Our Environment The Outlook for 1980

Our Environment The Outlook for 1980

Edited by
Alfred J. Van Tassel
Hofstra University

Lexington Books
D.C. Heath and Company
Lexington, Massachusetts
Toronto   London

Library of Congress Cataloging in Publication Data

Van Tassel, Alfred J
Our environment.

   1. Pollution—United States. 2. Refuse and refuse disposal—United States.
I. Title.
TD180.V36   363.6   72-7023
ISBN 0-669-84509-4

Copyright ©1973 by Hofstra University

All rights reserved. No part of this publication may be reproduced or transmitted in any form or by any means, electronic or mechanical, including photocopy, recording, or any information storage or retrieval system, without permission in writing from the publisher.

Published simultaneously in Canada.

Printed in the United States of America.

International Standard Book Number: 0-669-84509-4

Library of Congress Catalog Card Number: 72-7023

Contents

Chapter 1	Introduction	*Alfred J. Van Tassel*	1
	Part 1	Our Environment: Water	7
Chapter 2	A Methodology for the Statistical Study of Water Pollution	*Alfred J. Van Tassel*	9
Chapter 3	Lake Michigan	*Jay Parker*	25
Chapter 4	Lake Erie	*Robert Vollkommer*	51
Chapter 5	Puget Sound	*Donald G. Niddrie*	75
Chapter 6	San Francisco Bay	*Raymond A. Matern*	95
Chapter 7	Long Island Sound	*Robert T. Caponi*	127
Chapter 8	Hudson River	*Richard M. Martin*	149
Chapter 9	The Lower Mississippi	*Robert Schwartz*	175
Chapter 10	Savannah River	*Rogers R. French*	187
Chapter 11	Long Island Water	*Henry Lang*	211
Chapter 12	Water Recycling in Southern California	*James Carmichael*	237
Chapter 13	The Effects of Increased Production on the Oceans	*Frank D. Husson, Jr.*	257
Chapter 14	Summary of Findings Concerning Water	*Alfred J. Van Tassel*	281
	Part 2	Our Environment: Air	295
Chapter 15	A Methodology for the Study of Air Pollution in the United States	*Alfred J. Van Tassel*	297
Chapter 16	Air Pollution in Five Eastern Cities	*Peter L. Goffin*	311

Chapter 17	Air Pollution in Five Midwestern Cities *Frank Caulson*	335
Chapter 18	Air Pollution in Five Western Cities *George Fennimore*	367
Chapter 19	Global Pollution of the Upper Atmosphere *Carl A. Katz*	391
Chapter 20	Aircraft Noise Pollution and the U. S. Air Transport Industry Between 1970 and 1980 *Ronald Bruce*	417
Chapter 21	Summary of Findings on Air Pollution *Alfred J. Van Tassel*	445
	Part 3 Our Environment: Solid Waste Disposal	453
Chapter 22	Some Aspects of Solid Waste Disposal *John A. Burns and Michael J. Seaman*	455
Chapter 23	Solid Waste Disposal in Five Eastern Cities *Michael J. Seaman*	473
Chapter 24	Solid Waste Disposal in Four Eastern Cities *Howard W. Jaffie*	499
Chapter 25	Solid Waste Disposal in Five Western Cities *John A. Burns*	509
Chapter 26	Summary of Findings Concerning Solid Waste *Alfred J. Van Tassel*	547
	Part 4 Our Environment: Conclusions	555
Chapter 27	Conclusions *Alfred J. Van Tassel*	557
	Index	577
	About the Editor	591

Acknowledgments

The Hofstra University Library in all its departments rendered valuable assistance to the authors. The Documents Library, a depository for federal government publications, was especially helpful in pointing the way through the tangled thicket of government papers.

The volume would not have been possible without help from sources too numerous to detail in most instances. However, the chapters on solid waste disposal benefited enormously from access to computer printouts of the preliminary results of the 1968 National Survey of Community Solid Waste Practices. These invaluable data were made available through the cooperation of Robert Clark, Chief Urban Data Section, and Richard Toftner, Chief Planning Section, of the Bureau, Solid Waste Management, Cincinnati, Ohio.

The chapter on the lower Mississippi River, by Robert Schwartz, relied heavily on data supplied by Dr. Jay Gould from the files of his firm, Economic Information Systems, Inc., New York, N.Y.

Our Environment The Outlook for 1980

1 Introduction

ALFRED J. VAN TASSEL

The air is full of questions about the environment these days. Is our polluted air a major factor in lung cancer, as one recent investigation suggests? Will there be more or fewer waters fit for swimming, especially near the big cities? Will the mounting volume of garbage engulf us or prove a resource? Can we with impunity continue to use the oceans as an international dump of last resort? What lies at the root of the problem—too many people or too many goods? Or perhaps the fault lies in the way we produce the goods by new technologies more incompatible with nature, as Barry Commoner has suggested in his challenging book. [1]

This volume has an ambitious title, but it can hardly throw light on more than a few aspects of the conditions affecting our environment in 1980. However, it can and does address itself to the questions it covers in a particular way that should be useful. There is a large number of very necessary and useful books about the environment published every year, for example on the application of the principles of science and technology to combating pollution, or governmental reports mandated by an agency's responsibilities, or protests of outraged nature lovers who call attention to especially grievous assaults, or profound philosophical discussions of the relation of man to his universe. But there have been few attempts such as this one to view in a more or less consistent framework the outlook for the environment in a variety of localities representing different situations in all parts of the nation. The present study is characterized, as well, by the effort to gauge the state and outlook for the environment in quantitative terms wherever possible.

Lord Kelvin, the founder of thermodynamics, once said: "When you can measure what you are speaking about and express it in numbers, you know something about it. But when you cannot measure it, when you cannot express it in numbers, your knowledge is of a meager and unsatisfactory kind." [2] In obedience to Kelvin's dictum, we have sought throughout to "express it in numbers," fully conscious of the many pitfalls and the necessary qualifications on results. These latter are fully discussed in the methodology chapters 2 and 15 and in individual chapters as well.

This is the second quantitative study of environmental questions at Hofstra University to be directed by the present editor and published by Heath Lexington Books. It is also one in a long series of products of research seminars in the School of Business, known as Hofstra Yearbooks of Business, most of which have been published by the University. Hofstra's School of Business is one of the few that still requires a master's essay as one of the conditions for the Master of

Business Administration (MBA) degree, but such essays like these are prepared in research seminars rather than as unrelated efforts.

The earlier study, *Environmental Side Effects of Rising Industrial Output* [3], had the immense advantage of drawing for the basis of its projections on a common authoritative source. This was a monumental study prepared by the Ford Foundation's organization Resources for the Future, Inc. (RFF) entitled *Resources in America's Future, 1960 - 2000.* [4] This provided projections of output from 1960 to 2000 of virtually every commodity in the economy, done with professional skill and internal consistency. It was necessary to restrict, insofar as possible, the scope of such a far-ranging investigation and the RFF authors wisely made no more than passing reference to the possible side effects that might be associated with the enormous expansion of output their study envisioned. The type of question that arose on examining the RFF study was clear: If we produce all the gasoline and automobiles projected, what is the impact in terms of air pollution? If the expected expansion of output in paper products occurs, what is the effect in terms of stream pollution?

The RFF projections were used as the basis for fourteen chapters based on projections covering such subjects as water pollution including thermal and eutrophication, air pollution, strip mining, solid waste, pesticides, and air safety. While every effort was made to relate these to specific localities and though illustrations were mostly localized, the projections for the most part yielded only national or, at best, regional values for pollutants.

While the lack of specificity limited the value of the projections to some extent, the study did throw important light on many aspects of the problem. Among other things, it demonstrated that the rise in the scale of industrial output was the factor primarily responsible for raising water pollution questions, for example, to a level of commanding urgency. Contrary to Malthus, population grows relatively slowly and its contribution to water pollution in the United States might have been adequately controlled by improvements in treatment facilities except that geometrically increasing levels of industrial output without comparable improvements in treatment efficiency overwhelmed advances in municipal facilities.[a] In 1969, the General Accounting Office, the watchdog of Congress on the executive agencies, issued a report concluding that the Federal Water Pollution Control Administration had wasted $5.4 billion because it had spent that amount on pollution control facilities from 1957 to 1959 without effecting any improvement in water quality in many of the streams studied. The *New York Times* reports that: "In every case the accounting agency found that the efforts of Government were being overwhelmed by continued outpourings of industrial waste." [5] What the auditors found, in effect, was that we will have to run very fast indeed to stay in the same place. *Environmental Side Effects* documented this with respect to many aspects of pollution control.

In a generally favorable review in *Science,* H. J. Barnett and R. E. Kohn criticized the failure in the study to consider the political and economic processes by which counteracting forces to pollution come into play. [6] A

---

[a]To be sure, a growing population calls for increased output; the point is that the more dynamic increase has been in per capita consumption.

more serious shortcoming appeared to this author to be the limited localization of pollutional effects. It is not some abstract nation as a whole that is harmed by water pollution, but a specific, a particular stream. Attempting to focus on local situations presented formidable difficulties, especially from the standpoint of data availability, but also from the nature of the statistical estimating problem.

It is well known that projections for a large aggregate are more likely to be accurate in percentage terms than for a smaller, more narrowly defined target. Also, a comfortably distant time horizon of the year 2000 was employed in the first study. Such a framework gave room for very broad and general conclusions, frequently of a dramatic nature. But the results lacked concreteness, not because of the subjects studied, but due to the lack of focus. Moreover, organizing the study by types of pollution did not facilitate the examination, for example, of a polluted lake as a dynamic, interacting system in which the bacteria of municipal waste effluents are destroyed by dissolved oxygen whose supply is renewed from the air by wave action; where oxygen-demanding industrial wastes may enter to compete for the supply and the heated discharge of a power plant's cooling system enters to raise the temperature, accelerate life processes, and render the environment unsuitable for some species. Yet a study which focuses on just one form of pollution—municipal, or industrial, or thermal—will miss a good deal of the action. Similarly, with air pollution, it is necessary to examine the inputs from home and factory, aircraft, and motor cars in particular meteorological settings.

But we could hardly ignore the closely related questions of refuse disposal: if you burn it, you may contribute to air pollution, if you throw it in the water—to water pollution. Again, the effects are manifested in specific localities.

Sample of Waterways

All this argued in favor of a study of individual cities, of selected streams or lakes. Accordingly, the water studies were organized first in terms of bodies of water: lakes—Michigan and Erie; estuaries—Puget Sound, San Francisco Bay, Long Island Sound; rivers—the Hudson, the lower Mississippi, and the Savannah. Each of these was to be studied in terms of both municipal and industrial inputs, and since ours is a growing country with an expanding economy, an effort should be made to gauge the future magnitude of such inputs by projecting population and output trends. The target date was set as 1980 to make sure such projections did not take off into a distant and undisciplined future where vague expectations would suffice and some new technological magic might enter to solve all problems.

For Kelvin's measuring stick, the biological oxygen demand (BOD), for example, was selected, since it "expresses in numbers" the load placed on a waterway by either sewage or industrial wastes. BOD can be related roughly to either population or output and projected into the future.

Increasingly, the notion has begun to emerge that the long-term solution may be water recycling, and so two situations in southern California and on Long

Island were examined where this route has been given some thought, however diffident and tentative.

Finally, since all the waters end in the oceans eventually, it seemed well to look at the place where the buck finally stops. But for a study of the oceans, the opposite of our localized approach in other areas was called for.

For the studies of air pollution and solid waste disposal, fifteen cities were selected—five eastern, five midwestern, and five western. Although no pretense was made of selecting a "representative" sample, cities both north and south were included and one smaller city was selected for examination in each of the three broad regions. Like the oceans, the upper atmosphere has come to serve as a sort of international dumping ground of last resort, and one seminar member was given the challenging task of reviewing scientific thought on the subject. One other seminar member was interested in the effect of jet noise pollution on the air transport industry, and this subject is examined intensively in another chapter.

This volume, then, comprises for the most part a set of concrete and fairly narrowly focused studies. The authors were not permitted to evade responsibility for reaching conclusions about real places. Together it is hoped they can give some notion about how our country will fare in its battle against pollution in the next few years.

This is a far cry from the apocryphal tone of many recent environmental studies. Some others have used the techniques of economic and statistical projection to foretell the fate of mankind and the date—usually some time way off—of ecological doomsday. Typical of highly regarded environmental studies that are characterized by sweeping, long-range judgments is the recent widely publicized report of the "Club of Rome." [7]

Certainly there is little harm and perhaps much good in such activities. The work of the seer is pleasant, his hours are his own, and if you project far enough into the future nobody is likely to remember to check on the final outcome anyway. A sense of humor may be a useful tool in such enterprises, and nobody has given better warning against the risks of projection than Mark Twain. In his *Life on the Mississippi,* Twain describes how the Mississippi, a typical "meandering stream," had reduced its length by cutting through the bends. Thus, he concluded:

In the space of one hundred and seventy-six years the Lower Mississippi has shortened itself two hundred and forty-two miles. That is an average of a trifle over one mile and a third per year. Therefore, any calm person, who is not blind or idiotic, can see that in the Old Oölitic Silurian Period, just a million years ago next November, the Lower Mississippi was upward of one million three hundred thousand miles long, and stuck out over the Gulf of Mexico like a fishing-rod. And by the same token any person can see that seven hundred and forty-two years from now the Lower Mississippi will be only a mile and three-quarters long, and Cairo and New Orleans will have joined their streets together, and be plodding comfortably along under a single mayor and a mutual board of aldermen. There is something fascinating about science. One gets such wholesome returns of conjecture out of such a trifling investment of fact." [8]

It is hoped that the studies of our seminar yielded more than a "wholesome return of conjecture" and they certainly involved more than "a trifling investment of fact." The search for localized data was an arduous one and the steps involved in the long, drawn-out statistical process were quite demanding, and certainly educational. We may not always have succeeded in achieving meaningful figures bearing on local pollution problems. But it is certain that the only fully relevant information is that which bears on the specific areas in which problems must be solved. We were called upon to exercise our skills, in Wordsworth's words,

> Not in Utopia, subterranean fields,
> Or some secreted island, Heaven knows where;
> But in the very world, which is the world
> Of all of us,—the place where in the end
> We find our happiness, or not at all.

Notes

1. Commoner, Barry *The Closing Circle* (New York: Alfred A. Knopf, 1971), 326 pp.

2. *Bartlett's Quotations* (14th Edition), p. 723.

3. Alfred J. Van Tassel (Ed.), *Environmental Side Effects of Industrial Output* (Lexington, Mass.: D.C. Heath and Company, 1970), 548 pp.

4. Hans H. Landsberg, Leonard L. Fischman, and Joseph L. Fisher *Resources in America's Future* (published for Resources for the Future, Inc. by the Johns Hopkins Press, Baltimore, Md., 1963), 1,017 pp.

5. *New York Times,* March 13, 1972, p. 35.

6. *Science,* Vol. 172 (May 17, 1971), pp. 666-67.

7. *New York Times,* March 13, 1972, p. 35.

8. Mark Twain *Life on the Mississippi* (New York: Bantam Books), pp. 137-38.

**Part 1
Our Environment: Water**

2 A Methodology for the Statistical Study of Water Pollution

ALFRED J. VAN TASSEL

What is the net impact of the growth of population and industrial output on the nation's waters, considering also the steps being taken to overcome the effects of pollution? That is the general question to which the next eight chapters are addressed. To make the answer as concrete as possible the problem is attacked in terms of specific streams and bodies of water. Accordingly, this portion of the study will examine the outlook for water quality in 1980 of the following: Long Island Sound, Hudson River, Lake Erie, Lake Michigan, Puget Sound, San Francisco Bay, Lower Mississippi River, and Savannah River, a reasonably representative sample of the nation's major rivers, lakes, and estuaries.

Nature and Importance of Water Pollution

This section of the study deals with factors affecting the quality of water, principally in eight major waterways in various parts of the United States. It is hardly necessary to emphasize the importance of the subject. Water, the cradle of life's beginnings on earth, is still essential to human and other animal survival. A satisfactory supply of drinking water is a crucial prerequisite for any community.

A plentiful source of water of varying degrees of purity is likewise essential for many industrial processes, most frequently as a cooling or other heat exchange medium, but often as an essential ingredient or medium in chemical reactions as well. Water enters importantly into many processes for reduction of air pollution serving, for example, as the medium for removing such air pollutants as ammonia and sulfur dioxide, by solution in air scrubbing towers.

Finally, water is a basic constituent of the leisure plans of hundreds of millions of Americans who find recreation in swimming, skindiving, water skiing, boating, fishing, and sightseeing. The growing affluence of recent decades has meant that Americans have used the waters of their land more and become proportionately more conscious of the threat of water pollution.

There is, to be sure, a certain ambiguity to the term "water pollution." Aside from the distilled water of the laboratory, there is no "pure" water, and certainly not in the waterways of America, which have been time honored

---

The initiative in developing the practical application of the methodology outlined below was taken by Raymond Matern, Donald Niddrie, and Jay Parker, and this team was responsible for its detailed elaboration. This methodology was not applicable to chapters 11 through 13, which required independent approaches.—Ed.

channels for disposal of a variety of wastes. Whether or not such discharge is designated as pollution is a matter of definition. A sound pragmatic basis of classification of water relates quality to the use to which the water may be put. New York State, for example, has established five classes of fresh surface waters ranging from Class AA, which is suitable for use as a public water supply without filtration; Class A, suitable for public water supply after filtration; Class B, satisfactory for bathing and any less demanding usages; Class C, suitable for fishing; and Class D, satisfactory only for "natural drainage, agriculture, and industrial water supply." Associated with each of these classes are standards for such invisible qualities as concentration of dissolved oxygen, acidity, bacteria, toxic wastes, and heated liquids, as well as such more readily perceptible contaminants as odoriferous substances, solids, and oils [2: Table 4.3]. Similar usage-classification systems have been developed for salt waters as well. The main point is that such a classification system permits us to redefine water pollution as anything entering a waterway that lowers the class of permissible usage, i.e., from A to B, or C to D, etc. We may then distinguish between a water polluting discharge thus defined and waste water that is neutral with respect to effect on best usage of the receiving waterway.

The potential source of pollution most people think of first is municipal sewage, consisting principally of the water-borne output of the toilets and kitchens of the nation's cities and urban areas. The load from this source is proportional in the first place to the size of the urban population, which increased by 12 percent from 1960 to 1970 [4:17].

Municipal Sewage Treatment

The treatment of municipal sewage is perhaps the oldest and best understood, though by no means universally practiced, aspect of pollution control. A growing percentage of the throughput of municipal sewage systems is subject to some form of treatment before discharge into a waterway. Most widely practiced is *primary treatment,* in which the grosser solids are removed by screening and settling techniques, and many cities limit treatment to this phase. *Secondary treatment* involves the biological oxidation of organic matter to more stable form and is performed by such means as trickling filters, oxidation ponds, chlorination or activated sludge. An insignificant fraction of the nation's sewage is subject to various forms of *tertiary treatment* designed to further purify the effluent of the sewage treatment plant and render it fit for reuse at a higher level of use than would be possible without such treatment; e.g., water suitable only for irrigation might be made acceptable for swimming by tertiary treatment.

Disease-causing bacteria such as those found in human excreta constitute an obvious major target of a sewage treatment plant. Much of this contamination is removed in sewage sludge formed by the filtration and settling stages of primary treatment. Much of the destruction of bacteria is accomplished in the course of secondary treatment by oxidation, which figuratively speaking burns up the bacteria. It is possible to determine the amount of oxygen that will be required to destroy all bacteria and other organic materials in a unit of fluid and this is known as the BOD or biological oxygen demand. The concept of BOD, as defined later in this chapter, is one of the principal tools of this study.

Of course, purification by sewage treatment is never complete and there is some definable residual BOD of the effluent of the treatment plant. Hopefully, this residual BOD is met by the natural dissolved oxygen of the waterway into which the effluent is discharged. Such dissolved oxygen in a waterway will be replenished by solution from the air at a rate that depends most importantly on the relative temperatures of air and water and the extent of mixing of the two media. The dissolved oxygen in a waterway constitutes the first line of defense against contamination by harmful bacteria. But oxygen is removed, as we will see, by many other agencies besides bacteria.

Historically, Americans have relied heavily on dilution by natural waterways to render partially treated sewage effluent tolerable. Some communities still discharge raw sewage to nearby waterways and rely entirely on natural processes for purification. It is astonishing how many streams, lakes, and estuaries have survived such onslaughts without being rendered entirely unusable. However, there has been a slow growth in the extent and sophistication of sewage treatment facilities which has probably at least kept pace with the growth of population. However, the growth of industries that also release oxygen-demanding discharges to the nation's waterways has dwarfed that of population.

Industrial Wastes

It was noted earlier that urban population in the United States increased by 12 percent from 1960 to 1970. In the same period, U. S. industrial output rose by 54 percent. The rate of growth of output was especially high in those industries whose effluents exert higher than average pollutional effects. Thus, the level of output of the chemical industry increased in the decade 1960-1970 by 108 percent, rubber and plastic products (heavily weighted by the latter) by 107 percent, and paper and its products by 60 percent [4: Table 113]. Most industrial discharges do not add substantially to bacteriological pollution. Rather, many effluents of industrial processes have voracious appetites for oxygen and it is because of this that they contribute so importantly to pollution problems. Either such industrial discharges constitute a crushing burden on the oxidizing stages of a sewage treatment plant, or are likely to represent a serious drain on the dissolved oxygen of a waterway and thus leave it an easy prey to rapid multiplication by such bacteria as may enter it.

In general, the systematic treatment of industrial wastewater prior to discharge is limited and sporadic. To the extent such discharges are treated, it is usually in ordinary municipal sewage facilities. The exception is where it proves possible to recover a commercially salable by-product through the treatment of industrial wastewaters. To the extent that waste discharge regulations are strictly enforced, it would appear likely that such by-product recovery will prove profitable in an increasing number of instances, although by-product recovery is more likely to be profitable in large-scale operations than in small.

The petroleum and chemical industries have realized substantial profits from by-product recovery, but such a course is not always possible. As one chemical corporation head put it:

The frustrating truth is: we have hundreds of tons a day of waste—for which there is as yet no known use. We have spent a fortune searching for ways—either to use it or to dispose of it. But—I fear—in vain. So far, we can't burn it; we can't bury it [2:66].

For many firms faced with such an intractable problem, the solution has been to discharge waste to the municipal sewage system with the predictable result that improvements in the municipal system have been nullified by the increasing load as industrial output expands. Thus, the General Accounting Office reported to Congress in 1969 that much of the federal water pollution control program has resulted in the construction of treatment facilities that, because of pollution from other sources, have not had an appreciable effect in reducing pollution, or improving water quality, or upgrading possible uses of the waterways.

"We found that industrial wastes, which were responsible for much of the pollution, had precluded the realization of water improvement objectives, even on waterways where municipalities had constructed the facilities needed to provide adequate treatment for domestic wastes." [5]

One form of pollution that has been distinguished only recently as a possible threat of some importance is thermal pollution. In many industrial processes water is withdrawn from a waterway, used for cooling purposes, and returned to the waterway at an elevated temperature. This is known as thermal pollution and has achieved prominence recently because the electric power industry, perhaps the largest user of water for cooling purposes, has been one of the fastest growing sectors of the economy. Moreover, waste heat and therefore thermal pollution is greater per unit of power produced when conventional fuels are replaced by atomic fuels in generation, and an increasing percentage of future power is expected to come from atomic fuels.

The impact on the receiving waterway of thermal pollution is two-fold: (a) The artificially elevated temperature renders the waterway in the vicinity of the heated discharge unsuitable to, and sometimes fatal to, resident marine organisms; (b) Since oxygen is less soluble in water as temperatures rise, thermal pollution reduces the concentration of dissolved oxygen in the receiving waterway.

Eutrophication

Another form of pollution that has come into prominence only recently is eutrophication. Coming from the Greek *eu* meaning "well" and *trophein* meaning "to nourish," the term eutrophication is used in practice to refer to the state of overfertilization of a waterway that promotes excessive growth of aquatic vegetation, algae, and plankton. The excessive fertility of such eutrophic waters may arise from several sources: (a) the nutrients, principally phosphatic and nitrogenous, in the effluent of sewage treatment plants; (b) the same nutrients carried to the waterway by runoff from fertilized land or from livestock feedlots; (c) industrial wastes rich in compounds of nitrogen and phosphorus [2: Chap. 15].

Most noticeable of the growths induced by eutrophication is the blue-green algae that has marred the surfaces of so many American lakes in recent years. An explosive rate of algae growth occurs when a combination of required conditions is met, including: (a) minimum concentration of the required plant nutrients; (b) presence of other necessary building materials such as carbon; and (c) sufficient light to support photosynthesis. In vigorously flowing streams or tidal estuaries, dilution is usually sufficient to hold nutrient concentrations below requisite levels; the deep water of lakes will not support rapid growth because of lack of light, but the upper levels will. It is evident, then, that lakes are much more vulnerable to eutrophication than streams and estuaries. However, it is obvious that the criteria are relative as is the seriousness of the manifestation of eutrophication, so that generalization is difficult. The multiplicity of possible sources of excessive nutrition argues against faith in simplistic solutions such as a ban on phosphatic detergents. On the other hand, the evidence of growing nutrient concentrations even in rivers and estuaries should caution against complacency. Dilution—the favorite traditional American solution to pollution problems—may no longer suffice to overcome the problems of eutrophication in those waterways where no symptoms are presently perceptible.

The major tool to be employed in this study to measure pollution potential is the **BOD** of an effluent. How does eutrophication affect the dissolved oxygen supply of a waterway? Again, no simple answer is possible. However, it can be said in general that during the growth of algae or other vegetation, oxygen is added to the waterway, but when the vegetation dies, its decay may impose a crushing burden on the dissolved oxygen supply of the waterway.

It may be useful to illustrate the essential method followed here by considering the problem of projecting the likely change in water quality over a period of time in a small lake. Let us suppose that there are on the shores of a lake a sewage treatment facility that discharges its liquid effluent to the lake, an industrial plant that does likewise, and a power generating station that withdraws water for cooling purpose and returns it at an elevated temperature to the lake. Which of the pollutional effects from this complex will it be possible to measure for the present and project for the future, and what yardstick can we employ?

It is frequently quite feasible to measure the pollutional impact of the sewage treatment facility and industrial plant in terms of the biological oxygen demand (**BOD**) of the effluent. The BOD refers to the amount of oxygen that is required for the destruction of bacteria or for the decomposition of organic matter or for other chemical reactions induced by the contents of the effluent. Other aspects enter into a determination of overall water quality such as clarity, odor, presence or absence of solids and surface films, e.g., from oil. However, to a substantial degree, low water quality as reflected in a high BOD will be associated with other undesirable characteristics of the body of water. BOD, then, provides the most convenient yardstick for measuring present and projected water quality, and that fact is the foundation of the method employed here. BOD is usually expressed as five-day BOD, meaning the amount of oxygen required for the biological decomposition of organic materials to occur under standard conditions in a standard interval of five days at a temperature of 20 degrees centigrade. Thus, the BOD test is a useful procedure for calculating and standardizing the polluting strength of organic waste materials discharged to waterways.

Returning to our hypothetical study of a small lake, there is likely to be a recent test of its water quality to serve as a benchmark for projection into the future. Such a test will usually show the BOD per unit, but is almost certain to include levels for other indicators such as (a) dissolved oxygen and (b) coliform bacteria, each per unit of volume. The concentration of coliform bacteria, which are present in untreated human excreta, serves as a joint test of degree of bacteriological purity and of effectiveness of the treatment of the sewage discharged to the waterway. Concentrations of both dissolved oxygen and coliform bacteria are functions of the capacity of the waterway to purify itself by absorbing oxygen from the atmosphere, but also of the BOD of the effluents discharged to the waterway.

BOD as a Yardstick

What determines the BOD of the effluent discharged to our lake by the sewage treatment facility? First, the BOD of the raw, untreated sewage entering the sewage treatment facility (which has been found to be about 0.17 pound of BOD per person per day [1:318]; second, the effectiveness of sewage treatment. If cost is no consideration it is possible to purify sewage to drinking water quality, entirely eliminating all bacteria and other sources of BOD as well as undesirable colors or odors. For purposes of illustration, let us assume that the sewage treatment facility on the shores of our lake uses only primary treatment of entering sewage and thereby effects a 60 percent reduction in the BOD of the sewage it receives. Let us also assume that ten years later, the sewage treatment facility has been improved by additional stages of treatment so that the BOD of entering sewage is reduced by 88 percent, a value that corresponds to actual practice [2:94]. Finally, let us assume that the population served by the sewage treatment facility at the start of the period is 10,000, and 11,000 at the end of the ten years. Now, let P_t and $P_{(t+10)}$ represent population at start and end of period respectively; BOD_t and $BOD_{(t+10)}$ represent BOD per day of facility effluent start and end of period respectively; R_t and $R_{(t+10)}$ represent percentage reduction in BOD effected at beginning and end of period respectively. Then, given that BOD percentage is a constant C_p equal to .17 lb, we may write:

$$BOD_t = P_t \cdot C_p \, \frac{100 - R_t}{100} \quad \text{and}$$

$$BOD_{(t+10)} = P_t + 10 \cdot C_p \, \frac{100 - R_{(t+10)}}{100} \quad \text{or substituting values}$$

$$BOD_t = 10{,}000 \, (.17) \, \frac{(100 - 60)}{100} = 680 \text{ Pounds/day}$$

$$BOD_{(t+10)} = 11{,}000 \, (.17) \, \frac{(100 - 88)}{100} = 224 \text{ Pounds/day}$$

Thus, in this illustration, an increase in effectiveness of sewage treatment has more than offset the impact of a 10 percent increase in the sewered population.

Whether or not this would have resulted in an absolute reduction in the BOD of effluents discharged to our lake would depend on the contribution of the industrial plant we assumed was situated on the shores of the lake. Just as BOD attributable to sewage was proportional to population, so BOD attributable to industrial effluents will be proportional to output of finished product, ignoring effect of any change in degree of treatment in both instances.

In the case of the industrial plant discharging effluent to our lake, let us assume that there is only one product for the plant and that at the beginning of the period the output is 1,000 units per day resulting in an effluent with a BOD of 1 pound per unit produced per day. Let us assume that at the end of the period the output is 1,200 units per day with a BOD of 0.9 pound per unit produced per day. Now, let O_t and $O_{(t+10)}$ represent output in units per day at beginning and end of period respectively; BOD_t and $BOD_{(t+10)}$ represent BOD per day of industrial plant effluent at start and end of period respectively; CI_t and $CI_{(t+10)}$ represent BOD per unit of industrial plant output at beginning and end of period respectively, then we may write:

$$BOD_t = O_t \cdot CI_t \quad \text{and}$$

$$BOD_{(t+10)} = O_{(t+10)} \cdot CI_{(t+10)} \qquad \text{or substituting values}$$

$$BOD_t = 1{,}000 \cdot (1) = 1{,}000 \text{ pound/day}$$

$$BOD_{(t+10)} = 1{,}700 \cdot (0.9) = 1{,}530 \text{ pounds/day}$$

Thus, in this illustration, the large increase in industrial output has more than offset the slight reduction in BOD per unit of industrial plant output.

If we were to combine the figures in our two illustrations, we would find that the substantial increase in BOD generated per day by our lake's industrial plant has overpowered the considerable decrease in BOD generation by the lake's sewage treatment facility, with the result that there has been in our illustration a small increase over the period in BOD of the combined sewage and industrial effluent entering our lake. While the choice of figures used for this illustration was, of course, arbitrary, the fact is that something very much like this pattern seems to have been occurring in many of the nation's waterways. Sparked by widespread recognition of the need for control of water pollution and aided by grants of federal funds, there has been a nationwide push to construct new sewage treatment facilities and upgrade old ones. Moreover, population in most areas has increased relatively slowly. Not so industrial output, which rose during the 1960s in most sectors at a rate far in excess of population growth. Treatment of the waste effluent of industrial plants has proved intractable to major improvement in some industries and in general has not advanced in proportion to the increase in output. Our model, then, is not devoid of reality.

The use of the BOD yardstick permits us to relate the extent of pollution to the changing magnitude of sewered population and industrial output as modified by treatment. It is recognized that this does not take account of other pollutants that may enter the lake and affect its odor, appearance, or even toxicity. However, the first two of these are likely to be correlated with BOD levels; that is, a waterway with a high BOD level is likely to have bad odor and appearance as well. Toxicity is generally highly localized if mixing and dilution are adequate, and is, in any event, not readily susceptible to the statistical treatment that is the principal tool of this analysis.

The pollutional impact of the third establishment hypothesized for our lake—an electric power station—is likewise not amenable to straightforward statistical treatment. The power station adds no bacteria or other oxygen-consuming elements to the lake; instead, it affects the lake through what is known as *thermal* pollution, notably by discharging waters to the lake at temperatures significantly higher than those that would have obtained if the water had not been withdrawn for power station cooling purposes. The impact of such thermal pollution is two-fold: (a) It has a direct effect of varying intensity, usually adverse in the immediate vicinity of the discharge, sometimes raising the temperature to a level that kills fish and other marine organisms.[a] (b) Raising water temperatures tends to accelerate life processes and thus to exacerbate any bacteriological pollution that may pre-exist; oxygen is less soluble in water as the temperature rises so that oxygen that is the destroyer of bacteria and other organic pollutants may be expelled from solution by thermal pollution. Neither of these effects are readily quantifiable, nor easily linked in direct correlation with levels of power output. Accordingly our treatment of thermal pollution will be principally in qualitative terms.

The central purpose of this portion of our study is to apply the model outlined above for a hypothetical small body of water to some eight widely scattered major U. S. waterways. At its best, this approach should permit us to make projections of future BOD levels in various portions of such waterways, thereby providing quantitative gauges of approximate pollutional levels. The availability of the necessary data was the main factor limiting the degree of achievement of this goal. A good starting point for the discussion of this effort is to outline the data that would meet the needs most nearly ideally.

Ideal Data for Computation

For each such waterway, there is a geographically definable set of drainage basins that feed it. For each such drainage basin we should have the following:

---

[a] Usually, the elevated temperature from thermal pollution will simply repel fish and marine organisms that find it unsuitable, and, in a few cases, fish are attracted by the warmer water.

A. Respecting population.

1. Population for the starting year plus data for the prior decade, i.e., sufficient data to establish a trend that may be projected for the succeeding decade;
2. A BOD ratio for the untreated sewage generated per person (which may be assumed to remain constant);
3. A coefficient representing the reduction in BOD of untreated sewage characteristic of standard practice at the start of the period in the drainage basin. (The product of the starting-year population multiplied by ratio in 2 and coefficient in 3 would represent the BOD generation attributable to sewage for the basin in the starting year.)
4. A projection of population for the drainage basin a decade after the starting year derived from the data described in (1);
5. A coefficient representing the reduction in BOD of untreated sewage expected to be characteristic of standard practice at the end of the projection period in the basin. (The product of the population arrived at in (4) multiplied by 2; and 5 would represent the BOD level attributable to sewage for the end of the projection period, i.e., ten years after the starting year.)

B. Respecting industrial output.

1. For each industry, output in number of units of the drainage basin for the starting year plus data for the prior decade, i.e., sufficient data to establish a trend that may be projected for the succeeding decade;
2. A BOD ratio per unit of output representative of industry practice and degree of treatment as of the starting year. (The product of starting-year output from (1) multiplied by the coefficient in (2) would represent the BOD generation in the basin attributable to the industry in question during the starting year.)
3. A projection for each industry of output in units in the basin for the tenth year after the starting year.
4. A BOD ratio per unit of output representative of industry practice, and degree of treatment a decade after the starting year. (The product of tenth year output in (3) and the ratio in (4) would yield BOD generation in the basin attributable to the industry in question a decade after the starting year.)

A summation of the BODs attributable to all industries in the drainage basin plus that arising from general sewage treatment (i.e., due to population) could yield a reasonably satisfactory estimate of the BOD load in the starting year. The same summation derived from population and output projections for the tenth year after the starting year would yield a less secure estimate of the likely situation a decade later. A comparison of the two figures for a specific drainage basin would provide the best available estimate of the change in pollution potential of the effluents entering the waters of the drainage basin as measured by BOD *only*. The reason for the emphasis here is that the BOD level does not take account of other forms of pollution such as that due to toxic materials, oil films, suspended solids, etc. Moreover, the actual BOD level that eventually obtains will depend not only on the BOD potential of the entering effluents, industrial and other, but upon the natural capacity of the waterway to rejuvenate itself by absorption of additional dissolved oxygen. Nevertheless, the calculation of the change over time in the BOD potential of entering effluents is the figure likely to be of most interest in tracing the history of pollution of a waterway.

What data are likely to be actually available covering the waterways we have chosen to investigate? The sources generally employed in most of the chapters that follow will be described here; individual variations from the general in sources and procedures will be noted in the individual chapters.

BOD Coefficients

The most fundamental step in the process of arriving at an estimate of the BOD produced by each industry in each drainage basin was to arrive at a coefficient of BOD per unit of output in the starting year and ten years later. This also proved the most elusive task involved. The members of the seminar were not equipped to undertake the computation of such coefficients themselves. Fortunately, a major study by the Regional Plan Association covering waste management in the tri-state New York metropolitan region had brought together coefficients of waste load generated in pounds of BOD per unit of output for a number of industries and for municipal sewage treatment facilities. In the RPA study, such coefficients were principally the responsibility of Blair T. Bower, but drew heavily as well on the work of Allen V. Kneese, W. W. Eckenfelder, C. F. Gurnham, and D. H. Stormont [3]. BOD per unit of output figures were available for industry:

SIC Number	Description
2011	Meat packing plants
2013	Sausage and other meats
2026	Fluid milk
2033	Canned fruits and vegetables, etc.
2037	Frozen fruits and vegetables, etc.
2621	Paper except building
2631	Paperboard mills
2911	Petroleum refining

The industries listed include a number such as food processing, paper products, and petroleum refining that are principal contributors to the pollution of the nation's waterways, and a BOD input computed on the basis of these ratios will account for a substantial fraction of the total. However, it must be noted that absence of BOD ratios for the conglomerate known as the "chemical industry" and to a lesser extent for the metal-working industries constitutes one of the major shortcomings of the method employed here. Less damaging is the fact that the ratios are not specifically applicable to the years and regions studied. Such ratios change very slowly over time and should not change markedly in the relatively short time span of the present study. Also, techniques in most industries are fairly well standardized within the United States so that differences between regions in the ratios should not affect the results substantially. The most important factors are taken into account in the methodology of the present study, notably: (a) wide differences in ratios between industries representing major sources of industrial BOD in the waterways, and (b) changes over time in output per employee.

One of the most consistent and pervasive changes occuring in American industry is the persistent advance in output per worker. RPA estimated the change in output per worker for each industry for which it estimated a BOD coefficient [3]. These data were also drawn upon.

The unique feature of the present study is the effort to gauge the extent of the contribution to pollution of population and industrial growth, not in broad general terms for the nation as a whole, but with respect to specific bodies of water. Broad totals for the nation as a whole may be of some limited value if they tell something about the increase over a period in the magnitude of the pollutional problems to be overcome. Estimates of national totals may be useful as well in indicating which industries are the most important contributors to pollution. But the solutions as well as the causes of pollution will be found ultimately only in the study of particular waterways and, in some instances, only in the examination of specific portions of such waterways. Such studies to be fully effective must be multidisciplinary, requiring the services of scientific and engineering specialists, and are therefore beyond the resources available to this study. However, introductory studies of specific waterways employing the techniques of economic and statistical projection may serve to: (a) point to areas of imminent critical importance; (b) suggest the principal potential sources of pollution; and (c) provide a basis for estimate of the magnitude of the effort involved.

Accordingly, we have chosen at Hofstra to attempt a pioneering effort at a quantitative measure of the outlook for water pollution as measured by BOD for specific waterways. Fully mindful of the data deficiencies, we are aware that the present effort may be of most value to later investigators. However, to quote an earlier Hofstra study, we are also convinced that "pollution has become a matter of such commanding urgency because of the rise in scale of national output" [2:449], and that accordingly a study that reflects for specific localities the effects of this rise will be valuable even if results are unavoidably approximate. Given ratios of BOD per capita sewage effluent and per unit of industrial output, the next step was to locate the best sources of data on population and industrial output for the drainage basin involved.

Estimating BOD of Municipal Sewage

The search for and treatment of population statistics was straightforward. Each waterway studied usually drew upon a series of definable drainage basins. For the most important of these drainage basins, the individual counties involved were first identified. (It was felt that a county was the smallest unit for which it was practicable to obtain data.) The population for the counties comprising each drainage basin was totaled for 1959 and for 1969. The assumption was then made that the population change to 1980 would continue at the same annual rate that had prevailed from 1959 to 1969. Summing these data for all counties in each drainage basin yielded a 1969 population, and an annual rate of increase in the preceding decade. The 1969 population plus 11 times the annual rate of increase yielded a projected 1980 population for the drainage basin. The 1969 and 1980 population for each drainage basin multiplied by 0.17 pounds (the BOD per capita ratio for sewage from *Waste Management*) gave the 1969 and 1980 totals for BOD generated by sewage within the drainage basin in question.

It should be noted that this calculation refers to BOD *generated* only, and not to the BOD of sewage effluent discharged to the waterway involved. In other words, this calculation involves the BOD of that portion of the *inflows* of sewage treatment plants attributable to ordinary sewage. It does not take account of either the *reduction* in BOD arising from sewage treatment nor the *addition* to BOD associated with the effluent of industrial processes, whether discharged directly to the waterway or through municipal sewage treatment plants. The techniques used in estimating the contribution to overall BOD of *industrial* effluents will be discussed next.

The effectiveness of sewage treatment in reducing the BOD potential of ordinary, non-industrial sewage was not susceptible to generalized statistical treatment. Rather, account was taken of available information on the effectiveness of present sewage treatment facilities and of likely improvements in the next decade, to arrive in a variety of ways at an estimate, frequently qualitative, of the effect of the effluent of sewage treatment facilities on a waterway, the latter defined as narrowly and specifically as the particular data permit.

BOD of Industrial Discharges

The measurement and extrapolation of the BOD contribution of industrial effluents constitutes a much more complex task than that associated with comparable estimates concerning ordinary sewage. The per capita BOD ratio for ordinary sewage is believed to be fairly well applicable within the United States, but the BOD coefficient for industrial effluents is known to vary widely between industries and within an industry in accordance with differences in in-plant treatment prior to discharge. What must concern us here are the problems involved in estimating and projecting output in a specific drainage basin for a particular industry.

For this task, data from two publications of the U. S. Bureau of the Census were employed, to wit the *Census of Manufactures* and *County Business Patterns*.

The *Census of Manufactures* for 1967 supplies data on yearly average number of employees in each of these industries in that year, and this figure was used as a benchmark for estimating starting year and 1980 employment in each such industry. The state employment figure for an industry in the *Census of Manufactures* is an average of four quarterly figures and may accordingly be presumed to be free of seasonal variation. However, the most recent year for which *Census of Manufactures* data are available is 1967.

Country Business Patterns is published annually and the most recent issue available when the calculations in the present study were made was 1969. *County Business Patterns* gives employment by states and *by counties as well* for each of the four digit SIC industries listed for which BOD coefficients are available. However, the employment figure in *County Business Patterns* is not as representative as that in *Census of Manufactures* inasmuch as it is based on employment in a single week of the year, notably that week containing March 12, and is therefore subject to distortion due to seasonal variation. Accordingly, the benchmark figure for each state and industry for which computations were made was that given by the 1967 *Census of Manufactures*, which is presumed to be free of seasonal variation. This 1967 *Census of Manufactures* figure for each state and industry involved was then projected to 1969 and 1980 on the basis of trends in each such state industry as revealed by *County Business Patterns*. The procedure was to obtain employment for each state industry involved from *County Business Patterns* for the years 1959 and 1969 and from these data to derive a trend expressed in percentage annual change. This percentage annual change trend was applied to the state-industry employment figure from the 1967 *Census of Manufactures* to obtain state-industry employment figures for 1969 and 1980.

An output per employee coefficient for each industry involved was computed for 1969 and 1980 by straight line interpolation in the *Waste Management* data on this subject. Given output per employee coefficients for 1969 and 1980 and state-industry employment projections for the same years, it was a simple matter to compute for 1969 and 1980 the gross BOD generated in industrial effluents in each state by each of the industries for which BOD per unit of output coefficients were available.

In summary then, data by states on employment in the relevant industries was relied upon for projection purposes; data for the decade preceding the starting year were used to project employment and output in each industry a decade later. Then BOD coefficients for selected industries were applied to the state output figures to derive gross BOD projections.

Allocating State BOD to Drainage Basins

However, figures for BOD generation by states did not provide an adequate breakdown for the task at hand. What was required was BOD generation by *drainage basin* for each industry, rather than by state. The drainage basins for Lake Michigan, for example, comprised parts (though in some cases very small portions) of four states: Michigan, Wisconsin, Indiana, and Illinois. Clearly, a figure for a given industry stating gross BOD generation for the four states would be meaningless with regard to the pollution problem of Lake Michigan. However, a figure representing gross BOD generation from major sources for each of the *drainage basins* discharging into Lake Michigan would be highly pertinent to a study of its pollution problem. Fortunately, a set of data was available which permitted the allocation of state-industry BOD totals to specific drainage basins.

For a drainage basin wholly contained within a single state (the usual situation), the procedure for given industry was as follows: (a) A determination was made of the counties comprising the drainage basin; (b) Employment in the industry in question was determined for the starting year for each of the counties comprising the drainage basin; (c) Drainage basin employment for the starting year was computed by totaling the county figures from (b); (d) The ratio of drainage basin employment in the given industry to total state employment in the same industry was determined and the assumption was made that BOD generation by the industry within the drainage basin was proportional to the basin's share of state employment in the industry; thus, drainage basin BOD generation attributable to a given industry was determined.

It will be recalled that BOD for a drainage basin attributable to municipal sewage for a given drainage basin was determined earlier by basin 1969 and 1980 projected population figures. Now, the total BOD generation for each drainage basin is determined by summation for all industries and municipal sources for which data were available. Since such a computation was possible for municipal sewage and for a number of the industries contributing most significantly to BOD generation, the resulting drainage basin BOD figures were regarded as reasonably satisfactory measures, in most cases, of levels of starting year and projected year pollution potentials for each basin. In a few cases where there was no BOD coefficient for an industry known to be an important contributor to BOD, this omission is noted in the text. (See, for example, the discussion in Chapter 5 of the Calumet area of northwestern Indiana where the steel industry is an important contributor to the BOD load entering Lake Michigan.)

The computation of BOD inputs by drainage basin was crucial for the production of meaningful statistics in all of the waterways where a BOD estimate was attempted. In any waterway of consequence, the impact of an entering effluent of given BOD must be considered as, in some degree, temporary and geographically limited. In a river the effluent is continually diluted and its impact accordingly reduced by stream flow, and additional oxygen is dissolved to replace that removed at a rate varying with turbulence, temperature, and other natural conditions. A somewhat similar dilution and renewal of dissolved oxygen is associated with tidal movements in estuarine waters. Even in lakes where both mixing and dilution is slower than in rivers and estuaries, a slower but nonetheless certain oxygen renewal occurs in most instances, especially in the larger lakes.

The development of drainage basin statistics made possible a high degree of specificity in our projections of water quality outlooks. Not only were conclusions stated in terms of particular waterways, but with relation to defined sectors of those waterways. This is in contrast to the broad national generalizations to which we were limited for the most part in the predecessor study at Hofstra [2].

Conclusion

This completes our description of the methodology that was applied, with suitable variations depending on circumstances and data availability, in the next eight chapters. In addition to BOD data, each author brought to bear on his topic all the relevant information he could come across. In many instances this latter was abundant, since federal and state agencies have been busy studying the waterways. However, major emphasis was placed on the use of the additional data developed by the Hofstra tabulations. The methodology outlined in this chapter was not applicable to the three enormously important subjects covered in Chapters 10, 11, and 12. Specific techniques were developed for each by consultation between the author and the director of the study.

Notes

1. W. A. Hardenbergh and Edward B. Rodie *Water Supply and Waste Disposal* (Scranton, Pennsylvania: International Textbook Company, 1960), 513 pp.
2. Alfred J. Van Tassel (Ed.) *Environmental Side Effects of Rising Industrial Output* (Lexington, Mass.: D.C. Heath and Company, 1970), 548 pp.
3. *Waste Management* (New York: Regional Plan Association, 1968), pp. 94-96.
4. U. S. Department of Commerce, Bureau of the Census *Statistical Abstract of the United States,* 1971 (Washington: Government Printing Office, 1971).
5. Report to the Subcommittee on Air and Water Pollution of the Committee on Public Works, U. S. Senate by the Comptroller General of the U. S., *Administration of the Construction Grant Program for Abating, Controlling, and Preventing Water Pollution* 91st Cong., 1st Session (November, 1969) p. 15.

3 Lake Michigan

JAY PARKER

Water pollution problems have plagued numerous bodies of water in the United States for many years, to a greater or lesser degree, although only in the late 1960s did the matter become one of great public and private concern. A rapidly growing population coupled with a greatly accelerated increase in industrial output have lead to the discharging of vast amounts of pollutants to the nation's waterways, at the very time that the demand for water has even outstripped the rate of growth in the private and industrial sectors of the economy. Many new industrial facilities require more water for process and cooling purposes than did the older facilities, and rising disposable income and increased leisure time insure that the domestic demand for water will continue to rise at a rapid rate.

Lake Michigan was once thought to be almost immune to any sizable water pollution problem, because of its great size and assimilative capacity. This is no longer true; water pollution problems of great magnitude have manifested themselves in various local inshore areas in the western and southern portions of the lake. How badly is Lake Michigan affected by water pollution? How have rising industrial output, a larger population, and a greater demand for electricity contributed to Lake Michigan water pollution problems? What is the outlook for 1980? These are some of the questions that this chapter considers.

The scope of the investigation centers around the two major sources of water pollution—industrial plants and municipal sewage systems, and one potential source of water pollution—electric power generating plants. Both quantitative and qualitative data were collected, developed, projected, and analyzed. While much of the quantitative data lack a very high degree of precision due to the amount of estimation and projection involved, it is believed that useful comparisons can be made as to the relative contributions of the various sources of waste, and as to the impact on local areas. This study was predicated on the assumption that rising industrial output and a growing population will lead to greater waste disposal problems that must be overcome. Consequently, waste generation from the aforementioned sources was estimated in accordance with the methodology described in Chapter 2, related to present water quality, and projected ahead with 1980 as the target year. A comparison with present estimated and future projected waste loads for the most severely affected areas of the lake should yield significant indications of the magnitude of the problem, if not fully reliable trends of future water quality.

The Lake Michigan Watershed

Lake Michigan, the sixth largest fresh-water lake in the world, has a surface area of approximately 22,400 square miles, and a total drainage basin covering some 76,900 square miles [1: III, 1444]. All the rainfall, sewage, and industrial wastes from this basin enter Lake Michigan directly or via sewer pipes, rivers, streams, and creeks. Many of the rivers tributary to the lake are quite long, especially in

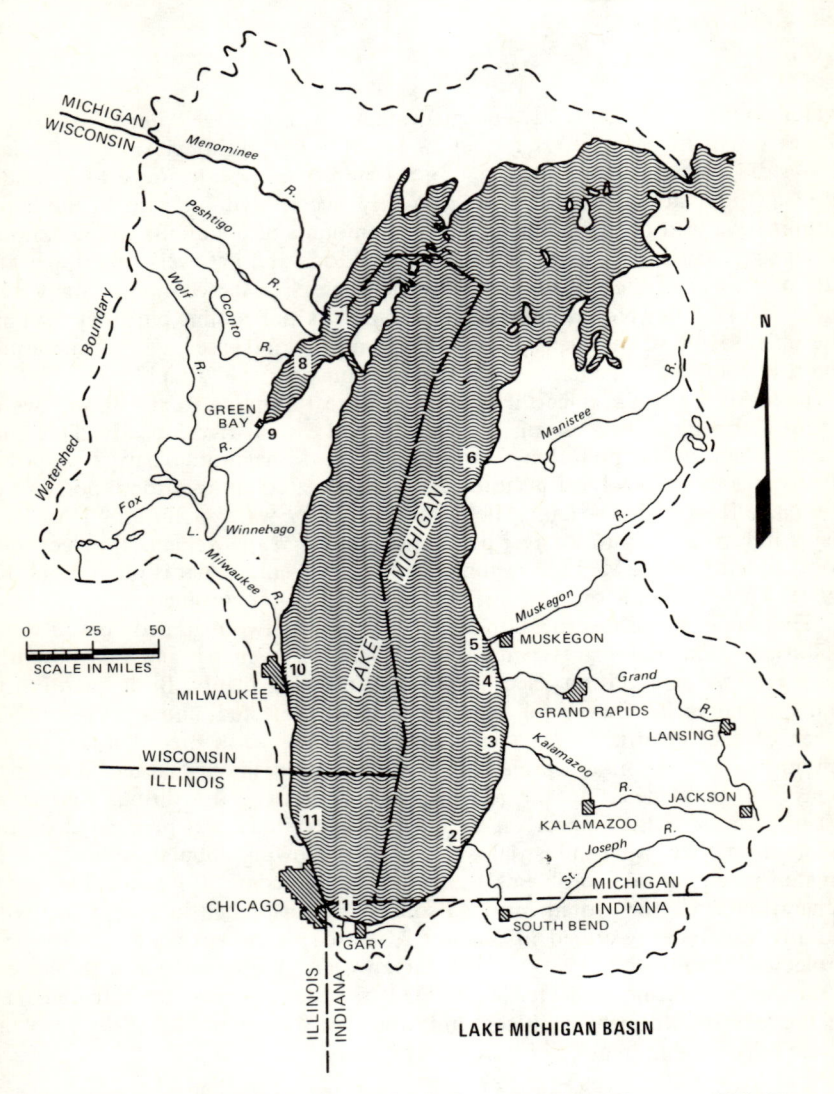

Figure 3.1. Lake Michigan and Its Tributary Basins. Source: U.S. Department of the Interior, Federal Water Pollution Control Administration, *Conference on the Matter of Pollution of Lake Michigan and Its Tributary Basins*, Chicago, Illinois, February 1968, Vol II, p. 527.

Michigan, so a good deal of purification takes place en route, sparing the lake itself of this oxygen-sapping chore. Four states share Lake Michigan waters, as figure 3-1 indicates. Michigan occupies the largest portion of the watershed and shoreline, and hence has a very direct interest in the water quality; but Wisconsin, by virtue of its heavy concentrations of manufacturing, shares this interest and is no less responsible for the water's quality. The magnitude of Indiana's contribution, partially offset by a smaller watershed and coastline, is increased by the heavy concentration of industry and population in the Gary-Hammond-East Chicago area, situated at the southern-most section of the lake. Illinois, although occupying a very small area of the watershed, is greatly concerned with Lake Michigan water quality, since the large concentration of industry and population in and around the Chicago Metropolitan area depends almost entirely on Lake Michigan as a water source. However, most of the water used in the Chicago Metropolitan area is not returned to the lake, but diverted south through the Illinois River [1: II, 528].

Industrial activity in the Lake Michigan watershed is diverse, with approximately 744,000 persons employed in manufacturing in 1969. The major water-using industries, including the following, employed some 154,000 persons, or about 21 percent of manufacturing personnel: food and kindred products, primary metals industries, and leather tanning are important in the Milwaukee area. The Gary-Hammond-East Chicago area of Indiana contains an especially heavy concentration of primary metals industries, and considerable chemical manufacturing and petroleum refining. Many pulp, paper, and paperboard mills are located in the watershed, primarily in Wisconsin, along the Fox River and other tributaries to Green Bay, and in Kalamazoo County, Michigan, along the Kalamazoo River [1: II, 529].

In 1970, an estimated 6,330,000 people lived within the Lake Michigan watershed boundary, and population projections indicate a 12.2 percent increase to approximately 7,105,000 persons by 1980. Major population concentrations are in the Milwaukee and Racine areas of Wisconsin, the Gary-Hammond-East Chicago area of Indiana, and in the St. Joseph and Grand River basins in Michigan. Although most of the Chicago Metropolitan area is not located within the Lake Michigan drainage basin, a large part of its 7,000,000 people use Lake Michigan for water supply and other purposes.

Magnitude of the Water Pollution Problem

In the procedure outlined in Chapter 2, the first step in assessing the polluting potential of a waste is to calculate its biochemical oxygen demand (BOD) in pounds. Because control of dissolved oxygen is important in water quality management programs, a means of measuring the oxygen-consuming potential of wastes is necessary [2]. The test commonly used for this purpose is the BOD test, which measures the amount of oxygen required for the biological decomposition of organic solids to occur under aerobic conditions at a standard time (usually five days), at a temperature of 20 degrees Centigrade [3]. Thus, the BOD test is a useful procedure for calculating and standardizing the polluting strength of organic materials discharged to waterways. BOD factors, in coefficient form, are available for a limited number of industrial processes and products. Care must be exercised in estimating gross BOD discharged from a particular industrial plant, since BOD coefficients, calculated on a per employee or per unit of production basis, can vary tremendously, depending on the type of production process used. For this reason, general BOD coefficients, not specifically calculated for an individual industrial plant but applied to that plant's output, may not result in an accurate, absolute measure of the polluting potential of the waste. Nevertheless, useful comparisons, on a relative basis, among industries can be made and the magnitude of the problem highlighted. A similar problem does not exist in regard to municipal domestic sewage. BOD calculations, on a per capita basis, have shown .17 pounds of BOD per person per day to be the mean figure, with a fairly small deviation around this average [4].

Although BOD, in terms of waste discharged, is often a reliable indicator of water quality, and is therefore widely used, many other toxic or otherwise undesirable substances and compounds find their way into waterways. These pollutants may pose serious problems, in that they are often not biodegradable or otherwise resist the natural purification process. In many cases they have been identified, and their concentrations in the receiving body of water determined, but to date few calculations have been performed to determine their rate of emission from the industrial plant or other source. Consequently, this investigation focuses on waste generation in terms of BOD and must necessarily exclude from consideration the very wide array of present and potential pollutants.

By far the greatest overall source of pollution to Lake Michigan is industrial waste, although in some areas of the lake municipal sewage discharge takes precedence. Other contributors to the pollution problem include land runoff, feed-lot runoff, harbor dredgings, pesticides, oil spills, and pleasure and commercial boating discharges. At times, the lake's shore may be cluttered with the dead alewives, a small, abundant fish which is afflicted occasionally with death on a massive scale.

The alewife situation vividly demonstrates how delicate the ecological balance of the Great Lakes system is. Not only has the alewife succeeded in crowding out almost all other desirable species of fish, but the massive annual die-offs create extremely serious and costly nuisance problems. This little fish, averaging about six inches in length, may have entered Lake Ontario via the St. Lawrence River from its natural habitat along the Atlantic Coast in the late 1800s. By 1890, it became the most abundant fish in Lake Ontario, and from there it migrated to Lake Erie, Lake Huron, Lake Michigan, and finally Lake Superior. The alewife is an extremely effective feeder, out-competing everything else, and has eliminated from Lake Michigan certain species of chub and substantially diminished populations of lake herring, smelt, yellow perch, and others. Lake Ontario, which should be highly productive in terms of a well-balanced and diverse fish population, instead contains over 99 percent alewife. In Lake Michigan, it is estimated that in excess of 90 percent of the fish flesh consists of alewife [1: II, 1099]. This is extremely unfortunate, because the alewife is a virtually useless fish. It is too small to appeal to the sportsman, is not good to eat, and commercially, as animal food, it has been only partially successful [5:43-44]. Furthermore, the massive die-offs have adversely affected municipalities, industries, and bathing beaches. One steel mill estimated it lost over $5 million in a ten-day period in April 1966, when cleaning screens on the water cooling system were unable to cope with alewife entering the intakes, even though the screening system removed over sixty tons of fish per day. One of Chicago's central district water filtration plants was forced to operate at reduced capacity during April 1965, when alewife caused breakdowns to 20 percent of the cleaning screens, which were handling ten tons of fish per hour [1: II, 1109-10].

No satisfactory solution to the alewife problem has yet been found. Prior to the introduction of the coho salmon to Lake Michigan, the alewife's predator was the lake trout, which itself was eliminated by the sea lamprey. The coho salmon was introduced to the lake in the early 1960s to help reduce the alewife population; but although it achieved great success both as a sport and commercial fish, its effect on the alewife was not as great as had been hoped. Moreover, scientists are not enthusiastic over these short-term solutions. They fear that an intensive stocking of predator species will not restore ecological balance, but may in fact cause greater instability because of cyclic interaction between the predator and the target species.

Table 3.1 indicates that the major water-using industries discharge about three times the amount of each class of waste as is discharged by the sewered population of the United States. Note the relative contribution in terms of BOD for the major industry groups. Chemical and allied products, an extremely complex and diverse industry, has the distinction of being the largest contributor, but the paper industry, smaller in size than the chemical industry, is characterized by an extremely concentrated effluent in terms of BOD.

Practically every industrial plant in the Lake Michigan watershed discharges some form of waste, unless treatment is 100 percent effective, but it is believed that a good estimate of present waste load generation and future projected waste load generation can be made by confining the analysis to the relatively few, but significant waste sources. This is also one of the limiting factors in the study, since BOD coefficients are available for relatively few industries.

In general, the procedure for calculating gross waste totals in terms of BOD was to first estimate total BOD by industry for each state in the Lake Michigan watershed, and then to apportion it to localized areas of the lake. The estimated gross industrial BOD for the selected industries is shown in table 3.2, and the projected 1980 totals are shown in table 3.3. Tables 3.2 and 3.3 were constructed using the sources and procedures of Chapter 2.

Municipal sewage poses a significant problem for many of the more populated local areas in the Lake Michigan drainage basin. However, much of the Lake Michigan watershed is occupied by fairly sparsely populated communities, many of which do not have integrated sewer systems, but rely on septic tanks. Consequently, at present, Lake Michigan is not substantially affected by the low concentrations of municipal sewage from these sparsely populated areas; but by 1980, many of these municipalities will have sewer systems, so it is desirable to examine the overall sewage effluent potential for the lake as a whole. Table 3.4

Table 3-1
Estimated Volume of Industrial Wastes Before Treatment—1964

	Wastewater Volume (billion gal.)	BOD (million lbs.)	Suspended Solids (million lbs.)
Chemical and Allied	3,700	9,700	1,900
Paper and Allied	1,900	5,900	3,000
Food and Kindred	690	4,300	6,600
Meat Products	99	640	640
Dairy Products	58	400	230
Canned and Frozen Vegetables	87	1,200	600
Sugar Refining	220	1,400	5,000
All others	220	670	110
Petroleum	1,300	500	460
Primary Metals	4,300	480	4,700
Blast Furnaces & Steel	3,600	160	4,300
All others	740	320	430
All Manufacturing	13,100	22,000	18,000
Sewered Population of U.S.	5,300[a]	7,300[b]	8,800[c]

Source: U.S. Department of the Interior FWPCA, COST OF CLEAN WATER, Detailed Analysis, Vol. 2 Jan. 1968.

[a] 120 million people times 120 gal. times 365 days.
[b] 120 million people times 1/6 lbs. times 365 days.
[c] 120 million people times .2 lbs. times 365 days.

Table 3-2
Estimated BOD[a] of Discharges of Plants in Selected Industries, Lake Michigan Watershed, 1969 (1,000 pounds)

SIC #[b]	Industry	Indiana	Michigan	Wisconsin	Total Lake Mich.
2011	Meat Slaughtering Plants	9,552	6,034	16,494	32,080
2013	Meat Processing Plants	557	1,906	847	3,310
2026	Fluid Milk	971	1,311	1,160	3,442
2033	Canned Fruits & Vegetables	9,156	11,970	30,880	52,006
2621	Paper Mills, except building	6,360	72,961	196,676	275,997
2631	Paperboard mills	23,442	66,446	29,850	119,738
2641	Paper coating and glazing	2,325	3,241	5,208	10,774
2911	Petroleum refining	13,690	4,491	–	18,181
3111	Leather Tanning	–	5,925	22,005	27,930

Source: Estimated using data on employment in relevant industries from U.S. Department of Commerce, Bureau of the Census, COUNTY BUSINESS PATTERNS and CENSUS OF MANUFACTURERS in accordance with the procedure outlined in Chapter 2.

[a] BOD—Biological Oxygen Demand.
[b] SIC—Standard Industrial Classification.

Table 3-3
Projected BOD[a] of Discharges of Plants in Selected Industries, Lake Michigan Watershed, 1980 (1,000 pounds)

SIC #[b]	Industry	Indiana	Michigan	Wisconsin	Total Lake Mich.
2011	Meat Slaughtering Plants	6,046	5,971	19,023	31,040
2013	Meat Processing Plants	472	2,678	1,066	4,216
2026	Fluid Milk	810	1,071	1,146	3,027
2033	Canned Fruits and Vegetables	11,654	31,777	76,885	120,316
2621	Paper Mills, except building	18,480	130,099	281,967	430,546
2631	Paperboard Mills	30,523	70,769	43,275	144,567
2641	Paper Coating and Glazing	7,087	2,373	7,020	16,480
2911	Petroleum Refining	10,812	6,890	–	17,702
3111	Leather Tanning	–	5,925	22,544	28,469

Source: Estimated using data on employment in relevant industries from U.S. Department of Commerce, Bureau of the Census, COUNTY BUSINESS PATTERNS and CENSUS OF MANUFACTURERS in accordance with the procedure outlined in Chapter 2.
[a]BOD—Biological Oxygen Demand
[b]SIC—Standard Industrial Classification

summarizes the estimated 1970, and the projected 1970 and 1980 populations and resulting BODs of municipal sewage discharges for the individual drainage basins and areas located within the overall Lake Michigan watershed. BOD was calculated on the basis of .17 pounds, per person, per day. The phosphate and nitrogen contribution is estimated to be three and six pounds per capita, per year, for the respective nutrients [6]. Phosphates and nitrogen are essential plant nutrients, and contribute heavily to a process known as eutrophication. Although, on average, perhaps only 65 percent or less of the total municipal BOD generated is discharged to Lake Michigan, some local areas have a much higher discharge rate, particularly the Milwaukee area, where municipal sewage is the leading source of water pollution.

Thermal Pollution

Thermal pollution has been defined as man-caused, deleterious changes in the normal termperature of water [7], and it differs from the more commonly thought of forms of pollution in that no foreign or toxic substances are discharged to a body of water; only the temperature is altered. Although many industries use water for cooling purposes, the electric power generating industry alone accounts for almost 80 percent of all cooling water used. Steam electric-generating plants use water for cooling condenser purposes, and the amount of water required depends on the type of plant, its efficiency, and the designed temperature rise within the cooling condensers. The temperature rise is usually

Table 3-4
Projected Population and BOD of Municipal Sewage Generated in Lake Michigan Drainage Basin, by Areas, 1970 and 1980 (Pounds per day)

	Calumet Area	Milwaukee Area	Green Bay Area	St. Joseph River Basin	Kalamazoo River Basin	Grand River Basin	Muskegon River Basin	Manistee River Basin	Total Lake Michigan
1970 Estimated Population	814,000	1,789,000	844,000	982,000	371,000	1,020,000	252,000	249,000	6,330,000
BOD 1,000 lbs./day	13.8	304.0	143.0	167.0	6.3	17.5	43.0	42.0	1076.0
Phosphates lbs./year (1,000 lbs.)	2,442.0	5,367.0	2,532.0	2,946.0	1,113.0	3,087.0	756.0	747.0	18,990.0
Nitrogen lbs./year (1,000 lbs.)	4,884.0	10,734.0	5,066.0	5,892.0	2,226.0	6,172.0	1,512.0	1,494.0	37,980.0
1980 Projected Population	997,000	2,016,000	865,000	1,143,000	407,000	1,179,000	276,000	222,000	7,105,000
BOD 1,000 lbs./day	170.0	343.0	147.0	194.0	6.9	20.0	4.7	3.8	1,207.8
Phosphates lbs./year (1,000 lbs.)	2,991.0	6,048.0	2,595.0	3,429.0	1,221.0	3,537.0	828.0	666.0	21,315.0
Nitrogen lbs./year (1,000 lbs.)	5,982.0	12,096.0	5,190.0	6,858.0	2,442.0	7,074.0	1,656.0	1,332.0	42,630.0

Source: Estimated using data on population and projected population from U.S. Department of Commerce, Bureau of the Census, COUNTY BUSINESS PATTERNS and CENSUS OF MANUFACTURERS in accordance with the procedure outlined in Chapter 2.

in the range of 10 to 20 degrees F., but nuclear steam electric plants require about 50 percent more condenser water, because of their less efficient operation, in terms of rejected heat [8].

It is anticipated that Lake Michigan will experience a more rapid rate of power production increase than the national average, because of the large volume of cooling water available and the rapid rate of industrial growth along its shores [9: II, 578]. Ten nuclear power plants are scheduled for completion along the shores of Lake Michigan by 1973, in addition to the eighteen major fossil-fueled plants currently operating, whose combined total capacity exceeds the nuclear power potential. In addition to the ten nuclear plants operating, being built, or planned, and in total to produce 7.1 million KWH of electricity, the AEC Reactor Development Division has estimated that an additional 6 million KWH of nuclear power will be needed for the Lake Michigan area by 1980. Plans for these plants and their sites have not yet been proposed.

The following formulae may be used to calculate the amount of heat in BTUs (British Thermal Units) rejected by fossil-fueled and nuclear plants, assuming no cooling devices are used [8].

For fossil-fueled plants: BTU/KWH = 3800 BTU times KWH
For nuclear plants: BTU/KWH = 6400 BTU times KWH

Table 3.5 shows the power capacity in kilowatts for all power plants, present and future, to be operable on the shores of Lake Michigan in the 1970s. Application of the above formulae indicates that all plants, operating at full capacity and not utilizing cooling devices, would discharge an estimated 76 billion BTUs per hour to Lake Michigan. The potential amount of heat rejected could be further increased by about 38.4 billion BTUs, if additional nuclear

Table 3-5

Major Fossil-Fueled and Nuclear Power Plants Located on Lake Michigan (Million kilowatts)

State	Operating Plants				Under Construction		Additional Power Projected	Total KWH	Total BTU (billion)
	Coal No.	KWH	Nuclear No.	KWH	Nuclear No.	KWH			
Illinois	3	2.128			2	2.100		4.228	21.5
Indiana	3	1.298			1	.515		1.813	8.4
Michigan	5	1.340	1	.070	3	2.808		4.218	22.5
Wisconsin	7	3.555			3	1.521		5.076	23.2
							6.000		38.4
Total	18	8.321	1	.070	9	6.944	6.000	21.335	114.0

Source: League of Woman Voters, Lake Michigan Inter-League Group, THERMAL POLLUTION, Chicago, Illinois, 1970, p. 1.

plants to produce 6 million kilowatts are constructed, as is estimated by the AEC Reactor Division.

While the overall long-range effects of thermal pollution are largely unknown, and are the subject of much current debate, it is agreed that in some instances heated water can be detrimental to fishes and aquatic life and can speed up certain undesirable biological processes. In addition, the value of water for drinking, recreational purposes, and industrial uses usually decreases at higher temperatures.

The magnitude of the Lake Michigan water pollution problem is certainly sizable and gross waste generated from all sources can be expected to increase significantly by 1980. Fortunately, remedies are available to combat the problem. These remedies may be thought of as falling into two general classifications: (a) adequate treatment facilities, employing sufficiently advanced levels of technology, and (b) legislative abatement programs.

Pollution Abatement — Programs and Techniques

Levels of Treatment Effectiveness

The strength and polluting potential of wastes discharged to Lake Michigan should not be substantially affected by any limitations in the state of present technology. Treatment approaching the 95 to 99 percent level of effectiveness is currently possible and feasible for the bulk of municipal and industrial wastes, including rejected heat. All indications point to the continuing trend of joint municipal and industrial waste treatment facilities, which should provide increasing economies of scale. Any level of purification can be obtained, but economic considerations usually determine the combinations of operations and the residual pollution potential of effluent discharged. Efficiencies are lowered when treatment plants are overloaded and part of the sewage bypasses the treatment facility. Primary treatment can be between 40 and 70 percent effective in removing suspended solids, putrescible matter, and bacteria. Secondary treatment commonly ranges from 80 to 95 percent effectiveness. Chlorination of the effluent can increase the destruction of bacteria to 99 percent or more. Advanced waste treatment (beyond secondary treatment) takes a variety of forms and can render the sewage almost pure by eliminating most of the previously unaltered compounds, such as phosphates and dissolved metals [10]. These efficiency rates refer primarily to municipal sewage treatment plants, but similar removal percentages can be attained by industrial plants. The most likely course of action is that industrial plants will increasingly provide some form of primary treatment, possibly removing some of the more toxic chemicals and metals not amenable to municipal treatment processes, and discharge the pre-treated effluent to municipal plants for further treatment. Obviously, the question is not, can we provide adequate waste treatment, but rather will we provide it, and if so, when? The Four-State Enforcement Conference indicated that Lake Michigan states will provide secondary treatment, including 80 percent phosphate removal, by 1972 for all major municipalities. If this is to be so, the question still to be answered is, what effect will this have on water quality in view of increased industrial output?

Waste heat rejected to Lake Michigan from electric generating plants can be reduced by the use of cooling ponds or cooling towers. Cooling ponds or reservoirs are the simplest method of cooling thermal effluent, and the least efficient in terms of air-water contact. Surface area requirements may average about two acres per megawatt output, so this system could not be used where land is unavailable or expensive [8]. There are many versions of cooling towers, and the terminology applied stems from basic differences in design or operation. The "wet tower" is the least expensive and may see widespread use as a cooling device along the shores of Lake Michigan. Strict thermal pollution standards were finally set by the Four-State Enforcement Conference, after years of debate and discussion, in March 1971. The thermal pollution standards accepted will prevent the discharge of any heated water into the lake that would raise the water temperature 3° F. above the normal temperature at a point 1,000 feet from the discharge pipe. To meet these standards, all shoreline plants, fossil-fueled and nuclear, now operating or under construction will have to upgrade their cooling facilities. Power company officials have estimated that it will cost a total of $280 million for all present and future plants to meet the new standards [11]. In March 1971, the Northern Public Service Company broke the power industry's solid front against the use of cooling devices and agreed to build a "wet tower" at its Bailly, Indiana, nuclear plant. Subsequently, the Consumers Power Company agreed to build a cooling tower at its South Haven, Michigan, nuclear plant.

Thus, in view of these recent developments, it is clear that thermal pollution on Lake Michigan will fall far below its full unchecked potential. This does not mean, however, that adverse effects arising from the reduced amount of heated water discharged will not materialize. Further research and time are needed to analyze and evaluate the possible and probable effects. In any case, the cost and scope of the undertaking are certainly not inconsequential.

The Four-State Enforcement Conference

An "enforcement conference" is a procedure called for under the Federal Water Quality Act of 1956. Its purpose is to get the federal and state authorities together to set water quality standards on a watershed basis. The need for water quality standards for Lake Michigan was perceived in 1965, the year that the public was first informed that this lake could conceivably suffer the same fate as Lake Erie. The Calumet or Two-State Enforcement Conference was convened late that year, in recognition of the distinct characteristics and special pollution sources of the southern basin. The Calumet Conference set standards for such matters as oil and turbidity, and deadlines for installment of treatment facilities by municipalities and industries [12]. In 1968, the first Four-State Enforcement Conference got underway. Its purpose was to hear testimony from the various pollution control agencies of each of the four states concerning pollution problems, legislation, pollution control techniques, and water quality goals, and decide upon common pollution standards at the termination of the conference. Of significance to this study are the following:

1. "The small quantity of oxygen normally dissolved in the water is perhaps the most important single ingredient necessary for a healthy, balanced aquatic life environment. The discharge of treated and untreated municipal and industrial wastes with their high concentrations of biochemical oxygen demand have caused oxygen depletion in many of the Lake Michigan tributaries and in some harbors. At present the main body of Lake Michigan has not yet evidenced signs of oxygen deficiency" [13:3324].
2. "Eutrophication is a threat now to the usefulness of Lake Michigan and other lakes within the basin. Feasible methods exist for bringing this problem under control" [13:3294].
3. "Evidence of severe bacterial pollution of tributaries has been found in the Fox River between Lake Winnebago and Green Bay, Wisconsin; in the Milwaukee River within Milwaukee, Wisconsin; in and downstream from the cities along the Grand River in Michigan and the St. Joseph River in Indiana and Michigan; and in the streams of the Calumet area, Illinois and Indiana. Although the bacterial quality of Lake Michigan is generally good in deep water, the water is degraded along the shoreline and in harbor areas" [13:3308].
4. "A contaminant entering directly into Lake Michigan, or dissolved in the water that feeds the lake, mixes with and eventually becomes an integral part of the lake water as a whole—regardless of the point of origin around the periphery or on the contributory watershed" [13:3309].
5. "The combined impact of siting many (nuclear) reactors on the shores of the lake poses a threat, particularly with respect to thermal pollution problems, and the accumulation of radionuclides in the ecosystem" [13:3347].

These general conclusions indicate that the lake has not yet been adversely affected by pollution, but that water quality degradation has been experienced in local inshore areas. Effluent discharged from municipal and industrial sources results in an increased rate of eutrophication, a lowered level of dissolved oxygen, and bacterial pollution. A number of general recommendations, including the following, were made and adopted by each of the four states as part of their individual abatement programs [13:3347]:

1. Advanced waste treatment should be provided by municipalities serving populations of 5,000 or more, or receiving over 800 pounds of BOD per day. This means at least 80 percent treatment effectiveness by December 1972.
2. By December 1972, adjustable overflow regulating devices should be installed on existing combined sewer systems to eliminate combined sewer overflow problems.
3. In conjunction with all urban reconstruction projects, combined sewers should be separated and prohibited on all new developments to eliminate pollution from combined sewers by July 1977.
4. Continuous disinfection should be provided for municipal effluents to eliminate dangerous quantities of coliform bacteria by December 1969.

5. Treatment facilities for industrial plants should be constructed by December 1972.
6. Industries should be encouraged to discharge treatable wastes to municipal sewers.
7. Industries not connected to municipal sewage systems and producing over 800 pounds of BOD per day should provide secondary treatment by December 1972.

Subsequent sessions of the Four-State Enforcement Conference, which meets annually, have continued to endorse these recommendations, subject to minor changes and alterations. The schedules call for widespread use of secondary and advanced waste treatment facilities by industries and municipalities by December 1972. By 1977, the pollution problems of combined sewers should be eliminated. Combined sewers collect both domestic sewage and rainwater. When there is heavy rainfall, the holding tanks and treatment facilities become flooded, much raw sewage bypasses the treatment facility, and flows directly to the receiving body of water, causing severe temporary pollution problems. Newly constructed sewer systems will have separate storm and sanitary sewers, but it will be quite some time before many of the older combined sewer systems can be separated.

The Localized Nature of Water Pollution

Because Lake Michigan is vast and its assimilative capacity is great, the overall water quality is still very good, particularly in the deeper regions of the lake, but pollution is becoming more and more apparent in the industrialized sections. Although pollution problems frequently manifest themselves in areas of concentrated population and industrial activity, they are not necessarily confined to those areas. One paper mill situated in an otherwise sparsely utilized area can have a tremendous impact on local water quality. Moreover, pollution problems cannot be assumed to be of a static nature. An examination and projection of historical trends may disclose significant shifts and changes in population and industrial activity, precipitating corresponding shifts in the pollution problem. Therefore, the approach used in this chapter is intensive in that it examines local water quality problems after the consideration just completed of the overall magnitude of the problem. Presently unaffected areas cannot be assumed to remain unaffected, unless projected waste loads and expected levels of treatment support this assumption. Consequently, our analytical technique calls for isolating individual drainage basins within the Lake Michigan watershed, and projecting future waste loads, levels of treatment, and water quality.

The entire Lake Michigan watershed is broken down into nine separate, natural drainage basins, or areas. Seven of these basins drain into major rivers, which flow into Lake Michigan. Two basins, or areas, the Calumet area of Indiana and the Greater Chicago area of Illinois, do not converge their wastes on major tributaries, but deliver the waste either directly to the lake or through various streams and creeks.

An estimation of gross waste loads in terms of BOD for each local area was made by apportioning the total gross waste for the four states as a whole, as previously calculated, to each local area for the relevant manufacturing industries. Briefly, the apportionment method is as follows:

1. Obtain total employees for each selected four-digit SIC industry group from *1969 County Business Patterns,* for each state.
2. Obtain similar employment data, as above, for each drainage basin (on a county basis).
3. Calculate basin employees as a percentage of state employees, by industry group. (#2 ÷ #1)
4. Apply percentage factor to total waste for each state (previously calculated) to obtain drainage basin waste.

The municipal sewage contribution was calculated by applying the .17 pounds per person, per day coefficient to the estimated population for each county in the drainage basin. Table 3.6 shows the estimated gross BOD generation for 1969 by drainage basin. Note the magnitude of the problem in the Green Bay and Milwaukee areas. The seriousness of the problem in the Calumet area of Indiana is somewhat obscured by the fact that BOD generation figures from steel mills, by far the largest industry in the area, are not available. The Greater Chicago area, which discharges most of its wastes to the Illinois River, is subject to special analysis, not conducive to the analytical methods used for the other drainage basins. Table 3.7 shows a projected 24 percent increase in gross BOD generation for the lake as a whole, with the Green Bay area bearing the brunt of the impact. What is the present water quality in some of these local areas, and what may we expect for 1980? Only an area by area analysis can properly consider these questions.

The Calumet Area of Indiana

The Calumet area of Indiana is highly industrialized and characterized by heavy industry, most of which is concentrated in the Gary-Hammond-East Chicago complex. Steel mills employ an estimated 53,000 persons, or about one-half of all manufacturing employees in the Calumet basin. Petroleum refining requires over 5,000 workers, and the chemical industry employs another 3,000 workers in various capacities. Wastes discharged either directly or indirectly to Lake Michigan include high levels of BOD, settleable and suspended solids, oil and grease, heavy metals, and other pollutants.

Table 3.6 indicates that gross BOD generation in 1969 was about 66 million pounds per year, but this figure does not include wastes from the major industry—steel. No BOD coefficients are available for steel mills, but Table 3.8 shows the amount of wastes generated per day by the major steel works in terms of suspended solids and oil. Since 1966, a water sampling program has been underway in the Calumet area. Table 3.9 summarizes the results and shows how the present water quality may be assumed to be affected by the aforementioned wastes. Results of the water quality tests show that generally there has been no overall improvement since 1966 [14]. An annual average of 90 percent of saturation of dissolved oxygen was set as the desired mean standard, and test results show that the actual amount of dissolved oxygen exceeded the 90 percent standard by a small degree, although there was no overall improvement over the years. Oil and grease, an important parameter associated with water treatment problems, indicated a deterioration in water quality. Threshold odor, another parameter important in water quality treatment problems, showed no

Table 3-6
Estimated BOD of Discharges of Municipal Sewage Treatment Plants and Plants in Selected Industries by Drainage Basins, Lake Michigan Watershed, 1969 (thousands of pounds)

SIC Number	Source	Calumet Area	Milwaukee Area	Green Bay Area	Kalamazoo River Basin	Grand River Basin	Muskegon River Basin	Manistee River Basin	St. Joseph River Basin	Total Lake Michigan
2011	Meat Slaughtering Plants	325	3,645	2,309	–	253	–	–	325	6,857
2013	Meat Processing Plants	75	407	56	–	282	–	–	254	1,074
2026	Fluid Milk	95	418	88	54	156	51	–	219	1,081
2033	Canned Fruits and Vegetables	–	4,725	4,138	–	–	–	5,697	2,143	16,703
2621	Paper Mills, except building	1,762	–	103,649	36,991	–	–	–	2,043	144,445
2631	Paperboard Mills	–	8,895	6,060	35,150	3,322	–	–	17,542	70,969
2641	Paper Coating and Glazing	588	375	4,234	681	–	1,215	–	–	7,093
2911	Petroleum Refining	12,540	–	–	–	853	–	–	–	13,393
3111	Leather Tanning	–	20,222	–	–	4,853	1,132	–	–	26,207
	Municipal Sewage (Total Population)	50,370	110,960	52,195	22,995	63,875	15,695	15,330	60,955	392,375
	Total Calculated BOD	65,755	149,647	172,729	95,871	73,954	18,093	21,027	83,481	680,197

Source: Estimated using data on employment in relevant industries from U.S. Department of Commerce, Bureau of the Census, COUNTY BUSINESS PATTERNS and CENSUS OF MANUFACTURERS in accordance with the procedure outlined in Chapter 2.

Table 3-7
Projected BOD of Discharges of Municipal Sewage Treatment Plants and Plants in Selected Industries by Drainage Basins, Lake Michigan Watershed, 1980 (thousands of pounds)

SIC Number	Source	Calumet Area	Milwaukee Area	Green Bay Area	St. Joseph River Basin	Kalamazoo River Basin	Grand River Basin	Muskegon River Basin	Manistee River Basin	Total Lake Michigan
2011	Meat Slaughtering Plants	206	4,204	2,663	206	–	251	–	–	7,530
2013	Meat Processing Plants	63	513	70	235	–	–	–	–	1,002
2026	Fluid Milk	79	413	87	179	44	128	41	–	971
2033	Canned Fruits and Vegetables	–	11,763	10,303	5,688	–	–	–	15,125	42,879
2621	Paper Mills, except building	5,119	–	148,597	3,643	65,960	–	–	–	223,319
2631	Paperboard Mills	–	12,896	9,227	18,683	37,437	3,538	–	–	81,781
2641	Paper Coating and Glazing	1,793	505	5,265	–	498	–	890	–	8,951
2911	Petroleum Refining	9,904	–	–	–	–	1,309	–	–	11,213
3111	Leather Tanning	–	20,717	–	–	–	4,853	1,132	–	26,702
	Municipal Sewage (Total Population)	62,050	125,195	53,655	70,810	25,185	73,000	17,155	13,870	440,920
	Total Calculated BOD	79,214	176,206	229,867	99,565	129,124	83,079	19,218	28,995	845,268

Source: Industrial—Projections made using data on employment in relevant industries from U.S. Department of Commerce, Bureau of the Census, COUNTY BUSINESS PATTERNS and CENSUS OF MANUFACTURERS in accordance with the procedure outlined in Chapter 2.

Municipal sewage—BOD projections based on usual method of estimation described in Chapter 2 and population projections for relevant areas from following:

R. Raja Indra, Michigan POPULATION HANDBOOK MICHIGAN DEPARTMENT OF HEALTH, Lansing, Nov. 1965, 120 pp.

INDIANA POPULATION PROJECTIONS, 1965-1985, Vol. 1, Bureau of Business Research, Graduate School of Business, Indiana University, Sept. 1966.

WISCONSIN POPULATION PROJECTIONS State of Wisconsin, Bureau of State Planning, Madison, April 1969, 145 pp.

ADVANCE REPORT ON 1970 CENSUS OF POPULATION IN WISCONSIN Document #BSP-IS-71-1(3) State of Wisconsin, Bureau of State Planning, Jan. 1971, 69 pp.

% increase 1969-1980: 24%.

Table 3-8
Discharge to Lake Michigan of Major Steel Mills in the Calumet Area, 1968 and 1970

Industry	Year	Flow MGD	Suspended Solids 1000 lbs./Day	Oil 1000 lbs./Day
Inland Steel Co.	1968	913	127.4	20.8
East Chicago	1970	910	75.0	12.0
Youngstown Sheet and Tube Co.	1968	324	74.3	15.7
East Chicago	1970	200	20.0	8.0
U.S. Steel Corp.	1968	537	198.0	44.0
Gary	1970	535	104.4	35.0
Total	1968	1,774	399.8	80.9
	1970	1,645	199.0	55.0

Source: WATER QUALITY IN THE CALUMET AREA, Conference on Pollution of Lower Lake Michigan, Calumet River, Grand Calumet River, Little Calumet River, and Wolf Lake, Illinois and Indiana Technical Committee on Water Quality, September 1970.

apparent change from 1966 to 1969, but it still exceeded the desired criteria, as did total phosphorus, an essential plant nutrient. In summary it appears that the overall water quality in the Calumet area in 1969 was, perhaps, fair. Some of the water quality parameters were within the acceptable limits, and a few others exceeded the criteria. Although the overall water quality throughout the basin area was basically uniform, individual water quality parameters did at times show marked deviations from station to station. Since some of the sampling stations were within several thousand feet of each other, it is evident that water quality problems can be extremely localized.

The future situation holds promise for some improvement, if not in water quality, then at least in terms of total wastes discharged to the basin area. Gross waste projections for 1980 indicate an increase of BOD generation of about 13 million pounds to about 79 million pounds, as shown in Table 3.7. However, these gross BOD figures are calculated before treatment is provided. Treatment at the 80 percent removal level or higher is the stated minimum requirement by 1980. Assuming overall treatment efficiency will be 80 percent (it may approach 90 percent), approximately 16 million pounds of BOD will still be discharged to Lake Michigan annually in the Calumet area. In addition, Table 3.8 indicates a large reduction in suspended solids, oil, and an assumed 50 percent reduction in BOD by 1970 for the major steel mills [14], a trend that could very well continue to 1980. The major oil refineries are scheduled to complete additional treatment facilities by 1971, but phosphorus removal is not contemplated until 1972, when it should be reduced by about 80 percent. Will the increase in treatment removal efficiency be enough to offset the increase in waste generation? The answer to this question must precede a realistic projection of water quality in the Calumet area for 1980. The present overall treatment efficiency is unknown, but if it were as high as 75 percent there would be no net reduction in

Table 3-9
Water Quality Parameters: Current Situation for Indicated Year

Parameter Year	Calumet Area (1) (1970)	Chicago Area (2) (1969)	Milwaukee Area (3) (1966)	Green Bay Area (4) (1966)
Coliform Bacteria	Yes	Yes	No	No
Fecal Bacteria	N.A.	Yes	No	No
Turbidity	N.A.	No	–	–
% Saturation Dissolved Oxygen	Yes	Yes	No	No
BOD	N.A.	N.A.	No	No
Oil and Grease	N.A.	Yes	–	–
Odor	No	Yes	–	–
Temperature	–	Yes	–	–
Total Phosphates	No	No	No	No

meets standard—Yes
does not meet standard—No
no data—

Sources:
(1) WATER QUALITY IN THE CALUMET AREA, Technical Committee on Water Quality Sept. 30, 1970, Conference on Pollution of Lower Lake Michigan, Calumet River, Grand Calumet River, Little Calumet River, and Wolf Lake—Indiana and Illinois.
(2) 1969 SURVEYS OF LAKE MICHIGAN OPEN WATER QUALITY AND LAKE BED, State of Illinois Sanitary Water Board, Report to the Governor and 76th General Assembly, June 1970.
(3) COMPREHENSIVE WATER POLLUTION CONTROL PROGRAM FOR THE LAKE MICHIGAN BASIN, Milwaukee Area, Wisconsin, U.S. Dept. of the Interior, FWPCA Great Lakes—Illinois River Basin Project, Region V, Chicago, Ill., June 1966.
(4) COMPREHENSIVE WATER POLLUTION CONTROL PROGRAM FOR THE LAKE MICHIGAN BASIN, Green Bay Area, Michigan and Wisconsin, U.S. Dept. of the Interior, FWPCA Great Lakes—Illinois River Basin Project—Region V, Chicago, Ill., June 1966.

wastes discharged to Lake Michigan by 1980, because the slight improvement to an assumed minimum of 80 percent removal would be offset by the rise in industrial output and, hence, total wastes generated. This 75 percent figure is probably unrealistically high, so one could conclude that based on the information analyzed there will be substantial, absolute reductions in wastes entering Lake Michigan from Indiana by 1980, although all of the sewage facilities will not be completed until 1977. Therefore, one could hazard a guess that water quality will not deteriorate, and may gradually improve.

The Greater Chicago Area of Illinois

The Illinois water basin area of Lake Michigan is located northwest of and adjacent to the Calumet basin of Indiana. There are only about six major industries discharging process wastes directly to Lake Michigan in the basin. All other industries discharge process wastes to sewer systems that flow to the Waukegan and North Chicago sewage treatment works [15]. All seven municipal sewage treatment plants, discharging to Lake Michigan in Illinois, are within the North Shore Sanitary District (NSSD), which is engaged in a $60 million construction program. When completed, the high quality effluent to be produced by the proposed facilities will be discharged to the Illinois River watershed, with no flow to Lake Michigan [15]. All of the municipalities in Cook County, adjacent to Lake Michigan, are served by the Metropolitan Sanitary District of Greater Chicago intercepter sewer system and sewage treatment facilities, which discharge to the Illinois River basin.

Consequently, Lake Michigan escapes receiving most of the vast amount of wastes generated in the Chicago Metropolitan area; and therefore, similar waste calculations in terms of BOD, utilized for the other areas, cannot be related to present water quality. Water quality data collected during 1969 indicated that, generally, Lake Michigan open water was of somewhat better quality than during 1968 [15]. Table 3.9 shows how various significant water quality parameters for which data were collected complied with the applicable standards in 1969, overall water quality was considered to have been fair. The raw water intakes at the municipal water filtration facilities that utilize Lake Michigan as their source of water supply were designated as the control points for monitoring the quality of Lake Michigan open waters. Five plants were selected for the 1969 monitoring program, and since the samples were taken directly from crib intakes, they were representative of open water [15].

The monitoring of shore waters reflects the influence of recent hydrologic conditions and localized human activities. In contrast, the monitoring of open waters reflects long-term changes resulting from activities and conditions within a major portion of the Lake Michigan watershed. The term "open water" is a relative term; the farther from shore that a water sample originates, the more representative it becomes of open water. It should be noted that the intakes of the water plants sampled during 1969 are not all located the same distance from shore. The average distances from shore of the eleven intakes at the five designated water treatment plants is about one mile. Normally, along the Illinois shore, mixing between shore waters and offshore waters occurs slowly and to a limited extent. Current velocity gradients and water quality gradients also exist along the edge of the lake. Due to these conditions, it is possible that differences in water quality may exist between plants due to the location of the water intakes within different current and flow patterns [15].

The 1980 water quality for the Illinois section of the Lake Michigan basin cannot be projected from the waste loads generated in the Chicago Metropolitan area. By 1980 all municipal wastes will flow to the Illinois River basin, and the six major industries currently discharging to Lake Michigan will have installed sufficiently effective treatment facilities. On the basis of these data alone, water quality could be expected to improve significantly. However, there is substantial mixing of water from the nearby Calumet area and the Illinois area, so that we may assume that water quality of the two adjacent areas will closely parallel each other, with, perhaps, the Illinois portion being of slightly higher quality.

The Milwaukee Area

The Milwaukee River drainage basin is located in southeastern Wisconsin, and accepts wastes from all of Milwaukee County and parts of Kenosha, Ozaukee, Racine, Sheboygan, Washington, and Waukesha Counties. The Milwaukee River collects most of the wastes generated in this area and discharges to Lake Michigan at Milwaukee. In addition, various minor streams collect wastes from the same drainage basin and discharge to Lake Michigan near the City of Milwaukee. Municipal waste effluents represent the largest source of pollution in the Milwaukee area [16]. The estimated 1970 population of the Milwaukee River drainage basin is approximately 1.8 million, as shown on Table 3.4, and the projection for 1980 is for slightly over two million. While these figures include the rural population in the outlying areas of the drainage basin, over one million people are concentrated in Milwaukee County alone. Industry in the area includes: heavy metals (24,000 employees), beverages (9,000), chemicals (5,000), leather tanning (4,000), and several paper mills and various canning plants.

Table 3.6 shows that in 1969, approximately 150 million pounds of waste in terms of BOD was generated in the drainage basin. This figure is expected to increase to about 176 million pounds per year by 1980, as a result of the increased population and, to a lesser degree, from increased paper mill and canning activity. Table 3.9 shows water quality parameters for Milwaukee harbor and adjacent waters, based on samples taken during 1962-1963, the latest available series of tests. The waters in and around Milwaukee harbor were considered to be extremely polluted in 1963. Milwaukee waters failed by sufficient margins to meet the standards set defining acceptability for municipal water use and recreational activity. Such water quality parameters as fecal coliform, BOD, and phosphates showed high concentrations and correspondingly low values for dissolved oxygen, all characteristic of municipal sewage pollution.

Since most of the polluting wastes originate from municipal sewage treatment plants, increased treatment effectiveness will have to precede water quality improvement. Moreover, any improvement in municipal treatment facilities will have to take the form of advanced waste treatment. There are nineteen municipal water treatment installations in the area, and all but two currently provide secondary treatment. The overall average removal efficiency is about 86 percent [16]. Advanced waste treatment may improve this removal efficiency to 95 percent or better. In view of the already high level of BOD removal, advanced treatment with removal efficiency of 95 percent or better is not likely to have a very great impact on water quality in terms of dissolved oxygen. Continuous disinfection, soon to be provided at municipal treatment plants, will reduce the level of fecal coliform. An estimated 3.2 million pounds of phosphorus per year is generated in the Milwaukee drainage basin [16]. Scheduled for completion by December 1972 are facilities designed to remove 80 percent of this essential plant nutrient, so a reduction in undesirable algae growth and odor may be expected.

From the data evaluated, 1980 water quality in the Milwaukee area is not likely to be much improved from the present situation. Industrial waste discharge is of minor importance in the area. The concentrated municipal population poses the greatest problem. Advanced municipal waste treatment will probably not substantially reduce the level of BOD discharged to Lake Michigan even though the present 86 percent removal rate is improved upon.

The Green Bay Area

Green Bay is approximately 118 miles long and 23 miles wide, with a mean depth of about 65 feet. The watershed tributary to Green Bay contains a total drainage area of about 15,000 square miles, or about one-third of the total Lake Michigan drainage basin. The Fox River, the largest stream in the Lake Michigan basin, heavily polluted with paper mill effluent, discharges to Green Bay at the City of Green Bay, Wisconsin, at the southern end of the bay. Other subsidiary drainage basins in the area are the Oconto, Peshtigo, and Menominee River basins [17].

The most significant source of waste in the Green Bay area are the pulp and paper mills, discharging their wastes to tributaries or the bay itself. Municipal waste treatment plant effluents are the second largest source of pollution in the Green Bay area [17]. Table 3.6 shows that an approximate total of 173 million pounds of BOD were generated in the combined Green Bay drainage basin in 1969. Pulp and paper mills channel 90 million pounds into the Fox River alone, and in total generate about 114 million pounds annually. The water shed had an estimated 1970 population of 850,000, of which about 63 percent was served by municipal sewers [17]. Table 3.6 indicates an approximate annual waste load in terms of BOD of about 52 million pounds for the entire population, or about 31 million pounds for the sewered population. The Green Bay watershed generates about 5 million pounds of phosphates annually, with about 2.4 million discharged to Green Bay from the Fox River [17]. As previously noted, 80 percent of the nutrient is scheduled to be affected by December 1972.

Present water quality in the Green Bay area is extremely poor, particularly in the southern end of the bay. Physical water quality parameters have indicated this to be the general condition of the bay, but additional indicators of water quality are the fact that the City of Green Bay obtains water from Lake Michigan rather than Green Bay itself, and bathing beaches in the area have been closed for more than twenty-five years. Table 3.9 shows the results of water quality tests made from March 1963 to April 1964, the latest available test results. All of the rivers and streams tributary to Green Bay show varying degrees of pollution, but the lower Fox River is in the worst condition, having at times little or no dissolved oxygen for distances exceeding twenty miles [17].

By 1980, BOD generation should increase substantially, with the paper industry accounting for most of the increase. Table 3.7 shows a projected total of 230 million pounds of BOD generated from industrial and municipal sources in 1980. Present municipal sewage treatment effectiveness is about 60 percent [17], but industrial treatment removal levels are substantially below this figure. By 1980, sewage treatment effectiveness should increase to 80 to 90 percent in terms of BOD reduction. It is anticipated that industrial waste treatment will greatly improve to approach the recommended 80 to 90 percent treatment level by 1980. Substantial recent progress has been made in the area of joint municipal-paper mill treatment facilities. The American Can Co. and Charmin Paper Products Co. have agreed to build a combined treatment plant with the Green Bay Metropolitan Sewage District, to be completed by September 1972, at an estimated total cost of about $41 million. This will be the first treatment plant in the United States to mix domestic sewage and sulfite pulp mill wastes [18]. As a result of these various factors, water quality in some sections of Green Bay could be expected to improve slightly, and once again become useful for some not-too-demanding purposes. While the southern portion of the bay, the most severely affected, may also experience water quality improvement, it seems unlikely that the degree of improvement will be great, so the water quality will probably remain of unsatisfactory quality.

Other Troubled Areas of Lake Michigan

Local areas along the eastern shore of Lake Michigan are plagued by water quality degradation similar to, but less severe than, those experienced by the Milwaukee and Green Bay areas. The sources of waste are industrial plants and municipal sewage treatment facilities, located in the drainage basins of the major rivers in Michigan. Table 3.6 shows the 1969 estimated and Table 3.7 shows the future projected waste loads in terms of BOD for the St. Joseph, Kalamazoo, Grand, and Muskegon River basins. Water quality in these areas at the mouth of each river is still good, but signs of impairment are beginning to show.

The St. Joseph River collects wastes from Michigan and Indiana, and the major pollutant is municipal sewage discharge. A total of 83 million pounds of BOD was discharged in 1969, of which an estimated 61 million pounds originated from municipal sources. Projected BOD totals for 1980 are about 100 million pounds. More effective municipal sewage treatment facilities will reduce significantly the amount of discharged sewage, and slightly improved water quality is to be expected by 1980.

The major waste sources in the Kalamazoo River basin are paper mills. They accounted for an estimated 72 million pounds of BOD in 1969, and this figure is expected to increase to about 104 million pounds by 1980. Corresponding figures for municipal sewage are 23 million pounds in 1969, and 25 million pounds in 1980. When adequate treatment facilities are installed at the paper mills, some degree of water quality improvement should become evident at the mouth of the Kalamazoo River, although current water quality is still good.

The principal source of pollution in the Grand River is municipal sewage. The 1,029,000 people living in the Grand River drainage basin generated some 64 million pounds of BOD in 1969. The projection for 1980 is for about 73 million pounds. The industrial contribution should increase from an estimated 4.5 million pounds to a projected 5.5 million pounds in 1980, a relatively insignificant quantity. Water quality improvement should result from increased effectiveness of municipal sewage treatment facilities.

The Muskegon River drainage basin collected about 16 million pounds of municipal BOD in 1969, compared with only 2.3 million pounds from industrial sources. Lake Michigan water quality at Muskegon, Michigan, is good and should improve slightly by 1980, assuming improved municipal sewage treatment facilities are installed.

In general, the Michigan portion of Lake Michigan has not experienced serious water quality problems to date. This is not true, however, of waters adjacent to the eastern section of Michigan. Much of the industrial activity and greatest population concentrations are centered around Bay City, Saginaw, Flint, and Detroit, outside the Lake Michigan drainage basin. In the Detroit area, the Detroit River and the River Rouge are heavily polluted, intensifying pollution problems on Lake Erie and Lake St. Clair. The state of Michigan is fortunate in that its major rivers tributary to Lake Michigan are quite long, so a good deal of assimilation and purification takes place en route to the lake. In addition, few industrial centers are located on or near Lake Michigan, in sharp contrast to the other three Lake Michigan states. This, of course, does not eliminate the Michigan pollution problem, but merely shifts the focus of attention, and justifiably so, to the tributary rivers and streams.

Lake Michigan 1980 Water Quality Summary

The offshore waters of Lake Michigan are now of high quality. They are just beginning to show slight, subtle changes in the direction of eutrophication. The quality of local inshore waters ranges from very good to extremely poor. The extent and primary cause of water quality degradation varies from area to area, around the lake, but municipal sewage discharge and industrial waste effluents, singly or in combination, are by far the largest contributors. Harbor dredgings, land runoff, and indiscriminate use of pesticides, thermal effluents, and the abundant alewife contribute, at times, significantly to local water quality problems. In the Indiana and Illinois basin areas, municipal and industrial wastes seem to be jointly responsible for the, at times, substandard water quality, although the overall water quality generally ranges from fair to good. In the Milwaukee area, although there is substantial industrial activity, the discharge of municipal sewage is the major problem, accounting for over two-thirds of the BOD currently generated. In the Green Bay area of Wisconsin, the paper industry is by far the leading cause of water pollution. The situation is some sections of Green Bay, particularly the southern portion, is extremely serious, with some water quality parameters, notably dissolved oxygen, nowhere near any possible lower acceptable limit.

In view of the present bleak situation in various areas of the lake, one might well wonder what the outlook is for 1980. Will there be further degradation, can we expect a leveling off of the deteriorating water quality, or is there hope for some improvement? The impatient observer may charge that not enough has been done, and that no real progress has been made. This, however, does not seem to be the case at all. It must be borne in mind that a larger population and rising industrial output, coupled with a greater demand for water, insure that greater gross waste loads will continue to be generated. To contend that our efforts to date have been largely ineffective and useless tends to dismiss the fact that we must deal with greater waste quantities, generated by rising industrial output and population increases. Viewed in this light, substantial progress has been made. But the relevant question is: has a suitable framework been laid to overcome the problems of rising industrial output, a larger population, and the interrelated ramifications? The facts indicate that progress has been made, and that we may look forward to further absolute improvements in Lake Michigan water quality.

It was shown in this study that the Lake Michigan basin population is expected to increase by about 12 percent from 1970 to 1980, and that total BOD generation for the selected industries in the Lake Michigan watershed is expected to increase by about 24 percent for the same period. Consequently, pollution control must not only strive to meet adequately the presently generated levels of waste, but must at least keep pace with the projected increase. Sufficiently advanced technology exists to reduce significantly the amount of wastes now being discharged to the lake. Pilot projects in various areas of the United States have indicated that the discharge of the essential nutrient phosphorus can be significantly reduced at reasonable cost. In Green Bay, Wisconsin, a $41 million joint municipal and sulfite paper mill treatment plant is being constructed, the first of its kind in the United States. The paper industry as a whole is spending some $321 million in 1971, up 110 percent from 1970, and representing 20 percent of its planned 1971 capital spending for pollution control [19]. In the Calumet area of Indiana, the large steel mills will have effected by 1971 a 50 percent reduction in discharged wastes through new treatment facilities. The petroleum refining plants are in the process of effecting similar reductions in net waste loads. Pursuant to new federal regulations, cooling towers or other cooling devices will be built by all electric power generating plants operating on the shores of Lake Michigan, and thereby greatly alleviating the possibility of thermal pollution problems. The cost will be enormous—$280 million for Lake Michigan alone. It should be noted in this vein that the Big Rock nuclear plant near Charlevoix, Michigan, although relatively small in size, has been operating since 1963 with no known adverse effects to the local environment. This points up that wholesale spending is not necessarily the answer to the pollution problem. The money must be effectively utilized.

Not to be forgotten is the fact that many of the pollution abatement programs conceived during the late 1960s are scheduled to become effective during 1971, 1972, and thereafter. The Lake Michigan Four-State Enforcement Conference meets annually to evaluate progress being made, and to report on the enforcement activities of the individual states. In general, the state agencies and the affected municipalities and industries are proceeding to implement control programs and to provide the necessary treatment facilities, pursuant to the recommendations developed during the first session in 1968. It is reasonable to assume that delays will be encountered and extensions granted for various reasons, but it does seem that by 1980, indeed before then, substantial progress will be made in the reduction of wastes discharged to the lake. Whether or not water quality will improve to any substantial degree is quite another question. There are so many interacting variables and ramifications involved that clear, long-term projections of water quality are difficult to make. For instance, to reduce the alewife population, the coho salmon was introduced, which became a tremendous success. However, this is leading to increased sport and commercial fishing with the attendant watercraft pollution problems, and the development of new marinas and boating facilities, which themselves alter the local aquatic environment. Since on Lake Michigan, water quality problems are quite local in nature, improvements will probably be of local character. Water quality in Green Bay is so degraded that even dramatic improvement may not be enough to render the water useful for recreational and municipal purposes. It is also not clear that plans call for construction of enough additional treatment facilities in the Milwaukee area to have any substantial effect on local water quality. These are the two most severely affected areas of the lake, but whether the projected, substantial reductions in waste discharged will be sufficient to overcome the already extremely serious pollution problems is questionable.

Public concern leading to effective legislation may be the single most important factor in determining Lake Michigan water quality in 1980. There is little question that the increasingly affluent Americans, faced with a greater amount of leisure time, will turn to recreation of all forms as an outlet for their energies and incomes, and the demand for water-oriented recreation is sure to increase significantly. Millions of vacationers annually visit Lake Michigan and several factors, including the planting of the coho salmon, the decreased availability of inland property, and the increased use of inland lakes [20], indicate that Lake Michigan water will experience a much greater use in the near future, not only by visitors, but by watershed residents. Thus, the increased demand for Lake Michigan water will come from that very segment of society that has the means, the influence, and the interest to stimulate sufficient public concern to lead to effective legislation and pollution control programs.

Notes

1. U. S. Department of the Interior, FWPCA, *Conference on the Matter of Pollution on Lake Michigan and its Tributary Basins,* Chicago, Illinois, Feb. 1968.

2. U. S. Department of the Interior, FWPCA, Great Lakes Basin, Chicago, Illinois, Jan. 1968, Water Quality Investigations, Lake Michigan Basin, *Physical and Chemical Quality Conditions.*

3. T. L. Willrich and N. W. Hines (Eds.), *Water Pollution and Abatement* (Ames, Iowa: Iowa State University Press, 1965).

4. *Waste Management* (New York, N.Y.: Regional Plan Association Inc., 1968).

5. *Water Pollution Problems of Lake Michigan and Tributaries,* Great Lakes Region, Chicago, Illinois, Jan. 1968, p. 43-44.

6. *Metropolitan Seattle Sewage and Drainage Survey,* Brown, Caldwell, March 1958, p. 232.

7. U. S. Department of the Interior, FWPCA, Feb. 1968, *Thermal Pollution*

– *Its Effect on Water Quality,* Administrative Report presented to the Water Pollution Control Advisory Board, Chicago, Illinois, p. 1.

8. U. S. Department of the Interior, FWPCA, Northwest Region, Pacific Northwest Water Laboratory, Corvallis, Oregon, Sept. 1968, *Industrial Waste Guide on Thermal Pollution.*

9. *U. S. Department of the Interior,* FWPCA, *Proceedings – Second Session in the Matter of Pollution of Lake Michigan and Its Tributary Basin,* Chicago, Illinois, Feb. 1969.

10. M. Fair Gordon and John C. Geyer, *Elements of Water Supply and Waste Disposal* (New York, N.Y.: John Wiley & Sons, Inc., 1958).

11. *New York Times,* March 26, 1971, "3 Lake Michigan States Agree to Thermal Pollution Standards."

12. *Bulletin of the Lake Michigan Federation,* Chicago, Illinois, Oct. 20, 1970.

13. U. S. Department of the Interior, FWPCA, *Conference on the Matter of Pollution on Lake Michigan and Its Tributary Basins,* Chicago, Illinois, Feb. 1969.

14. *Water Quality in the Calumet Area,* Technical Committee on Water Quality, Sept. 1970.

15. State of Illinois, Sanitary Water Board, *1969 Surveys of Lake Michigan Open Water Quality and Lake Bed,* June 1970.

16. U. S. Department of the Interior, FWPCA, Great Lakes-Illinois River Basins Project, Region V., *Comprehensive Water Pollution Control Program for the Lake Michigan Basin,* Milwaukee Area, Wisconsin, Chicago, Illinois, June 1966.

17. U. S. Department of the Interior, FWPCA, Great Lakes Region, *A Comprehensive Water Pollution Control Program,* Lake Michigan Basin, Green Bay Area, Chicago, Illinois, June 1966.

18. *Green Bay Press-Gazette,* "Waste Plant Agreement Signed," Jan. 1971.

19. *Business Week,* May 15, 1971, "Pollutors Raise the Cleanup Ante."

20. League of Women Voters, *Lake Michigan Basin Study: Michigan Section,* Jan. 1968, Detroit, Michigan, p. 26-27.

4 Lake Erie

ROBERT VOLLKOMMER

The lives of close to eleven million people inhabiting its shores are supported and enhanced by Lake Erie. On its waters may be seen coal- and ore-carrying lakers, ocean-going vessels, and fishing and pleasure boats of every shape and description. Along its shores and tributaries sprawl populous cities drawing millions of gallons of fresh water for their inhabitants and industrial complexes. A vast variety of agricultural products are raised on the fertile lands of the eastern plains and Ohio's Maumee River basin. Hundreds of recreational and resort areas that support many service facilities use Lake Erie as a focal point. This great industrial, agricultural, recreational, and commercial activity is indicative of man's progress in the region. Associated with such progress is the problem of pollution.

As decades have passed, changes have occurred beneath the lake's surface. Since 1890, a marked change in the fish population has been noted. Game fishing has declined as the population of fish better adapted to live in polluted environments has increased [4:137]. As population and industry have grown, yields of desirable fish have fallen. Thus, from 1900 to 1910 the catch of sturgeon fell from 1 million pounds annually to 77,000 pounds, and this had dropped further to 4,000 pounds by 1964. In the 1930s, the cisco harvest began to drop precipitously from an annual catch of 14 million pounds and never recovered; by the years 1960-1964, the annual cisco catch was 8,000 pounds. The total poundage of fish caught has not changed much since the start of the century, but now consists of less valuable species able to survive in the Lake Erie biological environment of today [2:96].

The changes in fish populations serve as a barometer signalling others of more direct impact on the population served by the Lake. Drinking water has developed odd tastes. Industrial and municipal intakes are plagued with algae and slime. An increasing number of bathing beaches are found unsafe for swimming. The pleasures of boating are marred by the evidences of pollution.

It is the purpose of this chapter to trace the relation between the increase in population and industrial output and the mounting pollution problems afflicting Lake Erie.

Description of Area[a]

Lake Erie is the twelfth largest lake in the world. The Lake Erie drainage basin is a generally low, level, clay-rich land formed by glacier abrasion and depositions.

---

[a]The description in this portion of the chapter relies mainly on two reports, one from the Federal Water Pollution Control Administration on Lake Erie and another from an international commission [14, 16].

Streams are sluggish and winding, carrying a high silt load, especially in the western part of the basin. Most streams also carry high waste loads.

Figure 4.1 helps to place the lake geographically. The names of subdrainage basins (southeast Michigan, Maumee River, etc.) correspond to the divisions employed in developing the basic statistics for the chapter as in tables 4.1 and Appendix table A4.1. However, near the end of the chapter we do make use of the three major basins (west, central, and east) for statistical purposes (see tables 4.3 and 4.4).

Located primarily in the eastern half of the drainage basin, Lake Erie's 9,940 square miles reaches depths of up to 216 feet, although the average depth is only 60 feet. It has a flat muddy bottom. The three major subbasins, western, central, and eastern, have geological significance, as is evident from figure 4.2. It will be noted from figure 4.2 that the western basin is the smallest as well as the shallowest. Because it is relatively shallow, the sunlight necessary to support growth of vegetation penetrates more readily than, for example, in the deeper eastern basin. Figure 4.2 also makes clear that the central basin covers the largest area of the three basins, and the eastern is the deepest.

The western basin has a large silt input and high algal productivity, making its water more turbid than the remainder of the lake. The polluted and silt-laden Detroit and Maumee Rivers flow into this basin. In the summer this basin

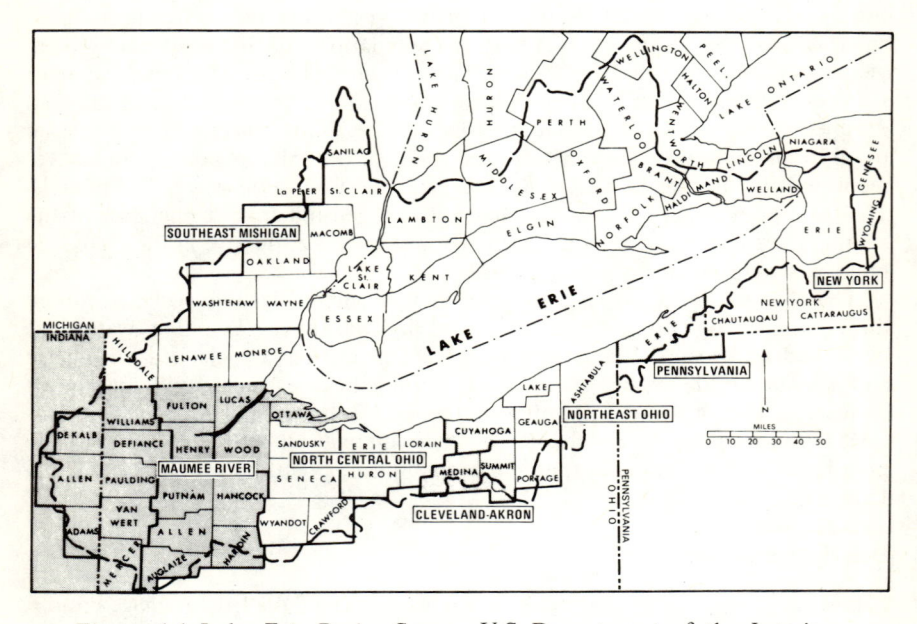

Figure 4.1. Lake Erie Basin. Source: U.S. Department of the Interior, Federal Water Pollution Control Administration, Great Lakes Region, *Lake Erie Report; A Plan for Water Pollution Control* (August 1968), p. 18.

Figure 4.2. Lake Erie Bottom Topography. Source: U.S. Department of the Interior, Federal Water Pollution Control Administration, Great Lakes Region, *Lake Erie Report; A Plan for Water Pollution Control* (August 1968), p. 19.

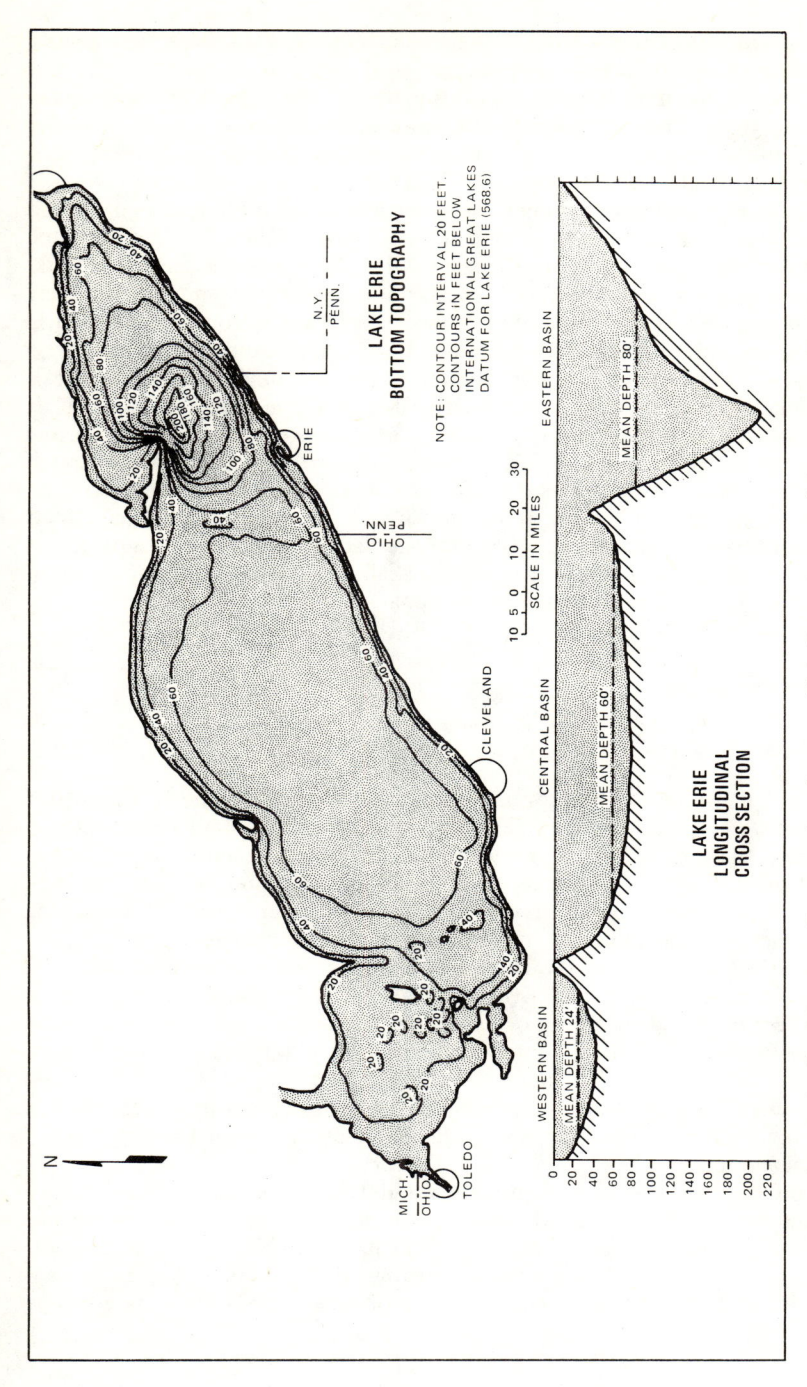

stratifies thermally, causing a rapid oxygen depletion toward the bottom, which has a drastic adverse effect on bottom organisms.

The central basin is the largest of the three basins. It is less turbid and biologically productive than the western basin.

Of the three basins the eastern basin is the deepest and has the clearest water and is least productive of vegetation [3:17-21].

The Lake Erie drainage basin in Canada consists of 5.6 million acres of land in Ontario. The United States portion covers 13.4 million acres distributed throughout five states: Indiana, Michigan, Ohio, Pennsylvania, and New York. The basin receives wastes either directly or indirectly from approximately sixty-five counties in the two countries (see fig. 4.1). The counties included in each subbasin employed in tables 4.1 and 4.3 to 4.9 are outlined in figure 4.1.

Figure 4.3 shows Lake Erie's tributaries. The width of the black lines gives a rough indication of the relative level of pollution at the points in question. It will be noted that the heavy black shading on some streams narrows as on the middle stretch of the Maumee or disappears altogether as on the middle stretch of the Raisin River. This is meant to show areas in which the natural recuperative powers of the stream have overcome partially or wholly pollutional effects further upstream. Figure 4.3 suggests that the Sandusky and the Chagrin Rivers are relatively unpolluted where they enter Lake Erie. The indication of degree of pollution in figure 4.3 is, of course, approximate.

Present Water Quality

Western Basin

The water quality of Lake Huron is crystal clear passing into the St. Clair River where the water quality remains generally good, as it flows into Lake St. Clair.

Lake St. Clair has pollution problems along its shoreline, but its open water is of good quality. Shoreline sources of pollution are the Clinton River, the Milk River Drain, and numerous storm water and sanitary sewer overflows. Swimming is sometimes endangered in this area, because of these discharges.

Lake St. Clair flows into the Detroit River. There are large concentrations of polluting discharges from this river and the whole southeast Michigan area, which affect the Michigan shores of Lake Erie. The water in this area, often unfit for body contact, is rendered suitable for municipal water supplies only with difficulty.

Southward from the Detroit River the quality of the water improves, but this improvement is interrupted by the Raisin River discharges of heavily polluted industrial waste water.

The discharges from the Maumee River and Toledo pollute the Maumee Bay area. This area is hazardous to health and impairs swimming and boating because of the heavy concentration of bacteria originating from Toledo's sewage treatment plant effluent, combined sewage overflow, and treatment plant bypassing. In all of Lake Erie the highest concentration of phosphorous and nitrate will be found in this area.

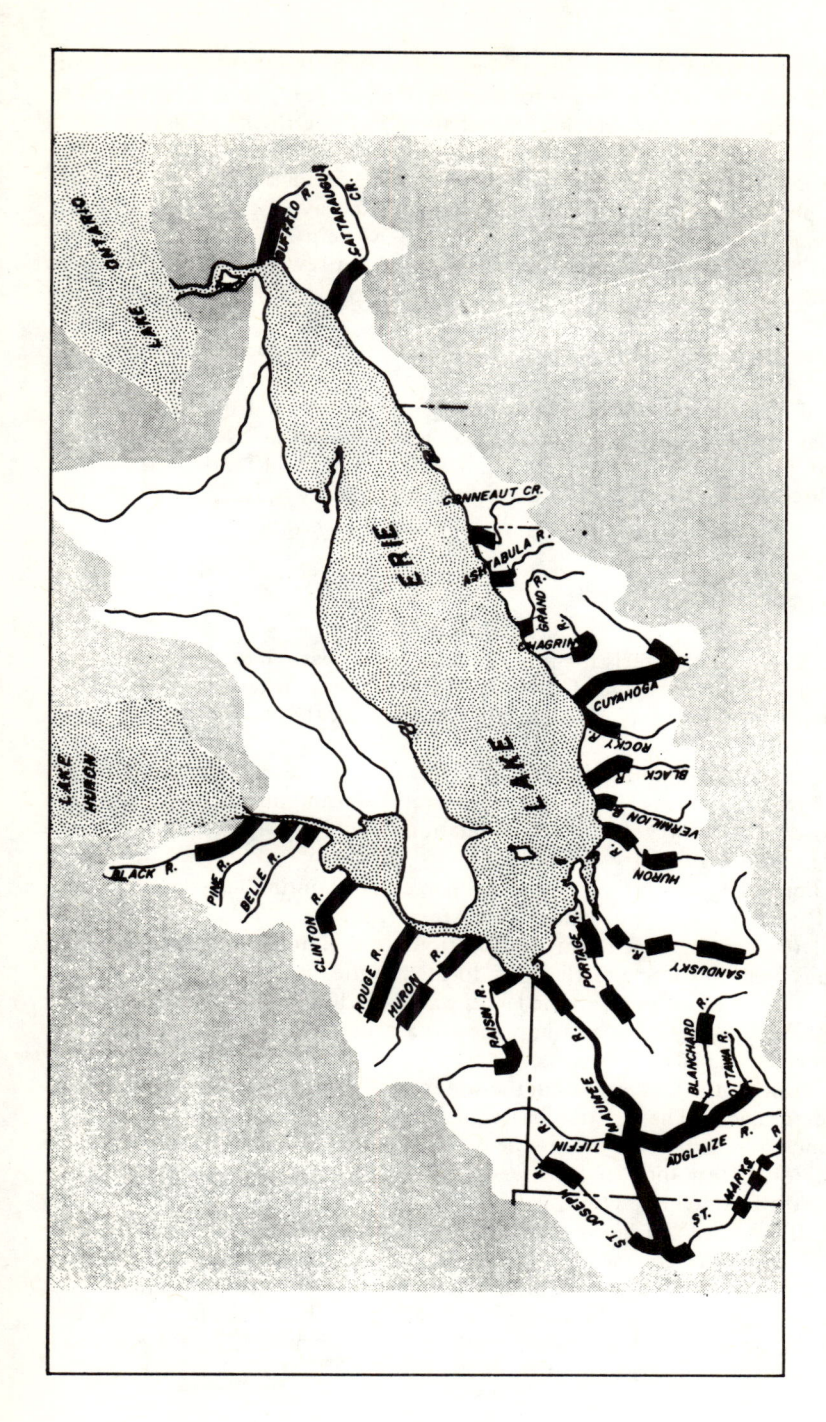

Figure 4.3. Heavily Polluted Tributaries to Lake Erie. Source: U.S. Department of the Interior, Federal Water Pollution Control Administration, Great Lakes Region, *Lake Erie Report; A Plan for Water Pollution Control* (August 1968), p. 6.

The western basin of Lake Erie is the shallowest of the three basins and thus readily admits the light necessary for growth of algae. It is rich in compounds of phosphorus and nitrogen, which are discharged in the effluent of sewage treatment plants serving a population of 5.9 million people (see table 4.1). The combination of light and abundant nutrients makes algae blooms a regular summer phenomenon. The death and subsequent decay of these algae places a heavy drain on the dissolved oxygen of the western basin, already heavily burdened by the oxygen demand of partially treated sewage effluent (see table 4.1, Southeast Michigan and Maumee) and industrial waste (see tables 4.2 and 4.3).

Because the western basin is relatively shallow, the surface area is high relative to volume aiding the absorption of oxygen from the atmosphere. However, when periods of stagnant air coincide with rapid algae growth and thermal stratification of lake waters, the oxygen deficit may reach dangerous levels [2:101-102]. The combined BOD attributable to population and industry in this basin is nearly three times as great as in either of the other two despite its small size (see tables 4.2 and 4.3).

Central Basin

The United States side feels the immediate effect of waste input in a band a half mile in width along its shore. This band widens and has higher concentrations of constituents in the vicinity of large municipalities located at the mouths of the tributaries.

There is contamination in Sandusky Bay due to sewage treatment plant effluents from Sandusky, combined sewer overflows, and untreated sewage from bay resort areas. However, this contamination is only considered serious in the bay waters fronting the city of Sandusky.

The immediate effects of waste discharges are continuously present from Rocky River to Willoughy, Ohio, which is part of the Cleveland metropolitan area. All the bathing beaches in this location are definitely polluted. The same conditions exist in Cleveland Harbour, located at the mouth of Cuyahoga River.

There is an improvement in the lake's water conditions as one moves into the East Lake-Willoughby area, and improves to a generally acceptable quality west of Erie, Pennsylvania.

Like those of the western basin, the waters of the central basin are also turbid and overenriched. The central basin receives the nitrogen and phosphorous compounds remaining in the effluent of sewage treatment plants serving a population of 2.9 million in 1966 (see table 4.1, North Central Ohio, Cleveland-Akron, Northeast Ohio). Other plant nutrients enter central basin waters from the runoff of fertilized lands of the intensively cultivated farms on both the U. S. and Canadian shores of Lake Erie. The central basin, like the western, is accordingly subject to excessive algae growth. Professor Barry Commoner cites data from the records of the Cleveland Waterworks that indicate an eight-fold increase in algae concentration of entering water from 1927 to 1945 and a further tripling of the algae count from 1945 to 1964 [2:107].

Table 4-1
Daily Estimated Municipal Treatment Plant Wastes Discharges in Population Equivalents for Seven Major United States Sub Areas in the Lake Erie Basin

Sub Area	Population Equivalents Before Treatment 1966[a] Millions	BOD[b] Million lbs.	Treated by Secondary Facilities Percentage	Treatment Efficiency[c] Percentage	Population Equivalents After Treatment 1966 Millions	Residual[d] BOD–1966 (Million lbs)[d] Million lbs.	Population Equivalents Before Treatment (Projected)[e] Millions	Residual BOD 1980 (Projected) Without Improved Facilities[f] Million lbs.	Residual BOD 1980 (Projected) With Improved Facilities[g] Million lbs.
Southeast Michigan	4.8	0.8	90	28	3.5	0.6	6.1	0.7	0.13
Maumee	1.1	0.2	79	78	0.2	–	1.3	0.1	0.02
North Central Ohio	.4	0.1	–	64	0.2	–	0.5	–	–
Cleveland Akron	2.4	0.4	64	74	0.6	0.2	3.0	0.2	0.05
Northeast Ohio	0.1	–	–	40	0.1	–	.1	–	–
Pennsylvania	0.4	0.1	70	88	0.1	–	.5	–	–
New York	0.2	–	39	59	0.1	–	.3	–	–
Total	9.4	1.6	68.4	50	4.8	0.8	11.8	1.0	0.2

[a]Population equivalents before treatment–1966
[c]Treatment efficiency percent–1966
Source: U.S. Department of The Interior Federal Water Pollution Control Administration, Great Lakes Region, LAKE ERIE REPORT A PLAN FOR WATER POLLUTION CONTROL, August 1968, p. 55.

[b,d,f,g]BOD .17 pounds per person
Source: U.S. Department of The Interior Federal Water Pollution Control Administration, THE COST OF CLEAN WATER SUMMARY REPORT, Volume I (Washington, D.C.: Government Printing Office, 1968), p. 21.

[e]Population equivalents before treatment–1980 (Projected)
Source: U.S. Department of Commerce, Bureau of The Census, POPULATION ESTIMATES OUR POPULATION 1970-2020 (Washington, D.C.: Government Printing Office, 1968), p. 21.

Table 4-2
BOD of Waste Discharges After Treatment in Lake Erie by Basin, Attributable Population and Industry, Late 1960s (millions pounds BOD per year)

	Attributable to		
Basin	Population	Industry	Total
West[a]	227.1	82.1	309.2
Central[b]	53.4	16.8	90.2
East[c]	8.2	72.3	80.5

Source: Population—1966 population equivalents (after treatment) for each basin from U.S. Dept. of Interior LAKE ERIE REPORT, (August 1968) p. 55, multiplied by 62.05 pounds BOD per person per year (i.e., 0.17 pounds per person per day multiplied by 365). Industry—Totals for 1969 basin from Appendix to Chapter 4.

Note: Extent of treatment of waste "Attributable to Population" is as shown in LAKE ERIE REPORT and Table 4-1. Extent of treatment by industry is as assumed by REGIONAL PLAN ASSOCIATION WASTE MANAGEMENT (New York, 1968: Regional Plan Association) 102 pp.

[a]West includes Southeast Michigan and Maumee River.
[b]Central includes all Ohio except Maumee River.
[c]East includes Pennsylvania and New York.

Table 4-3
BOD of Waste Discharges Before Treatment in Lake Erie by Basin, Attributable Population and Industry, Late 1960s and 1980 (millions of pounds of BOD per year)

	Attributable to					
Basin	Population		Industry		Total	
	1966	1980	1969	1980	1960s	1980
West	363.0	455.1	82.1	114.2	445.1	569.3
Central	181.8	225.3	16.8	32.3	198.6	257.6
East	36.0	48.0	72.3	95.0	108.3	143.0

Source: Population—1966 population equivalents (before treatment) for each basin from U.S. Dept. of Interior LAKE ERIE REPORT, (August, 1968) p. 55, multiplied by 62.05 pounds BOD per person per year. Projected to 1980 by multiplying 1966 figure for each basin (West, Central, East) by ratio of sum of values in column *e* Table 4-1 for basin to sum of values for basin in column *a* of Table 4-1. Industry—Totals for 1969 and 1980 for each basin from Appendix to Chapter 4. Definition of basins as in Table 4-2.

Eastern Basin

In this portion of the United States we find the least serious of waste discharges except at Erie, Pennsylvania and Buffalo, New York. At Erie, pollution is a result of combined storm and sanitary sewer overflows and treatment plants effluents, plus the considerable oxygen demand of paper mill discharges.

The main pollution source at Buffalo is the Buffalo River, which contains industrial and municipal wastes. On the Canadian side of the Lake we find untreated sewage discharges at Port Rowan, Ontario.

Because the eastern basin is deeper than the other two, it is less liable to suffer oxygen deficits even in the summer period of thermal stratification. Also because of its depth and volume, pollutants are more diluted, as are the nutrients that lead to eutrophication. Also, the deeper waters are denied the sunlight needed to promote the growth of algae. For all these reasons water quality in the eastern basin tends to be higher than in the central and western basins. For many of the same reasons any impairment of water quality in the eastern basin might be expected to take longer to correct.

Sources of Erie Pollution

The primary dilemma for both the present and future usefulness of Lake Erie is caused by nutritional overenrichment and its derivative problems. The eutrophication is taking from the lake its necessary oxygen supply, a supply essential for the preservation of animal life. Such high nutrient concentrations are not particularly harmful in themselves; however, they have the synergistic effect of stimulating an excess growth of aquatic plants.

In the western basin and south shore of Lake Erie there are concentrations of phosphorous of twenty or more times the amount needed to actuate algae blooms. The aquatic plants create a paradox in that they constitute a vital part of the food chain, yet an overabundance will have detrimental effects.

While some observers hold that a gradual increase in nutrient concentrations is an inevitable accompaniment of the aging of a lake, it is generally agreed that in the absence of man-made wastes the time required for a lake the size of Erie to reach its present state would be immeasurably long. Lake Erie is more susceptible to such an aging process, because in comparison to the other Great Lakes, Erie has always been in a more advanced stage of eutrophication due to its shallowness, warmth, and high natural sedimentary input from its inflowing tributaries. Each of these factors contributes to the loss of its vital dissolved oxygen. Studies made forty-five years ago recorded no bottom areas seriously depleted of oxygen, but more recent investigation, since man's flagrant misuse of this natural water supply, indicates the occurrence of serious oxygen deficiencies. Approximately 25 percent of the total lake bottom water area is nearly devoid of oxygen during the thermal stratification of the summer.

Moreover, the oxygen deficit is preserved for the future by the accumulation of oxygen demanding wastes in bottom deposits as algae bloom and die off. "Thus," as Professor Commoner notes, "instead of Lake Erie forming a waterway for sending wastes to the sea, it has become a trap that is gradually collecting in its bottom much of the waste material dumped into it over the years—a kind of huge underwater cesspool!" [2:104]

A major source of nutrients entering Lake Erie is the sewage treatment plants. Alleviation of one part of the pollution problem has allowed others to go unabated. By using the microbial oxidation method at the sewage treatment plants, organic waste matter is converted into inorganic substances. It had been assumed that these inorganic wastes would place little or no demand on the dissolved oxygen supply. This was an incorrect assumption. Some inorganics, especially phosphorus, are most detrimental [16]. Such compounds discharged into the lake are reconverted with other elements into cellular matter which in turn places a high demand on the available oxygen.

Much of the fertilizing elements entering Lake Erie drain from lands on both U. S. and Canadian shores. Since forest cover has been replaced by farms and urban developments, the amount of land subject to fertilization has increased. Moreover, recent decades have witnessed an increase in per acre use of chemical fertilizers both for farms and home lawns and gardens.

This chapter has emphasized the dangers of eutrophication as the key to the pollution problems of Lake Erie. The annual cycle of growth and decay of algae is the crucial factor responsible for the persistent oxygen deficit, although the deficit is intensified by the oxygen demands especially of industrial wastes, but also of sewage effluents. Such an emphasis on the deficit of dissolved oxygen should not obscure the very real dangers of bacteriological pollution. The close connection between the two forms of pollution is evident because the dissolved oxygen in the lake is its first defense against dangerous bacteria.

Bacteria concentration levels in waste input streams constitute a direct health hazard in the major population centers of Detroit, Cleveland, Toledo, and Buffalo. A major source is raw municipal sewage discharged directly into the lake, especially from storm and sanitary combined sewers. Even effluents of treatment facilities contain high bacterial concentrations. Approximately one-third of the United States-Lake Erie shore is either continually or intermittently fouled by such contamination. The correction of such problems is far more straightforward than the problem of eutrophication. Disinfection of waste streams is merely a stopgap measure and cannot replace the need for adequate sewage treatment facilities [16:35].

Another source of direct discharge of raw or inadequately treated sewage to Lake Erie is from the extensive fleets of commercial and pleasure craft. Although such discharges are not believed to be a significant factor in overall water quality they are locally damaging, especially in harbor areas. "The bacteriological and nutrient pollution load from commercial vessels on the lake is equivalent to the raw sewage of 1,200 persons for eight months of the year or a permanent population of 800. The pollution contribution from pleasure crafts is equivalent to the raw sewage of a permanent population of 5,500" [16:57]. The Water Quality Improvement Act of 1970 calls for the installation of devices to prevent the discharge of inadequately treated sewage from boats.

Municipal Pollution

Fifty years ago in the Lake Erie basin, when sewage was discarded directly into waterways, the natural process of self-purification was capable of treating and disposing of the wastes with no residual effects on water quality [3:4]. Accompanying the growth in population, however, is an ever-increasing volume of sewage wastes, so large that natural purification is incapable of handling its processing. To aid in such processing, municipal wastes are handled by a system of sewage process plants.

The quality of water discharged by these process plants is directly dependent upon the efficiency of the method of treatment. In primary treatment facilities approximately 35 percent of the organic pollutants are removed. In such a treatment, solids are separated from process streams solely by filtration. Such means of purification completely disregards the solubility of the particular waste in water. A rudimentary example of the effect of solubility may be demonstrated by mixing five pounds of sodium chloride (table salt) in one gallon of cold water. Upon filtration, approximately two pounds of solid salt would be recovered; the remainder would be in solution and simply pass through the filtering mechanism. A similar result may occur when filtering organic wastes, the amount of pollutants passing through depending on that particular contaminant's solubility.

A secondary treatment, consisting of a biological degradation of pollutants, may remove up to 90 percent of organic impurities from the sewer streams. As noted in table 4.1, secondary waste treatment is neither uniform nor total throughout the Lake Erie basin.

In discussing water treatment efficiency, an examination of the oxygen concentration present in the lakes and tributaries into which the treated effluent is discarded is of paramount importance. The life and death of any body of water is directly dependent upon its ability to maintain a certain concentration of dissolved oxygen. It is the dissolved oxygen that sustains aquatic life either directly or indirectly as an integral part of the oxygen-carbon dioxide cycle. If small amounts of sewage are dumped into streams, natural biological degradation eliminates the waste and life is not affected. If, however, excessive amounts find their way into streams, rivers, or lakes, all available oxygen may be consumed and death and decay result. It is the biological oxygen demand, BOD, that a treatment system effluent requires of that body of water into which it is discarded, that is the criterion by which the treatment efficiency is determined. If the effluent has a high concentration of organic pollutants following treatment, it is said to have a high BOD and the treatment facility is operating at a low efficiency.

In 1966 approximately 9 million people in the United States portion of the Lake Erie basin discharged partially treated sewage into the lake and its tributaries. This waste was treated by 63 primary and 155 secondary municipal treatment plants [16:35]. Table 4.1 shows that in 1966 only 68.4 percent of the municipal sewage in the Lake Erie basin received secondary treatment. In consequence, sewage treatment plants in the basin discharged effluent equivalent in BOD to the untreated waste of a population of 4.8 million people. (It should be noted that this does not imply a bacteriological pollution of anything like this magnitude.) Projecting to 1980 and assuming no increase in the extent or efficiency of treatment, the effluent of U. S. municipal sewage plants discharged to Lake Erie would have a BOD equivalent to the raw sewage of a population of 5.9 million. Projecting, however, the construction of secondary treatment facilities by 1980 so that all sewage may be treated, and assuming the overall efficiency to be the hoped-for 90 percent, the oxygen demand of the sewage discharge may be greatly reduced. A breakdown of specific areas may also be found in table 4.1. The comparison is most striking when noting the projected oxygen demand in 1980, before and after suggested improved facilities, a difference by a factor of five. It is not clear, however, that such improvements will be in widespread use by 1980.

Unlike the U. S. shore of Lake Erie, which is heavily populated, the Canadian portion of the lake's drainage basin is relatively sparsely populated and devoted to recreation and agriculture. Table 4.4 compares the sewage effluent entering Lake Erie from U. S. and Canada in terms of population equivalents and BOD. By either standard, the Canadian contribution from municipal sewage to the pollution of Lake Erie is negligible compared to that of the United States. Thus, table 4.4 indicates that the BOD from U. S. sewage effluents was 33 times as great as that from Canada.

Industrial Pollution

The Lake Erie basin industrial growth is surpassing its population growth. The basin accounts for 25 percent of the total production of the five states [16:27]. In the *Lake Erie Report* "Industrial wastes are defined as those spent process wastes associated with industrial operations which are discharged separately and not in combination with municipal wastes" [16:55].

The largest producers of industrial waste on the Lake Erie basin, excluding electric power production, are the steel chemical, oil, paper, food producing, and other allied industries. "Slightly more than half of these industries have inadequate treatment" [16:7]. There is some form of waste discharge from every industrial plant in the Lake Erie basin. This chapter, however, is limited to a few, since general BOD coefficients are only available for a small number of industries.

The estimate of the BOD of industrial wastes made in this chapter was performed using substantially the same methods as those described in Chapter 2. The unit of estimate in this chapter was always initially a sub-basin (Southeast Michigan, North Central Ohio, etc.) and the first step was to identify the counties comprising the sub-basin. (In the Lake Erie basin no sub-basin extended over state lines.) Thereafter, the procedure called for a straightforward application of the methods, techniques, and coefficients described in Chapter 2. Appendix table A4.1 was constructed using these methods, and tables 4.2 and 4.3 were derived from it.

Table 4.3 shows a 41 percent increase in the gross industrial BOD by 1980 for the whole Lake Erie basin. Table 4.3 also shows the amount of increase for each of its drainage basins, and Appendix table A4.1 supplies additional detail.

The 1980 projections are based on the assumption that industrial waste treatment conditions are the same as in 1969. Therefore, with increased industrial output, greater amounts of industrial wastes will be generated.

Table 4-4
Comparative Analysis of United States vs. Canada Total Municipal Wastes Discharged into Lake Erie

	Population Equivalent Before Treatment—1966 Millions	BOD Million Lbs.	Population Equivalent Before Treatment—1980 (Projected) Millions	Residual BOD 1980 (Projected) Without Improved Facilities Million Lbs.
United States	9.4	0.8	11.8	1.0
Canada	0.022	0.0024	0.029	0.0031

Source: U.S. Data: Table 4-1. Canadian data from the International Joint Commission Report of the Commission, POLLUTION OF LAKE ERIE, LAKE ONTARIO AND THE INTERNATIONAL SECTION OF THE ST. LAWRENCE RIVER, Volume 2-Lake Erie (U.S. and Canada: The International Joint Commission, 1969), pp. 9 and 193-194.

Southeast Michigan Basin

Nearly half of Lake Erie's industrial waste comes from this basin (according to the *Lake Erie Report* [16]), but less than a third of the BOD generated by our selected industries. In this area is Lake St. Clair, whose waters are polluted by the Clinton and Milk Rivers. Here we also find the Detroit River, which starts as a fairly clean stream; but by the time it enters Lake Erie it is the basin's largest waste carrier.

Great contributors to industrial waste in the upper Detroit River are manufacturers of steel, paper, and automobiles, along with the chemical industries.

Last but not least in this area is the Raisin River, whose industrial pollution is caused by paper mill waste.

Maumee River Basin

Wastes from this area affect the entire lake, but to a lesser degree than the Southeast Michigan basin. This basin has the Maumee River flowing through it, which receives waste from the Auglaize River, which carries the effluents of a bottling plant and two packing companies. It is further degraded in its confluence with the Ottawa and Blanchard Rivers. Into the Ottawa River flows the wastes of the Sohio's refinery, chemical, and petrochemical plants.

In the Lower Maumee River, the river receives wastes of oil and phenols produced by steel and oil companies.

North Central Ohio Basin

In this area of the lake, the major source of pollution is the Sandusky River, whose principal industrial waste contributors are rubber and sugar companies.

Other polluting rivers are the Black and Portage, whose source of industrial waste comes from oil brines, steel, meat packing, and canning industries.

Cleveland-Akron Basin

Industrial pollution caused by the rubber industry comes from the Cuyahoga River in the Akron area. The lower portion of this river is contaminated in the Cleveland area by the steel industry.

Northeastern Ohio Basin

Major industrial pollution comes from the Grand and Ashtabula Rivers, the source of which is the chemical industries located at Fairport, Painsville, and Ashtabula.

Pennsylvania Basin

Wastes discharges are less serious except for locations at Erie, Pennsylvania. The huge Hammermill paper plant pollutes the Erie area waters.

New York Basin

The main source of pollution is the Buffalo River, which receives industrial waste discharged into it by the Bethlehem Steel Company, Donner-Hanna Coke, and Mobil Oil. Another polluting stream is Cattaraugus Creek, whose source of industrial wastes are from Monarch Tanning Company and Peter Cooper Corporation [16:37-52, 72-79].

Again, the Canadian side of Lake Erie, because of its recreational and agricultural nature, imposes an insignificant oxygen demand on the lake from its industries.

The *Lake Erie Report* recommends that, if the conditions of the lake are to be reversed:

1. The iron and steel industry should install and maintain waste reduction facilities to levels that will comply with federal water quality standards.
2. The heavy chemical industry should install and maintain waste reduction facilities to control their discharges of suspended and dissolved solids.
3. The food-producing and other allied industries should install and maintain waste reduction facilities to control the discharges of suspended solids, oxygen-demanding substances and color so that the flow from such plants meet state requirements including those for the total allocated oxygen demand of the lake [16:14-15].

Combined Municipal and Industrial Pollution

Table 4.2 presents a summary of the BOD of waste discharges into Lake Erie, attributable to population and to industry. Data on BOD of population wastes are derived from table 4.1, which in turn came principally from the *Lake Erie Report*. This involved the combination of figures relating to sub-basins into broader totals for the three major basins whose bottom contours are illustrated in figure 4.2. There is little forcing involved in this grouping; the western boundary of the eastern basin lies close to the Ohio-Pennsylvania line, and only the North Central sub-basin is seriously split by a major basin boundary. The massive inputs of the Southeast Michigan and Maumee River sub-basins overwhelm statistically any slight error involved in allocating all of the North Central Ohio's totals to the central basin.

The industry data of table 4.2 are taken from Appendix table A4.1, where they are shown for sub-basins. These industry estimates were made using the data sources and procedures discussed in Chapter 2 and are subject to all the numerous and important qualifications there. In terms of the BOD of pollutants entering Lake Erie itself in contrast to a figure representing BOD arising *somewhere* in the drainage basin, a further qualification must be noted. Much pollution originating in the headwaters of a tributary stream entering the lake, especially if expressed in terms of BOD, may be eliminated by the time it reaches the lake through oxygenation; this is suggested in a number of cases by figure 4.4.

Pollution Situation by Basins

A number of conclusions concerning the pollution problem of Lake Erie emerge quite clearly from the data of table 4.2. It is evident that the western basin, the shallowest and from that standpoint most subject to eutrophication, also suffers under by far the heaviest burden of BOD. Most of this arises from wastes attributable to population; the inadequately treated sewage effluent of the City of Detroit was a big factor, but that of Toledo added importantly to the load as well. The BOD of industrial wastes was also highest for the western basin of the three divisions. Clear evidence of the mounting deterioration of the western basin is shown in figure 4.4, which depicts the spread of sludgeworms, a pollution-resistant bottom organism in the western basin.

Table 4.2 shows a very modest BOD for industrial discharges in the central basin, which undoubtedly understates the magnitude of the problem very seriously. No account could be taken in estimating the BOD of industrial wastes attributable to a number of important industries; notably the rubber, chemical, and steel industries, all substantial contributors to industrial pollution in the central basin. The Cuyahoga River, notorious as the one that caught fire, is not as relatively clean as table 4.2 might suggest.

Figures on the Eastern basin in table 4.2 are interesting for two reasons: first, it is the only basin in which industrially originating BOD is most important, actually overwhelmingly so; second, the very small BOD attributable to municipal wastes in the Eastern basin reflected a much smaller population than the other basins, but also a very much higher average level of treatment efficiency than in the other basins. The paper industry accounts for nearly three-fourths of the BOD generated by the selected industries in the Eastern basin (see Appendix table A4.1).

The Future of Lake Erie

Table 4.3 indicates that Lake Erie's pollution problems are not likely to go away by themselves in the next few years. It represents, in a sense, the magnitude of the obstacles that will have to be overcome if BOD pollution is to be eliminated. The BOD totals for both population and industrial wastes are much higher in table 4.3 than in 4.2. This is so because the calculations were made assuming no treatment of population waste in either period covered and no improvement in treatment of industrial waste. Neither assumption has any relation to the facts and the table is, in no sense, a set of estimates. Since there is no possible way of

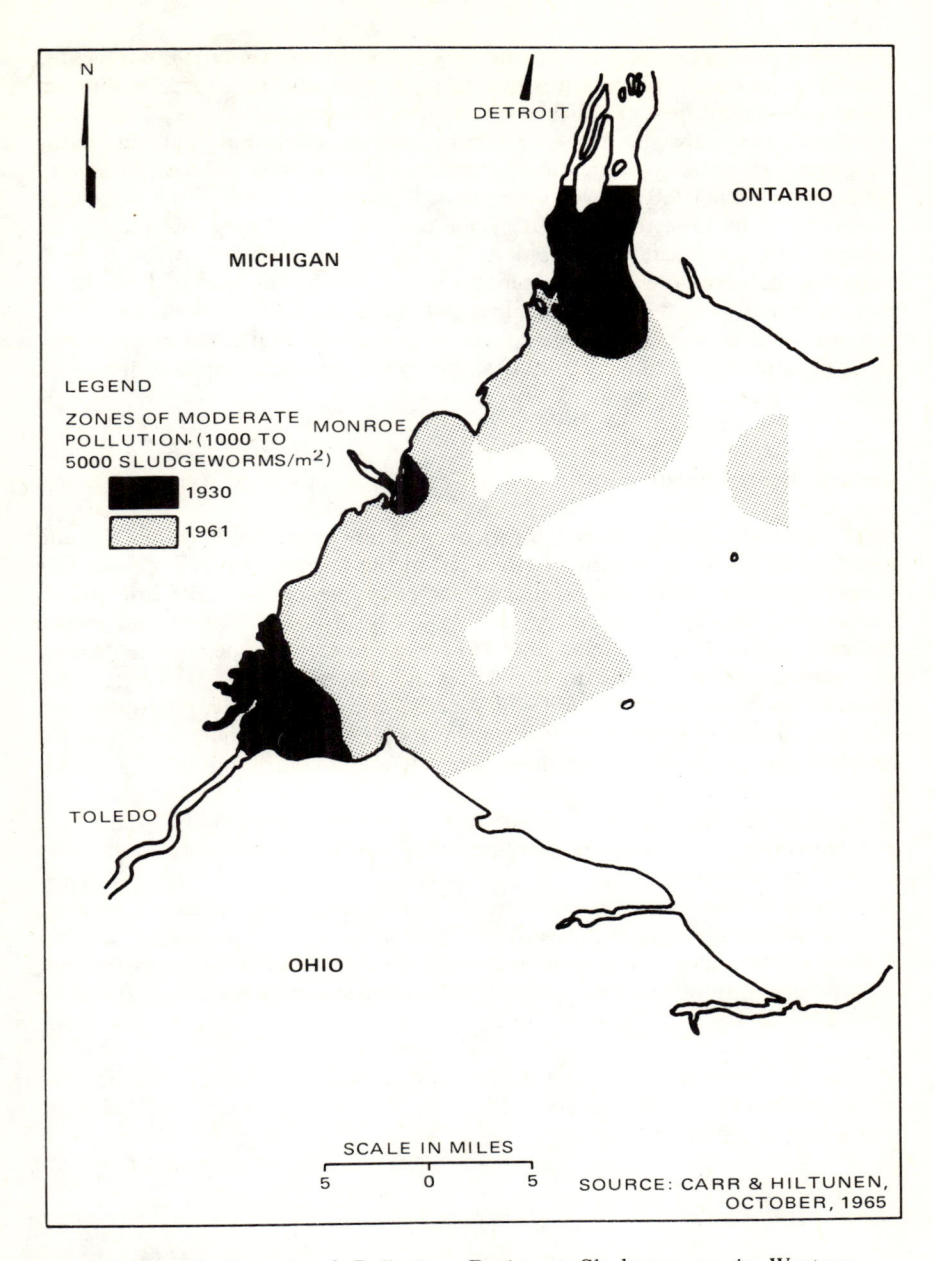

Figure 4.4. Spread of Pollution Resistant Sludgeworms in Western Basin, Lake Erie, 1930 to 1961. Source: U.S. Department of the Interior, Federal Water Pollution Control Administration, Great Lakes Region, *Lake Erie Report; A Plan For Water Pollution Control* (August 1968), p. 34.

estimating how much treatment in either sector will improve by 1980, it seemed wisest to make the calculation solely to reflect the rise in *potential* pollution, that is before treatment at either the starting or terminal date.

Looked at in this way, table 4.3 leads to several conclusions: (1) considering the general inadequacy of treatment during the 1960s there will have to be very rapid progress by 1980 to overcome the additional load arising from expanding population and industrial output; (2) moreover, overcoming BOD loads by conventional treatment methods does little or nothing to reduce the flow of nutrients that promote the eutrophication that lies at the root of Lake Erie's problems; and (3) the volume of other pollutants such as discoloration, odors, and toxics usually rise in some rough proportion to BOD, and conventional sewage treatment methods usually deal with such pollutants haphazardly or not at all.

Consequences of Pollution

What are the consequences of the sort of pollution suggested by tables 4.2 and 4.3? One of the most significant effects is on the quality of life. The *Lake Erie Report* carries a fifteen-page recital of the consequences for Lake Erie and its tributaries [16:37-52], a dreary catalog of blighted landscapes and waters rendered unfit for most human uses. However, most of this deplorable description applies almost entirely to the shores and the tributaries of the lake; the open waters of the lake a few miles offshore remain, surprisingly, relatively clean and even potable. To be sure, it is on the shores and the tributaries where the people are to be found and their lives adversely affected in a number of ways:

1. Most of the municipalities on the lake draw their water supply from it and increasingly heroic measures may be needed to eliminate objectionable tastes or odors at some seasons;
2. Swimming and other water contact sports are barred at many beaches, frequently those nearest the centers of population (see fig. 4.5);
3. People who can afford sizable boats may be able to escape such restrictions, but frequently only after traversing harbor and shore areas made unattractive by pollution.

There can be little doubt that there would be extensive dividends in the form of better quality of life for the residents of the basin if a successful cleanup campaign could be mounted.

Thermal Pollution

Most people think of water pollution as caused by wastes that have been discharged or washed into rivers or lakes from homes or businesses, oil and chemicals from industry, or dirt and debris from city streets or country fields. The effects of pollution can frequently be seen or smelled.

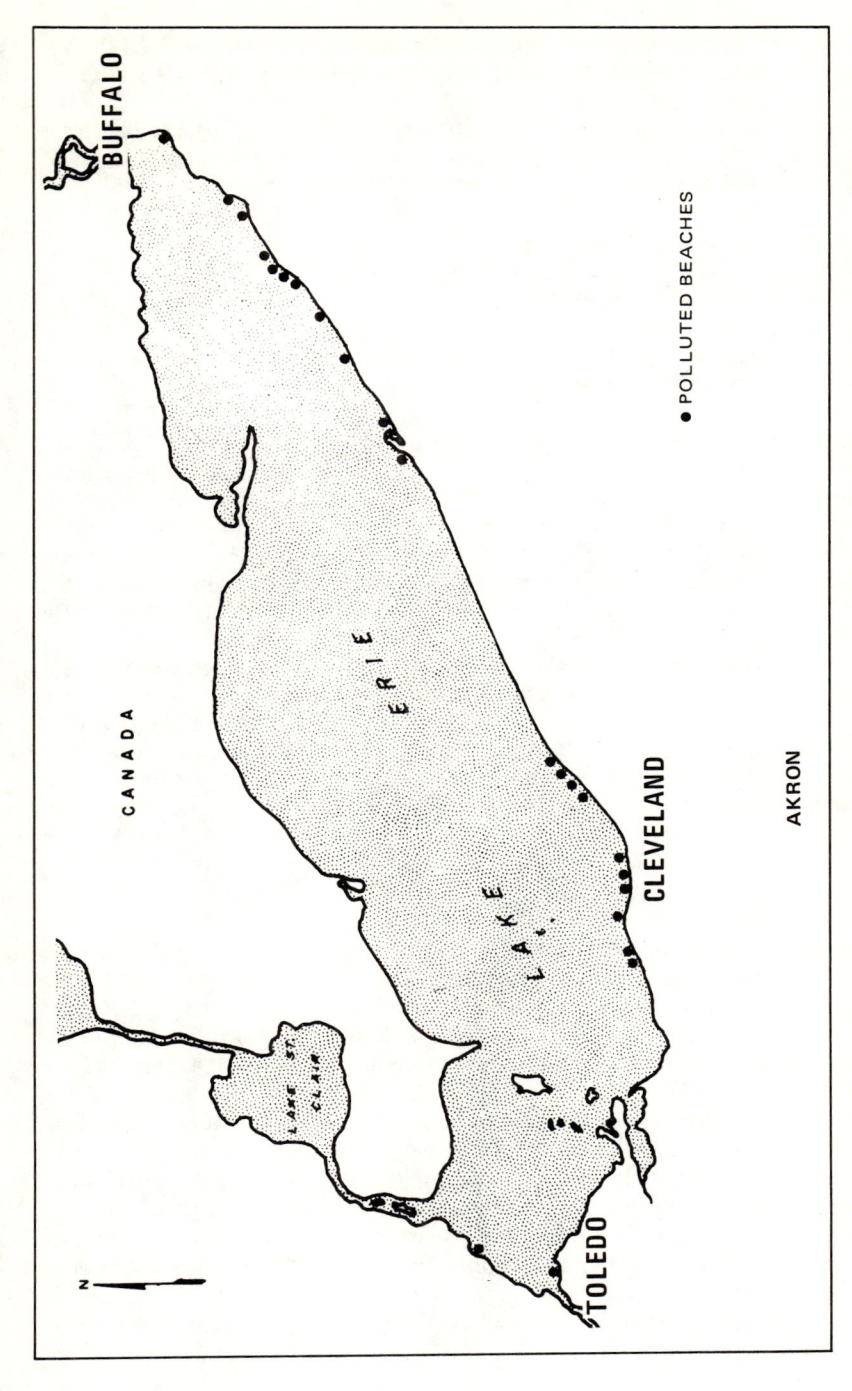

Figure 4.5. Polluted Beach Areas in Lake Erie. Source: U.S. Department of Interior, Federal Water Pollution Control Administration, Great Lakes Region, *Lake Erie Report; A Plan for Water Pollution Control* (August 1968), p. 3.

A far more insidious type of pollution, however, is heat, because even though biologically pure, water discharged at elevated temperatures into a stream may have serious harmful effects [12].

A temperature change of three or four degrees can under certain conditions have serious effects upon fish and aquatic life. This addition of heated water to rivers and lakes results in:

1. A decrease in dissolved oxygen content,
2. An increase in the toxicity of existing pollutants,
3. An increase in water turbidity,
4. An increase in algae growth,
5. An increase in the metabolic rates of fish and other organisms, which changes or interferes with their reproductive cycle [18].

Great amounts of heat are produced by industry, which needs water for cooling. Since the electrical power industry is a major user of water for cooling purposes, its projected growth can illustrate the problem ahead. Energy in the United States has increased at a faster rate than population, and electric power is the form of energy whose use is increasing most rapidly. For 1950 the per capita consumption of electric power was 2,000 kilowatt hours. In 1968 this increased to 6,500, and by 1980 it should be about 11,500 kilowatt hours [19:3-4] (see table 4.5).

The major producers of waste heat in the lake are, again, the power generation facilities in the United States portion of the lake basin. Eleven power plants are now responsible for this discharge. The 1966 capacity of the plants was at least 7500 megawatts [14:249]. Production is expected to increase to at least 16,000 megawatts by 1980, as indicated in table 4.6. Currently, there are no electric power plants on the Canadian shore of Lake Erie.

Conclusion

Eleven million people inhabit and are supported by Lake Erie. For them, it is a source of industry, recreation, agriculture, and commercial activities. It is a heritage that should be preserved and enjoyed by future generations.

The Lake Erie drainage basin is low and level with sluggish and winding streams. It has three major sub-basins—western, central, and eastern—whose water quality ranges from fair to very poor.

The primary dilemma for both the present and future usefulness of Lake Erie is caused by increasing eutrophication, namely nutritional overenrichment and its derivative problems. The eutrophication is taking from the lake its necessary oxygen supply, a supply essential for the preservation of animal life.

These high nutrient concentrations are reflected in excessive growths of aquatic plants whose overabundance cause detrimental effects on the lake's vital dissolved oxygen.

Source of fertilizing nutrients are many. Chemical fertilizers are a source that enriches algae growth by drainage from the land. Since 1950 there has been a

Table 4-5
U.S. Electric Utility Power Statistics Relating to Population and Consumption

	1950	1966	Estimated 1980
Population (Millions)	152	202	235
Total Power Capacity (Millions of KW)	85	290	600
KW capacity/person	0.6	1.4	2.5
Power consumed per person per year (KW-Hr.)	2,000	6,500	11,500
Total Consumption (KW-Hr.)	325 billion	1.3 trillion	2.7 trillion

Source: Joint Commission on Atomic Energy Hearings, ON ENVIRONMENTAL EFFECTS OF PRODUCING ELECTRIC POWER Phase I, October 28-31 and November 4-7, 1969.

Table 4-6
Electric Utility Power Statistics Relating to Population and Consumption in the U.S. Portion of the Lake Erie Basin

	1966	Estimated 1980
Population (millions)	11.5[a]	14.5[b]
Total power capacity (million of KW)	16.1	36.3
KW capacity/person	1.4	2.5
Power consumed per person per year (KW-HR.)	6,500	11,500
Total Consumption (KW-HR.)	7.5 billion	16.7 billion

[a]1966
Source: The International Joint Commission, Report of the Commission POLLUTION OF LAKE ERIE AND LAKE ONTARIO AND THE INTERNATIONAL SECTION OF THE ST. LAWRENCE RIVER (U.S. and Canada: The International Joint Commission, p. 9.

[b]1980
Source: U.S. Department of Commerce Bureau of THE CENSUS POPULATION ESTIMATES AND PROJECTIONS 1970 TO 2020 (Washington, D.C.: Government Printing Office, 1970), p. 8. Joint Committee on Atomic Energy Hearings on ENVIRONMENTAL EFFECTS OF PRODUCING ELECTRIC POWER Phase I, October 28-31 and November 4-7, 1969.

trend to hard chemicals, those being resistant to biological breakdown. Because of recent government concern, major agriculture chemical manufacturers are starting to develop products that degrade more readily, reducing the future chances of their drainage from land.

Vessel discharges are locally damaging. However, under the *Federal Water Quality Improvement Act of 1970,* standards of performance are being established for marine sanitation devices designed to prevent the discharge of inadequately treated sewage from boats.

The principal source of pollution is municipal wastes. In 1966 approximately nine million people in the United States portion of Lake Erie discharged partially treated sewage into the lake and its tributaries. This waste was treated by primary and secondary treatment plants, leaving the waste-equivalent of 4.8 million people being discharged into the lake. The estimated 1980 population will be a little greater than 11 million people; using present methods of waste disposal the result will be an equivalent of raw sewage of 5.9 million people. However, projecting the construction of improved facilities to an overall 90 percent efficiency would leave a residual discharge equivalent to 1.2 million people in 1980.

Industrial waste is the second largest source of pollution. Every industrial plant in the Lake Erie basin has some industrial waste discharge. At present a little less than 50 percent of these industries have adequate treatment facilities.

Production by 1980 will expand to the extent that there will be an estimated 41 percent increase in gross industrial BOD, in the selected industries discussed in this chapter, if there is no future improvement in industrial waste disposal facilities.

Various industries are improving those facilities. This improvement can be further stimulated if each basin state establishes uniform industrial pollution controls defining pollutants and their acceptable reduction before discharge.

Appendix

See table A4-1 on page 74.

Notes

Books

1. Donald E. Carr, *Death of Sweet Water* (New York: W. W. Norton & Company, Inc., 1966), 257 pp.

2. Barry Commoner, *The Closing Circle,* (New York: Alfred A. Knopf, 1971), 326 pp.

3. J. H. Dales, *Pollution, Property and Prices* (Canada: University of Toronto Press, 1968), 109 pp.

4. Frank Graham, Jr., *Disaster By Default – Politics and Water Pollution* (New York: M. Evans & Co., Inc., 1966), 256 pp.

5. Louis Klein, *River Pollution III Control* (London, England: Butterworths, 1966), 484 pp.

6. George A. Nikolaieff, *The Water Crisis* (New York: The H. W. Wilson Company, 1967), 192 pp.

Government Publications

7. U. S. Department of Commerce Bureau of Census, *County Business Patterns; Indiana, Michigan, Ohio, New York, Pennsylvania 1959, 1964, 1969* (Washington D.C.: Government Printing Office, 1959, 1964, 1969).

8. U. S. Department of Commerce, Bureau of Census, *Census of Manufactures 1958, 1963, 1967* (Washington D.C.: Government Printing Office, 1958, 1963, 1967).

9. U. S. Department of Commerce, Bureau of the Census, *Population Estimates; Our Population–1970 to 2020* (Washington D.C.: Government Printing Office, 1970).

10. U. S. Department of the Interior, Federal Water Pollution Control Administration, *The Cost of Clean Water – Summary Report* Volume 1 (Washington D.C.: Government Printing Office, 1968), 39 pp.

11. U. S. Department of the Interior, *About Boats and Pollution 1970* (Washington, D.C.: Government Printing Office, 1970).

12. U. S. Department of the Interior, Federal Water Pollution Control Administration, *Heat Can Hurt, Better Water for America* (Chicago, Illinois: Office of Public Information, 1969).

13. Joint Committee on Atomic Energy Hearing, *On Environmental Effects of Producing Electric Power* Phase I, October 28-31 and November 4-7, 1969.

14. The International Joint Commission, Report of the Commission, *Pollution of Lake Erie, Lake Ontario and The International Section of the St. Lawrence River* (U. S. and Canada: The International Joint Commission, 1969, Volume II, Lake Erie), 316 pp.

15. Regional Plan Association, *Waste Management* (New York, 1968: Regional Plan Association), 102 pp.

16. U. S. Department of the Interior, Federal Water Pollution Control Administration, Great Lakes Region, *Lake Erie Report; A Plan for Water Pollution Control* (August 1968), 107 pp.

Miscellaneous Publications

17. League of Women Voters, Lake Erie Basin Committee, *Lake Erie? Requiem or Reprieve* (Cleveland, Ohio: League of Women Voters of Cleveland, 1966), 50 pp.

18. League of Women Voters, Lake Erie Basin Committee, *Lake Erie Letter, Thermal Pollution – Another Threat to Lake Erie* (Bay Village, Ohio: League of Women Voters, 1969).

19. Dean E. Abrahamson, *Environmental Cost of Electric Power* (New York: Scientists' Institute for Public Information, 1970), 36 pp.

Table A4-1

Gross BOD Generated by Industries in the Lake Erie Drainage Basin, by Industry and Sub-Basin, 1969 Estimated 1980 Projected (Millions of Pounds of BOD per Year)

SIC #	Name	Southeast Michigan		Maumee River		North Central Ohio		Cleveland-Akron		Northeast Ohio		Pennsylvania		New York		Lake Erie Total	
		1969	1980	1969	1980	1969	1980	1969	1980	1969	1980	1969	1980	1969	1980	1969	1980
2011	Meat Packing Plants	6.3	11.0	5.3	5.3	–	–	1.0	1.2	–	–	3.2	3.3	0.5	0.3	16.3	20.1
2013	Sausage and other meats	4.3	6.1	1.3	2.7	0.2	0.5	0.6	1.2	–	–	0.3	0.7	1.3	2.6	8.0	13.8
2026	Fluid milk	1.1	1.0	0.7	0.6	0.2	0.2	1.4	1.4	0.1	0.2	0.2	0.3	0.7	0.5	4.3	4.2
2033	Canned fruits, vegetables, etc.	1.2	3.6	6.2	15.0	2.3	7.6	0.8	2.4	–	–	2.1	5.3	5.9	11.1	18.5	45.1
2037	Frozen fruits, vegetables, etc.	–	–	2.3	5.0	0.6	1.4	2.6	5.7	–	–	–	–	2.4	10.8	8.0	22.9
2621	Paper except building	22.9	28.1	–	–	–	–	3.1	5.2	–	–	44.9	51.8	3.1	3.6	74.1	88.7
2631	Paperboard mills	16.3	17.3	4.8	6.6	–	–	–	–	–	–	–	–	5.7	3.0	26.8	27.0
2911	Petroleum refining	2.7	3.8	6.7	7.9	–	–	3.7	5.4	–	–	2.1	1.6	–	–	15.2	18.6
	Total	54.9	70.9	27.3	43.2	3.4	9.7	13.2	22.4	0.1	0.2	52.8	63.0	19.5	31.9	171.2	241.4

Source: Estimated using data on employment in relevant industries from U.S. Department of Commerce, Bureau of the Census, COUNTY BUSINESS PATTERNS and CENSUS OF MANUFACTURERS in accordance with the procedure outlined in Chapter 2.

5   Puget Sound

Donald G. Niddrie

Steadily rising industrial output and rapid population growth has exacted a heavy toll on the quality of the water environment in many of our lakes, rivers, and estuaries. A considerable portion of Puget Sound has, thus far, succumbed to these forces. The people of Puget Sound have already spent millions of dollars to reverse the deleterious effects created by factors inherent in continuous growth. The principal objective of this chapter is to weigh the possible balance of these forces for the year 1980. Will the balance be in favor of the forces in opposition to the pollution problem, or will the continuing upward trends in population and output perpetuate the present imbalance in favor of these potentially deleterious forces? This is the crucial question that must be answered.

It is well known that pollution problems generally do not manifest themselves over a wide area. This is especially true in Puget Sound because it is a vast body of water with tremendous natural assimilative capacity. Consequently, this study will be restricted to those areas in the sound where significant pollution problems have already appeared.

Description of the Puget Sound Region

The Puget Sound region is located in the extreme northwest corner of the continental United States in the State of Washington. The basic geography was formed by glacial action that created an elongated depression bounded on the east by the extremely rugged Cascade Mountain Range and on the west by the Olympic Mountains. The sound lies on a north-south axis between these natural boundaries. Gradually, tidal action and natural erosion carved out the multitude of bays, coves, and inlets that characterize this body of water. The majority of the land areas surrounding the sound is heavily wooded rolling hills interspersed with fertile river valleys.

Puget Sound is one of the deepest salt water basins in the United States, with depths ranging from 930 feet in the northern portions to 550 feet in the south. Total surface area is about 650,000 acres with shorelines extending some 1,330 miles. The total drainage basin encompasses a land area of about 11,000 square miles with drainage contributing a mean flow of 50,000 cfs (cubic feet per second) to the sound, 80 percent of which enters via the major rivers [1:250]. The tidal currents are strong but irregular throughout the sound, but the dominant hydrographic feature is the net seaward flow of less saline surface waters and the net inward flow of heavier marine waters. These flows enter and exit through the Straits of Juan de Fuca and Georgia in the north [2:12]. This creates a natural flushing action of the wastes entering the Puget Sound basin.

The five counties that extend along the eastern shore of the sound are, by far, Puget Sound's most heavily industrialized and populated area. From north to south they are: Whatcom, Skagit, Snohomish, King, and Pierce Counties. The principal community in this five-county region is Seattle, which began its development in the early 1850s with the arrival of the first permanent settlers to the Puget Sound region [1:46]. Since that time, the population has spread north and south with the development of the port cities of Everett and Tacoma in Snohomish and Pierce Counties respectively. The principal centers of population and employment in the two northern counties are Bellingham, in Whatcom County, and Anacortes, in Skagit County.

Historically, the economy of the entire Puget Sound basin has been closely related to its nearby waters, forests, and fields. However, in recent years the products coming from these sectors have been overshadowed by developments in the aerospace field. Boeing Company, one of the world's largest aerospace equipment manufactures, has its corporate headquarters and a large part of its plant located in the Seattle metropolitan area. Approximately one-third of all manufacturing workers in the State of Washington are employed in the aerospace sector [3:19]. The aerospace field has provided the impetus for the tremendous growth experienced in the five-county region over the past ten to fifteen years. However, growth in the aerospace sector has declined in recent years. Offsetting the decline in aerospace employment and production has become a considerable problem, but recent growth in metals, machinery, chemicals, petroleum refining, and other diversified manufacturing has helped to strengthen and broaden the economic structure. Growth in nonmanufacturing has also been significant. The area now serves as the financial and trade center for a large part of the Pacific Northwest [3:1-27].

Water Pollution in Puget Sound Today

The water pollution problems that exist in Puget Sound today are generally confined to the bays, harbors, and tributary streams in the immediate environs of the sound's major population and industrial centers of concentration. There is no evidence that water pollution, in terms of the conventional parameters, has spread to the open water areas or to the majority of the areas that are still sparsely populated. The vast assimilative capacity of the sound has, thus far, maintained these waters in a pollution-free state.

The extent of the water quality degradation that has occurred has been thoroughly documented by the state's Department of Ecology in its *Implementation and Enforcement Plan for Water Quality Regulations* published in September 1970. At that time, the only significant pollution problems in the sound were found to be concentrated in the five-county region bordering on the eastern shore.

Although 58 percent of the state's population was residing in this region in

1969, only a small portion of the degraded water quality conditions can be attributed to the domestic wastes generated by this source. A very large percentage of the domestic wastes received a minimum of primary treatment prior to final discharge into the sound or its tributary streams. Table 5.1 summarizes the condition of the major water bodies in each of these counties as of September 1970. The most serious pollution problem that can be related directly to the domestic waste discharges of the population was found to be excessive amount of coliform bacteria. Thirteen of the eighteen watercourses cited in table 5.1 were below the acceptable limits established for their respective classifications by the Department of Ecology. As a result many of the public bathing beaches located in these areas were closed. In most cases the problem was one of inadequate disinfection of treated effluent rather than the indiscriminant discharge of raw sewage.

Low dissolved oxygen concentrations were found in many of the cited watercourses. The inner harbors and the lower reaches of the tributary streams

Table 5-1
Water Quality in Selected Areas of Puget Sound

County	Watercourse	Water Quality Parameters						
		Total Coliform	Dissolved Oxygen	Temp.	pH	Turbidity	Toxic or Deleterious	Aesthetics
Whatcom	Bellingham Bay	X	X	S	S	X	X	X
	Bellingham Bay, Inner Harbor	X	X	S	X	X	X	X
	Outer Bellingham Bay	S	X	S	S	S	X	S
	Nooksack River from mouth to River Mile 4	X	X	S	S	S	X	X
	Guemes Channel, Padilla, and Samish Bays	S	X	S	S	S	X	S
Skagit	Skagit Bay	S	S	S	S	S	S	S
	Skagit River from mouth to Burlington	X	S	S	S	S	X	S
	Possession Sound, Port Susan, Saratoga Passage	S	S	S	S	S	S	S
Snohomish	Everett Harbor	S	X	S	S	X	X	X
	Everett Harbor, Inner	X	X	X	X	X	X	X
	Snohomish River from mouth to latitude 47° 56' 30"	X	S	N	S	X	X	X
	Elliott Bay	X	X	S	S	S	X	X
King	Duwamish River	X	X	S	S	X	X	X
	Lake Washington Ship Canal	X	X	N	X	X	X	X
	Commencement Bay	X	X	S	S	X	X	X
Pierce	Commencement Bay, Inner	X	X	S	S	X	X	X
	Port Industrial and City Waterways (Tacoma)	X	X	S	S	X	X	X
	Puyallup River from mouth to River Mile 1	X	S	S	S	N	X	X

Legend: X = Unsatisfactory
S = Satisfactory
N = Natural

Source: State of Washington, Department of Ecology, IMPLEMENTATION AND ENFORCEMENT PLAN FOR WATER QUALITY REGULATIONS, (n.p.: n.p., 1970), pp. 43-53.

were hardest hit in this category. Serious deterioration of water clarity (turbidity) and the impairment of aesthetic values were found in all counties except Skagit and only two of the watercourses cited herein did *not* exhibit a toxic or deleterious condition. On the brighter side, abnormal water temperatures and acidity values (pH) were not a problem in most areas. Only the inner harbors in Bellingham and Everett and the Lake Washington Ship Canal had abnormal pH values. Unsatisfactory temperature conditions or thermal pollution was found only in the inner harbor at Everett. The already-degraded water quality in many of these areas along the shores of the sound portend a more serious condition in the future if the pollution is not checked.

The Magnitude of the Existing Pollutional Load

Quantifying the magnitude of the existing pollutional load is severely hampered by the general lack of equivalent unit pollution generation rates or pollutional coefficients for uncommon liquid waste streams. Biochemical oxygen demand (BOD), the oxygen-depleting characteristic of liquid wastes, is the only reliable magnitudinal indicator that can be safely used to equate the broad field of waste-producing agents in our society. BOD, however, is a good if not perfect overall indicator of the total waste load generated by most industries and the population. It is recognized that there are many other contaminants contained in industrial and domestic waste streams that also have a degrading effect on the water environment, but these are not suited for broad generalizations as to relative contribution to the total pollution load. In any event, unsatisfactory conditions in other respects are usually associated with high BOD levels. Presented in table 5.2 are 1969 estimates of the BOD generated by the population and a carefully selected group of industries located in the five-county region. Although this table does not and could not include all the industry groups operating in this region, it is felt that the most significant contributors to the pollution problem, in terms of BOD generation, are included. The Standard Industrial Classifications of the industries included in table 5.2 are identified below.

SIC Code	Industry
2011	Meat Slaughtering Plants
2013	Meat Processing Plants
2015	Poultry Dressing Plants
2026	Fluid Milk
2033	Canned Fruits and Vegetables
2037	Frozen Fruits/Vegetables
2611	Pulp Mills
2621	Pulp and Paper Mills
2631	Paperboard Mills
2911	Petroleum Refining

Calculations from the 1963 Census of Manufactures indicate that food and

Table 5-2
Estimated Gross BOD Load for the Population and a Selected Group of Industries, 1969 (BOD expressed in millions of pounds annually)

Source	Whatcom		Skagit		County Snohomish		King		Pierce		Total by Source	Percentage of Five County Total	Percentage of State Total
	BOD	%	BOD	%	BOD	%	BOD	%	BOD	%			
Population	5.09	3.1	3.26	2.3	15.99	5.1	71.58	82.6	25.14	43.8	121.06	16.0	58.0
SIC 2011	.14	–	.11	–	.18	–	4.05	4.7	1.91	3.3	6.39	.8	54.5
SIC 2013	.01	–	–	–	–	–	.33	.4	.01	–	.35	–	87.5
SIC 2015	.01	–	.11	–	.02	–	.65	.8	.26	.5	1.05	.1	89.0
SIC 2026	.07	–	.07	–	–	–	.43	.5	.04	–	.61	–	49.2
SIC 2033	1.89	1.1	.37	–	–	–	3.10	3.6	–	–	5.36	.7	34.6
SIC 2037	5.00	3.1	6.75	4.8	5.49	1.7	6.52	7.5	2.71	4.7	26.47	3.5	80.0
SIC 2611[a]	142.44	87.5	127.17	91.4	263.41	84.5	–	–	–	–	533.02	70.2	64.5
SIC 2621	6.97	4.3	–	–	26.99	8.7	–	–	20.47	35.7	54.43	7.2	48.5
SIC 2631	–	–	–	–	–	–	–	–	6.54	11.4	6.54	.9	9.7
SIC 2911	1.19	.7	1.37	.9	.06	–	–	–	.21	.4	2.83	.4	100.0
Total	162.81	–	139.21	–	311.83	–	86.66	–	57.29	–	758.11	100.0	–
Percentage of Five County Total	21.5	–	18.4	–	41.1	–	11.4	–	7.6	–	100.0	–	–

Source: Estimated using data on employment in relevant industries from U.S. Department of Commerce, Bureau of the Census, COUNTY BUSINESS PATTERNS and CENSUS OF MANUFACTURERS in accordance with the procedure outlined in Chapter 2.

Legend: BOD is expressed in millions of pounds annually. Blank spaces in the BOD columns indicate that the county had no employment in the SIC category. Blank spaces in the percentage columns indicate an insignificant percentage of the total.

[a]See Appendix A.

kindred products, paper and allied products, and petroleum and coal products accounted for over 70 percent of the total industrial water intake and over 73 percent of the total water discharged in the State of Washington [9]. The four-digit SIC categories listed above are the main constituents of these broader SIC groups.

The overwhelming fact revealed by table 5.2 is the tremendous relative and absolute contribution of the pulp, paper, and paperboard group. This, of course, has long been well known by Washington State authorities and was thoroughly documented by the Washington State Enforcement Project's report entitled *Pollutional Effects of Pulp and Paper Mill Wastes in Puget Sound,* published in March 1967. In summary, the report attributed the major part of the degraded water conditions in Whatcom, Skagit, and Snohomish Counties to the waste discharges of the pulp and paper producers in these areas [2]. Of the selected sources considered for these counties it is estimated that the pulp and paper mills generated a minimum of 91 percent of the total BOD load. No other single source contributed more than 5.1 percent of the total. It is obvious that improvement of the water quality in these three counties must begin with abatement of the pollution created by the pulp and paper group.

In King and Pierce Counties the population contributes the bulk of the BOD that is generated. Domestic wastes accounted for 83 percent of the total in King County and 44 percent in Pierce County. This reflects the fact that King and Pierce are the state's most heavily populated counties with nearly 48 percent of the total population. In 1969 the population of King County was estimated at 1,153,618 while Pierce County contained 405,146 people. The total for the state was estimated at 3,365,000 [5]. Despite the very heavy concentration of the total state population in these two counties, its actual contribution to the existing degradation of the watercourses in this area of the sound is almost nonexistent. The reason for this is that effective sewerage systems have been developed to serve a very large percentage of the domestic population residing in these counties.

The cornerstone of the pollution control efforts in King and Pierce Counties rests on the pollution experienced in the Lake Washington basin after World War II. Lake Washington has long been King County's most precious recreational asset. Rapid proliferation of the urban population in this natural drainage basin created a tremendous sewage disposal problem, which culminated in 1958 with the realization that the lake was unsafe for swimmers and was heading for a rapid death. Actually, the community had made a very commendable effort to avoid pollution. Ten secondary sewage treatment plants had been constructed to serve twenty-three of Seattle's satellite cities and sewer districts. The problem was not with bacteria count but rather with excessive nutrient loading. Algae, stimulated by the nutrient content of the sewage effluent entering the lake and its tributary streams, ran its life cycle at an accelerated rate. The result was a lake deficient in dissolved oxygen, cloudy and foul smelling, and marred with slime and scum [8].

An extensive study was undertaken to solve the problems of King County and particularly Lake Washington. This culminated in the acceptance of a comprehensive sewage disposal plan for the entire Metropolitan Seattle Area. The ultimate aim of the plan was to divert all waste discharges away from the Lake Washington basin and to provide adequate treatment for *all* wastes in the metropolitan area prior to final discharge into Puget Sound or its immediate tributaries. The results have been no less than spectacular so far. Twelve years and roughly $145 million later, the Municipality of Metropolitan Seattle (METRO) now provides sewage disposal for an area in excess of 290 square miles [6:2]. An excerpt from METRO's *Annual Report on Operations* tells the story.

Since removal of the sewage effluent from the lake, the recovery of Lake Washington has been phenomenal. The phosphorus content of the lake has diminished from a high of 72 parts per billion (ppb) in 1963 to 1965 period to a minimum of 22 ppb in 1969, a level comparable to that in 1933 when measurements were first initiated. Correspondingly, aesthetic conditions have improved markedly with disappearance of some of the nuisance forms of algae and surface scums and odors that plagued the lake as late as 1965. Transparency of the water has increased to a depth of 12 feet, an increase of 400% from the minimum of 2½ feet. The rehabilitation of the lake has exceeded all expectations and has been described as one of the most remarkable water pollution control achievements in the past decade [6:22-23].

Similar comprehensive studies were conducted for the Metropolitan Tacoma Area. The objective in both Pierce and King Counties was to take full advantage of the vast assimilative capacity of Puget Sound. This would minimize treatment requirements to the primary level for the majority of the wastes generated and thus reduce construction and operating costs. METRO's system has been designed for stage construction and will eventually handle the wastes of practically the entire county. The treatment facilities in the system have been designed to accommodate the population at saturation levels with integrated plans to upgrade the level of treatment if that should become necessary.

The sewage disposal systems in King and Pierce Counties are operating satisfactorily. The degraded water conditions that still exist are generally confined to the heavily industrialized harbor areas. Here the problem has been getting the industrial entities connected to the municipal system. In most cases the industrial polluters are being required to intercept with the municipal systems. Where this is neither feasible nor desirable, the Department of Ecology has set forth specific treatment requirements together with a time schedule for their expected completion.

Washington State Water Pollution Abatement Program

Waste disposal is only one of the many important water uses in the State of Washington. The regulation of this important water use, so as to prevent the impairment of the water environment for other possible beneficial uses, is the responsibility of the Department of Ecology. The department, formerly the Pollution Control Commission, was established in 1945 and is empowered to prohibit the pollution of all Washington waters, coastal and inland, by prescribing and enforcing rules and regulations for the disposal of waste.

Water Quality Standards

The department had adopted a system that allows assignment of specific quality standards for each watercourse within the state. This is accomplished by utilizing a scheme of watercourse classification where the characteristic uses and the acceptable water quality criteria for these uses are fitted into a single class designation. Each body of water in the state has been so designated.

Waste Treatment and Control Requirements

In order to attain water quality that is in compliance with the standards, the State of Washington has adopted a set of minimum waste treatment and control requirements applicable to all units discharging wastes to its surface waters. Generally, secondary treatment is required of all entities. Where the waste discharge is to salt water, primary treatment is allowed if an engineering study demonstrates that this will preserve or enhance existing water quality conditions. Advanced waste treatment is required when the waste discharge is to a lake or its feeder streams or is within a national park or forest. In no case is less than primary treatment with adequate disinfection and outfall facilities acceptable. Commercial and industrial operations *must* connect to a municipal system where at all feasible; adequate pretreatment is required for these wastes [4:55].

The Department's definition of primary treatment, secondary treatment, advanced waste treatment, and adequate disinfection is as follows:

Primary Treatment . . . the removal of settleable and floatable solids from the waste flow followed by adequate disinfection of the effluent.

Secondary Treatment . . . the removal of settleable and floatable solids from the waste flow and the application of additional waste treatment processes to attain a minimum of 85 percent removal of BOD and 90 percent removal of suspended solids, followed by adequate disinfection of the effluent.

Advanced Waste Treatment . . . secondary treatment followed by further reduction of BOD, suspended solids, MPN and/or nutrients to a level determined to be adequate.

Adequate disinfection has been obtained when the treated effluent exhibits a median value of total coliform organisms of 2400/100 ml or less with no more than 10 percent of the samples exceeding 10,000/100 ml. . . . In no case shall the disinfected effluent cause the degradation of the receiving water quality below the criteria established for the waterway [4:93].

Implementation and Enforcement Plan

To ensure that the waste treatment and control requirements are adhered to, the department relies on various implementation and enforcement tools at its disposal. Domestic waste discharges are controlled by the issuance of directives and orders, and the review and approval of engineering reports and designs for treatment plant construction. State legislation gives the department authority to approve the design for treatment plant construction. In the case where existing facilities no longer provide adequate treatment, the director of the department issues written directives requiring the appropriate modifications.

Commercial and industrial waste discharge control is accomplished by the issuance of waste disposal permits, which are classified temporary or permanent. A temporary permit is one that is issued for a limited and specified period, during which necessary improvements are made in order to comply with department regulations. A permanent permit is issued, for a term of five years, when the treatment and control facilities meet the standards established by the department. Only permanent type permits are issued to new industrial and commercial facilities. All existing permits are reviewed, prior to their expiration, by the director of the department. The permit holder is notified, in writing, as to needed improvements necessary to obtain a new temporary or permanent permit at a reasonable time prior to the expiration of his original permit.

Miscellaneous waste discharges and non-point sources of pollution, not covered under the domestic, commercial, or industrial categories, are controlled by the issuance of regulatory orders and directives. Examples of such potential sources of pollution include: agricultural waste waters, marina and vessel wastes, storm runoff, dredging and dredge spoils, log storage, soil erosion, river impoundments, and animal feedlots. Methods of control for these potential sources of pollution have not yet been determined. The Department of Ecology indicates that studies will be undertaken to assess the magnitude and significance and the best method of control for each of these non-point sources of pollution [4:89]. Appropriate implementation measures will be taken upon completion of the studies.

Surveillance Program

The department's surveillance program consists of inspection of all treatment and control facilities, the monitoring of waste discharge characteristics, and the monitoring of receiving water quality. Compliance or noncompliance with departmental directives, orders, regulations, and waste disposal permits is ascertained by this method. The surveillance program has resulted in a rather thorough documentation of waste dischargers, and receiving water quality in the State of Washington. The *Implementation and Enforcement Plan for Water Quality Regulations,* published in September 1970, lists *all* waste dischargers in the state whose facilities require improvement and compares the water quality of each watercourse with the quality criteria adopted for its class.

Enforcement Program

The ability of the department to enforce its regulations has been strengthened by Washington State legislation that allows the department to levy civil penalties and issue regulatory notifications, orders, and directives. A civil penalty of $100 per day may be levied against any person who violates the terms of his waste discharge permit, discharges without a permit, or who otherwise pollutes the waters of the state.

Future Water Quality in Puget Sound

Water quality in the presently undeveloped areas of Puget Sound will not be degraded in 1980 or at any other point in the future. Even if these areas experience substantial industrial and population growth in the next ten years, there will be no degradation of the watercourses. The reasons are simple. New commercial and industrial operations *must* now incorporate adequate treatment facilities or interception plans in the construction design as only permanent-type waste disposal permits will be issued to these facilities. Thus, potential pollution problems can be accommodated prior to the actual construction of the facility. Urban development will not be allowed without provisions for adequate treatment facilities and/or connections to existing adequate treatment and collection systems. The Department of Ecology's control over treatment plant construction design will ensure that the treatment level and capacity of new facilities maintain or enhance the existing water quality in the receiving system. In general, the relatively undeveloped areas will have the advantage of increased emphasis on matters pertaining to the environment; uncontrolled and environmentally harmful development seems sure to be avoided.

In the areas that have already experienced substantial development, namely the five-county region along the eastern shore of Puget Sound, the future water quality is not as clear, although there is certainly reason for some restrained optimism in certain locations. Presented in table 5.3 are 1980 estimates of BOD for the five-county region. The technique employed to make the 1980 estimates for the industrial sector assumes that gross BOD generation rates per unit of throughput will remain constant throughout the next decade. In other words, changes in the manufacturing processes, should they occur, will not significantly increase or decrease the amount of BOD generated per unit of product produced. In this sense the industries under investigation here are considered to be technologically frozen in that they will be unable to reduce BOD per unit of output through changes in the manufacturing process and must turn to external reduction of the waste load. Thus, growth in output will be shadowed by a proportionately equal growth in the quantity of wastes or obstacles to be overcome by external treatment. Table 5.3 projects the Washington State output trends experienced over the last ten years to 1980 for the selected industry groups operating in the five-county region.

There are far too many variables and intangibles involved in making population forecasts on a county by county basis. Sophisticated methods have certainly not justified their worth in the past. For this reason no attempt has been made to estimate growth rates for the five counties in this region. However, to place things in perspective, the assumption has been made that the absolute growth experienced in each of the counties over the last ten years will be duplicated in the next ten. Since this was generally a period of rapid growth the estimates are

Table 5-3
Estimated Gross BOD Load for the Population and a Selected Group of Industries, 1980 (BOD expressed in millions of pounds annually)

Source	County										Percentage Total by Source	Compound of Five County Total	Compound Annual Growth Rate
	Whatcom		Skagit		Snohomish		King		Pierce				
	BOD	%	BOD	%	BOD	%	BOD	%	BOD	%			
Population	6.35	2.5	4.96	2.3	26.28	5.5	100.52	75.0	35.04	40.2	173.50	14.9	3.3
SIC 2011	.22	—	.17	—	.28	—	6.41	4.8	3.02	3.5	10.10	1.0	4.3
SIC 2013	.02	—	—	—	—	—	.40	—	.02	—	.44	—	1.7
SIC 2015	.02	—	.18	—	.03	—	1.04	1.0	.42	—	1.69	.2	4.4
SIC 2026	.11	—	.10	—	.01	—	.62	—	.06	—	.90	—	3.6
SIC 2033	4.14	1.6	.81	—	.27	—	6.78	5.1	.14	—	12.14	1.0	7.7
SIC 2037	13.94	5.6	18.29	8.5	15.32	3.2	18.18	13.6	7.57	8.7	73.30	6.3	9.7
SIC 2611[a]	213.89	85.9	188.73	87.8	402.62	84.0	—	—	—	—	805.24	69.2	3.8
SIC 2621	8.87	3.6	—	—	34.32	7.2	—	—	26.03	29.9	69.22	5.9	2.2
SIC 2631	—	—	—	—	—	—	—	—	14.65	16.8	14.65	1.3	7.6
SIC 2911	1.44	—	1.66	.7	.08	—	—	—	.25	—	3.33	.3	1.5
Total	249.00	—	214.90	—	479.21	—	133.95	—	87.20	—	1164.25	—	3.9
Percentage of Five County Total	21.4	—	18.5	—	41.2	—	11.5	—	7.5	—	100.00	—	—

Source: Estimated using data on employment in relevant industries from U.S. Department of Commerce, Bureau of the Census, COUNTY BUSINESS PATTERNS and CENSUS OF MANUFACTURERS in accordance with the procedure outlined in Chapter 2.

Legend: BOD is expressed in millions of pounds annually. Blank spaces in the BOD columns indicate that the county did not contain the SIC category. Blank spaces in the percentage columns indicate an insignificant percentage of the total.

[a]See Appendix A.

likely to be high. The 1980 estimates of BOD projected for the population in table 5.3 reflect this assumption.

Whatcom and Skagit Counties

The future quality of the surface waters in Whatcom and Skagit Counties will depend almost entirely on the success or failure of the efforts to reduce and maintain the residual waste discharges of the pulp and paper industry at a level significantly below that which presently exists. All other point sources of pollution existing in these counties are insignificant in comparison to the magnitude of the waste discharges of this group. Realization of any improvement in the water quality, with the exception of coliform bacteria, can only be expected if the contaminants generated by the pulp and paper sources are adequately treated prior to their final disposal in the Puget Sound receiving system.

In addition to the Georgia-Pacific pulp and paper operations in Bellingham and the Scott Paper Company's pulp mill in Anacortes, the Department of Ecology has cited thirty-five other private entities with unacceptable discharges. However, it is safe to assume that these polluters will not pose a pollution problem at the end of the decade. Twenty-three must merely convey their wastes to existing treatment systems. Ten must upgrade their existing facilities, while the remaining two must modify or expand presently adequate facilities to accommodate future production capacity. The latest completion date for these improvements is 1973 [4:59-87].

The only problem that merits serious attention is the matter of the pollutional load that could be generated by pulp and paper producers. Present plans for abatement call for a minimum of primary treatment of *all* pulp and paper wastes, improvement of existing outfalls, and in the case of the Georgia-Pacific operations in Bellingham, removal of 80 percent of the Sulfite Waste Liquor (SWL) load [4:67, 79]. Will this be enough to insure that the already degraded water environment in the vicinity of these mills will be returned to an acceptable state in the years ahead? The answer lies in the expected magnitude of the gross waste load and the abatement requirements that have been set forth.

As late as 1969, waste treatment at the Scott and Georgia-Pacific mills consisted primarily of screening and in-plant control [4:67, 79]. For all practical purposes the residual waste load and the gross waste load generated were the same. Since Georgia-Pacific and Scott constitute the entire pulp and paper operations in Whatcom and Skagit Counties, the waste load estimates in tables 5.2 and 5.3 apply directly to these two companies. In terms of BOD, the estimates of table 5.2 suggested that Georgia-Pacific discharged over 149 million pounds of waste to Bellingham Bay while Scott added another 127 million pounds of BOD to Guemes Channel in 1969. By 1980, the projections of table 5.3 indicate, the Georgia-Pacific operation could be generating an annual gross waste load of 222 million pounds of BOD. At the same time, Scott's annual gross waste load could very well be approaching 189 million pounds.

The Department of Interior had estimated that primary treatment of pulp and paper mill wastes reduces between 10 and 50 percent of the BOD load depending on the process employed and the operational efficiency [7:33]. Furthermore, it has been determined from the same source that about 60 percent of the BOD load generated in the sulfite process emanates from the strong digester wastes containing the Sulfite Waste Liquor (SWL). Combining the abatement plans for these two sulfite mills, the 1980 estimates of the gross BOD loads, and the treatment efficiency levels expected by the Department of Interior for primary treatment facilities results in an estimated residual BOD load that could range from a low of 107 million pounds to a high of 165 million pounds annually for the Georgia-Pacific operation at Bellingham and a low of 92 million pounds to a high of 141 million pounds for the Scott Mill at Anacortes.

What do the above figures portend for 1980 in these two northern counties? The estimated magnitude of the residual discharges, in terms of BOD, will *not* be significantly below those that presently exist in either Whatcom or Skagit Counties. Thus, the unsatisfactory dissolved oxygen concentrations in Bellingham Bay and Guemes Channel (see table 5.1) will still be unsatisfactory in 1980. On the brighter side, the toxic condition in Bellingham Bay should be eliminated by the 80 percent SWL removal requirement established for the Georgia-Pacific operations. The primary treatment requirement should improve the impaired aesthetic values in the Bellingham Bay system.

The great, but unchallenged, capacity of the Guemes Channel to assimilate wastes may be seriously tested by the residual waste load of the Scott Mill in 1980. The Department of Ecology has obviously banked on this capacity by not requiring Scott to remove a percentage of its SWL load. If the Department is right, the citizens of Anacortes should continue to witness reasonably good water quality conditions.

In summary, Whatcom and Skagit Counties should see some improvement in the water quality conditions of their environmental waters by 1980. However, serious attention must be paid to the actions taken by the pulp and paper companies in their attempt to conform to the rules laid down by the pollution control authorities.

Snohomish County

Snohomish County faces very serious problems in the years ahead. The problems will not emanate from growth in population, but rather from the industrial sector. Only 5.5 percent of the total estimated BOD load for 1980 is attributable to the assumed population of 360,000. It is estimated that the selected industries will be contributing nearly 453 million pounds of BOD annually, as shown in table 5.3 for this county. This represents an increase of 157 million pounds over the 1969 estimates for the industrial sector, as shown in table 5.2. Based on a per capita annual BOD generation rate of 70 pounds, the increase in the industrial sector is the equivalent of adding 2,240,000 to the county's population. Over 93 percent of the expected increase is attributable to the pulp and paper operations in the county. The problem is aggravated by the fact that the pulp and paper producers are all concentrated in the immediate vicinity of Everett Harbor and the lower reaches of the Snohomish River. Scott, Weyerhauser, and Simpson-Lee operate facilities in this area.

Simpson-Lee operates a relatively small kraft pulp and paper mill on the Snohomish River some ten miles upstream from its mouth [2:259]. The size of the operation and the processes employed in the manufacture of its pulp and paper tend to minimize the impact of the effluent discharged to the Snohomish. Scott and Weyerhauser operate sulfite-type pulp and paper mills in the Everett Harbor area. Weyerhauser also operates a kraft mill on the lower reaches of the Snohomish [2:256]. Heretofore, the waste disposal practices of Snohomish County's pulp and paper industry have been primitive. Only screening and in-plant control procedures were utilized at the primary facilities of Scott, Weyerhauser, and Simpson-Lee [4:79, 80, 85]. As a result, almost the entire gross load generated entered the sound.

The treatment requirements prescribed for the pulp and paper producers in Snohomish County are generally the same as those described earlier for Georgia-Pacific's operations in Bellingham. Based on similar treatment efficiency assumptions, it is estimated that the residual BOD of the Snohomish County mills in 1980 will range between 197 million to 308 million pounds annually. This is certainly *not* a significant reduction from the 1969 estimates, which were presented earlier in table 5.2. The residual discharge at the lower end of the range is equivalent to a raw waste discharge of nearly 2,800,000 people. When one considers that this discharge will be dumped into a relatively small area, the hope of improving the water conditions is all but destroyed.

The people of Snohomish County must take a long, hard look at the situation they face. The total wastes generated by the sources considered in this chapter by 1980 will be equivalent to a population of nearly seven million people. This is more than two times the existing size of the population in the entire state. A concerted effort will have to be made by industry and citizens alike in order to bring the pollution problem under control.

King County

King County is well prepared to handle the expected increases in the pollutional load projected for 1980. The total BOD load from the sources analyzed in this study will contribute an additional 50,000 pounds a day over the estimates made for 1969. This should be easily absorbed in the municipal system. The only area that might cause some concern is the potential deleterious effects of industrial waste discharges on the operating efficiency of the treatment facilities. METRO officials are well aware of this potential problem and have, in fact, experienced some difficulties already. In May 1969 an industrial waste engineer was hired to begin work on METRO's industrial waste program. The emphasis of the program is being placed on the elimination of all toxic wastes at the source of entry into the municipal system [6:22]. Headway is already being made in this effort. A sampling program was begun in December 1969 to analyze the waste discharges of the area's major metal finishing industries, including the Boeing Company.

The samples are being analyzed with the help of a sophisticated atomic absorption spectrophotometer [6:22]. In addition to the detailed analysis of waste discharges, METRO has developed a comprehensive receiving water quality control program for all environmental waters that may be affected as a result of its operations. About 630 monitoring stations have been established in the sewerage area, some of which automatically feed data to a computer for analysis [6:22-24].

The degree of sophistication and vigor already exhibited by METRO in attacking the pollution problem and the success with which it has met, should result in materially improved water quality in the next ten-year period. "The White House Council on Environmental Quality, in its first annual report, cited Seattle's METRO project as one of the two outstanding antipollution success stories in the nation" [8]. There appears to be no reason to dispute this evaluation. The balance of forces in King County have shifted in favor of a cleaner water environment.

Pierce County

Although the projected waste load in Pierce County is the smallest of any of the five counties selected for investigation, the future water quality in this county is uncertain. Despite the fact that the population contributes the bulk of the waste load, this source will not present a problem. The increase of 27,000 pounds of BOD per day projected for the assumed 1980 population of 480,000 should be easily assimilated into the existing waste treatment systems. On the other hand, rising output in the industrial sector will result in an additional 20 million pounds of BOD annually by 1980. Much of this will again come from pulp, paper, and paperboard sources. Substantial increases will also be posted by the meat slaughtering and frozen fruits and vegetables industries.

The Department of Ecology cited thirty-three commercial and industrial firms with inadequate treatment facilities in the county. Twenty-five of these were located in the City of Tacoma. The remainder were scattered throughout Pierce County [4:59-87]. Among the list were three constituents of the pulp, paper, and paperboard industries. The pulp, paper, and paperboard firms are all being required to upgrade their treatment facilities. As in other areas of the sound, this group was merely utilizing screening and in-plant control procedures. The upgraded treatment requirements should result in significant reductions in the residual discharges of these facilities.

There are other industries, outside those selected for analysis in this chapter, which may be a considerable factor in Tacoma's Commencement Bay area. The department also cited six chemical companies and three metals firms in its *Implementation and Enforcement Plan*. It would seem that further investigation into the pollution problems created by these other manufacturing groups is warranted.

Conclusions

Water pollution in Puget Sound in 1980? Unlikely, except for those areas where the pulp and paper industry reigns supreme. The Puget Sound region has two key advantages over other areas in our great country. One, it is only in the early stages of development; and two, the sound itself has vast assimilative capacity. When water pollution came to the forefront of our thinking some five to ten years ago, the State of Washington was already well down the road to solving the problem. The fact that a state agency in the United States has tabs on *all* of its waste dischargers is something just short of phenomenal. Such is the case in Washington. Its industrial waste discharge permit system, which has been in effect for quite some time, is now or will be shortly adopted by the federal government. All things considered, the State of Washington is probably the most advanced state in the union when it comes to water pollution control. Because of these reasons, the odds are in favor of a cleaner water environment in the Puget Sound region.

This chapter shows that serious attention, however, should be placed on planning for growth in industrial output and growth in population if we are to succeed in combating the pollution problem. Emphasis should be placed on predicting and planning for the growth in industrial output. The wastes generated by the population per se do not present any unique problems to combat. All that counts here is setting a plan in motion that will insure that all domestic wastes are gathered and properly treated prior to returning them to nature. On the other hand, wastes generated in the industrial sector can and indeed have presented unique treatment problems. By predicting the level of output we can, at least, prepare the methods of control that will be necessary to maintain residual waste discharge below the assimilative threshold of the receiving system. Certainly, with first-hand knowledge at the source of potential pollution problems, we should be able to develop systems that are better suited to quantify the level of output and the concomitant level of gross waste generated than the method employed in this chapter.

It has been shown in this chapter that, despite substantial increase in the output of a selected group of industries, in most cases, the effort to control pollution should be rewarding. However, we must not fall into the trap that has plagued our society since its inception of pollution control programs. And that is that we let growth outpace the efforts to control the concomitant wastes. We must control growth and plan for it if we expect society, *as a whole,* to benefit from it.

Appendix

BOD coefficients were available in *Waste Management* for all SIC categories *except* 2611, pulp mills. Several studies revealed, however, that this industry was a particularly important contributor to the pollution problems in Puget Sound. Thus, it was necessary to develop a BOD coefficient for pulp mills (SIC 2611) in order to assess the relative and absolute impact of this industry on the present and future water quality of Puget Sound.

Several authoritative sources were consulted in an attempt to derive a reliable BOD coefficient for pulp mills. Valuable insight into the problem was provided in *The Cost of Clean Water: Industrial Waste Profile No. 3 – Paper Mills,* a 1967 publication by the Federal Water Pollution Control Administration. The above source revealed that the magnitude of BOD generated by pulp mills varied widely with the particular pulping process employed and the degree of technological advancement of the particular mill. *Probable average values* (not truly statistical averages) derived in *The Cost of Clean Water* series ranged from 480 pounds of BOD per ton of pulp produced in older sulfite pulp mills to 70 pounds of BOD per ton of pulp for newer unbleached sulfate (kraft) mills. The generation rates for individual mills used to produce the *probable average values* ranged as high as 700 pounds of BOD per ton of pulp produced.

A joint federal and State of Washington investigation of Puget Sound pulp and paper mill wastes found BOD generation rates from pulp sources ranging from about 40 pounds of BOD per ton to 1,065 pounds per ton. Results of the investigation are found in *Pollutional Effects of Pulp and Paper Mill Wastes in Puget Sound,* published in 1967 by the Federal Water Pollution Control Administration. This study gave BOD generation rates and pulp production rates for *all* mills in the Puget Sound area.

The wide dispersion of rates from mill to mill and process to process makes the use of an "average" figure meaningless except where the population is large. As a result, a ballpark figure of 450 pounds of BOD per ton of pulp produced was used to estimate the statewide BOD load. After apportioning the statewide figure to individual counties by percentage of Washington State employment, the county figures were adjusted to reflect variations between the ballpark pulp mill BOD coefficient used to determine the statewide BOD load and the actual BOD coefficients existing in Puget Sound pulp-producing counties. Where more than one source existed, a weighted average figure was used based on individual mill BOD generation rates and pulp production rates. Table A5.1 shows the calculation of the county adjustment factors.

See table A5.1 on following page.

Table A5-1

County	Source(s)	BOD Coefficient (lbs/ton of pulp)	Average Daily Output (tons)	Weighted Average Coefficient (lbs/ton of pulp)	County Adjustment Factor
Whatcom	Georgia-Pacific (Sulfite)	519	527	519	519/450 = 1.15
Skagit	Scott Paper (Sulfite)	989	138	989	989/450 = 2.2
Snohomish	Weyerhauser (Sulfite)	1,065	304		
	Weyerhauser (Kraft)	92	417		
	Scott Paper (Sulfite)	989	828		
	Simpson-Lee (Kraft)	46	271	655	655/450 = 1.45

Source: U.S. Department of Interior, Federal Water Pollution Control Administration, Pollutional Effects of Pulp and Paper Mill Wastes in Puget Sound (n.p.: n.p., 1967). Tables 6-1, 18-1, 23-1, 23-3, 23-5, 23-6, pp. 23-264.

Notes

1. Brown and Caldwell Civil and Chemical Engineers, *Metropolitan Seattle Sewerage and Drainage Survey,* Report to the City of Seattle, King County and the State of Washington, May 19, 1958 (Seattle: Brown & Caldwell, 1958).

2. U. S. Department of the Interior, Federal Water Pollution Control Administration, *Pollutional Effects of Pulp and Paper Mill Wastes in Puget Sound* (Washington, D.C.: Government Printing Office, 1967).

3. State of Washington, Department of Commerce and Economic Development, Business and Economic Research Division, *Regional Economies of Washington State,* 1969.

4. State of Washington, Department of Ecology, *Implementation and Enforcement Plan for Water Quality Regulations,* 1970.

5. Revised 1969 estimates by the State of Washington, Bureau of Vital Statistics, Population and Research Section dated 6/29/70 received via correspondence.

6. Metropolitan Engineers, *Report on Operations: Municipality of Metropolitan Seattle, January 1, 1968 - December 31, 1969,* 1970.

7. U. S. Department of the Interior, Federal Water Pollution Control Administration, *The Cost of Clean Water,* Vol. III, *Industrial Waste Profiles No. 3 – Paper Mills* (Washington, D.C.: Government Printing Office, 1967), Table 6, p. 33.

8. *New York Times,* September 18, 1970, p. 30.

9. U. S. Department of Commerce, Bureau of the Census, *Census of Manufactures: 1963* (Washington, D.C.: U. S. Government Printing Office, 1966), Vol. I, *Summary and Subject Series,* pp. 10-62 - 10-63.

6 San Francisco Bay

RAYMOND A. MATERN

The Geographic and Economic Environment

The San Francisco Bay System

The area of the San Francisco Bay system under consideration extends from the eastern tip of Chipps Island where the San Joaquin and Sacramento Rivers join, westward and southward to the mouth of Coyote Creek near San Jose. This includes a total distance of some 85 miles. The Golden Gate is halfway between San Jose near the Bay's southern end and Antioch at the north and is the bay's only connection with the ocean [1:3-1].

There are eighty tributary individual drainage basins to the bay seaward of Chipps Island. The actual watershed tributary area to the northern arm of the bay system landward of Chipps Island is about 45,420 square miles [2:23].

Much of San Francisco Bay is very shallow, the average depth being approximately 20 feet. The consequences of this are important due to the fact that wind-generated waves disturb the floor of the bay and contribute to the high turbidity of the water. The depth of the bay is also an important factor in the surface reaeration in the overall oxygen balance of the bay [1:3-2].

The Delta System

The Delta area is triangular in shape and extends from Chipps Island on the west, to Sacramento on the north, and to the Vernalis gauging station on the south.

This area is the focal point for the transfer of surplus water for drainage from the Central Valley entering the Bay-Delta system. The Delta's channels are currently being used for the transfer of needed water. It is proposed that, in the future, there be the construction of a peripheral canal around the eastern edge of the Delta region. This will be done to transfer high-quality Sacramento River water to the south [1:3-2].

The River System of the Central Valley

The San Joaquin and Sacramento river basins comprise nearly 40 percent of the total area of the State of California. They are the principal fresh water source inflows into the Bay-Delta system. As use within the Central Valley grows, the net Delta outflow will diminish.

Economic Environment

The economy of the Bay-Delta area is a diverse one. Due to the geographic location, the economy of the area relies to a great extent on maritime trade and related activities. It remains the distribution and wholesale center for the region. A stock exchange and Bank of America's main headquarters are located in San Francisco, thus making it the western financial capital of the United States.

The most important industries are publishing and printing, the manufacture of food products, fabricated metal products, chemicals, furniture, and machinery. The specific industries that contribute substantially to the water pollution problems of the area will be discussed fully in the following pages.

Statistics released in 1967 for the nine-county Bay region show that although this area is only 4.5 percent of California's total area, it accounts for 23 percent of the population and retail trade of the state, 25 percent of the effective buying income, 31 percent of wholesale sales, 30 percent of the state's bank deposits, and 54 percent of the waterborne commerce. The area's transit facilities also contribute to the economy of the area with the San Francisco International Airport ranking fifth in air passenger traffic in the United States [4].

Pollution Abatement Programs

San Francisco is unique due to its early involvement in pollution abatement. As early as the mid-1950s citizens of the community were beginning to recognize general problems of pollution. Landfill projects were slowly reducing the total Bay area and therefore not only causing serious consequences to the water quality of the bay, but destroying the natural beauty of the area. The latter result of the landfill project probably initiated citizen involvement. As time progressed though, people began to realize that this action also caused serious water quality problems.

Many clubs, organizations, and citizens concerned about the future of the bay started to apply pressure on both local and state legislators for positive action. Due to this pressure and to the general political popularity of the general subject matter as an issue for voters, the State of California and the localities involved with the Bay area are taking positive steps and proposing dynamic programs to help solve this problem. Two of the major important programs will be discussed below.

University of California Program

The University of California, on July 1, 1970, submitted a report of recommendations and findings to the State Water Resources Control Board, thereby complying with the terms of their agreed contract.

They concluded that the major water quality problems in San Francisco Bay resulted from toxicity, coliform bacteria, floatables arising from oil and grease, and biostimulation resulting in excess algae growth. To complicate the problem, these four major problem areas are those in which there is generally the least reliable quantitative data, especially with respect to toxicity, biostimulation, and floatables. They emphasized that it was essential that sufficient quantitative data be developed in these problem areas so that rational planning would become possible.

The report dealt with the concepts of biostimulation and toxicity and the buildup of such concentrations in the Bay area. It admitted that although nitrogen and phosphorous concentrations might be high, they were not the critical factors presently stimulating the production of algae growths. The report recommended that further study of these factors be conducted with an emphasis on toxicity. Such an analysis is currently being conducted for the State Water Control Board by the Departments of Fish and Game and Water Resources and the Sanitary Engineering Laboratory of the University of California.

Their recommendations as to the treatment of waste discharges, receiving water problems, and sewerage planning seem to be mostly theoretical in nature. They call for more study, intensive monitoring facilities, sampling systems, and regional agencies to help accumulate accurate and reliable data [2:1-85].

San Francisco Bay-Delta Water Quality Control Report

The San Francisco Bay-Delta Water Quality Control Program was created by the state legislature in 1965. Its purpose was to determine the need for, and the feasibility of a comprehensive waste collection and disposal system serving the Bay-Delta area, as directed by the legislature. A report of their findings was published in March of 1969.

This report differs from the University of California report in that it offers concrete solutions for the alleviation of the water pollution problems of the area. A description of this program follows.

After careful study by the State Water Resources Control Board, it was determined that the following action would be essential for instituting any effective water quality control system.

1. Take steps to alleviate the most immediate and serious water quality problems.
2. Consider the various dilution capabilities of the different parts of the Bay-Delta area.
3. Design a program with maximum flexibility.
4. Design for the possibility of regional and local wastewater reclamation.
5. Design a means of disposing unreclaimable wastewaters.
6. Achieve by means of the foregoing the maximum realization of beneficial uses of the Bay-Delta area waters [2].

The time span of the recommended program is fifty years. Paramount attention, however, is focused on the ensuing five- to ten-year period. This is true because, in order to get the most serious water quality problems solved rapidly, one must focus attention on the relatively short-term problems. This is not to say that long-range goals of the system are not important. Even at this early planning stage, there must be long-range goals in order to give some meaning to the shorter-term objectives. Realistically, however, one cannot plan fifty years in advance with accuracy. There are so many variables involved that accuracy would be almost impossible.

The recommended system consists of three different phases. *Phase one* includes the formation of various regional agencies capable of implementing the recommended plans. The general design of this phase involves the export of treated wastewaters from the extremities of San Francisco Bay to the central regions of the bay. Construction of this would be completed by 1980. *Phase two* is to be completed by 1990 and includes in its plan a regional wastewater treatment plant near Redwood City discharging to an ocean outfall system. *Phase three,* to be developed after 1990, is presented as a guide rather than a specific recommendation. This plan has two basic options built into it because of the difficulty of estimating progress so far in the future.

Phase One Program

Since this chapter deals with projections to 1980, it will concentrate on phase one. Its main proposals for construction are as follows:

1. A South Bay interceptor to collect treated wastewater effluents from existing municipal and industrial plants between Milpitas and Redwood City. In 1980, these effluents would be discharged through an outfall and diffuser in the deepest water of the South Bay sector opposite Redwood City.
2. An interceptor from Antioch to Oakland to collect treated wastewater effluents from existing municipal and industrial plants and discharge in Central Bay through a large diffuser section just south of San Francisco-Oakland Bay Bridge.
3. An interceptor from Calistoga to Vallejo to collect treated wastewater from existing plants would discharge to Carquinez Strait in 1975. In 1980 a regional treatment plant, RP-1, with an initial capacity for secondary treatment and filtration of 25 mgd would be constructed at Vallejo and local treatment facilities would be taken out of service.
4. An interceptor from Sonoma County to a regional treatment plant at San Rafael. This system would collect and treat wastewaters between Sonoma and Greenbrae, south of San Rafael. The regional treatment plant would provide chemical treatment and filtration and would discharge in the Bay south of the Richmond-San Rafael Bridge.
5. An interceptor system (Alameda-Redwood City interceptor) from the Livermore Valley and the east shore of the bay would discharge locally-treated wastewaters to the bay, sharing a common outfall with the Milpitas-Redwood City interceptor [3:85].

A significant improvement in the major problem areas would be effected with the inplementation of the phase one system, thereby eliminating wastewater discharges from the South Bay, San Pablo, and Suisun Bay areas. It would also provide the base for continued development of any of several alternative water quality control systems with no significant loss in investment.

There is no change in the capital requirement estimates for the implementation of phase one due to the options available in exercising phases two and three. Table 6.1 shows the recommended phase one cost summary.

As can be seen in table 6.1, the capital cost projections for phase one of the system are divided several ways. Breakdowns are given as to the costs of the

Table 6-1
San Francisco Bay Area Sewage System

Recommended Phase One Cost Summary			
Capital Expenditures (Millions of 1969 Dollars) Elements	By 1970	1970- 1974	1975- 1979
Conveyance		44	317
Treatment			
Municipal		47	94
Discrete Industrial	91	4	65
Land and Right of Way		14	27
Contingency		21	87
Engineering and Administration		18	78
Total	91	148	668
Average Annual Operation and Maintenance Costs (Millions of 1969 Dollars)			
	by 1970	1970- 1974	1975 1979
Conveyance			0.3
Treatment			
Municipal		10.1	12.9
Discrete Industrial		5.0	6.3
Total		15.1	19.5
Total, Five Years		75.5	97.5
Grand Total (5 year increment)			
Municipal		195	669
Discrete Industrial	91	29	96
Accumulated Grand Total	91	315	1080

Source: REPORT OF THE STATE WATER RESOURCES CONTROL BOARD ON SAN FRANSISCO BAY-DELTA WATER QUALITY CONTROL PROGRAM, Kerry W. Mulligan, chairman (Sacramento, California: 1969), pp. 9-10.

regional system facilities and the in-plant industrial treatment facilities. A distinction is also made between costs in current dollars as compared with future escalating costs. Within this context the total capital costs for the phase one system in 1969 dollars is $907 million with $747 million allocated to regional system facilities and $160 million allocated to in-plant industrial treatment facilities. In estimated future costs this figure comes to a total of $1,473 million, an estimated increase of 62.4 percent [1:10-4]. The projected phase one capital costs for assumed improvement districts total $1,255 million for the various districts involved.

The average annual operation and maintenance costs for the years 1970 through 1979 are divided equally between the five-year periods. Total annual costs for the 1970-1974 period in regard to the operation and maintenance of the system amount to $15.1 million. The costs for the 1975-1979 period come to an annual total of $19.5 million. Therefore the estimated total costs for the operation and maintenance of the phase one system over the ten-year period reaches $173 million.

Whether developmental or operational cost projections, one economic fact of life must be considered, namely, inflation. Table 6.2 estimates the projected costs when inflation and its effects are considered.

It is currently estimated, therefore, that due to inflation the capital requirements for phase one will be increased from an uninflated $907 million in terms of 1969 dollars to $1,473 million throughout the ten-year period, an estimated increase of 62.4 percent. Similar results could be expected for capital outlays relating to the annual operational and maintenance costs of phase one.

Phase Two Program

Construction between 1980 and 1990 for phase two would include three major features. By 1985 the Antioch-Oakland interceptor would be extended across San Francisco Bay and south along the west shore, thus collecting wastewaters from San Francisco and other cities in this area. The terminating point would be

Table 6-2
San Francisco Bay Area Sewage System—Escalated Phase One Developmental Capital Costs

	By 1970	1970-1974	1975-1979	Total
Regional System Facilities	0	$183	$1,072	$1,255
In-Plant Industrial Treatment Facilities	$97	5	116	218
Total	$97	$188	$1,189	$1,473

Source: REPORT OF THE STATE WATER RESOURCES CONTROL BOARD ON SAN FRANSISCO BAY-DELTA WATER QUALITY CONTROL PROGRAM, Kerry W. Mulligan, chairman (Sacramento, California: 1969), pp. 10-4.

near the treatment plant at Redwood City. By 1985 a regional wastewater treatment plant near Redwood City would treat the wastewaters from the Antioch-Redwood City interceptor and would be expanded to treat those from the Milpitas-Redwood City interceptor in 1990. The third plan is to have an interceptor serving Livermore and southern Alameda County that would connect to the regional treatment facility in 1985, thereby utilizing the interim outfall initially constructed for the Milpitas-Redwood City interceptor for the balance of the bay crossing [3].

The major purpose of the phase two system is to provide a disposal system to the ocean. This is the goal mainly because the ocean has dilution capability of at least four times that of the Central Bay. The recommended system would reduce the 1970 BOD loading to the Bay-Delta area by only 30 percent by 1980, but this figure by 1990 would increase to an estimated 85 percent. The total relative toxicity discharge to the Bay-Delta area would increase slightly by 1980 under this system; however, by 1990, it would be reduced by more than 40 percent below the levels of the 1970 discharge figures [3].

Phase Three Program

Because of the difficulty of planning fifty years in advance, there are only two system alternatives complementing the previous two phases. One alternative assumes that a significant demand for reclaimed wastewater will be evident around 1990. The other assumes that this demand will not exist and that protection of the estuarine environment will necessitate the discharge of most of the wastewaters to the ocean. These two alternatives compose the recommended third phase of the San Francisco Bay-Delta Water Quality Control Plan.

Cost projections for the second and third phase of the recommended system differ, depending upon which option is used. Current estimates indicate that the disposal option would be the most expensive to construct and maintain, although the maintenance expense is slightly lower than the reclamation option. Complete cost summaries of both options can be found in the appendix of this chapter.

With any cost projections, as indicated, the question of inflation must be considered. Earlier, inflation's effect on phase one was cited. These projections must be considered and viewed as a mere guide to the ultimate cost of the entire system. It should be recognized that there is no method by which one may predict the inflation rate ten years hence with any degree of accuracy. Therefore economic conditions and uncertainties being as they are, it is impossible for one to arrive at a correct approximation of cost figures even further in the future. Thus, complete objectivity is impossible. The individual reader must make his own determination as to the accuracy of the current inflation projections and make his own subjective judgment on these figures in an effort to come up with approximations of what the cost projections will ultimately be.

Trends of Population and Employment Concentrations

Population Projections

When trying to estimate the extent of a pollution problem one must consider population statistics, for obvious reasons. The twelve-county region of the Bay-Delta area is not unlike the rest of America. As is shown in table 6.3, this area will have, in some cases, dramatic growth projections to cope with.

Marin, Yolo, and Sonoma Counties show the most spectacular growth projections. This is due to the small populations in these areas today. Conversely, not unlike most other major cities, the City of San Francisco is near its saturation point as far as population density is concerned, and a slow population growth rate may be expected.

Employment Concentrations

As in each major city in the United States, San Francisco has a greatly diversified industrial economy. Add to this the eleven surrounding counties making up the total Bay-Delta area, and there is even more diversification. For this reason we will only concentrate on the major industrial pollutors.

The counties under consideration include the same list as was stated for the projected population growth figures. These counties make up the entire Bay-

Table 6-3
San Francisco Bay Area—Projected Population Growth (in thousands)

County	Population 1965	Population 1970	Population 1980	Percentage Change
Marin	40	220	310	675.0
Yolo	80	120	280	250.0
Sonoma	180	220	510	183.3
Solano	150	200	290	94.4
Sacramento	650	780	1,080	66.2
Santa Clara	840	1,070	1,320	57.1
Napa	80	100	120	50.0
San Joaquin	280	310	400	42.8
Contra Costa	500	590	710	42.0
San Mateo	500	550	660	32.0
Alameda	1,020	1,110	1,340	31.3
San Francisco	750	790	800	6.6
Total	5,070	6,060	7,820	

Source: REPORT OF THE STATE WATER RESOURCES CONTROL BOARD ON SAN FRANSISCO BAY-DELTA WATER QUALITY CONTROL PROGRAM, Kerry W. Mulligan, chairman (Sacramento, California: 1969), pp. 4-7.

Delta region. The industries chosen in this study are considered from the four-digit SIC code classification, therefore giving a micro-analysis. The SIC code classifications will be listed below. In future pages, therefore, some references will be made only to the SIC code number.

Industry	SIC Code
Meat Slaughtering Plants	2011
Meat Processing Plants	2013
Poultry Dressing Plants	2015
Fluid Milk	2026
Canned Fruits and Vegetables	2033
Frozen Fruits and Vegetables	2037
Cane Sugar Refining	2062
Distilled Liquor Except Brandy	2085
Paper Mills Except Building	2621
Millwork Plants	2631
Paper Coating and Glazing	2641
Sanitary Food Containers	2654
Petroleum Refining	2911

The statistical method used in this study is as follows. Briefly, we arrived at statewide industrial statistics for California for specific industries. With this data, one could arrive at an apportionment factor that could be applied at a county level. Through interpolation with existing statistics, the total BOD and input/output statistics statewide could be determined. With this information for the years 1969 and 1980, it became possible to focus on the county level for a specific industry and arrive at the total percentage that an individual county contributed by way of pollution, the total BOD load for that county and industry for the years 1969 and 1980, and the same figures for the entire twelve-county area under consideration.

An examination of the general characteristics and possible problem areas of the twelve-county Bay-Delta area is helpful for the individual to get an overall perspective of the area. Following this general analysis will be an examination of categories of wastewaters and individual counties that may present problems in 1980.

Table 6.4 shows the percentage of the total state BOD load that the twelve-county area contributes. From these data one can immediately determine the industries that are important in the Bay-Delta area relative to the remainder of the state. These specific industries will be reviewed individually. It should be noted, however, that what is most important is the total BOD load for the Bay-Delta region; therefore, not only the major industries will be reviewed, but also the industries that contribute the greatest amount of BOD for the area.

Categories of Wastewater

There are basically four major categories of wastewaters in the Bay-Delta system: (1) industrial wastes, (2) agricultural wastes, (3) municipal wastes, and

Table 6-4
Bay-Delta Percentage of Total Statewide BOD

100.0%	Cane Sugar Refining	2062
82.1%	Canned Fruits and Vegetables	2033
54.1%	Millwork Plants	2631
51.6%	Distilled Liquor Except Brandy	2085
46.1%	Paper Coating and Glazing	2641
35.2%	Paper Mills Except Building	2621
32.9%	Meat Processing Plants	2013
30.3%	Petroleum Refining	2911
29.6%	Meat Slaughtering Plants	2011
29.1%	Sanitary Food Containers	2654
23.1%	Fluid Milk	2026
17.8%	Frozen Fruits and Vegetables	2037

Source: Estimated using data on employment in relevant industries from U.S. Department of Commerce, Bureau of the Census, COUNTY BUSINESS PATTERNS and CENSUS OF MANUFACTURERS in accordance with the procedure outlined in Chapter 2.

(4) natural runoff. There is some unavoidable overlapping in these categories. For example, a portion of industrial wastes may be discharged through municipal systems. However, it is useful to categorize the different sources to organize presentation and discussion of wasteload projections.

Industrial wastes are those discharged by all industrial operations. These include both discharges through industrial and municipal systems. In many sources though, the phrase "discrete industries" is used. This category includes industries that dispose of their wastes solely through industrial systems.

Agricultural wastes are due to drainage from agricultural land. These wastewaters may be caused by subsurface drainage that is collected in a drainage system or excess irrigation flows returned to the surface streams. The area of greatest concern with respect to agricultural wastes is that of the drainage of the Sacramento and San Joaquin Valleys. The drainage results in an almost uncontrolled discharge of agricultural wastes to these rivers and their tributaries. Agricultural waste load projections are based mainly on future land use and water requirements, water reuse potential, cropping patterns, and the concentrations of constituents in the waste flows.

Municipal wastes are those wastes discharged by the local nonindustrial community.

Natural runoff results from water running off the land during a snow melt or after a rain. In urbanized areas these waters usually contain large quantities of BOD, solids, nutrients, nitrates, oils, and greases. Eventually a large majority of these wastewaters will be mixed with municipal wastewater systems.

Major Contributing Industries

Cane Sugar Refining – 2062

As the previous table indicates, the entire industry of cane sugar refining is within the Bay-Delta region. Moreover, the entire statewide industry exists in the county of Contra Costa. The present waste load generated for cane sugar is two pounds per ton of product. The production of cane sugar is a relatively straightforward process, with essentially one product resulting.

The apportionment factor, therefore, is 1.0 per unit or 100 percent, due to the fact that all the state's industry is in one area. With this factor multiplied by the total 1969 BOD load for the state, the figure in pounds of BOD is 3,031,000 pounds in the Bay-Delta area. Projecting this to 1980, the figure decreases to a total of 2,653,000 pounds of BOD. There are possible changes projected in the process of cane sugar refining for the year 2000. These changes, however, are not likely to materially distort the above statistical data.

Canned Fruits and Vegetables – 2033

The statistics generated show that 82.1 percent of the California canned fruits and vegetable output originate in the twelve-county area of the Bay-Delta region. The bulk of this industry originates in Santa Clara county, which contributes 33.6 percent; Alameda county, which contributes 22.2 percent; and the county of San Joaquin, with 12.9 percent of the total industry for the area. Therefore, these counties generate 68.7 percent of the area's 82.1 percent portion.

Unlike the cane sugar refining industry, the total BOD load for the Bay-Delta region will increase between now and 1980. Specifically, total BOD load was 80,189,000 pounds in 1969, as compared to a projected 153,623,000 pound load in 1980. In this particular process, there is a predominance of vegetable production in the given area with some hope of improving efficiency in the distant future. Thus, as it stands today and for 1980, the canned fruits and vegetables industry remains a serious problem.

Paperboard Mills – 2631

For this industry, 60 pounds of BOD results from each ton of product according to *Waste Management* [5]. Contra Coast contributes 35.7 percent of the total 54.1 percent allocated to the twelve-county area. To a lesser extent other contributing counties include Alameda, San Joaquin, and Santa Clara.

For the paperboard mills, total BOD will likely increase some 21,815,000 pounds statewide over the eleven-year period of 1969 to 1980. On a county area basis for the Bay-Delta region, total BOD is expected to increase from

Figure 6.1. THE SAN FRANCISCO BAY REGION—showing points referred to in text of Chapter 6. These include small towns as well as major cities. Source: Atlases and text of Chapter 6.

20,619,000 pounds in 1969 to 32,420,000 pounds in 1980. There is some hope that this figure for 1980 will be somewhat reduced. It has been estimated by *Waste Management* that for the year 2000 there will be only 25 pounds of BOD per ton of product as compared to 60 pounds in 1969. As for 1980, an estimate would be speculation as to the accuracy of this data on possible improvement within the industry, and within the context of this given time span.

Paper Mills, Except Buildings – 2621

Although contributing only 35.2 percent of total statewide BOD, this amount as related to the Bay-Delta region is of great importance. The projected statewide increase in total BOD between 1969 and 1980 is an enormous 149.8 percent. Total statewide BOD was 233,984,000 pounds in 1969 with a projected total of 583,367,000 pounds for 1980. County statistics of the Bay-Delta region reveal that total BOD jumps from 82,362,000 pounds to 205,344,000 pounds over the eleven-year period. The counties contributing these BOD loads are Contra Costa and San Joaquin, each contributing 17.6 percent.

There is some hope for these dismal statistics, however. According to *Waste Management,* currently for every 80 tons of product produced, 100 pounds of BOD results. Due to such factors as process modification, recovery of materials, and by-product production, by the year 2000 this figure should be reduced by 60 percent. Therefore some of this improvement should come between now and 1980. Even so, this particular industry presents a serious problem.

Frozen Fruits and Vegetables – 2037

This industry also does not contribute much produce in this area as compared to statewide totals, but its BOD load is significant. A total of 17.8 percent of the state's total comes from the Bay-Delta area. For 1969 total BOD load was 5,731,000 pounds with a projected total of 14,428,000 pounds for 1980; therefore a total percentage increase of 151.6 percent.

Process modification within the next ten years does not seem likely. Accordingly one should not expect the above estimates to be reduced due to such modifications.

Petroleum Refining – 2911

The Bay-Delta region contributes some 30.3 percent of the state's total refining. Twenty-nine percent of this total is refined in the county of Contra Costa. A total 1969 BOD load was 11,641,000 pounds as compared to an estimated figure of 13,646,000 pounds in 1980.

Slight improvement can be expected through process modification and new technical knowledge. Thus the 1980 projections will likely be somewhat overstated.

Distilled Liquors – 2085

Although contributing over 50 percent of total statewide production, the distilled liquor industry as related to BOD loads in the Bay-Delta area presents little problem. Total BOD will likely increase from 366,000 pounds in 1969 to 572,000 pounds in 1980.

Paper Coating and Glazing – 2641

The twelve-county Bay-Delta area contributes some 46.1 percent of the total statewide production. Alameda and Santa Clara are the largest producers on a countywide basis, contributing 28.9 percent and 13.1 percent respectively.

There is a projected total BOD load increase of 145.5 percent statewise over the eleven-year period 1969 to 1980. This is compared to a 147.4 percent increase for the twelve-county area over this time period. Total BOD for this area is expected to increase from 1,634,000 pounds to 4,013,000 pounds during the eleven-year period. However, due to some projected process modification this figure should be somewhat reduced by 1980.

Major Trouble Areas

Based on the foregoing statistics, a closer examination will reveal that there are four counties that are the major contributors to the pollution problem. In order of importance they are: Contra Costa, San Joaquin, Santa Clara, and Alameda. Figures to demonstrate this fact are shown in table 6.5 and are based on the large industrial polluters, namely SIC codes 2033, 2062, 2621, 2631, 2641, and 2911.

It is interesting to note that total BOD for the industries included in table 6.5 and for the twelve-county Bay-Delta region are 215,330,000 pounds in 1969 and a projected 441,056,000 pounds in 1980. Therefore one can easily see that the industries listed above along with the counties listed above are truly the areas of greatest concern. In percentages, the four counties accounted for 89 percent of the twelve-county total in 1969 and are projected to account for 88 percent in 1980. Looking at a map, one finds that the counties of Contra Costa, Alameda, and Santa Clara surround San Francisco Bay from its central section to the southern portion of the bay. We therefore can see why the poorest water quality

Table 6-5
BOD Contribution of Principal Industrial Sources[a] in Four Area County are Having Major Pollution Problem, 1969 Estimated and 1980 Projected

County	1969 Total BOD	1980 Total BOD
Contra Costa	73,780,000 lbs.	148,588,000 lbs.
San Joaquin	55,962,000 lbs.	130,044,000 lbs.
Santa Clara	35,696,000 lbs.	68,746,000 lbs.
Alameda	25,050,000 lbs.	47,261,000 lbs.
Totals	190,488,000 lbs.	394,639,000 lbs.

Source: Estimated using data on employment in relevant industries from U.S. Department of Commerce, Bureau of the Census, COUNTY BUSINESS PATTERNS and CENSUS OF MANUFACTURERS in accordance with the procedure outlined in Chapter 2.

[a]Industries included are SIC Codes 2033, 2062, 2621, 2631, 2641 and 2911.

of the bay is in the southern regions. Here the dissolved oxygen concentrations often fall to zero in late summer. If this situation is left to continue, the problem would expand to other regions of the bay where dissolved oxygen concentrations are consistently above 80 percent of saturation [2:69].

Specific county by county BOD projections for the industries under consideration are summarized in table form in the appendix of this report.

Water Quality Statistics

Water Quality Characteristics

This section will consider bay areas in relation to certain minimum water quality standards, thus spotlighting the problem areas. Where there are problems, more detailed descriptive information will be supplied.

Table 6.6 identifies acceptable and unacceptable water quality zones. The criteria of what actually is considered acceptable and what is not are based on various national, state, and regional guidelines. It must be emphasized that these are the minimum acceptable standards. Even with the standards at this level, in acceptable areas, there will still be considerable room for improvement.

Water Quality Problems Within the Bay-Delta Region

Table 6.6 shows that there are certain areas where improvement should be sought. In addition to the parameters noted in the table, other problem parameters will now be discussed.

Table 6-6
Acceptability of Water Quality

Parameters	South Bay	Lower Bay	Central Bay	North Bay	San Pablo Bay	Suisun Bay
pH (Hydrogen Ion Concentrations)	A	A	A	A	A	A
Dissolved Oxygen	A	A	A	A	A	A
Transparency	A	U	A	U	U	U
Dissolved Oxygen Concentrations	U	A	A	A	A	A
Coliform Bacteria	U	U	U	U	U	U

A = Acceptable
U = Unacceptable
Mean Values Used

Source: E.A. Pearson, P.N. Storrs, and R.E. Selleck, A COMPREHENSIVE STUDY OF SAN FRANCISCO BAY, REPORT TO THE STATE WATER CONTROL BOARD, Berkeley, Calif., July 1, 1979 (Berkeley Calif.: University of California Press, 1979), p. 38.

Hydrogen Ion concentrations are by and large fairly good throughout the Bay-Delta region. The same would appear for dissolved oxygen; however since these statistics are mean statistics, the whole story is not told. It is generally recognized that dissolved oxygen quantities should not fall below 5 mg/l. The lowest mean statistic is in the South Bay area where this figure is 5.1. However, in the late summer this figure falls to 0.7 thereby creating a water quality problem. A parameter associated with dissolved oxygen, namely dissolved oxygen saturation, does show this poor quality where the dissolved oxygen saturation percentages fall to a mean of 55 percent and a low of 9.3 percent. Suisun Bay would be characterized as a problem area for this parameter due to a low reading of 65 percent. Generally speaking, dissolved oxygen saturation percentages should be above 80 percent.

By far the worst water quality statistics in table 6.6 is that of coliform bacteria. It is by no means acceptable in any part of the Bay-Delta region. Coliform bacteria should not exceed a reading of 1000 MPN/100 ml. Coliform bacteria concentrations ranged from 10 to more than 2 million bacteria per 100 ml. Again the highest concentrations and those with the greatest range were found in the South Bay area [2:39]. Various studies conducted in San Francisco Bay between 1959 and 1968 show that coliform bacterial levels exceeded the Public Health standards as related to water contact sports. These coliform levels are so high that sport shellfishing is not safe, and commercial harvesting is no longer permitted. In order to abate this situation domestic wastewaters must be treated and wastes from watercraft controlled [2].

Indications are that the bay will be in serious trouble regarding toxicity if the present situation is not reversed. Between the years 1963 and 1968 there were some thirty-one reported fish kills, eleven of which were the result of wastewater discharges or spills. The causes of the other incidents have gone unexplained. It seems likely, though, that at least a portion of these unexplained incidents were due indirectly to wastewater discharges [1:3-12].

Another problem confronting both the Bay and Delta regions is the large population of biostimulants and algae. This is of particular concern in the Delta region. Tests have shown that nitrogen and phosphorous concentrations range from ten to one hundred times greater in the Delta than is found necessary for substantial growths of algae. The highest concentrations are in the vicinity of Stockton and the San Joaquin River. In San Francisco Bay enrichment is detected along the shores and in the tidal reaches of some of the bay's tributaries [1:3-11].

As a result of the foregoing water quality problems, annual mean transparency in San Francisco Bay is about 2 feet. The water is most turbid in the Suisun Bay area where the transparency ranges from 0.5 to 2 feet [2:37].

Projections of total nitrogen concentrations are shown in figure 6.2. There seem to be two problem areas that warrant attention. The concentrations of nitrogen are presently of highest magnitude in the South Bay area. Here the concentrations are expected to increase a total of 41.2 percent over the fifteen-year period. A more dramatic change will occur in the San Pablo and Suisun Bay areas. Here the expected increase in total nitrogen is estimated at a growth of 230.5 percent. Therefore immediate attention should be directed toward both

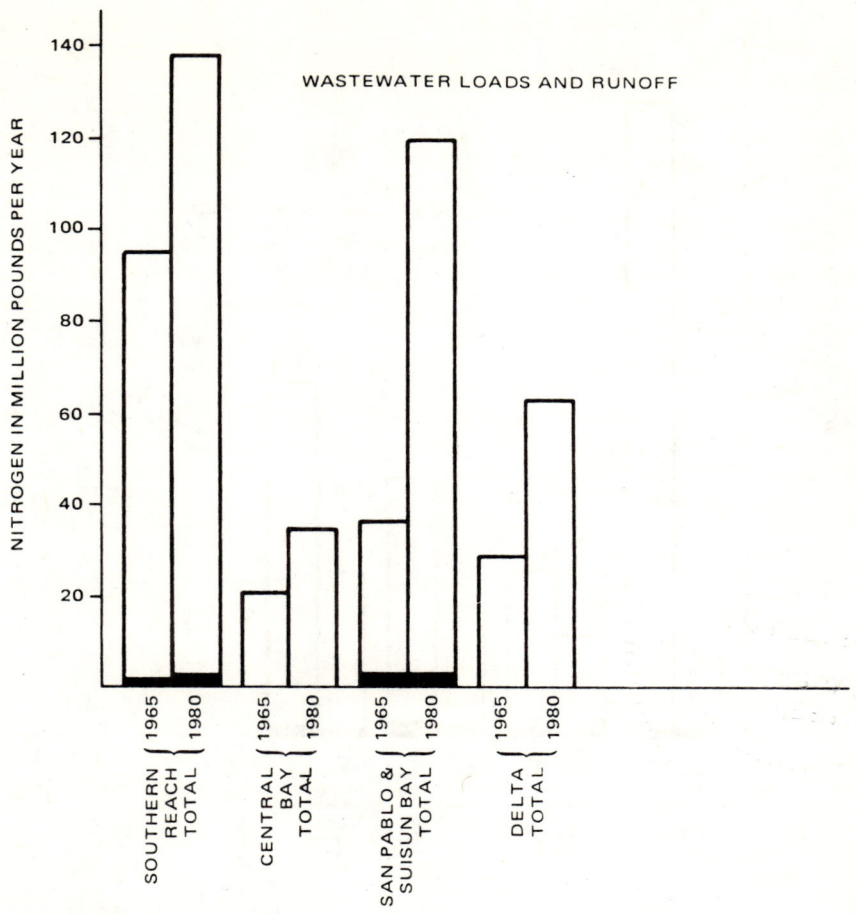

Figure 6.2. Nitrogen Compound Content of Wastewater Entering San Francisco Bay (By Selected Year and Area). Source: *Report of the State Water Resources Control Board on San Francisco Bay-Delta Water Quality Control Program*, Kerry W. Mulligan, chairman (Sacramento, California: 1969), pp. 4-14.

these areas, for even in 1980 total nitrogen in the South Bay area will be an estimated 15.1 percent higher than in the San Pablo and Suisun Bay area.

As was indicated in the previous pages, BOD load concentrations as related to specific industries are highest in the southern region of the bay. Figure 6.3 includes BOD load from municipal and industrial sources and natural runoff. A 37.7 percent increase in BOD for the South Bay area is estimated. The other

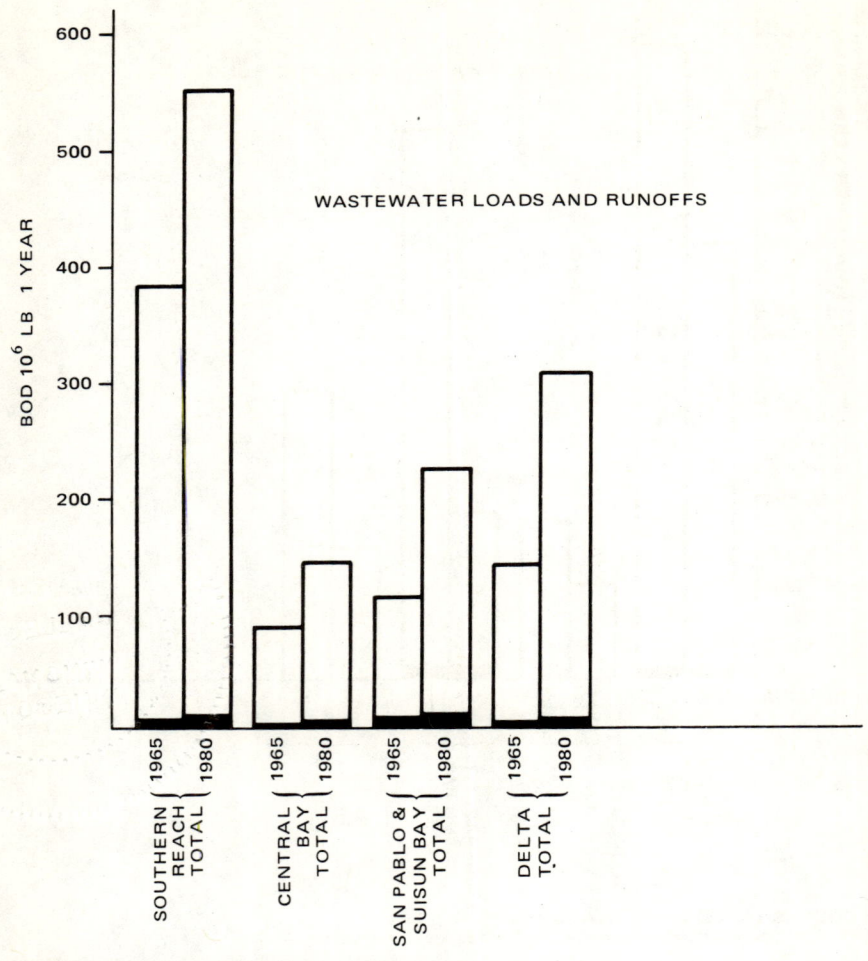

Figure 6.3. BOD Load of Wastewaters Entering San Francisco Bay (By selected year and area). Source: *Report of the State Water Resources Control Board on San Francisco Bay-Delta Water Quality Control Program*, Kerry W. Mulligan, chairman (Sacramento, California: 1969), pp. 4-14.

sectors of the Bay-Delta area do not constitute near the problem as the South Bay area does. To illustrate this, one can see that in the Delta area, for example, total projected BOD for 1980 is some 23.1 percent lower than the established 1965 BOD for the South Bay area.

Sewage Treatment Plants

An integral part of any abatement program is the concentration and efficiency of the sewage treatment plants in the area. A study by the San Francisco Bay Conservation and Development Commission in July of 1967 revealed that there were some sixty-six major sewage treatment plants near San Francisco Bay. Almost all of these plants empty their effluent into the bay or into streams flowing into the bay. These plants serve more than 3.7 million persons, which is approximately 85 percent of the population [6]. A map of the location of these major treatment plants along with the level of treatment is shown in figure 6.4.

San Francisco has historically lagged in sewage treatment. There was no separation of storm drains from sewers, as called for by good practice. As late as 1969 even in dry weather, San Francisco was often in violation of state water quality requirements. Turbidity, muddiness, and floatable solids were often visible at the outfall points in the bay. Since sewage and drainage share the same pipes, in wet weather storm water and waste water merge and flow into the treatment plants together. The plant reaches capacity as the volume of water builds up. At this point an overflow valve is opened and the sewage flows into the bay untreated [7].

Progress is being made, however. Recently in San Mateo an installation of an odor-free facility that incinerates down to an antiseptic sand was installed. The process is a natural gas fired, electrically-powered furnace. Solid materials are dumped in the top and cycled down a five-leveled design past successive hearths by large rotating rakes until they emerge at the bottom as a sterile dust. In a similar installation at Lake Tahoe, water emerges that meets the U. S. Public Health standards for drinking. Preliminary indications are that this process is efficient and it is certainly evident that it does the necessary job. Therefore some real sense of optimism would appear to be justified [8].

General Problems of San Francisco Bay-Delta Area

At this point it seems well to review the problems of San Francisco generally, since the foregoing has been an in-depth analysis of specific water quality problems of the area.

There has been general indication and agreement that the rate of overall population and economic growth of the Bay area will continue to exceed that of the United States as a whole. Therefore, it seems obvious that an agressive abatement program must be instituted in order for the problem to just remain at the present magnitude, to say nothing of making progress on current problems.

Another general problem is the multiplicity of small districts throughout the Bay area dealing with sewage and sewage treatment activities. More often than not, these districts, each with an independent board of directors, often do an adequate job within their own boundaries, yet without relation to the needs and problems of the adjoining areas. There are instances where the desire for local independence or the lack of leadership from any other source has resulted in greater eventual cost than would have been the case otherwise.

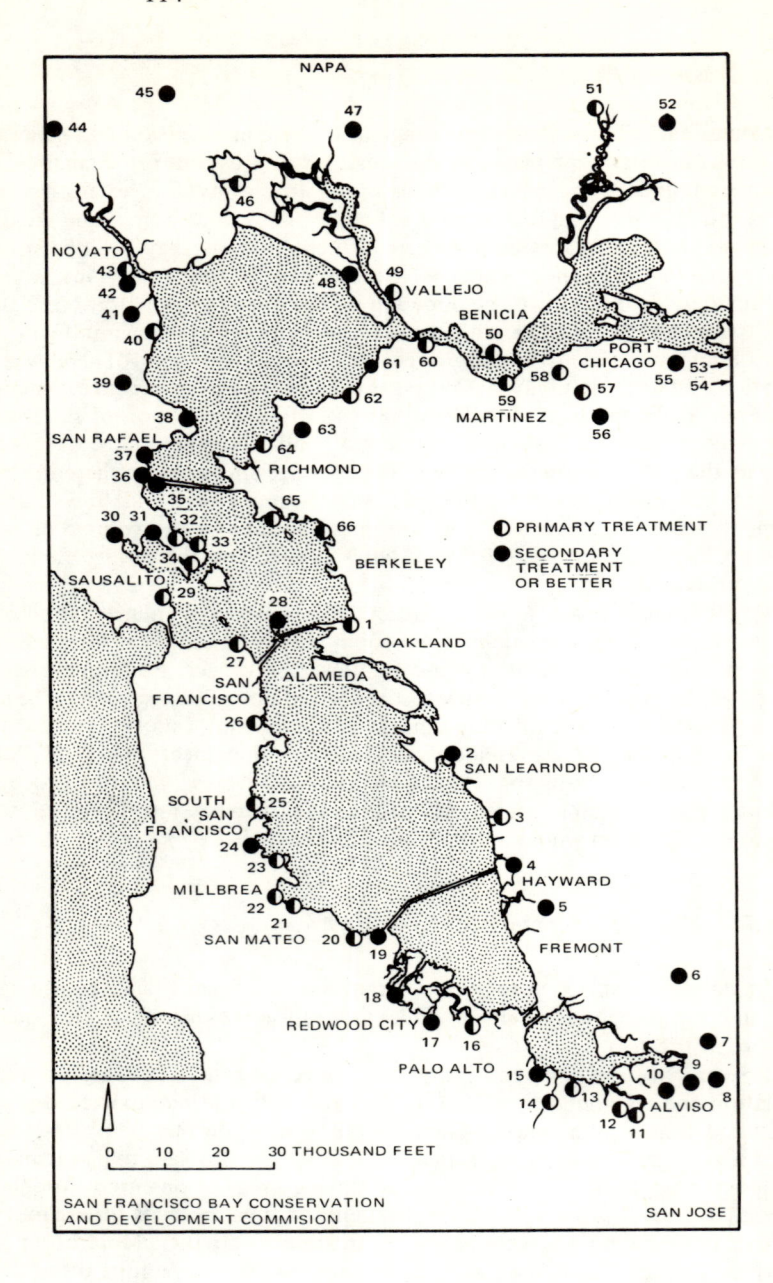

Figure 6.4. Major Publicly-Operated Sewage Treatment Plants Near San Francisco Bay. Source: Regional Water Quality Control Board, Sewage Plant Operators.

In relation to this, there seems to be an absence of an effective unifying force to bring all these various plans and proposals together. There seems to be significant documentation of this fact from local newspapers in the Bay-Delta area. Resulting from this fact is a waste of money, misdirected objectives, and inefficiency.

Conclusions and Recommendations

It is not the intention of this chapter to recommend a unique plan to save San Francisco Bay. Millions of dollars have been spent by various agencies to study this problem. As a result of these studies much more will be spent to combat the water pollution problem of San Francisco Bay. Therefore, it is outside the scope of this chapter to recommend a revolutionary new program for abatement.

In an effort to crystalize the main thrust of the findings, this section will deal with a number of important key facts regarding the water pollution problems of the Bay-Delta region. The likely consequences of these facts will also be given.

           Fact: The dangers of water pollution were recognized early in the Bay-Delta area.
Consequence: Citizens concerned about the Bay's future applied various amounts of pressure on local authorities, thus bringing about corrective legislation and detailed commissioned studies.

           Fact: Little quantitative information is available on such factors as biostimulation and toxicity.
Consequence: Until future studies tackle this problem, water quality in relation to biostimulation and toxicity is likely to deteriorate at an increasing rate.

           Fact: The area of poorest water quality is that of the South Bay region of San Francisco Bay.
Consequence: Unless improved in the near future, the poor quality of water will migrate northward, thereby affecting other regions of the bay.

           Fact: Overall lack of unity among the various Bay-Delta municipalities.
Consequence: Wastes of time, money, and resources leading to a continued deterioration of water quality.

           Fact: Quantitatively, coliform bacteria is the worst water quality problem in the Bay area.
Consequence: Coliform bacteria will continue to be a major problem unless strict attention is paid to municipal sewage and the problem of discharges from watercraft, for these are the two principal sources of coliform bacteria.

Under the proposed recommended system, total expenditures by 1980 are projected to be over a billion dollars on sewage treatment and control of industrial pollution. Despite these expenditures, however, the likelihood at best is for only a slight improvement in water quality by 1980, and most probably maintenance of the level of overall quality of water in the bay that exists today. If there was better organization the water quality could be moderately improved by 1980; that is to say, if there were no more delays in the implementation of the recommended program. It would seem highly unlikely that there will be no future delays in the system and, therefore, at best, only slight improvement in water quality is foreseen.

Figures have shown that total expenditures on industrial treatment plants will be greater than on municipal plants; nevertheless, the biggest threat to pollution of the bay will probably come from the industrial sector. In accordance with this, it would seem prudent that great emphasis be placed on industrial treatment in the counties of Contra Costa, San Joaquin, Santa Clara, and Alameda. As was pointed out earlier, these counties are the major trouble areas in regard to industrial pollution. If such a concentrated effort were initiated and maintained in these counties, the corresponding waters of the bay would show appropriate improvement.

Perhaps the biggest immediate problem to hurdle is not any specific water quality problem of the area, but the problem of bureaucracy. The recommendations of the San Francisco Bay-Delta Water Quality Control Board are sound in theory, but unless they get off the drawing boards water quality will not improve. There just isn't enough time for future studies, evaluation of such studies, and finally action. We cannot worry if the recommendations of the board are completely sound. The fact is that it is the most comprehensive program to date and that immediate implementation of the program should begin with a minimum of bureaucracy. Any errors in planning can be corrected after the plan is initiated.

After careful study and examination, the final conclusion is that in one way or another the water pollution problem of the Bay-Delta area will be solved. The time horizon, however, may have to be greatly expanded. It should be noted that as time is lost for one reason or another, the problem becomes more difficult to solve and probably more expensive due to inflation. Unless there are technological breakthroughs that are unforeseen today it seems unlikely that the problem will be even close to being solved by 1980. There is just too much to accomplish and too little time to accomplish it.

In trying to qualify the chances of success of pollution abatement, based on the foregoing investigation, it would seem prudent to say that by 1980 there will have been enough progress in the fight for clean water to continue on, but not enough progress made for unchecked optimism.

Appendix

See tables starting on following page.

Table A6-1
Recommended System Cost Summary (Disposal Option)

Capital Expenditures (millions of 1969 dollars) Elements	By 1970	Phase One 1970-1974	Phase One 1975-1979	Phase Two 1980-1984	Phase Two 1985-1989	Phase Two 1990-1994	Phase Two 1995-1999	Phase Three 2000-2004	Phase Three 2005-2009	Phase Three 2010-2014	Phase Three 2015-2019	Total
Conveyance		44	317	303	28	170	307					1,169
Treatment												
Municipal		47	94	29	78		50	28	30			356
Discrete Industrial	91	4	65		46		35		35			276
Land and Right of Way		14	27	(9)	11	6	13	1	5			68
Contingency		21	87	69	23	35	76	6	7			324
Engineering and Administration		18	78	62	21	31	66	5	6			287
Total	91	148	668	454	207	242	547	40	83			2,480
Average Annual Operation and Maintenace Costs (millions of 1969 dollars)												
Conveyance			0.3	2.0	3.8	4.2	5.1	6.6	6.8	7.2	7.4	
Treatment												
Municipal		10.1	12.9	24.2	20.1	21.1	24.0	25.5	27.2	29.2	31.5	
Discrete Industrial		5.0	6.3	10.2	8.7	10.1	11.4	9.2	10.1	11.2	12.3	
Total		15.1	19.5	36.4	32.6	35.4	40.5	41.3	44.1	47.6	51.2	
Total, Five Years		75.5	97.5	182.0	163.0	177.0	202.5	206.5	220.5	238.0	256.0	1,818
Grand Total (5 year increment)												
Municipal		195	669	585	280	369	658	200	217	182	195	3,540
Discrete Industrial	91	29	96	51	90	50	92	46	86	56	61	758
Accumulated Grand Total	91	315	1,080	1,716	2,086	2,505	3,255	3,501	3,804	4,042	4,298	4,298

Source: REPORT OF THE STATE WATER RESOURCES CONTROL BOARD ON SAN FRANCISCO BAY-DELTA WATER QUALITY CONTROL PROGRAM, Kerry W. Mulligan, chairman (Sacramento, California: 1969), pp. 9-14.

Table A6-2
Recommended System Cost Summary (Reclamation Option)

Capital Expenditures (millions of 1969 dollars) Elements	By 1970	Phase One 1970-1974	Phase One 1975-1979	Phase One 1980-1984	Phase Two 1985-1989	Phase Two 1990-1994	Phase Two 1995-1999	Phase Three 2000-2004	Phase Three 2005-2009	Phase Three 2010-2014	Phase Three 2015-2019	Total
Conveyance		44	317	303	28	173	28	95				988
Treatment												
Municipal		47	94	29	75	71	22	57	25			420
Discrete Industrial	91	4	65		46		44		43			293
Land and Right of Way		14	27	(9)	9	(6)	5	6	4			50
Contingency		21	87	69	22	50	11	31	6			297
Engineering and Administration		18	78	62	20	45	10	28	6			267
Total	91	148	668	454	200	333	120	217	84			2,315
Average Annual Operation and Maintenance Costs (millions of 1969 dollars)												
Conveyance			0.3	2.1	3.8	4.1	5.0	5.3	5.8	6.1	6.3	
Treatment												
Municipal		10.1	12.9	24.2	20.1	21.1	28.5	33.1	35.4	38.1	40.7	
Discrete Industrial		5.0	6.3	10.2	8.7	10.1	10.1	10.7	12.6	13.8	15.1	
Total		15.1	19.5	36.5	32.6	35.3	43.6	49.1	53.8	58.0	62.1	
Total, Five Years		75.5	97.5	182.5	163.0	176.5	218.0	245.5	269.0	290.0	310.5	2,028
Grand Total (5 year increment)												
Municipal		195	669	585	273	460	243	409	248	221	234	3,537
Discrete Industrial	91	29	96	51	90	50	95	53	106	69	76	806
Accumulated Grant Total	91	315	1,080	1,716	2,079	2,589	2,927	3,389	3,743	4,033	4,343	4,343

Source: REPORT OF THE STATE WATER RESOURCES CONTROL BOARD ON SAN FRANCISCO BAY-DELTA WATER QUALITY CONTROL PROGRAM, Kerry W. Mulligan, chairman (Sacramento, California: 1969), p. 9-10.

Table A6-3
Summary Statistics of Countywide Apportionments

SIC Code 2011	Alameda	Contra Costa	Sacramento	Marin	Napa	San Francisco
Meat Slaughtering						
Total Percentage	.0180	.0168	.0118	.0008	.0008	.0586
1969 Total BOD Load	327	305	214	14	14	1064
1980 Total BOD Load	496	463	325	22	22	1614

SIC Code 2013	San Joaquin	Santa Clara	Solano	Sonoma	San Mateo	Yolo	Totals
Total Percentage	.0134		.0450	.0040	.1058	.0210	.2960
1969 Total BOD Load	243		817	73	1922	381	5374
1980 Total BOD Load	369		1239	110	2914	578	8152

SIC Code 2013	Alameda	Contra Costa	Sacramento	Marin	Napa	San Francisco
Meat Processing						
Total Percentage	.1349		.0520			.1108
1969 Total BOD Load	476		184			391
1980 Total BOD Load	724		279			595

	San Joaquin	Santa Clara	Solano	Sonoma	San Mateo	Yolo	Totals
Total Percentage	.0100	.0147		.0009	.0059	.3292	
1969 Total BOD Load	35	52		3	21		1162
1980 Total BOD Load	54	71		5	32		1568

Table A6-3 (cont.)

SIC Code 2026	Alameda	Contra Costa	Sacramento	Marin	Napa	San Francisco		
Fluid Milk								
Total Percentage	.0526	.0044	.0458	.0008	.0004	.0598		
1969 Total BOD Load	168	14	146	2	1	191		
1980 Total BOD Load	228	19	199	3	2	260		
	San Joaquin	San Mateo	Santa Clara	Solano	Sonoma	Yolo		Totals
Total Percentage	.1032	.0008	.0328	.0064	.0052	.0088		.2310
1969 Total BOD Load	42	2	105	21	17	28		737
1980 Total BOD Load	57	3	142	28	23	38		1002

SIC Code 2033	Alameda	Contra Costa	Sacramento	Marin	Napa	San Francisco		
Canned Fruits								
Total Percentage	.2261	.0442	.0550			.0008		
1969 Total BOD Load	21909	4369	5438			50		
1980 Total BOD Load	41455	8268	10289			150		
	San Joaquin	San Mateo	Santa Clara	Solano	Sonoma	Yolo		Totals
Total Percentage	.1286	.0070	.3364	.0006	.0130	.0140		.8212
1969 Total BOD Load	12714	692	32259	59	1285	1384		80189
1980 Total BOD Load	24057	1309	62932	112	2432	2619		153623

SIC Code 2037	Alameda	Contra Costa	Sacramento	Marin	Napa	San Francisco
Frozen Fruits						
Total Percentage	.0040	.0040	.0056	.0008		.0040
1969 Total BOD Load	249	249	349	50		249
1980 Total BOD Load	627	627	878	125		627

	San Joaquin	San Mateo	Santa Clara	Solano	Sonoma	Yolo	Totals
Total Percentage	.0100	.0008	.0628				.1784
1969 Total BOD Load	623	50	3912				5731
1980 Total BOD Load	1568	125	9851				14428

	Alameda	Contra Costa	Sacramento	Marin	Napa	San Francisco
SIC Code 2062						
Cane Sugar						
Total Percentage		1.0000				
1969 Total BOD Load		3031				
1980 Total-BOD Load		2653				

	San Joaquin	San Mateo	Santa Clara	Solano	Sonoma	Yolo	Totals
Total Percentage							1.0000
1969 Total BOD Load							3031
1980 Total BOD Load							2653

	Alameda	Contra Costa	Sacramento	Marin	Napa	San Francisco
SIC Code 2085						
Distilled Liquor						
Total Percentage				.1300		.2560
1969 Total BOD Load				92		368
1980 Total BOD Load				144		572

	San Joaquin	San Mateo	Santa Clara	Solano	Sonoma	Yolo	Totals
Total Percentage		.1300					.5160
1969 Total BOD Load		92					368
1980 Total BOD Load		144					572

Table A6-3 (cont.)

SIC Code 2621	Alameda	Contra Costa	Sacramento	Marin	Napa	San Francisco	
Paper Mills							Totals
Total Percentage		.1760					.3520
1969 Total BOD Load		41181					82362
1980 Total BOD Load		102672					205344

	San Joaquin	San Mateo	Santa Clara	Solano	Sonoma	Yolo	
Total Percentage	.1760						
1969 Total BOD Load	41181						
1980 Total BOD Load	102672						

SIC Code 2631	Alameda	Contra Costa	Sacramento	Marin	Napa	San Francisco	
Millwork Plants							Totals
Total Percentage	.0530	.3570					.5410
1969 Total BOD Load	2020	13606					20619
1980 Total BOD Load	3176	21394					32420

	San Joaquin	San Mateo	Santa Clara	Solano	Sonoma	Yolo	
Total Percentage	.0530		.0780				
1969 Total BOD Load	2020		2973				
1980 Total BOD Load	3176		4674				

SIC Code 2641	Alameda	Contra Costa	Sacramento	Marin	Napa	San Francisco	
Paper Coating							
Total Percentage	.2890	.0160				.0060	
1969 Total BOD Load	1024	57				21	
1980 Total BOD Load	2517	139				52	

123

SIC Code 2654 — Food Containers

	San Joaquin	San Mateo	Santa Clara	Solano	Sonoma	Yolo	Totals
Total Percentage	.0160	.0030	.1310				.4610
1969 Total BOD Load	57	11	464				1634
1980 Total BOD Load	139	26	1140				4013

	Alameda	Contra Costa	Sacramento	Marin	Napa	San Francisco
Total Percentage	.0650	.0240				.0550
1969 Total BOD Load	543	201				460
1980 Total BOD Load	812	300				687

SIC Code 2911 — Petroleum Refining

	San Joaquin	San Mateo	Santa Clara	Solano	Sonoma	Yolo	Totals
Total Percentage	.0270	.0100	.1100				.2910
1969 Total BOD Load	227	84	919				2434
1980 Total BOD Load	337	125	1374				3635

	Alameda	Contra Costa	Sacramento	Marin	Napa	San Francisco
Total Percentage	.0025	.2990	.0005			
1969 Total BOD Load	97	11536	19			
1980 Total BOD Load	113	13462	26			

	San Joaquin	San Mateo	Santa Clara	Solano	Sonoma	Yolo	Totals
Total Percentage		.0010					.3030
1969 Total BOD Load		39					11691
1980 Total BOD Load		45					13646

Source: Estimated using data on employment in relevant industries from U.S. Department of Commerce, Bureau of the Census, COUNTY BUSINESS PATTERNS and CENSUS OF MANUFACTURERS in accordance with the procedure outlined in Chapter 2.

Table A6-4
Employment, Output/Input BOD Projections

SIC Code	Population Trends			Output/Input Per Employee		BOD Coefficient	Total BOD (1000)	Output/Input Per Employee	BOD Coefficient	Total BOD (1000)
	1967	1969	1980	1969		1969	1969	1980	1980	1980
2011	5,900	6,128	7,782	247		12	18,163	295	12	27,548
2013	4,000	4,204	5,326	70		12	3,531	84	12	5,368
2026	5,300	5,134	4,219	519		1.2	3,197	858	1.2	4,343
2033	21,400	21,400	21,429	154		30	98,868	291	30	187,075
2037	8,400	8,952	11,991	232		30	62,305	439	30	156,867
2062	2,000	1,938	1,194	782		2	3,031	1,111	2	2,653
2085	400	418	517	3.4		0.5	710	4.3	0.5	1,111
2621	12,800	14,716	25,254	159		100	233,984	231	100	583,367
2631	1,572	1,596	1,728	398		60	38,112	578	60	59,927
2641	1,000	1,144	1,935	62		50	3,546	90	50	8,707
2654	2,500	2,532	2,603	66		50	8,355	96	50	12,494
2911	9,800	9,480	7,723	37		110	38,583	53	110	45,025

Source: Estimated using data on employment in relevant industries from U.S. Department of Commerce, Bureau of the Census, COUNTY BUSINESS PATTERNS and CENSUS OF MANUFACTURERS in accordance with the procedure outlined in Chapter 2.

Notes

1. *Report of the State Water Resources Control Board on San Francisco Bay-Delta Water Quality Control Program,* Kerry W. Mulligan, Chairman (Sacramento, California: 1969).

2. E. A. Pearson, P. N. Storrs, and R. E. Selleck, *A Comprehensive Study of San Francisco Bay,* Report to the State Water Control Board, Berkeley, Calif., July 1, 1970 (Berkeley, California: University of California Press, 1970).

3. *The Nation's Estuaries: San Francisco Bay and Delta, Calif.* Hearings Before a Subcommittee on Government Operations, House of Representatives Ninety-First Congress (Washington, D.C., 1969).

4. "San Francisco," *Colliers Encyclopedia,* 1967, XXIV, 402.

5. *Waste Management* (N. Y.: Regional Plan Association, 1969).

6. San Francisco Bay Conservation and Development Commission, Report of the Commission, *Pollution – Water Pollution and San Francisco Bay* (San Francisco, California, 1967), p. 32.

7. Thomas Powell, "Skiing, Swimming Curtailed in Bay," *New York Times,* Oct. 8, 1969, p. 1.

8. "New Weapon in War On Pollution," *New York Times,* May 1, 1970, p. 5.

7 Long Island Sound

ROBERT T. CAPONI

What are the factors that could cause a deterioration of the sound's quality? Pollutants from municipalities, institutions, industries, federal installations, combined overflows, recreational boating, and discharges of oil and petroleum products are all potential threats to the sound. Thermal discharges and disposal of dredged spoils also contribute to the pollution of Long Island Sound.

The sound is affected by those municipalities that immediately border the body itself and a larger area consisting of counties close to the sound. The areas immediately bordering the sound will have the greatest impact on the sound, whereas the adjacent areas have a lesser but still significant impact.

In 1960 the population for those municipalities affecting the sound, excluding New York City, was about two million. This total population is expected to increase to approximately three million by 1980. Industrial projections were made utilizing employment figures for individual firms, then projecting them on state or national growth patterns. Engineering consultant firms were used in two specific Connecticut plants.

Present Overall Water Quality

Long Island Sound is a tidal estuary located between Long Island, New York, and Connecticut, an area of approximately 930 square miles with an average depth of 60 feet. The sound has an entrance to the Atlantic Ocean through Block Island Sound and the Race. (The Race is a stretch of water between Fishers Island and Plum Island.) Entrance to the sound from the west is by way of the East River at Throgs Neck.

The quality of the water in Long Island Sound improves as one moves from the New York City border eastward to Block Island Sound and from shore and harbor areas to the center of the sound. The Connecticut River Basin (11,265 square miles), the Housatonic River Basin (1,950 square miles) and the Thames River Basin (1,470 square miles) are the three major contributors of fresh water to the sound. An additional 1,300 square miles of drainage area is supplied by smaller streams in Connecticut and New York.

The highly developed urban areas surrounding the sound have caused pollution problems to the sound's harbors and embayments. Water quality in some of these areas does not meet the approved standards as set by state, interstate, and county agencies (see fig. 7.1).

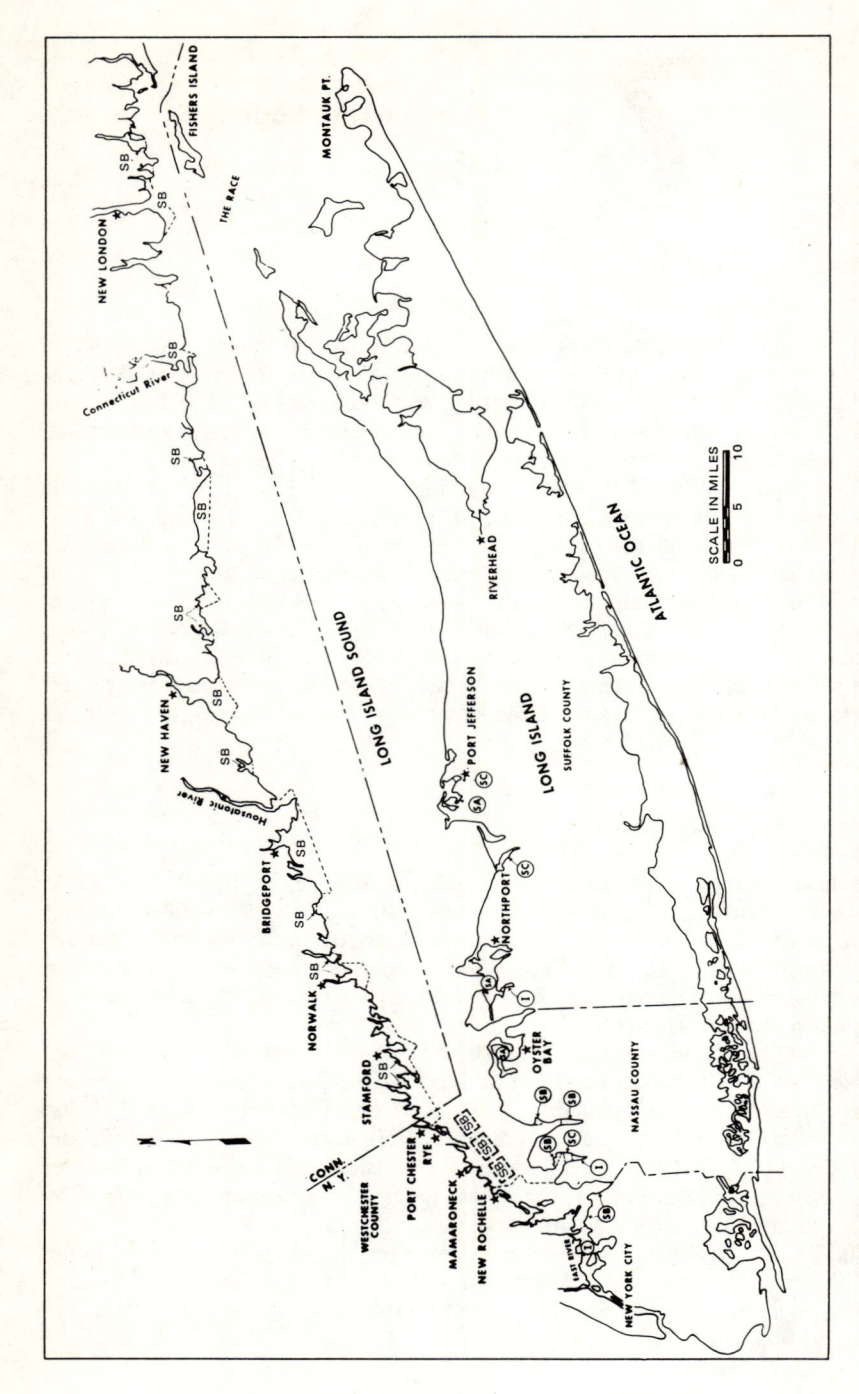

Figure 7.1. Quality Classification of Waters of Long Island Sound. Source: Federal Water Pollution Control Administration, *Report on the Water Quality of Long Island Sound, November 1969.*

Temperature, dissolved oxygen, and bacteria standards have been established for measuring water quality. Nutrient standards have not been established, but one of the principal pollution problems confronting the sound is excessive concentrations of nutrition which are responsible in turn for excessive growths of algae. The excess growth of the latter may lead to population explosions of species that feed on the vegetable species in question. Water quality is measured against the established standards and assigned a water quality classification as a result of the comparison. The standards include minimum levels for water temperature, dissolved oxygen, and bacteria densities.

Adequate levels of dissolved oxygen are necessary to support fish and other forms of aquatic life. If the oxygen demand in a body of water exceeds the available dissolved oxygen, a condition exists that causes the water to become foul smelling. The higher the water classification the greater the dissolved oxygen content required.

A primary source of bacteria in water comes from raw and inadequately treated domestic waste. Total coliform densities have traditionally been used as an indication of human contamination. Water quality standards specify levels of total coliform densities that cannot be exceeded. Total coliform densities for the highest classified waters in the sound are approximately 14 percent of those established for the lowest classified waters.

Water movement in Long Island Sound is influenced by the inflow of salty Block Island Sound water, the inflow of fresh water streams and groundwater, and by tides, winds, and the earth's rotation. An inflow of water from the East River affects the western portion of the sound near the New York Metropolitan Area. The dilution of marine waters can be related to the circulation pattern of Long Island Sound. The less dense surface waters move eastward, being replaced by more dense bottom waters at Block Island Sound. This type of system has the effect of enriching the waters of the estuary because the bottom water is generally richer in organic and inorganic material. Bottom water entering at Block Island Sound decreases in speed from almost six miles per month near New Haven to almost zero at the western end of the Sound. On the other hand, surface water flow increases from west to east averaging almost nine miles per month near New Haven.

Sources of Pollution

The foremost danger to the sound is from population increases. As population grows the load on municipal treatment facilities increases. Thus, unless such facilities are improved, water quality will deteriorate.

A second major potential problem arises from increased industrial output and waste associated with production. Although there aren't any industrial waste discharges directly into the sound, there are at least 137 into embayments and estuaries. Only five are located in New York and the remainder are located in Connecticut. A variety of contaminants including oxygen-consuming materials, oil, grease, solvents and toxic substances, chromic and other acids, cyanide and

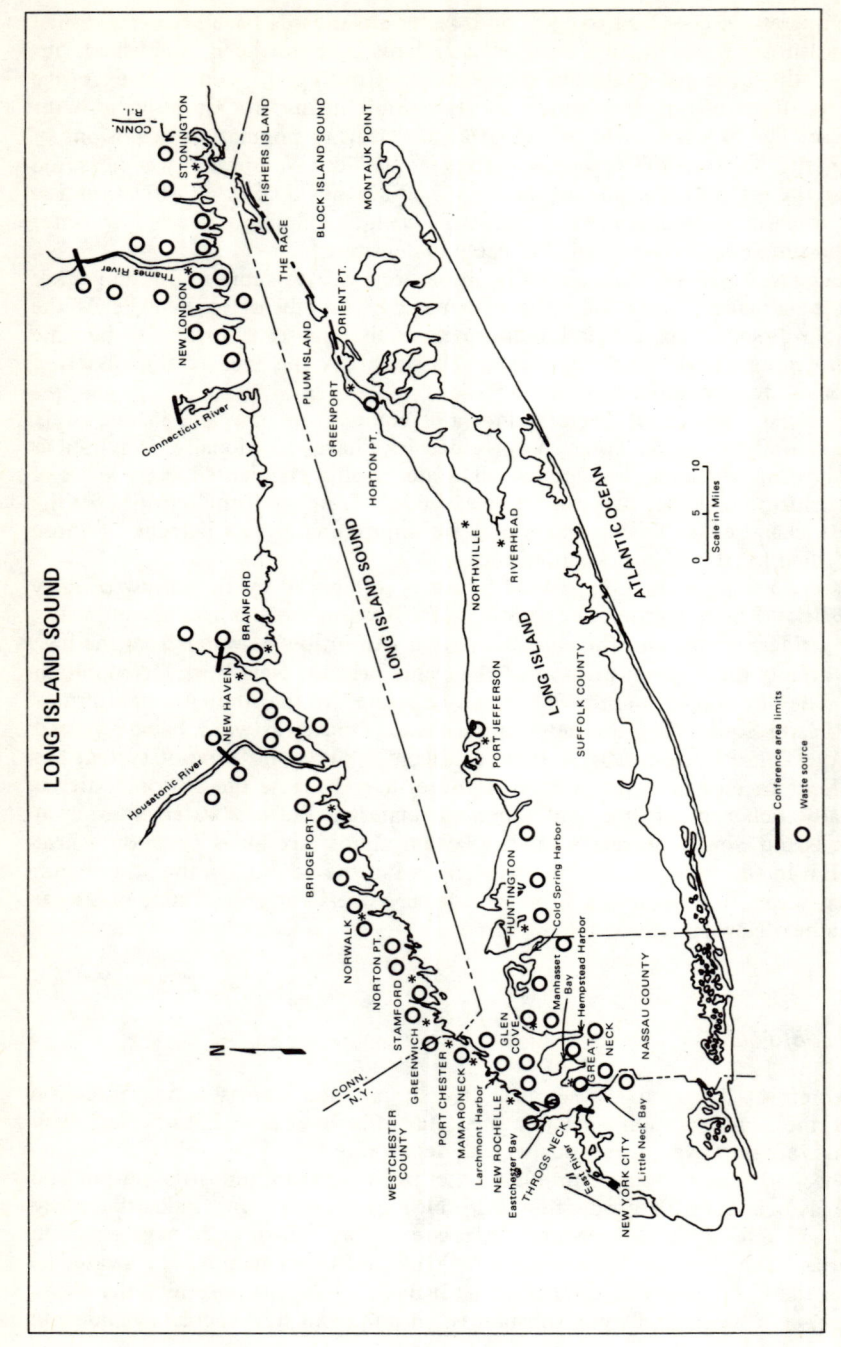

Figure 7.2. Municipal and Institutional Waste Sources. Source: 1971 Environmental Protection Agency, *Report on the Water Quality of Long Island Sound*, Figure B-1.

alkali is usually processed by municipal treatment plants before entering the sound.

Contaminants also enter the sound from electric generating plants, boating activities, and dredging activities.

Projections

Sixty sewage plants emptying into the sound or connecting waters serve approximately 1,300,000 people discharging approximately 190 million gallons per day (MGD) of waste water into the open waters of Long Island Sound or its embayments (see table 7.1). The population is expected to increase to more than 1,400,000 by 1980. Unless improvements are made to existing facilities and new plants are added, it will not be possible to handle the projected increased waste load without adding seriously to contamination.

Industry located on the sound is concentrated in a few categories. Industrial wastewaters result from the manufacture of food and kindred products, primary metals products, chemicals and allied products, and textiles. The effect of industrial waste is more apparent in harbors and embayments, the same places where municipal wastes are emptied, than in the middle of the sound.

Seventeen fossil-fuel steam electric generating plants producing 2,930 megawatts (Mwe) use the waters of Long Island Sound for cooling purposes. In 1967, 2,100 MGD of water from the sound was used by electric generating plants. Increased water temperature from these plants can upset the delicate ecological balance of the sound.

The projected increase in population and industry will require greater amounts of electricity. Proposed new power plants will add 7,177 megawatts per year by 1980. The estimated cooling water requirements to meet these future demands will exceed 5,700 MGD. The possible adverse effects to the sound from this proposed heated water must be a consideration in the development of any new power plants.

The sound is used extensively for recreational boating as attested to by the presence of nearly 240 marinas. In 1967, residents of Westchester, Nassau, and Suffolk counties owned more than 120,000 private boats, not including sail or row boats. The number of motor boats is projected to increase to approximately 183,000 by 1980. About 65,000 boats are based in Connecticut and operate on the sound. The number of boats in Connecticut is expected to increase to approximately 100,000 by 1980.

To the above vessels must be added countless other boats using the sound for weekend or extended visits. Waste from boating activities constitutes another major source of pollution.

Connecticut

Water quality in New London Harbor during a 1969 survey indicated that a substandard condition existed. The dissolved oxygen and fecal coliform densities

Table 7-1
Municipal, Private and Institutional Waste Sources, Long Island Sound

Name	Receiving Waters	Estimated Population Served	Average Daily Flow (MGD)[1]	Degree of Treatment
Bridgeport—West Side	Cedar Creek	109000	25.90	Primary
Bridgeport—East Side	Bridgeport Harbor	47000	10.60	Primary
Stratford	Housatonic River	40000	7.80	Primary
Shelton—Main	Housatonic River	10000	1.00	Primary
Shelton—Route 8	Housatonic River	—	1.00	Secondary
Derby	Housatonic River	8500	.85	Primary
Milford (Devon)	Housatonic River	—	—	Individual
Milford (Town Meadows)	Milford Harbor	10000	1.60	Secondary
Milford (Harbor Plant)	Milford Harbor	4000	.60	Secondary
Milford (Gulf Pond Plant)	Gulf Pond	6000	2.40	Secondary
West Haven	New Haven Harbor	40000	5.70	Primary
New Haven—Boulevard Plant	New Haven Harbor	63100	13.40	Primary
New Haven—East St. Plant	New Haven Harbor	67100	15.40	Primary
New Haven—East Shore Plant	New Haven Harbor	35000	6.80	Primary
North Haven	Quinnipiac River	16000	5.00	Secondary
East Lyme—Rocky Neck State Park	Lons Island Sound	—	.02	Primary
East Lyme—Niantic St. Farm	Bride Brook	—	.03	Secondary
East Lyme—Camp Dempsey	Niantic River	—	.32	Primary
Waterford—Seaside St. San.	Long Island Sound	—	.03	Primary
New London—Trumbull St. Plant	Thames River	32500	3.20	Primary
New London—Riverside Plant	Thames River	2600	20	Primary
Waterford	Thames River	—	—	Individual

Montville	Thames River	–	–	Individual
Norwich & Yantic Interceptor	Shetucket & Thames R.	28500	1.50	Primary
Ledyard–Lifetime Homes	Williams Brook	–	.08	Secondary
Groton City–Fort St. Plant	Thames River	10000	.85	Primary
Groton City–Branford Court	Bakers Cove	800	.03	Secondary
Groton Town–Fort Hill Homes	Mumford Cove	4800	.30	Secondary
Stonington–Mystic	Mystic River	–	.05	Individual
Stonington–Boro	Stonington Harbor	–	.12	Individual
Greenport	Long Island Sound	3000	.15	Primary
Port Jefferson	Port Jefferson Harbor	2000	1.20	Primary
Kings Park Hospital	Smithtown Bay	9500	.80	Secondary
Northport	Northport Harbor	6000	.18	Primary
Huntington	Huntington Harbor	34700	1.40	Secondary
Quantitative Biological Lab	Cold Spring Harbor	40	.008	Secondary
Oyster Bay	Oyster Bay Harbor	6000	1.20	Secondary
Glen Cove–Morgan Estates	Long Island Sound	250	–	Primary
Glen Cove	Hempstead Harbor	25000	4.80	Secondary
Roslyn	Hempstead Harbor	3000	.40	Secondary
Port Washington	Manhasset Bay	25000	2.60	Secondary
Great Neck SD #1	Manhasset Bay	14000	2.50	Secondary
Great Neck Village	Manhasset Bay	9000	1.00	Secondary
Belgrave Sd.	Little Neck Bay	15000	1.42	Secondary
Orchard Beach, NYC	Long Island Sound	Seasonal	–	Septic Tank
City-Hart Island, NYC	Long Island Sound	6000	1.00	Primary
New Rochelle	Long Island Sound	72000	12.00	Primary
Mamaroneck	Long Island Sound	95000	15.50	Primary
Shenrock Shore Club, Rye	Long Island Sound	Seasonal		Septic Tank

Table 7-1 (cont.)

Name	Receiving Waters	Estimated Population Served	Average Daily Flow (MGD)[1]	Degree of Treatment
American Yacht Club, Rye	Long Island Sound	Seasonal		Septic Tank
Blind Brook	Long Island Sound	23000	2.10	Primary
Port Chester	Byram River-Port Chester Harbor	27000	4.70	Primary
Greenwich	Greenwich Harbor	42000	7.00	Secondary
Stamford	Stamford River	60000	10.40	Primary
Stamford–Indian Ridge	Stamford River	500	0.5	Secondary
Darien	Long Island Sound	6500	1.10	Primary
Norwalk	Norwalk River	55000	9.00	Primary
Norwalk (6th Tax Dist.)	Norwalk River	—	—	Septic Tank
Westport	Saugatuck River	5000	1.00	Secondary
Fairfield	Long Island Sound	30000	4.00	Secondary

[1] Millions gallons daily.

did not meet established standards. The Thames River flows by New London and Groton before emptying into the sound. A study of the Groton municipal treatment plant was done by Metcalf and Eddy Engineer Consultants in 1967. They projected that by the year 2020 the per capita per day level for biochemical oxygen demand (BOD) and suspended solids will be .20 pounds and .22 pounds respectively upon its receiving waters. Groton's estimated sewer-served population is expected to increase to 16,250 by 1980 [20:16].

In addition to the population waste, industrial waste is produced by Chas. Pfizer and Co. and General Dynamics Electric Boat Division. Pfizer, employing approximately 2,000, represents the major threat to the Thames River because of the quantity and type of effluent from its pharmaceutical production. Pfizer has also had permission for the last fifteen years to dump a cake-like brown substance from the production of antibiotics directly into the sound. Pfizer claims the material that is dumped into the sound, about three miles offshore, is a nontoxic material. Pfizer has also maintained that approximately 1,000 employees will be laid off if the Army Corps of Engineers cancels the permit and an alternative land site cannot be found.

The cost of upgrading the primary treatment plant in Groton to a secondary treatment plant with city waste treated to 90 percent influent BOD removal was estimated at $2,340,000. An additional $62,000 annually will be needed to operate and maintain the new plant [20:16].

The Connecticut River is a second major river having an effect on the sound. The Connecticut River is the disposal area for a sewered population of approximately 360,000; all levels of treatment are at the primary level except for two locations which receive no treatment. Most treatment plants flowing into the Connecticut River are operating close to designed flow capacity. The municipal plant at Hartford accounts for approximately 75 percent of the sewered population. The Hartford plant is working above its designed flow capacity.

It is expected that the Connecticut River will handle the effluent for more than 420,000 by 1980. Unless the level of treatment afforded the sewage is upgraded, at a minimum, to the secondary level the BOD of wastes released on the already overburdened Connecticut River will be enormously increased with inevitable adverse results for the sound.

The Connecticut River also acts as a source of cooling water for the Connecticut Yankee Atomic Power Plant. During plant operation, surface water temperature differed from bottom temperature by as much as 8° Centigrade. Prior to plant operation the maximum difference was 3° C. During periods of plant shutdown the difference in temperature reverted to pre-operative conditions. The difference was noticeable even during short periods of shutdown. Dissolved oxygen, phosphate, and nitrate levels were higher after the introduction of heated water. The increase in nutrients may be attributable to increased land runoff and sewage added upstream [10].

The sound has a moderating effect on the temperature of the Connecticut River. During the summer months, the water from the sound is cooler than the Connecticut River and therefore has a cooling action. In winter the sound is warmer, thus heating the waters of the Connecticut River.

Bridgeport represents one of the areas on the sound of lower water quality.

Two municipal treatment plants located in Bridgeport serve an estimated 156,000 population. Both plants are primary treatment plants with a combined designed flow of 32 MGD. The West Side Plant is averaging 25.9 MGD, better than 43 percent above the designed level. The plants may be handling waste from approximately 182,000 people by 1980. Work is in progress on the East Plant to convert it to an activated sludge treatment plant that will remove 95 percent BOD and suspended solids. Similar work is progressing at the West Side plant with the $13,300,000 project scheduled for completion during 1972.

There is a heavy concentration of industry on the Mill River (Bridgeport-Fairfield Area). Federal Paper Board Co., Inc., discharges raw sanitary sewage into the city sewage system, while New Haven Board and Carton (Simkens Industries Inc.) discharges partially treated industrial waste directly to the river. Simkens' industrial sewage averages approximately 1.6 MGD currently with as much as 1.8 MGD projected for 1980. Federal Paper Board averages 2.5 MGD currently, projecting a rise to more than 2.6 MGD by 1980. Analysis of present waste discharged by both companies shows Simkens Industry generates 160 tons annually of suspended solids and BOD. Federal Paper Board Co., Inc., discharges 500 tons of suspended solids and 250 tons of BOD annually. Assuming no change in the level of treatment, Simkens and Federal Paper Board will be emitting a combined load of 720 tons of suspended solids and 440 tons of BOD annually by 1980.

The estimated investment required to update treatment plants to handle the industrial load of these two plants ranges from $4,604,000 to $7,327,000.

Beechmont Dairy in Bridgeport, which employed 100 in 1969, is capable of expanding to a point where they may be producing approximately 25 tons of BOD annually. Pepsi Cola, also located in the Bridgeport area, is expected to bottle more than 2.4 million cases of soft drinks annually. Finally Tilo Company could be processing as much as 91,000 short tons of asphalt annually by 1980. The effluent of these three companies must be handled through company treatment facilities or municipal plants.

Norwalk has a primary treatment plant serving 55,000 people. Population projections to 1980 will bring that total to 64,000. A $5 million expansion plan calls for an average design flow of 30 MGD at 95 percent removal of BOD and suspended solids. Additional improvements are estimated to cost $3 million, totaling $8 million for the modernization program [8:25].

In the Norwalk area, Pepperidge Farms bakery and Metallized Products are potential polluters. Pepperidge Farms could be processing more than 51 million pounds of flour annually in 1980. An organization the size of Metallized Products (35 employees) growing at a rate equal to the industry's growth rate between 1958 and 1966 would produce approximately 54.5 tons of BOD annually by 1980.

The potential quantity of pollutants that could be produced by just these two plants gives us an indication of the obstacles to be overcome to maintain water quality in the sound near Norwalk.

During 1969 approved bacteria and dissolved oxygen standards were violated in Stamford Harbor, Norwalk Harbor, the Housatonic River, and Milford Harbor. Bacterial levels were also high in the Housatonic River, suggesting that the river receives discharges of inadequately treated municipal waste. The degradation includes both the river and the area adjacent to its mouth. A 1970 report by the Water Resources Commission, State of Connecticut on Water Quality Standards indicates that problems of quality exist in the Connecticut River, Housatonic River, Byram River, Stamford Harbor, Norwalk Harbor, Bridgeport Harbor, New Haven Harbor, and Mill River. From Branford to the Connecticut-Rhode Island state line shoreline waters, excluding the Thames River, were of high quality. In the Thames River, the dissolved oxygen and nutrient standards were violated. The bacterial levels of the Thames were exceeded in Long Island Sound only by waters in the western end. This condition represents a definite hazard for recreational users.

Long Island

The northern shore of Eastern Long Island from Lloyd Harbor to Orient Point discharges little waste to the waters of Long Island Sound. Industry is negligible and the population discharging its waste to the sound is small.

The Port Jefferson Sewer District provides primary treatment for approximately 2,000 people. Long Island Lighting Company has a plant at Port Jefferson that provides septic tank treatment for its employees. Port Jefferson is planning to upgrade its present primary treatment plant to activated sludge treatment and expand the capacity from its present 1.5 MGD level to 6 MGD. In conjunction with the improvements, work is in progress at the LILCO plant to divert its flow to the Port Jefferson plant.

Northport is another area worthy of consideration. The municipal plant at Northport provides 6,000 people with Imhoff tank treatment. However, Northport plans a .3 MGD aeration treatment plant to handle future needs. The plant is designed for 90 percent BOD reduction and total suspended solids removal [8:24].

Long Island Lighting Company (LILCO) has a power plant at Northport that utilizes the sound for cooling water. The heated water from the plant does have some effect on the organisms in the area. The same variety of organisms does not appear to be as prevalent within 1.5 km of the plant as beyond that range [7:85]. However, the fish population found in the area of the LILCO plant did not differ significantly from those found elsewhere on the sound. The same observation was made of the bird population in the vicinity of the plant.

Another potential trouble area is in Shoreham, Long Island. LILCO is proposing a 820,000 kilowatt atomic generating plant to serve nearly one million families. The plant was originally scheduled to start commercial service during the spring of 1975. About 600,000 gallons per minute of water will be taken from the sound to be utilized as cooling water. The water will be returned to the sound about 3,000 feet from the plant, entering the sound waters via a perforated pipeline. LILCO maintains the returning water will not heat the surface waters by more than 1.5 degrees during the summer months, thus complying with New York State requirements pertaining to thermal pollution.

Based on the effects of thermal pollution at Northport, it does not appear that the Shoreham plant will cause damage to the fish and bird population in the surrounding area, but environmental opponents have nevertheless mounted an intensive campaign against the plant which at a minimum is expected to delay opening.

Continuing westward along the north shore of Long Island, Hempstead Harbor is the next area deserving investigation. The Glen Cove Morris Avenue plant provides secondary treatment for 25,000 people. The plant averages 4.8 MGD even though it was designed for 2.7 MGD. At projected per capita waste loads estimated by the Regional Plan Association and assuming a population increase of 1,500 by 1980, the Glen Cove plant will be required to handle an additional 870 to 1,740 tons of waste annually, about a 6 percent increase over present levels.

Chemco Photo Products in Glen Cove treats its effluent prior to discharge to the sewer system. Projections to 1980 show Chemco's effluent increasing to as much as .017 MGD.

Glen Cove is taking steps to combat pollution from projected municipal and industrial waste loads through upgrading of its Morris Avenue plant. The modernization program will improve the efficiency of the plant to the tertiary level from a secondary level. The Morris Avenue plant will also handle waste from the Morgan Island plant in Glen Cove, which is being converted to a pumping station.

LILCO has a generating plant at Glenwood Landing, which is also located on Hempstead Harbor. The LILCO plant presents the chance of oil spills from oil tankers and discharges heated water to the harbor. The plant does have three septic tanks to handle sanitary sewage from the plant.

Because of the restrictive harbor flow, Hempstead Harbor could easily become seriously polluted unless the sources of pollution are checked.

The next bay area proceeding westward is Manhasset Bay. Port Washington Sewer District, which is located on the bay, serves 25,000 people with secondary level of treatment. The plant processes 2.6 MGD, having room for expansion to 3.0 MGD, its design flow level. Port Washington may be called on to serve only 2,000 additional people in 1980. The additional population would mean an additional waste load ranging between 1,160 tons and 2,320 tons annually.

Two plants located at Great Neck, Long Island, serve 23,000 people at a secondary level of treatment. A small population increase of 1,400 people over the ten-year period will yield between 812 and 1,624 additional tons of waste annually.

One future project under consideration in Nassau County calls for a sewer district that will serve the entire length of Nassau's north shore. The plans call for increasing capacities of existing plants and expansion of collection districts. Plans contemplate the consolidation of the districts. At the same time, a survey is under study calling for the expansion of the Belgrave Sewer District to 4 MGD as part of the North Shore Plan. The Great Neck plant is being expanded to 8 MGD at a cost of $4,300,000.

Manhasset Bay's largest problem appears to be its proximity to the East River. Water quality violations were found in the western end of Long Island Sound. Recorded dissolved oxygen levels were lower in the Hutchinson River near the New York City line than further east and off the mainland at a point near Orchard Beach. Total coliform densities also decrease as one moves from Throgs Neck, where the East River meets the sound, to Orchard Beach. Total coliform densities from Throgs Neck to Orchard Beach were two and one-half times the monthly median value established by the standards. Bacterial concentration measured in 1970 were two to ten times those observed in 1969. This indicates a recent contamination by inadequately treated human or animal feces. Nutrient levels are two to ten times higher than levels elsewhere in the sound. The above water quality conditions indicate the effect of waste sources originating from the metropolitan New York City area, which discharge to the East River.

Nutrients

The most pronounced feature of western Long Island Sound, particularly in recent years, has been the increase in nutrient concentrations such as that just noted. This arises not only from the vast sewage effluent of New York City, whose treatment facilities are frequently overloaded, but also from those of Nassau and Westchester counties. These treatment plants, such as those from Oyster Bay to Great Neck in Nassau County, provide what has always been regarded as reasonably adequate treatment, as do most of those about to be described on the Westchester shore. But the best of these treatment plants, serving noticeably affluent communities, do not have facilities capable of effecting removal of any substantial portion of the nutrients, other than that portion removed as solids. Thus effluents rich in phosphorus and nitrogen flow from these large and growing treatment plants into the sound. The predictable result is an accelerated growth of vegetable and animal plankton in the sound, one of the manifestations of the phenomenon of eutrophication. While the consequences of such eutrophication are many and varied, marine scientists see an inverse correlation between nutrient concentration and water quality. Thus M. Grant Gross, research director of the Marine Sciences Research Center at the State University at Stony Brook, New York, noted that he used measurement of nutrient concentrations to indicate "instantaneous water quality at the time you're there." The reason a high concentration of nitrogen compounds is associated with lower water quality, according to Dr. Gross, is that the nitrates stimulate such a rapid growth of marine organisms that they cannot survive on the food available. The death of the organisms in large quantities and their decomposition deprives fish of the oxygen required for survival [29:19].

One happy consequence of this for Long Island fishermen is the almost explosive population growth of the mossbunker, which feed on the brown algae whose growth is stimulated by the nutrients. Bluefish that feed on the mossbunker have also been active residents of the sound for the past several years. Less happy are the occasional massive fish kills during the height of summer when a combination of the calm weather and the decay of algae leave thousands of dead mossbunker floating in the bays and harbors.

There is nothing in sight to change the prospect of a tendency in the western sound to higher nutrient concentrations. The ebb and flow of the tides exert a powerful flushing action on the sound which holds down concentrations in open water, but diminishes in force in the harbors and embayments in all areas and especially as one moves westward. Thus, the dangers of excessive nutrient build-up increase as one approaches the more densely populated shores of the sound's western end.

Westchester

Westchester County, adjoining Connecticut, has heavily populated areas along the sound at Port Chester, Rye, Mamaroneck, Larchmont, and New Rochelle. All of the Westchester shore is used extensively for boating.

A ten-year study, conducted between 1951 and 1961 by William F. O'Connor, reported the bacterial count in 1960 was at its highest point in twenty years. The previous five years showed a continuous increase in the count. The increase was attributed to increased shore and harbor activity. Samples taken on Mondays were consistently higher than recorded during the remainder of the week.

A 1969 study showed bacteria levels near the New Rochelle and Mamaroneck outfalls were greater than the established standard for contact recreation. A 1969-1970 New York State Department of Environmental Conservation study showed an improvement in water quality west of Mamaroneck Harbor. However, high densities of bacteria continued to persist.

Port Chester Harbor is the first harbor in New York after leaving Connecticut. Surveys taken in 1968 and 1969 show that the waters in Port Chester Harbor violate approved water quality standards. In 1970 extensive floating trash, aquatic plant growth, and a slight oil slick were observed. During 1968, the Byram River, which flows through Port Chester Harbor, was found to contain high bacteria levels that would indicate that inadequately treated municipal wastes enter the Byram River waters. The Byram River was visibly polluted by oil and suspended matter and a distinct oily odor was present. At several locations, gas bubbles were surfacing and oil was observed on its banks. A 1961 U. S. Public Health Service report determined the Byram River through Port Chester Harbor was suitable only "for transportation of sewage and industrial wastes without nuisance and for power, navigation, and other industrial uses . . ." [14:14].

A 6.0 MGD primary treatment plant serves 27,000 people in the City of Port Chester. The plant is operating at full capacity and during many periods has been operated in excess of its designed capacity. The Interstate Sanitation Commission has rated the level of treatment unsatisfactory for the removal of settleable and suspended solids. On many occasions the effluent has contained more settleable and suspended solids than the influent.

The plant was seldom in good operating condition, resulting in heavy accumulations of grease and sludge. Discoloration and suds on the Byram River are also a result of ineffective treatment facilities at Port Chester. Further evidence of the ineffectiveness of the plant is seen in the fact that its effluent has a higher BOD count than the influent and the residual chlorine content has rarely been in excess of .5 parts per million (ppm).

Port Chester plans to begin construction of a secondary treatment plant in the fall of this year with completion scheduled for the summer of 1973. The new plant will handle waste from the Blind Brook Plant and submit it to thermal oxidation.

Port Chester Harbor is in poor condition as a result of industrial waste and the inefficient municipal treatment plant. The condition will continue to deteriorate unless the secondary plant is completed and stricter controls are maintained on industrial and municipal discharges.

The smallest of the Westchester municipal treatment plants is located at Blind Brook. It serves a population of 23,000, removing 90 percent of the settleable solids but only 50 percent of the suspended solids. The effluent is released into Long Island Sound about one mile southeast of Rye Beach after 10 to 40 percent BOD has been removed and the residual chlorinated to .5 ppm. The capacity is 5 MGD, affording it capacity to treat waste from a population projected to 30,000 by 1980.

Sewage from a population of 95,000 is treated at Mamaroneck's primary treatment plant. The 15.5 MGD flow removes 70 percent suspended solids and 90 percent settleable solids during treatment. Between 30 and 60 percent BOD is removed before the residual, carrying .5 ppm chlorine, is released just outside Mamaroneck Harbor. A planned increase of plant capacity to 70 MGD would produce an additional 54.5 MGD. The resultant BOD, suspended solids, and settleable solids daily load would more than triple. However, an upgrading to the secondary level of treatment would remove 10 to 20 percent more suspended solids and 25 to 50 percent more BOD from the total load.

New Rochelle treatment plant serving 75,000 people removes almost all settleable solids and about 60 percent of the suspended solids. At present the plant is processing 12 MGD at a 30 percent BOD removal efficiency. New Rochelle is to be upgraded to a secondary treatment level at an approximate cost of $15 million. Under a New York State Pollution Abatement order, New Rochelle was scheduled to complete construction by January 1, 1972.

Gries Reproducer Co. in New Rochelle employed 539 in 1969 with that figure projected to increase to 839 in 1980. Based on industry employment data and employees' annual output, Gries should be producing 16,780 tons of castings in 1980. The present waste from Gries enters the sewer system and is treated by the municipal plant before the final effluent is emptied into the sound.

Seven-Up Bottling Co. is projected to produce 24,000 cases per employee by 1980. The projection is based on a 1.9 percent increase in output annually per production employee. The effluent will be treated by a municipal treatment plant before final discharge.

Recreational Boating and Shellfishing

There are approximately 183,000 acres suitable for shellfish production in Long Island Sound. Of the 183,000 acres there are 76,000 acres closed to shellfish harvesting due to the pollution of the sound's waters.

Closure of shellfish grounds has resulted in a reduction of commercial shellfish activities. Commercial fishers are being forced to relocate their beds in deeper waters with results being a slower shellfish growth and more expensive operation. Seed and market shellfish are still raised in closed areas, then transplanted under the supervision of the State Health Department to certified areas. The shellfish are released for direct marketing after examination to ensure compliance with proper standards. Branford, New Haven Harbor, Housatonic River, Darien, and Norwalk areas in Connecticut are examples of closed areas being used to raise shellfish by this procedure. Westchester and Nassau County also use this method.

Six hundred acres of shellfish beds that are available between October 1 and April 30 are lost for the remainder of the year because of pollution from summer homes. The best shellfish area on the east coast is located inside the chain of islands in Norwalk Harbor. Shellfish taken from this area are the best in size, development, and flavor.

The Norwalk estuary, including parts of Darien, contributes about $1 million worth of shellfish annually.

Five oyster companies in the vicinity of Greenwich, Stamford, and Darien have been forced out of business because of pollution to the area waters. Prior to 1920 oyster companies located on the sound did more than $5 million in business. It has been estimated by the Connecticut Shellfish Commission that during the past fifty years, losses in excess of $500 million have resulted from water deterioration caused by pollution. Pollution from sludge has caused the closing of 1,000 acres of productive shellfish-growing area in New Haven Harbor and the Quinnipeac River.

The Westchester Shore Area has been a major shellfish-producing area in the past. During the last twenty years its waters have been restricted for this purpose. There are approximately 7,500 acres of onshore waters that do not meet required standards for shellfish harvesting, but have extensive population of clams. Harvesting is restricted to firms that transplant to designated clean areas further east.

By 1970, the water quality in the western portion of Westchester County had improved sufficiently to justify a recommendation for a monitoring program looking to the day when the taking of shellfish from the area for direct marketing purposes might again be permitted. This area would add approximately 5,000 acres of underwater lands for shellfish harvesting [11:2, 10].

Boating

There were in 1967 approximately 185,000 boats registered to residents in the area of Long Island Sound. This figure was estimated to increase at a 5 percent rate until 1970, then continue at 3 percent annually to 1980.

The amount of revenue contributed to the area from recreational boating activities is estimated in the region of $642 million annually. The projected increased boating activity on the sound will force this amount upward to approximately $850 million annually by 1980. Projections are made on present cost per boat annually expended by the projected boat population by 1980. The cost consists of estimates of the value of boats operated on the sound and related expenses associated with owning a boat.

The discharging of untreated human waste to the sound leads to a health hazard in areas of human contact recreation and harvesting of shellfish. Litter discharged from boating activities can be detrimental to the bacteriological and chemical quality of the receiving waters.

Motor boats cause pollution by unburned gasoline and petroleum spillage. Spillage is also inevitable in marinas where gasoline is sold. The possibility of a major oil spill is always present because of human or mechanical failure. When these pollutants are accumulated in restricted harbors and embayments, the effect on the water quality shows up quickly.

Legislation

The Interstate Sanitation Commission has been active in the water pollution field since 1936. One of the first activities by the Commission took place in 1937 when a majority of the waters were classified.

In 1965 New York, New Jersey, and Connecticut agreed that secondary treatment be required on all Commission waters. The Commission received a court order requiring Port Chester, New York to remove industrial wastes within four months and provide the necessary treatment to meet the Commission standards within two years.

New York, New Jersey, and Connecticut, in 1967, agreed to secondary treatment abatement for all their municipal plants by 1972 [8: Appendix B, pp. 2, 7, 8].

Connecticut and New York enacted legislation allowing the Commission to set new classifications for its waters. New Jersey also approved the new classifications which became effective April 15, 1971.

Each state has the authority to issue Pollution Abatement Orders as part of its regulatory authority. The order requires plans and dates of improvement on existing facilities.

November 1965 saw New York's citizens vote approval of a $1.7 billion or more Pure Waters Program to clean up the waters of New York State. A major element of the program provided for construction of municipal treatment plants. "Under this program, municipalities which commence construction by March 31, 1972, can qualify for a 30 percent state grant and a state-guaranteed underwriting of an additional 30 percent, if necessary to prefinance the minimum federal share. The municipality pays the balance of 40 percent and under a constitutional amendment, local government obligations for sewage treatment works incurred until 1973 are wholly exempted from constitutional debt limits" [12:1].

Industry is offered a tax incentive to build its own wastewater treatment plants. The industry can take a net operating loss on its state income tax in the year the expenditure is made. Upon completion of the facility, it becomes tax exempt from property taxes and other special levies [12:4].

Once a violator has been identified, if he does not cooperate with the state, the state can initiate penalty proceedings or refer the case to the Attorney General for legal action. As of 1969, twenty-eight penalty assessments had been initiated and twenty-five cases were referred to the Attorney General.

The state also requires that boats have a provision for holding tanks.

Connecticut has passed recent legislation similar to New York. Connecticut's *Clean Water Act of 1967* contains the following provisions:

1. Authorization of "150 million in bonds to be issued as needed for grants to municipalities."
2. Grants up to 30 percent for the separation of storm waters from sanitary sewer systems when such separation will eliminate a substantial source of pollution.
3. Authorization for the state to prefund to the municipalities the federal grants to which they would be entitled under the federal program as well as advances to cover the cost of preliminary engineering reports and project planning.
4. Industrial waste treatment plants are exempt from local property taxes, from state use and sales taxes, and are permitted an accelerated write-off against the state corporation tax [2].

Conclusions

Long Island Sound is a valuable natural resource which is used extensively for recreational purposes while contributing significantly to the economic position of the area. The shellfish and recreational industries provide a substantial source of income to the area. A deterioration of the quality of the sound will lead to a deterioration of the area's standard of living.

Pollution, particularly in the harbors and embayments, has an adverse effect upon the fish, aquatic life, and aesthetic value of the sound. The shellfish industry has already sampled the effects of poor water quality. Many residents in the immediate vicinity of the sound have chosen their homes because of the beauty and recreation offered by the sound. The property value around the sound would decline rapidly if the sound could no longer be used for boating, bathing, and other recreational purposes.

The major contributor of bacteria pollutants to the waters of the sound is from the discharge of raw and inadequately treated domestic wastes. As population increases in the area of the sound, the volume of waste will increase proportionately. Connecticut presently accounts for a majority of the population whose effluent is disposed of in the sound. By 1980 New York counties will account for the majority of the waste attributable to population and thus must be prepared to handle the projected increased waste load. The discharge of inadequately treated municipal and industrial wastes has been the major cause for the low levels of dissolved oxygen in some harbors and embayments. The main body of the sound doesn't show signs of low level dissolved oxygen at the present time.

Connecticut is the home of most industry located on the sound but has fallen behind New York in updating its municipal treatment plants. Connecticut must move rapidly to upgrade its plants and convince industry to do likewise to prevent further degradation of the sound. New York must also continue to improve its municipal plants and maintain a close surveillance of industrial activities.

Nutrient levels are high in some harbors and embayments causing algal blooms. The harbors and embayments are of lower water quality for a number of reasons: the water is shallow and therefore doesn't have the capability to self-purify as the deeper waters, the harbors and embayments are used as outfall areas for municipal and industrial waste, and thirdly, boating activities concentrate in these areas adding inadequately treated waste and unburned fuels. The deteriorating effect that boating activities have can be demonstrated by the continuous increase in the bacteria count along the Westchester shoreline, particularly on Monday mornings.

The water quality in the western end of the sound is lower than that in the east. The foremost reason for this condition is the low quality of the East River, an adjoining tidal estuary, which carries many pollutants into the sound from New York City.

Areas of lower quality can also be found at the mouth of the Thames, Connecticut, and Housatonic Rivers on the Connecticut shoreline and the Byram River in New York. The low quality of each of these rivers has a direct effect on the sound.

The overall quality of the sound will depend directly upon the quantity of pollutants allowed to be deposited in the sound plus the degree of treatment prior to discharge.

Notes

Reports

1. William A. Boyd, *Connecticut River Ecological Study: Water Temperature 1966-1969,* Report to the Connecticut Water Resources Commission (Essex, Connecticut: Essex Marine Laboratory, April 1970).

2. Clean Water Task Force of Connecticut, *Connecticut's Clean Water Act of 1967—An Analysis* (Hartford, Connecticut: 1967).

3. Connecticut Labor Department, *Directory of Connecticut Manufacturing and Mechanical Establishments 1970,* 1970.

4. Connecticut Labor Department – Employment Security Division, *Labor Market Letter,* October 1970.

5. Connecticut State Department of Finance and Control, Office of State Planning, *Population and Employment Projections for Regions and Towns,* January 1970.

6. Charles D. Hardy, "Hydrographic Data Report: Long Island Sound – 1969," Marine Sciences Research Center, State University, Stony Brook, N.Y., January 1970.

7. George Hechtel and Erwin Ernst, "Biological Effects of Thermal Pollution, Northport, New York," Marine Sciences Research Center, State University, Stony Brook, N.Y., January 1970.

8. Interstate Sanitation Commission, "Report on the Water Pollution Control Activities and the Interstate Air Pollution Program, 1968-1970," a report to the Governors of New York, New Jersey, and Connecticut.

9. Long Island Lighting Co., *The Environment* (Mineola, N.Y.: 1970).

10. Merriman, Daniel, et al., "Connecticut River Ecological Study Semi-Annual Reports #7-11," a report to the Water Resources Commission of the State of Connecituct.

11. New York State Department of Environmental Conservation, Division of Marine and Coastal Resources, Bureau of Environmental Control, "Westchester Shore Shellfish Growing Area #55 1970 Report," Ronkonkoma, N.Y., 1970.

12. New York State Department of Health, *Pure Waters Progress—1969*, 1969.

13. New York State Department of Health, Water Resources Commission, *Surface Waters of Nassau County — Official Classifications*, 1965.

14. New York State Department of Health, Water Resources Service, *Upper East River and Long Island Sound Drainage Basins*, (Albany, New York, 1964).

15. New York State Office of Planning Coordination, *Demographic Projections for New York State Counties to 2020 A.D.*, (Albany, New York: Office of Planning Coordination, August 1969).

16. New York State Office of Planning Coordination, *Long Island Water Resources*, January 1970.

17. Alfred Perlmutter, "Ecological Study—1969 — Shoreham Nuclear Power Station," Long Island Lighting Company, July 1970.

18. "Recreational Boating Registration Statistics," National Association of Engine and Boat Manufacturers, May 1970.

19. "Report on the Industrial Waste Survey — City of Stamford, Connecticut." Report furnished by Charles A. Maguire and Associates, Engineers of Boston, 1969.

20. "Report to the City of Groton, Connecticut Sewer Authority on the Secondary Treatment of Sewage." Report furnished by Metcalf and Eddy Engineering Consultants, March 6, 1968.

21. "Report Upon Facilities for Treatment of Waste Waters From Industries Located on the Mill River." Report furnished by Goodkind, O'Dea, Fay, Spofford and Thorndike Engineers of Boston, 1969.

22. U.S. Department of the Interior, Federal Water Pollution Control Administration, *Report on the Water Quality of Long Island Sound*, (Washington, D.C.: U.S. Government Printing Office, 1969).

23. U.S. Department of the Interior, *National Estuarine Pollution Study* (Washington, D.C.: U.S. Government Printing Office, 1970).

24. Water Quality Office — Northeast Region, Environmental Protection Agency *Report on the Water Quality of Long Island Sound*, March 1971.

Periodicals

25. "A Healthy Environment — Everyone's Concern," *LILCO Almanac* No. 16 (May 1970), p. 2.

26. "Hysteria Over Heads," *Time* 97, 18 (May 3, 1971), p. 57.

Newspapers

27. "Layoffs Seen in Ban on Waste in Sound," *New York Times,* April 15, 1971, p. 38.

28. Harry Pearson, "LILCO Alters A-Plant Design," *Newsday,* September 23, 1970, pp. 3, 27.

29. Sylvia Carr, "Pollution Scores on Six Harbors," *Newsday,* February 7, 1972, p. 19.

8 Hudson River

RICHARD M. MARTIN

The Hudson River begins in upper New York State, on Mt. Marcy, in a two-acre pond called "Lake Tear in the Clouds." From this pond, it flows generally southward for approximately 315 miles, becoming one of the most important watersheds in North America. The drainage area is estimated at about 12,400 to 14,500 square miles and encompasses a human population of about 12 million people.

The Hudson River is an area of important historical associations. The area is exceptional as a recreational resource as well as an area of unsurpassed natural beauty. It is also a drain for industrial wastes and domestic sewage. The economic, industrial, and recreational potential of this river has not yet been fully realized. Unless care is exercised, unwanted pollution may destroy much of its usefulness.

The drainage basin covers most of the northern and eastern parts of New York State and small areas of Vermont, Connecticut, Massachusetts, and New Jersey.

Mountainous terrain accounts for 48 percent of the Hudson drainage basin, cultivated lands for 42 percent, and the remaining 8 percent by urban locations.

The Hudson Valley is generally viewed, physiographically, as made of of four longitudinal segments, two through mountains and two through lowland relief. The first segment begins at the source of the river, 4,293 feet up on the side of Mt. Marcy, and includes its passage through the Adirondack region. At the lower extremity of this region, it joins its principal tributary, the Mohawk River.

The second portion begins at the Federal Dam and Lock at Troy, New York, and extends southward to the northern extremities of the Hudson Highlands, in Dutchess and Ulster Counties.

The third segment of the river valley is called the Hudson Canyon. It is contained between the Catskill Mountains on the west bank and the Taconic Mountains on the east bank. This portion extends southward to, roughly, Stoney Point in Rockland County.

From that point, the fourth section begins and extends south of the Highlands, to the Atlantic Ocean [1:7-8].

Average precipitation is about 43 inches, of which 24 inches appear as actual runoff at the mouth of the Hudson. As the freshwater flow diminishes during the summer months, the salt water from the ocean "pushes" upstream from near the Tappan Zee Bridge in Westchester County in the spring, reaching sometimes as high as Troy, New York, in the late fall, over 150 miles from the mouth of the Hudson River.

Tidal action has a mean range of 2.9 to 4.4 feet in Upper New York Bay [2:7]. Tidal influence in the Hudson River extends upstream about 153 miles to the Federal Dam at Troy, New York, and affects the entire area. As sea water pushes up the river at flood tide, it tends to form a wedge of denser saline water at the bottom of the river bed. Because the Hudson is more than 100 feet deep in places, this saline water usually remains near the bottom. The salinity of the river increases from May to November. In November there is usually an added flow of fresh water before the cold temperatures of midwinter reduces the flow. In general, salinity remains high throughout the winter and the spring thaw brings about a massive dilution, usually in March or April.

The northward motion of its tides during some periods apparently reverses the river flow, because of the massive tidal flow, and dwarfs the summer freshwater flow.

During the summer of 1956, dye was added to the river at different sites for fourteen tidal cycles. At Troy, the northern-most extremity of tidal action, the tidal influence was 3 miles, but the velocity of net movement of the dye mass downstream was a mile and one-half for each tidal cycle, or about 20 miles in a week. However, below Kingston, New York, there was no net movement of the dye downstream. The overall effect was to produce a movement of brackish water, but little effective exchange [3:60-82].

This important study left several clear implications that have great importance in the matter of water pollution and eutrophication of the Hudson. Apparently the Hudson acts like a huge brackish lake rocked back and forth (north and south) by the tide. The fresh water inflow is sufficient for only about 0.3 to 2.5 percent to be exchanged each day. This will cause the effluents and nutrients discharged into the Hudson to be recirculated during the dry months between the sediments, water, and biota until eutrophic conditions occur [4:8-15]. Only the high spring runoffs provide a flushing volume of water necessary to prevent an accumulation of pollutants and eventually eutrophication.

The channel of the river can be traced well out into the Atlantic Ocean, and at the northern extremity of the tidal action at Troy, the river bottom is still four feet below sea level.

The depth of the Hudson up to Troy, some 150 miles, opened up the Hudson to deep draft vessels for a greater distance than any other river on the eastern seaboard. This enabled the early settlers to penetrate into the interior along this valley more deeply than any other European settlement. Another important factor in the early settling along the Hudson is the high salt content of the harbor and estuary. The salinity keeps it relatively ice free in winter.

Nutrient Loadings

The sparsely populated, agricultural Mohawk Valley watershed, stocked with farm animals and using fertilized land, provides a large nutrient input. This is reflected in the high nutrient levels of Mohawk River waters as compared to other American Rivers [5:41-58]. Of the amounts recorded, about 17 percent could be attributed to natural runoff, the remainder to artificial sources.

The human population of about 12 million people in the Hudson watershed produce 69.7 million kg. of nitrogen and 6.2 million kg. of phosphorus as personal waste in a year, much of which will eventually be carried out to sea [6:7]. Besides agricultural wastes, other nutrient wastes, such as from detergents, the meat and dairy industries, yeast and paper production, are contributors to the nutrient load.

In 1965, municipal waste discharges north of Yonkers provided 1.6 percent (spring) to 16 percent (summer) of the volume of river flow [6:7-8]. In New York harbor, the proportion was much higher, due to the population density in the New York metropolitan area.

The nutrient content of the Hudson River water is a reflection of how much it is used. Samples of river water taken at Indian Point (43 miles from New York Harbor), showed the phosphorus content ranged from 9.5 to 2.5 ug-atoms/liter and at the southern tip of Manhattan, a high of 12 ug-atoms/liter was recorded [7]. To put this in proper perspective, the generally accepted value for phosphorus content in unpolluted water is 2.8 ug-atoms/liter.

Nutrient loadings are an important consideration when considering eutrophication in the Hudson River. The present method of treating sewage in the Hudson Valley is to discharge a liquid effluent as the end product. It is very likely that this procedure does little to reduce nutrient levels of the Hudson River. It is even possible that nutrients, thus made available in a more soluble form, may make the situation worse than it would have been as more and more sewage treatment plants come into operation. Nitrates and phosphates develop algae blooms, leading to an increased oxygen demand to support the photosynthesis that creates organic matter.

Economic Growth

New York State's economic growth to the year 1980 should be closely aligned with rapid social and technological change. The entrance of the postwar generation into the social and economic picture, rising affluence, higher levels of education, and new job opportunities should be important in driving New York State's economy upward.

Many of the decisions made in the late 1960s and in 1970 will play an important role in the development of New York State in 1980. Looking ahead is essential, in order to try to establish programs to strengthen the state's grip on the problem of water pollution of the Hudson River and its major tributaries.

Population

New York State's population will pass 20 million by 1980. At the present growth rate, annual increases will amount to 200,000 people per year. The state at this rate should add approximately 2 million people to its rolls over the next decade.

The most important observation is that the state's two great areas, the greater New York City region (New York City, Nassau-Suffolk, and Mid Hudson planning and development regions) and Upstate New York are expected to grow at similar rates over the decades ahead. Thus, the greater New York City region will continue to account for roughly two-thirds of the state population in the forseeable future [8:7].

In order to view the population expansion into 1980 and judge its impact on the Hudson and Mohawk Valleys, this study is divided into four geographical areas. The Upper Hudson, Mohawk Valley, Mid Hudson, and Lower Hudson will be the general classifications for the Hudson drainage basin. The upper Hudson consists of Albany, Essex, Rensselaer, Saratoga, Schenectady, Schoharie, Warren, and Washington counties of New York State.

The Mohawk Valley area consists of Fulton, Herkimer, Montgomery, and Oneida counties. The New York counties of Columbia, Dutchess, Greene, Orange, Putnam, and Ulster comprise the Mid Hudson area. The Lower Hudson encompasses Rockland, Westchester, the five boroughs of New York City (Bronx, Brooklyn, New York, Queens, Richmond), and a limited section of New Jersey, including Bergen, Essex, Hudson, Middlesex, Morris, Passaic, Somerset, and Union counties.

In this manner, each section of the drainage basin will be viewed individually, with its own peculiarities and problems.

The projections of the New York State Office of Planning Coordination were used for this study with some slight adjustments. As the 1970 census figures became available, they were compared to the statistical estimates made by the Office of Planning Coordination. After a comparison, some slight adjustments were made and projected to 1980.

For the selected counties of New Jersey, the projections used were those of the New Jersey Department of Conservation and Economic Development, Division of Economic Development, Bureau of Research and Statistics. Again, slight adjustments were made to conform with the 1970 census data.

There are many factors, which will have various weights, affecting the characteristics of the population in the Hudson drainage basin. Some of these factors are social attitudes towards size of family, preferences for locations in terms of living and working environments, economic conditions, and technology changes.

Much of New York State's growth in population since World War II has been the direct result of a very large increase in the birth rate, with a steady rate of deaths. The period from 1946 to 1965 showed that the birth rate was double the rate of deaths for those twenty years [9:59]. This alone accounted for over 70 percent of the state's increase in population during that time.

In terms of housing and land distribution, it is expected that a pronounced impact will be felt in urbanizing the land areas immediately adjacent to New York City proper. This will have a dramatic effect on the so-called Hudson Corridor. This is the area directly north of New York City on both sides of the Hudson River. This area includes Orange, Rockland, Westchester, and Putnam Counties in New York.

Tables 8.1 and 8.2 show the expected growth patterns. The most significant areas of population growth are the Mid Hudson and Lower Hudson areas. There is an expected increase of almost 1,655,000 people in the Lower Hudson region and 332,000 in the Mid Hudson region. This growth by far overshadows the increases projected for the Upper Hudson and Mohawk Valley. The projected growth in the lower Hudson area between 1970 and 1980 is 1,655,400 compared to 191,900 for the Upper Hudson and Mohawk Valley combined. Thus, the growth in population in the lower Hudson area, already the most seriously afflicted with pollution problems, is expected to be over eight times as great as in the northern areas where the problem is less severe.

The large increase in population projected for the Mid Hudson region suggests that pollution control officials in that area will have to work to achieve advances. Nearly two-thirds of the expected population increase on the Lower Hudson area of New York and New Jersey is in the suburban counties of Rockland and Westchester in New York and Bergen, Essex, Middlesex, Morris, and Somerset in New Jersey. This is in line with recent national trends that have shown a slowdown in migration from rural areas to metropolitan, accompanied by increased movement from urban centers to suburbia. These trends as well as the sharp rise in projected Mid Hudson population would seem to support the theory that there will be a rise in the "Hudson Corridor" population above New York City.

Table 8-1
Population Projections by County in the Hudson River Drainage Basin (in thousands)

	1950	1960	1970	1980 (Projected)
New York State	14,830.2	16,782.3	18,190.7	20,756.8
Upper Hudson				
Albany	239.4	272.9	285.6	352.4
Essex	35.1	35.3	34.9	37.1
Rensselaer	132.6	142.6	152.5	186.2
Saratoga	74.9	89.1	121.6	127.9
Schenectady	142.5	152.9	160.9	181.6
Schoharie	22.7	22.6	24.7	23.3
Warren	39.2	44.2	49.4	54.6
Washington	47.1	48.5	52.7	49.6
Mohawk Valley				
Fulton	51.0	51.3	52.6	51.0
Herkimer	61.4	66.4	67.4	73.6
Montgomery	59.6	57.2	55.8	54.5
Oneida	222.9	264.4	273.0	331.6

Table 8-1 (cont.)

	1950	1960	1970	1980 (Projected)
Mid Hudson				
Columbia	43.2	47.3	51.5	54.9
Dutchess	136.8	176.0	222.2	317.9
Greene	28.7	31.4	33.1	35.7
Orange	152.3	183.7	220.5	394.4
Putnam	20.3	31.7	56.6	81.6
Ulster	92.6	118.8	141.2	172.4
Lower Hudson-New York				
Rockland	89.3	136.8	229.9	307.4
Westchester	625.8	808.9	891.4	1,138.2
Bronx	1,451.3	1,424.8	1,472.2	1,591.4
Kings	2,738.2	2,627.3	2,601.8	2,670.6
New York	1,960.1	1,698.8	1,524.5	1,478.2
Queens	1,550.8	1,809.6	1,973.7	2,181.3
Richmond	191.6	222.0	295.4	381.4
Lower Hudson-New Jersey				
Bergen	539.1	780.2	897.1	1,053.7
Essex	905.9	923.5	932.3	998.5
Hudson	647.4	610.7	609.2	612.1
Middlesex	264.8	433.8	583.8	842.8
Morris	164.3	261.6	383.4	571.9
Passaic	337.1	406.6	460.7	520.5
Somerset	99.0	143.9	198.3	283.3
Union	398.1	504.2	543.1	620.4

Sources:
(1) U.S. Dept. of Commerce, Bureau of Census, U.S. CENSUS OF POPULATION, 1950, 1960, 1970 (Washington, D.C.: U.S. Government Printing Office, 1950, 1960, 1970).
(2) N.Y.S. Office of Planning Coordination, DEMOGRAPHIC PROJECTIONS FOR NEW YORK STATE COUNTIES TO 2020 A.D., (August, 1969).
(3) New Jersey Dept. of Conservation and Economic Development, 1969 POPULATION ESTIMATES FOR NEW JERSEY, (Trenton, 1969), pp. (not numbered).

Table 8-2
Percentage Distribution of Population by County in the Hudson River Drainage Basin (in thousands)

	Change 1950-1960		Change 1960-1970		Projected Change 1970-1980	
	Number	Percentage	Number	Percentage	Number	Percentage
New York State	1,952.1	13.2	1,408.4	8.4	2,566.1	14.1
Upper Hudson						
Albany	33.5	14.0	12.6	4.7	66.8	23.3

Table 8-2 (cont.)

	Change 1950-1960		Change 1960-1970		Projected Change 1970-1980	
	Number	Percentage	Number	Percentage	Number	Percentage
Essex	.2	0.6	−.6	−1.9	2.2	6.3
Rensselaer	10.0	7.5	9.9	7.0	33.7	22.0
Saratoga	15.8	19.0	35.5	36.6	6.3	5.1
Schenectady	10.4	7.3	8.0	5.3	20.7	12.8
Schoharie	−.1	−0.4	2.1	9.4	−1.4	−5.6
Warren	5.0	12.2	5.4	12.3	5.2	10.5
Washington	1.4	2.8	4.2	8.8	−3.1	−5.8
	76.2		77.1		130.4	
Mohawk Valley						
Fulton	.3	0.6	1.3	2.6	1.6	3.0
Herkimer	5.0	8.1	1.0	1.6	6.2	9.1
Montgomery	−2.4	−4.0	−1.3	−2.4	−1.3	−2.3
Oneida	41.5	18.6	8.6	3.3	58.6	21.4
	44.4		9.6		61.5	
Mid Hudson						
Columbia	4.1	9.6	4.1	8.9	3.4	6.6
Dutchess	39.2	28.7	46.2	26.3	95.7	43.0
Greene	2.7	9.1	1.7	5.6	2.6	.8
Orange	31.4	20.7	36.8	20.0	173.9	78.8
Putnam	11.4	56.2	24.9	78.7	25.0	44.1
Ulster	26.2	28.3	22.4	18.9	31.2	22.0
	115.0		136.1		331.8	
Lower Hudson-New York						
Rockland	47.5	53.2	93.1	68.1	77.5	33.7
Westchester	183.0	29.3	82.5	10.2	246.8	27.6
Bronx	−26.4	−1.8	47.4	3.3	119.2	8.0
Kings	−110.8	−4.0	−25.4	−1.0	68.8	2.6
New York	−261.8	−13.4	−173.7	−10.2	−46.3	−3.0
Queens	258.7	16.7	164.1	9.1	207.6	10.5
Richmond	30.4	15.9	73.4	68.1	86.0	29.1
	120.6		2.6		759.6	
Lower Hudson-New Jersey						
Bergen	241.1	44.7	116.9	15.1	156.6	17.4
Essex	17.6	1.9	8.8	0.7	66.2	7.1
Hudson	36.7	−5.7	−1.5	−0.2	2.9	.4
Middlesex	169.0	63.8	150.0	34.6	259.0	44.3
Morris	97.3	59.2	121.8	46.6	188.5	49.1
Passaic	69.5	20.6	54.1	13.3	59.8	12.9

Table 8-2 (cont.)

	Change 1950-1960 Number	Change 1950-1960 Percentage	Change 1960-1970 Number	Change 1960-1970 Percentage	Projected Change 1970-1980 Number	Projected Change 1970-1980 Percentage
Somerset	44.9	45.3	54.4	37.8	85.5	43.1
Union	106.1	26.7	48.9	7.7	77.3	14.2
	782.2		543.4		895.8	

Sources:
(1) New York State Division of the Budget, NEW YORK STATE STATISTICAL YEARBOOK-1970, (April, 1970), p. 48.
(2) U.S. Bureau of Census, 1970 CENSUS OF POPULATION, Advance Report PC (VI)-34, (January, 1971), p. 3.
(3) U.S. Bureau of the Census, U.S. CENSUS OF POPULATION: 1970, NUMBER OF INHABITANTS, Final Report PC(1)-A32, New Jersey, Table 10 (Washington, D.C.: U.S. Government Printing Office), pp. 32-20 to 32-23.

Economic Outlook

The economy should reach new heights during the 1970s and into 1980 in the Hudson and Mohawk River valleys. However, only a healthy and growing national economy can insure this growth. Many changes will undoubtedly take place. New industries will be born, and established ones will decline. Due to advances in communication and scientific discovery, technological changes should occur even more rapidly during the 1970s than in the past. However, it is not the purpose of this chapter to predict changes in technology, but only to try to discern the trends that will have an effect on New York and New Jersey's industrial growth as related to the Hudson River.

It is expected that the large increases in employment during the 1970s will occur in the service industries. Manufacturing in many sectors is not expected to grow and the number of manufacturing employees may even decline. The reasons for the decline in the manufacturing sector are as follows:

a. Growth rates in the past have not been clearly established; the number of manufacturing establishments has not shown significant rise in the past twenty years.
b. Technological change, stimulated by research and development, will make manufactured goods easier to produce with fewer employees.

The following industries are expected to provide increases in employment: electric machinery, transportation equipment, instruments, and printing. Declines are expected in food processing, apparel, textile, leather, and primary metal industries. It will be noted that in terms of future water pollution in the Hudson basin, the industrial outlook is promising. The expected increases in output are in such durable goods sectors as machinery where water pollution effects per unit produced are less than in production of such nondurables as food and leather.

Manufacturing in the Hudson and Mohawk Valleys has been declining in importance for a number of years. There seems to be a shift in the manufacturing mix from the slower-growing nondurables towards the faster-growing durables. Nondurables' production is concentrated in the heavily populated Lower Hudson area, while durables' production looms larger in the less populated upstate areas.

Industrial Projections

The magnitude of the future pollution impact on the Hudson basin of household sewage can be judged in terms of population projections. The future pollutional impact of industry is likewise a function in some sense of the size of the industrial population. However, in the latter case, a simple head count of establishments is not adequate because the impact varies enormously from one industry to another. The polluting effects of the effluent of a single paper mill, for example, may be the equivalent of the sewage of a sizable city. In contrast, the discharge of a large factory manufacturing computers may have a negligible polluting effect.

Accordingly, as a first step in measuring the present pollutional effects of industry in the Hudson basin on its waters, the industrial population of the area was studied by references to industrial directories for the year 1958, 1964, and 1971 [10]. For each of the twenty-one counties of New York State that are in the Hudson River drainage basin, and for each industry, the number of establishments in each of the above years was determined. These plants were classified into ten size groups depending upon the number of employees and organized by industry according to the standard industrial classification. Although the resulting tabulations covering 8,774 industrial establishments were invaluable in providing a profile of area industry, it was felt that no analytical purpose would be served by their reproduction here. These tabulations provided much of the basis for the description of industrial composition of areas that follows.

Upper Hudson and Albany Area

The Upper Hudson region, with about 5 percent of the state's employment, has a diversified industrial base, with state government and trade in Albany, electrical machinery in Schenectady, and "soft goods" nondurables manufacturing in Troy. In this region, the Albany area should be the one of the largest growth into 1980. Albany's trading area extends to the Canadian border and also encompasses Vermont and Massachusetts.

One of the principal industrial sources of water pollution in the Upper Hudson region is the paper industry. Papermaking firms employ 11 percent of the region's manufacturing employees. The area, with forests for raw materials and an abundance of water for power, transportation, and processing, was a particularly advantageous site for the first pulp and paper mills in New York State. Paper manufacturing is concentrated in Saratoga and Warren counties, particularly in the former, where over 40 percent of the industry's jobs are located.

However, many of these mills are among the first producing paper and pulp in New York State. Not surprisingly, they are now being out-produced by more efficient and modern plants in other parts of New York and the United States. Several recent announcements have indicated the closing of several of these old mills in Saratoga County in the early 1970s. These closings should produce a significant decline in the paper and pulp industry and a corresponding decline in the amount of lignin effluent discharged into the Hudson basin.

The apparel industry is fourth in importance in this area, and is centered in Rensselaer County. The fibers from this industry represent an important form of water pollution.

Stone, glass, and clay manufacturing employ 8 percent of the area's workers. The industries manufacture a variety of products, such as abrasive paper and cloth, pressure-sensitive tape, brake linings, blocks, insulating materials, cement, and bricks.

Much of the area's food processing is located in Albany County. One large meat packing firm and a large baking firm employ a good share of the workers in this industry. This industry, along with the paper, glass, and stone industries, constitutes a good portion of the industrial water polluters in this region.

The electrical and nonelectrical machinery groups comprise the largest industries in terms of total number of employees in this region. Approximately three out of every ten manufacturing employees are employed in these two industries in the Upper Hudson region [11:1-11]. However, their water pollution effects are small.

Mohawk Valley Region

This region has been traditionally known for the manufacture of gloves, carpets, and machinery. The leading industry, electrical machinery, is concentrated in Oneida County. The leather goods industry is principally located in Fulton County. The nonelectrical machinery industries are mainly centered in the Utica-Rome areas of Oneida County.

The textile, primary metals, and food industries employ about one-fourth of the manufacturing force in the Mohawk Valley [12:1-8]. The textile industry is primarily the manufacture of rugs and carpets. Another major employer is the dyeing and finishing of knit cloth. The primary metal metal industry consists mainly of processing the copper and brass in the forms of sheets, strips, tubings, rods, and forgings. In the food products, the principal employer is engaged in producing gum, cereals, baby food, and cough drops. Other food concerns produce beer and ale, meat products, instant coffee, tea, and dessert preparations.

Also important to the pollution of the Mohawk River is its agricultural industry. Dairying is by far the most important agricultural activity in the Mohawk Valley area. In 1964, area farms sold products valued at nearly $57 million, of which more than three-fourths were from dairy products [9:146-47].

Fulton County has a large number of plants, particularly connected with the leather industry. However, the bulk of the plants are small and have less than fifty employees. Oneida County is the largest in terms of industrial output and contains by far the greatest portion of large manufacturing plants that contribute industrial effluents.

Mid Hudson Region

The Mid Hudson Region is one of the fastest-growing areas of the state industrially, as well as in population. For industry, it is ideally located between New York City and the Upper Hudson region. It is growing in both the residential areas as well as in commerce and industry. Research and development activity is a leading activity with many large firms establishing R & D facilities in this area. The most significant industrial growth has been in the machinery-producing area. This industry is concentrated in the counties of Orange, Ulster, and Dutchess. The electrical machinery industry includes data processing equipment and electronic components. Nonelectrical machinery includes processing equipment for the food, chemical, and process industries, as well as milking machines and allied dairy equipment. The R & D activities and machinery-producing industries do not contribute significantly to pollution of the Hudson.

The apparel industry is characterized by many small firms. One-half of this industry is located in Orange County. Textile producers, and printing and publishing firms rank third and fourth respectively in number of employees in the Mid Hudson region. Textile firms are quite numerous, but small to medium in size. In the printing and publishing field, there are several large firms that print children's books, business forms, books, and newspapers.

In general, most of the industrial activity in this region is of the light, nonbasic type, and can be generally characterized as noncontributory in terms of industrial pollution. Despite the concentration of industry in Orange, Dutchess, and Ulster Counties, most streams are unmarred by industrial pollution. There are a few notable exceptions, as witnesses to Modena Creek will testify.

Lower Hudson Region

This area centers on New York City as its hub, with Rockland and Westchester Counties to the north and northwest, and New Jersey to the west. There is a high degree of economic interdependence between the sections, with high levels of commutation in both directions. This region contains about two-thirds of New York State's population and a somewhat larger percentage of its economic activity.

New York City, which contains the bulk of the population in the area and is the economic center of gravity, remains the financial, commercial, and communications capital of the nation. Accordingly, while practically every conceivable industry is represented in the city, economic activity is highly oriented toward non-goods production, while manufacturing activity emphasizes such nondurables as apparel. Almost all industrial waste generated in New York City is disposed of through the municipal sewage system. That part of the municipal sewage effluent that is discharged to the Hudson constitutes an enormously flagrant source of pollution. However, it is important to emphasize that much more of the New York City municipal sewage effluent is discharged in other New York waters such as Long Island Sound, East River, New York Harbor, and the bays along the south shore of Long Island. Much of this discharge nevertheless adversely affects the quality of the Lower Hudson River through the effects of tidal action. Whatever its source, inadequately treated sewage rather than industrial waste is the major contribution of New York City to the pollution of the Lower Hudson.

It is not so clear that the same thing is true of the other major population centers in the region, the New Jersey counties draining into the Lower Hudson and New York Harbor. Though many of the New Jersey communities discharge incompletely treated sewage effluent, there are also serious direct discharges by industrial polluters. New Jersey is a center for chemical manufacturing and associated industries as well as petroleum refining and the wastes of these establishments have serious pollutional effects that are carried into the Lower Hudson, again in part, by tidal action.

Westchester and Rockland counties, in New York State above New York City, have been the scene of rapid industrial expansion during the last two decades. However, very little of this expansion was in the "heavy" or basic industries that are traditionally the major polluters, such as the paper or steel industries. The main industries of the Lower Hudson region in New York are commercial, apparel, printing and publishing, miscellaneous manufacturing, and food products. Westchester County wastes are discharged to Long Island Sound as well as the Hudson River.

Pollutional Discharge into the Hudson and Mohawk Rivers

Water quality in the Hudson River system is affected by numerous discharges of untreated and inadequately treated wastes from municipalities and industries from the upper reaches in the Adirondack Mountains to the Narrows in New York Harbor. The purpose of this section is to summarize the estimated and projected waste loads entering the Hudson from municipal and industrial sources in the previously designated sections of the river.

Wastes are discharged into the Hudson and Mohawk by municipalities and industrial plants throughout their lengths. Although the per capita daily sanitary waste load is fairly constant (except perhaps in New York City), the amount contributed by the municipality is dependent not only on its population but also on the amount and quality of its private disposal systems, the degree of treatment provided, as well as seasonal variations such as population fluctuations and storm water runoff.

When organic material is discharged into a waterway, river biota assimilate this matter into their life cycle of synthesis and respiration. The dissolved oxygen in the river is consumed to support both processes. Because the amount of dissolved oxygen required to degrade a given amount of a certain waste is both determinable and constant, this oxygen requirement will be used for expressing the strength of any waste. The oxygen requirement is called the biochemical oxygen demand and will be hereafter referred to as BOD. The BOD will be expressed in terms of pounds of BOD discharged.

The estimates and projections of BOD loads were made following the general methodology described in Chapter 2.

For the BOD load attributable to population, this involved multiplying the estimated or projected population by the generally suggested average daily per capita BOD contribution of 0.17 pounds per capita per day [13:102-103]. This is the coefficient for the sewage and the same source cites smaller coefficients for per capita BOD contributions of treated sewage effluents. However, the detailed knowledge of present and future sewage treatment facilities was not available to permit application of the coefficients for treated sewage. Accordingly, it was necessary to content ourselves with estimates and projections of the potential contribution to the BOD load of the population.

Population estimates and projections were obtained from the federal and state sources cited in the footnotes of table 8.1 For each region, the 1980 projected population multiplied by the above-cited coefficient of 0.17 pounds per capita per day, or 62.05 pounds per capita per year, yields the BOD attributable to population in table 8.3.

Table 8-3*
Projected Liquid Waste Generated in the Hudson Drainage Basin, 1980

	Lbs. of BOD (in thousands)
Upper Hudson Region	
Population	62,838.0
Industrial	32,495.5
Mohawk Valley Region	
Population	31,688.8
Industrial	11,482.3
Mid Hudson Region	
Population	65,580.6
Industrial	7,629.0
Lower Hudson Region	
Population	946,367.8
Industrial	216,450.2

*This table is derived from a summary of tables 8-4-8.11 covering outputs in each of the four Hudson regions.

Determining the quantitative factors for the industrial waste discharges necessitated a multifaceted approach. Several governmental agency and independent studies were available. Many of these surveys were compiled during the last five years and pinpointed virtually every known industrial polluter actually discharging directly into the Hudson River system at various specific locations. However, it is important for this study that each of the regions of the Hudson drainage basin be viewed in its entirety. In this manner, a gross BOD can be calculated and assigned to each region in the study.

A previous study by the Regional Plan Association [14:95-96] established a procedure for this type of analysis, and Chapter 2 details its application here. Using certain selected industries, a coefficient of liquid waste generated was extrapolated using existing data. These data were generally based on the existing technology and knowledge of water utilization practices in given industries. The industries selected are listed by the Standard Industrial Classifications (SIC):

SIC	Description
2011	Meat Packing Plants
2013	Meat Processing Plants
2015	Poultry Processing Plants
2026	Fluid Milk and Related Products
2033	Canned Fruits and Vegetables
2037	Frozen Fruits and Vegetables
2062	Refined Cane Sugar and By-Products
2085	Distilled, Rectified, and Blended Liquors
2261	Finished Cotton Broadwoven Products
2262	Finished Manmade Fibers and Silk
2621	Papermill Products
2631	Paperboard Mill Products
2641	Coated and Glazed Paper
2654	Sanitary Food Containers
2911	Petroleum Products

These industries are most appropriate to the Hudson drainage basin. In each region of the Hudson most, if not all, of the fifteen selected industries play an important role in the water quality of that region.

Table 8.3 brings together the end results of this enormously drawn-out and complex procedure for projection to 1980 of the pollution potential of wastes discharged to the Hudson. The pollution potential in accordance with the methodology set forth in Chapter 3 is expressed in terms of BOD of waste discharges, with a distinction made between that attributable to the sewage generated by the population and that portion arising from discharges by plants in the selected industries.

An examination of table 8.3 reveals the broad outlines of the Hudson pollution problem. The bulk of the BOD comes from wastes generated in proportion to the population which accounted in total for all regions for more than four times the BOD of industrial wastes. This ratio was highest in the Mid Hudson region where population accounted for more than eight times the BOD of industry, lowest in the Upper Hudson and Mohawk Valley, and in-between in the Lower Hudson where enormous populations were concentrated. The problem of pollution of the Hudson is most intractable in the Lower Hudson region, which generated more than five times the potential BOD than arose from the discharges of all the other regions of the river combined. Striking as this may appear, it is clear that this comparison obviously understates the severity of the problem in the Lower Hudson because oxygen dissolved from the atmosphere meets much of the relatively light and dispersed demands of the Mid Hudson region and above, while only a tiny fraction of the overwhelming oxygen demand in the Lower Hudson can be met from this source.

Upper Hudson Region

In the upper reaches of the Hudson to Albany County, there is a concentration of old paper mills that contribute the major portion of industrial waste to the Hudson. By the time the Hudson reaches Cohoes, where the Mohawk River empties into it, the area has become what has been characterized as an "open sewer" [15:94]. Fortunately for the pollution problem, the number of paper mills has been declining steadily [16:37] and several recently have been designated as "too inefficient" and will soon be closed.

Tables 8.4 and 8.5 give the wastes generated in the Upper Hudson region. In this region, the amount of industrial waste is more than half the amount of waste attributed to population as contrasted with smaller proportions elsewhere. The reason for this is the large part played by paper mills. The liquid waste in the Upper Hudson is primarily generated by industrial plants until it reaches the city of Albany, New York. The large paper mills are dispersed up the Hudson above Albany about 50 miles. This distance gives the assimilative capacity of the Hudson a chance to act before the wastes reach Albany.

Table 8-4
Waste Generation Upper Hudson Region—Population

County	1980 Pop. Estimated	Lbs. BOD Generated (thousands)
Albany	352.4	21,866.4
Essex	37.1	2,302.0
Rensselaer	186.2	11,553.7
Saratoga	127.9	7,936.2
Schenectady	181.6	11,268.3
Schoharie	23.3	1,445.8
Warren	54.6	3,387.9
Washington	49.6	3,077.7
Total		62,838.0 lbs. of BOD

Source: Table 8-1 values multiplied by 62.05 pounds BOD per capita per year (i.e., 0.17 lbs. BOD per day times 365).

Table 8-5
Waste Generation Upper Hudson Region—Industrial

SIC Industrial	Lbs. of BOD Generated (in thousands)
2011	510.5
2013	756.2
2015	491.2
2026	240.4
2033	2,372.6
2037	395.6
2062	0
2085	0
2261	1,245.8
2262	618.5
2621	20,679.4
2631	4,968.7
2641	196.2
2654	20.4
2911	0
Total	32,495.5 lbs. of BOD

Sources:
(1) U.S. Dept. of Commerce, Bureau of Census, COUNTY BUSINESS PATTERNS, FIRST QUARTER 1959, Part 3A, Middle Atlantic States (Washington, D.C: U.S. Government Printing Office, 1961), pp. 4-8.
(2) U.S. Dept. of Commerce, Bureau of Census, COUNTY BUSINESS PATTERNS—1969 NEW YORK, Vol. CBP-69-34 (Washington, D.C.: U.S. Government Printing Office, July 1970), pp. 4-8.
(3) U.S. Dept. of Commerce, Bureau of Census, COUNTY BUSINESS PATTERNS-1969-NEW JERSEY, CBP-69-32 (Washington, D.C.: U.S. Government Printing Office, 1970), pp. 4-7.
(4) U.S. Dept. of Commerce, Bureau of the Census, 1967 CENSUS OF MANUFACTURES, Vol. 11, Industry Statistics, part 1, Major Groups 20-24 (Washington, D.C.: U.S. Government Printing Office, 1968).
(5) U.S. Dept. of Commerce, Bureau of the Census, 1967 CENSUS OF MANUFACTURES, Volume 11, Industry Statistics, Part 2, Major Groups 25-33 (Washington, D.C.: U.S. Government Printing Office, 1968).

Mohawk Valley

In the Mohawk region, in contrast to the Upper Hudson, industrial discharges play a much less important role than do municipal discharges. The population and industrial waste loading are shown in tables 8.6 and 8.7 respectively. As the Hudson River flows southward, the effects of population waste generation become an increasing factor. Industrial waste arises mostly from food processing and paper products in this region.

In the Mohawk region, a great deal is being done with the problem of municipal sewage treatment by the state and local governmental agencies.

Table 8-6
Waste Generation Mohawk Valley Region—Population

County	1980 Pop. Estimated	Lbs. BOD Generated (in thousands)
Fulton	51.0	3,164.5
Herkimer	73.6	4,566.8
Montgomery	54.5	3,381.7
Oneida	331.6	20,595.8
Total		31,688.8 lbs. of BOD

Source: Table 8-1 population values multipled by 62.05 pounds BOD per capita per year (i.e., 0.17 pounds BOD per day times 365).

Table 8-7
Waste Generation Mohawk Valley Region—Industrial

SIC Industry Category	Lbs. of BOD (in thousands)
2011	485.6
2013	670.2
2015	424.2
2026	213.0
2033	2,102.9
2037	350.6
2062	0
2085	0
2261	956.6
2262	454.9
2621	4,344.1
2631	1,043.8
2641	412.2
2654	4.2
2911	0
Total	11,482.3 lbs. of BOD

Sources:

(1) U.S. Dept. of Commerce, Bureau of Census, COUNTY BUSINESS PATTERNS, FIRST QUARTER 1959, Part 3A, Middle Atlantic States (Washington, D.C.: U.S. Government Printing Office 1961), pp. 4-8.

(2) U.S. Dept. of Commerce, Bureau of Census COUNTY BUSINESS PATTERNS—1969 NEW YORK, Vol. CBP-69-34 (Washington, D.C.: U.S. Government Printing Office, July 1970), pp. 4-8.

(3) U.S. Dept. of Commerce, Bureau of Census, COUNTY BUSINESS PATTERNS-1969- NEW JERSEY, CBP-69-32 (Washington, D.C.: U.S. Government Printing Office, 1970), pp. 4-7.

Table 8-7 (cont.)

(4) U.S. Dept. of Commerce, Bureau of the Census, 1967 CENSUS OF MANUFACTURES, Vol. 11, Industry Statistics, Part 1, Major Groups 20-24 (Washington, D.C.: U.S. Government Printing Office, 1968).

(5) U.S. Dept. of Commerce, Bureau of the Census, 1967 CENSUS OF MANUFACTURES, Vol. 11, Industry Statistics, Part 2, Major Groups 25-33 (Washington, D.C.: U.S. Government Printing Office, 1968).

Mid Hudson Region

The Mid Hudson region is an area of fairly good water quality characteristics. The municipal discharges are scattered and not particularly heavy. The industrial discharges are quite light, in comparison, with a few notable exceptions such as food processing and textiles. Tables 8.8 and 8.9 show the population and industrial wastes generated. It is agreed among water quality people that this is the area where the Hudson "cleans" itself by its natural assimilative capacity before it flows into the New York Metropolitan region.

As previously discussed, this Mid Hudson region is part of the "Hudson Corridor" and consequently is an area expected to substantially increase in population during the 1970s. This is also the area where the water is fairly clean and has its greatest possibilities for recreational activity as well as potential water supply. For these reasons, much careful consideration must be given to improving sewage treatment facilities in this area to meet demands of a rising population.

Table 8-8
Waste Generation Mid Hudson Region—Population

County	1980 Pop. Estimated	BOD (in thousands)
Columbia	54.9	3,406.5
Dutchess	317.9	19,725.7
Greene	35.7	2,215.2
Orange	394.4	24,472.5
Putnam	81.6	5,063.3
Ulster	172.4	10,697.4
Total		65,580.6 lbs. of BOD

Source: Table 8-1 population values multiplied by 62.05 pounds BOD per capita per year (i.e., 0.17 pounds BOD per day times 365).

Lower Hudson Region

The metropolitan area or Lower Hudson region is of concern to the entire country, and has been given much national media attention. The estimates shown in tables 8.10 and 8.11 give ample evidence of the cause for concern. However, the magnitude of the figures sometime tends to lull the mind to complacency. One must venture out into New York Harbor to appreciate the

Table 8-9
Waste Generation Mid Hudson Region—Industrial

SIC Industrial Category	Lbs. of BOD (in thousands)
2011	361.1
2013	498.4
2015	312.5
2026	158.4
2033	1,563.7
2037	260.7
2062	0
2085	0
2261	912.1
2262	452.8
2621	712.4
2631	1,711.9
2641	676.0
2654	7.0
2911	0
Total	7,629.0 lbs. of BOD

Sources:
(1) U.S. Dept. of Commerce, Bureau of Census, COUNTY BUSINESS PATTERNS, FIRST QUARTER 1959, Part 3A, Middle Atlantic States (Washington, D.C.: U.S. Government Printing Office, 1961), pp. 4-8.
(2) U.S. Dept. of Commerce, Bureau of Census, COUNTY BUSINESS PATTERNS-1969 NEW YORK, Vol. CBP-69-32 (Washington, D.C.: U.S. Government Printing Office, July 1970), pp. 4-8.
(3) U.S. Dept. of Commerce, Bureau of Census, COUNTY BUSINESS PATTERNS-1969-NEW JERSEY, CBP-69-34 (Washington, D.C.: U.S. Government Printing Office, July 1970), pp. 4-7.
(4) U.S. Dept. of Commerce, Bureau of the Census, 1967 CENSUS OF MANUFACTURES, Vol. 11, Industry Statistics, Part 1, Major Groups 20-24 (Washington, D.C.: U.S. Government Printing Office, 1968).
(5) U.S. Dept. of Commerce, Bureau of the Census, 1967 CENSUS OF MANUFACTURES, Vol. 11, Industry Statistics, Part 2, Major Groups 25-33 (Washington, D.C.: U.S. Government Printing Office, 1968).

Table 8-10
Waste Generation Lower Hudson Region—Population

County	1980 Pop. (Estimated)	Lbs. BOD (in thousands)
New York State		
Rockland	307.4	19,074.2
Westchester	1,138.2	70,625.3
Bronx	1,591.4	98,746.4
Kings	2,670.6	165,710.7
New York	1,478.2	91,722.3
Queens	2,181.3	135,349.6
Richmond	381.4	23,665.9
		604,894.4
New Jersey State		
Bergen	1,053.7	65,382.1
Essex	998.5	61,956.9
Hudson	612.1	37,980.8
Middlesex	842.8	52,295.7
Morris	571.9	35,486.4
Passaic	520.5	32,197.0
Somerset	283.3	17,578.7
Union	620.4	38,495.8
		341,473.4
Total		946,367.8 lbs. of BOD

Source: Table 8-1 population values multiplied by 62.05 pounds BOD per capita per year (i.e., 0.17 pounds BOD per day times 365).

deplorable condition of the water in this area. In this region, the blame and the responsibility becomes an interstate matter, with both New York State and New Jersey guilty as major contributors. The problems in this area are enormous and will require a great deal of consideration, effort, and money.

New Jersey becomes a significant contributor of both industrial and municipal pollutants starting at about mile point 9 from the Battery in New York Harbor. However, the large scale of discharges from the tributaries such as the Passaic River, Kill van Kull, and Newark Bay more than make up for the relatively short distance in which discharges occur. As indicated previously, the Lower Hudson region is a complex problem in itself and is a matter of national concern. The computations that resulted in tables 8.10 and 8.11 do more to suggest the complexity and magnitude of the problem than to provide a basis for detailed analysis of the area.

Only a fraction of the sewage generated in New York City and the New Jersey counties listed in table 8.10 is discharged into the Hudson. However,

Table 8-11
Waste Generation Lower Hudson Region—Industrial

SIC Industrial Category	Lbs. of BOD (in thousands) New York	Lbs. of BOD (in thousands) New Jersey	Total
2011	5,731.7	138.8	5,870.5
2013	7,781.3	5,984.5	13,765.8
2015	5,101.9	1,309.0	6,410.9
2026	2,502.3	701.8	3,204.1
2033	24,674.6	7,133.9	31,808.5
2037	4,108.9	7,085.4	11,194.3
2062	2,503.5	−0−	2,503.5
2085	750.0	844.1	1,594.1
2261	7,007.6	3,678.7	10,686.3
2262	8,229.1	32,195.0	40,424.1
2621	6,360.2	2,922.9	9,283.1
2631	15,240.1	37,327.7	52,567.8
2641	603.5	809.9	1,413.4
2654	63.3	1,194.4	1,257.7
2911	1,394.9	23,071.2	24,466.1
	92,052.9	124,397.3	216,450.2 lbs.

Sources:
(1) U.S. Dept. of Commerce, Bureau of Census, COUNTY BUSINESS PATTERNS, FIRST QUARTER 1959, Part 3A, Middle Atlantic States (Washington, D.C.: U.S. Government Printing Office, 1961), pp. 4-8.
(2) U.S. Dept. of Commerce, Bureau of Census, COUNTY BUSINESS PATTERNS—1969 NEW YORK, Vol. CBP-69-34 (Washington, D.C.: U.S. Government Printing Office, July 1970), pp. 4-8.
(3) U.S. Dept. of Commerce, Bureau of Census, COUNTY BUSINESS PATTERNS—1969—NEW JERSEY, CBP-69-32 (Washington, D.C.: U.S. Government Printing Office, 1970), pp. 4-7.
(4) U.S. Dept. of Commerce, Bureau of the Census, 1967 CENSUS OF MANUFACTURES, Vol. 11, Industry Statistics, Part 1, Major Groups 20-24 (Washington, D.C.: U.S. Government Printing Office, 1968).
(5) U.S. Dept. of Commerce, Bureau of the Census, 1967 CENSUS OF MANUFACTURES, Vol. 11, Industry Statistics, Part 2, Major Groups 25-33 (Washington, D.C.: U.S. Government Printing Office, 1968).

because of complex tidal actions in the waters of New York Harbor and its tributaries, sewage effluents entering waters some distance from the Lower Hudson may affect its quality. Moreover, wastes discharged from vessels have a great effect, direct if they are in the Lower Hudson, indirect if elsewhere in the waters of New York Harbor.

The figures shown as total BOD loadings in each region are indicative of the region, but cannot be regarded as exact quantitative measurement. Total figures throw light on a region in its entirety and provide a perspective or insight into each area. Individual point loadings are necessary also to reflect the effects on water quality in a particular area and the contributions it will have to the surrounding estuary. In understanding the water quality for a particular region, the assimilative capacity of the body of water must be given consideration also.

Government Regulation

In December 1964, Governor Nelson Rockefeller announced the Pure Waters Program a plan that was basically designed to improve the quality of water bodies of New York State during the early 1970s. This program was officially adopted by the New York State Legislature in the spring of 1965. In the election in November 1965, the voters approved a $1 billion bond issue for the construction of water treatment plants throughout the state. The target date to begin construction was 1972.

The Division of Pure Waters was created within the New York State Department of Health to implement the Pure Waters Program. It is essentially the duty of Pure Waters to formulate and maintain water quality practices and provide effective planning of water supply and waste disposal. It is also within their jurisdiction to conduct a water quality monitoring program throughout the state.

In devising water quality standards, the Division of Pure Waters has developed and submitted water quality standards to the Federal Water Pollution Control Administration under the Water Quality Control Act of 1965, which were subsequently approved. As part of maintaining these water quality standards, the previously mentioned surveillance program is carried out.

During the third session of the conference held concerning the pollution of the interstate waters of the Hudson River, a statement by Governor Rockefeller, read by Dr. Hollis S. Ingram, Commissioner of the New York State Department of Health [24:27-30], outlined an ambitious six-point program that was undertaken as part of the Pure Waters Program. Part of the program includes 164 waste treatment construction projects that were then under construction or completed and an additional 175 projects in the design and planning stages. One of these projects is the "North River" project, which is needed to treat sewage on the upper west side of Manhattan. However, this project has been a highly controversial political issue and is still far from settled.

Conclusions

What will be the quality of Hudson River waters as we approach 1980? A major conclusion of this chapter must be that there can be no one answer. It is fatuous to speak in broad general terms of the Hudson's being polluted or unpolluted. The river's capacity for rehabilitation, for restoration of its dissolved oxygen defenses renders any such simple answer unsatisfactory and meaningless. Just as the description and quantitative analysis had to be conducted in terms of specific sectors, so must be our statement of conclusions.

Upper Hudson River

The outlook for 1980 in the Upper Hudson region is for some improvement in water quality. Population is projected to rise by a modest 15 percent between 1970 and 1980, and with the State Pure Waters Program some upgrading of municipal facilities is to be expected. Industry, which might be expected to contribute about a third of the BOD loading (see Table 8.3), may actually fall well below this portion in part due to improved treatment facilities, but also due to the phasing out of some obsolete paper-producing facilities.

Mohawk Valley

Like the Upper Hudson region, the Mohawk Valley is expected to witness a relatively moderate population increase of population of about 62,000 or 14 percent (see table 8.1). Since this area has been the scene of intensive efforts to improve municipal sewage treatment facilities, progress should be evident by 1980. Industry, which is projected to account for about 37 percent of the BOD loading by 1980, is not expected to change this. Industrial developments in the area are not such as to alter this outlook.

Mid Hudson Region

By the Mid Hudson region the natural assimilative capacity of the river has restored water quality to a fairly good level. The nature of the present and likely future development of industry, which is expected to account for only 10 percent of the BOD load by 1980, is such that it presents few pollution problems. The major threat to an improved water quality outlook by 1980 might arise from a failure to improve treatment facilities sufficiently to accommodate the rapid increase in population. An increase of 332,000 or nearly 46 percent is projected between 1970 and 1980.

It will also be necessary to monitor the nutrient content of Mid Hudson water carefully to give adequate warning of danger from eutrophication which might threaten because of continued nutrient formation and sluggish flow-through of fresh water during much of the year.

Lower Hudson

Only the most incurable optimist could hope for any substantial improvement of water conditions in the Lower Hudson by 1980. The sewage effluent of 15 million people mostly treated to some degree, but some raw, contributes directly through discharge to Hudson waters or indirectly through tidal action to the river's problems. Any substantial progress in improving the conditions requires as a minimum the construction of the North River sewage treatment plant, plans for which have been bogged down in apparently insoluble political controversy. Even if this were accomplished, the Lower Hudson would still be heavily polluted by the discharge of many industrial polluters, particularly New Jersey enterprises. Accordingly, federal and interstate cooperative action, always difficult to achieve, will be required. Even international cooperation on the part of the fleet that discharges wastes in some of the waters of New York Harbor will be required to effect a really satisfactory situation.

These considerations mean that prospects are bleak for a readily discernible improvement in conditions in the Lower Hudson and New York Harbor by 1980. However, a substantial beginning might well be made by then. First of all, about 80 percent of the waste that contributes even indirectly to the BOD load in the Lower Hudson is traceable to population. Federal and state funds are available to aid the communities involved in improving facilities and there are few unsolved technical problems. The federal pressures on industrial sources of direct pollution of New York Harbor are just beginning and should begin to show results by 1980.

Coordinated Program for the Hudson

There are powerful reasons for coordinated programs of attack on pollution problems, especially those of the Lower Hudson. Unless rapid progress is made in reducing the BOD load on the Lower Hudson, tidal action may overcome progress in the Mid Hudson. This argues for: (a) giving New York City and New Jersey communities massive federal and state help in cleaning up their portions; and (b) making as rapid progress as possible in cleaning up New Jersey industrial sources. Finally, such a cooperative program should include constant monitoring for signs of eutrophication.

The rewards for cleaning up the Hudson would be so enormous as to be almost incalculable. Consider the recreational values for the millions, on or near its shores, of a Hudson fit to swim in from the headwaters to the mouth.

Notes

1. Quirk, Lawler, and Matusky, *Hudson River Water Quality and Waste Assimilative Capacity Study*, (December, 1970), pp. 4-8.

2. U. S. Dept. of Health, Education and Welfare, *Report on Pollution of the Hudson River and its Tributaries* (Washington, D.C., September 1965).

3. Hudson River Valley Commission, *Hudson River Ecology, Symposium* (October 4-5, 1966), pp. 60-82.

4. Dept. of Scientific and Industrial Research, *Effects of Polluting Discharges on the Thames Estuary*, Technical Paper No. 11 (London, September 1964).

5. J. H. Feth, "Nitrogen Compounds in Natural Waters — A Review," *Water Resources Research* 2, 1 (1966).

6. G. P. Howells, M. Eisenbud, and T. J. Kneip, "The Biology of the Hudson River," *Development of a Biological Monitoring System and a Survey of Trace Metals, Radionuclides and Pesticide Residues in the Lower Hudson River* (New York, October 1969).

7. New York State Dept. of Environmental Conservation, *Computer Printout* (1966).

8. New York State Office of Planning Coordination, *New York State Economic Outlook for the Seventies* (Albany, 1964).

9. New York State Division of the Budget, *New York State Statistical Yearbook — 1970* (April 1970).

10. New York State Dept. of Commerce, *Directory of Manufacturing and Mining Firms in New York State* (1958), pp.
New York State Industrial Directory, (1964), pp. U29-U192.
New York State Industrial Directory, (1971), pp. GG-1-GG-264.

11. New York State Dept. of Commerce, *Capital District — Business Fact Book, 1967-1968.*

12. New York State Dept. of Commerce, *Mohawk Valley Area — Business Fact Book, 1967-1968.*

13. E. B. Besselievre, *Industrial Waste Treatment* (New York: McGraw-Hill, 1952).

14. Regional Plan Association, *Waste Management-Generation and Disposal of Solid, Liquid, and Gaseous Wastes in the New York Region,* Bulletin 107 (New York, March 1968).

15. Robert H. Boyle, *The Hudson River* (New York: Norton, 1969).

16. Alfred J. Van Tassel, (Ed.), *Environmental Side Effects of Rising Industrial Output* (Lexington, Mass.: D.C. Heath & Co., 1970).

17. U. S. Dept. of Commerce, Bureau of Census, *1967 Census of Manufactures,* Vol. II, Industry Statistics, Part 2, Major Groups 25-33 (Washington, 1968).

18. U. S. Dept. of Commerce, Bureau of Census, *County Business Patterns,* First Quarter 1959, Part 3A, Middle Atlantic States (Washington, 1961).

19. U. S. Dept. of Commerce, Bureau of Census, *County Business Patterns — 1969 New York,* Vol. CBP-69-34 (Washington, 1970).

20. U. S. Dept. of Commerce, Bureau of Census, *County Business Patterns — 1969 New Jersey,* Vol. CBP-69-32 (Washington, 1970).

21. U. S. Dept. of Commerce, Bureau of Census, *U. S. Census of Population, 1950* (Washington, 1951).

22. U. S. Dept. of Commerce, Bureau of Census, *U. S. Census of Population, 1960* (Washington, 1961).

23. U. S. Dept. of Commerce, Bureau of Census, *U. S. Census of Population, 1970* (Washington, 1971).

24. U. S. Dept. of Interior, *Conference in the Matter of Pollution of the*

Interstate Waters of the Hudson River and its Tributaries – New York and New Jersey (Washington, June 1969).

25. New York State Office of Planning Coordination, *Demographic Projections for New York State Counties to 2020 A.D.* (August 1969).

26. New Jersey Dept. of Conservation and Economic Development, *1969 Population Estimates for New Jersey* (Trenton, 1969).

27. New York State Dept. of Environmental Conservation, *Water Quality Surveillance,* 390 pages.

9 The Lower Mississippi

ROBERT SCHWARTZ

The Mississippi River has been described in many ways and many times as the backbone of America both physically and functionally. In geographical position the river begins as far north as Lake Itasca in Minnesota. From there it winds its way through the heart of the United States, eventually emptying into the Gulf of Mexico some 2,500 miles away. The river drains an area of 1.2 million square miles stretching from the Appalacian Mountains on the east to the Rocky Mountains on the west.

From the time La Salle floated down the river in 1673 as well as eons prior, the Mississippi has played an important role in the development of our country. It has served as drinking water for both human and animal consumption, as an irrigation system to foster the growth of food crops, as an easily accessible transportation route, as a source of recreation, and as a readily available prime site for industry.

Unfortunately the river has come to serve one other purpose since the advent of man and his industry that is already causing the eradication of the above-mentioned uses. That purpose is to receive the large volume of wastes generated by man and his ever-increasing modern industrial society.

The basic purpose of this chapter is to examine this society, the wastes it generates, and the effect of these wastes on the lower portion of the river. It will be a statistical analysis using data compiled by various government agencies. It will include an analysis of the present level of pollution and an extension of the present growth and pollution trends to the year 1980.

Water pollution in the Mississippi River from St. Louis to the Delta can be broken down into two basic categories: natural pollution and man-made pollution.

Natural pollution in the form of sedimentation has always been a problem on the river. This sedimentation problem is an outgrowth of the natural physical phenomenon of rapid water flow that results in the constant erosion of the banks and bottom of the river coupled with land drainage. It serves to give the Mississippi its characteristic of "muddy" and creates the tremendous delta that exists at the mouth of the river as 400 million tons of silt are washed down each year.

A United States Geological Survey Study of suspended sediment discharges at St. Louis for the period 1948 to 1958 showed discharges ranging from 4,250 to 7,010,000 tons per day. The average discharge was 496,000 tons per day.

A U. S. Department of Agriculture estimate of sediment yield for the lower Mississippi River showed a high of 8,210 tons per square mile per year, a low of

1,560 tons, and an average of 5,200 tons. These figures are the highest for any river in the United States.

New flood control methods and better agricultural techniques are being used to help correct this problem, but the task is substantial. For the future it appears that this type of pollution will remain at its present level which, while creating certain problems in water use, is not a terminal threat to the river.

The above types of "natural" pollution previously the major source of problems have long since been replaced by the advent of man into the Mississippi River valley. Lured on by the promises of rich lands for production and a swift river for transportation of those products, in the beginning mostly agricultural, men swarmed along its banks bringing waste after waste. As the towns and cities sprang up along its banks more and more sewage systems were dumping directly into the river. All along the way raw sewage and human wastes were dumped directly into the river without the thought of treatment. The amazing fact is that this same waste basin supplied the major cities with their drinking water. People remained, however, relatively unconcerned.

The old adage that "dilution solves all pollution" was the keynote that sparked the growth of industries of all kinds, a growth that has continued to skyrocket. As late as 1964, for example, St. Louis, Memphis, Vicksburg, Natchez, and New Orleans were still pouring out raw waste materials into the river while at the same time continuing to use that very same source as the origin of its drinking water. As the population increased and industry grew, the waste loads continued to elevate.

To analyze and quantify this pollution the wastes can be divided into two basic types: industrial wastes and municipal wastes. In order to calculate the effect of these wastes and determine what the future water quality of the river will be, a quantitative analysis must be made. To do this the river has been divided into six sections, with a water quality measuring station at the southern point of each section. These stations are located at St. Louis, Missouri; Cape Girardeau, Missouri; West Memphis, Arkansas; Vicksburg, Mississippi; St. Francisville, Louisiana; and New Orleans, Louisiana.

In using this method of segmenting the river there is one point that should be noted; that is that pollutants in one segment of the river or in a tributary do not affect the pollution levels in the next segment. While this is not entirely true, it is a fairly accurate assumption due to the ability of the river to clean itself given some distance before the entry of more pollutants.

It must be determined which geographical areas are contributing primarily to the pollution in the river. After an analysis of the region, 97 counties have been chosen that deposit wastes directly into the Mississippi. Figure 9.1 shows a map containing this area.

The next step is to quantify the waste loads being delivered into the river by these counties. This is done by calculating the amount of pollutants delivered by each major industry and city within the given segment of the river.

The municipality category relates primarily to the nonindustrial sewage waste discharged by the many cities along the river. The best measure of this type of waste is to relate it directly to population. Table 9.1 shows the urban population of the six regions for 1950 and 1960 with projections through 1980.

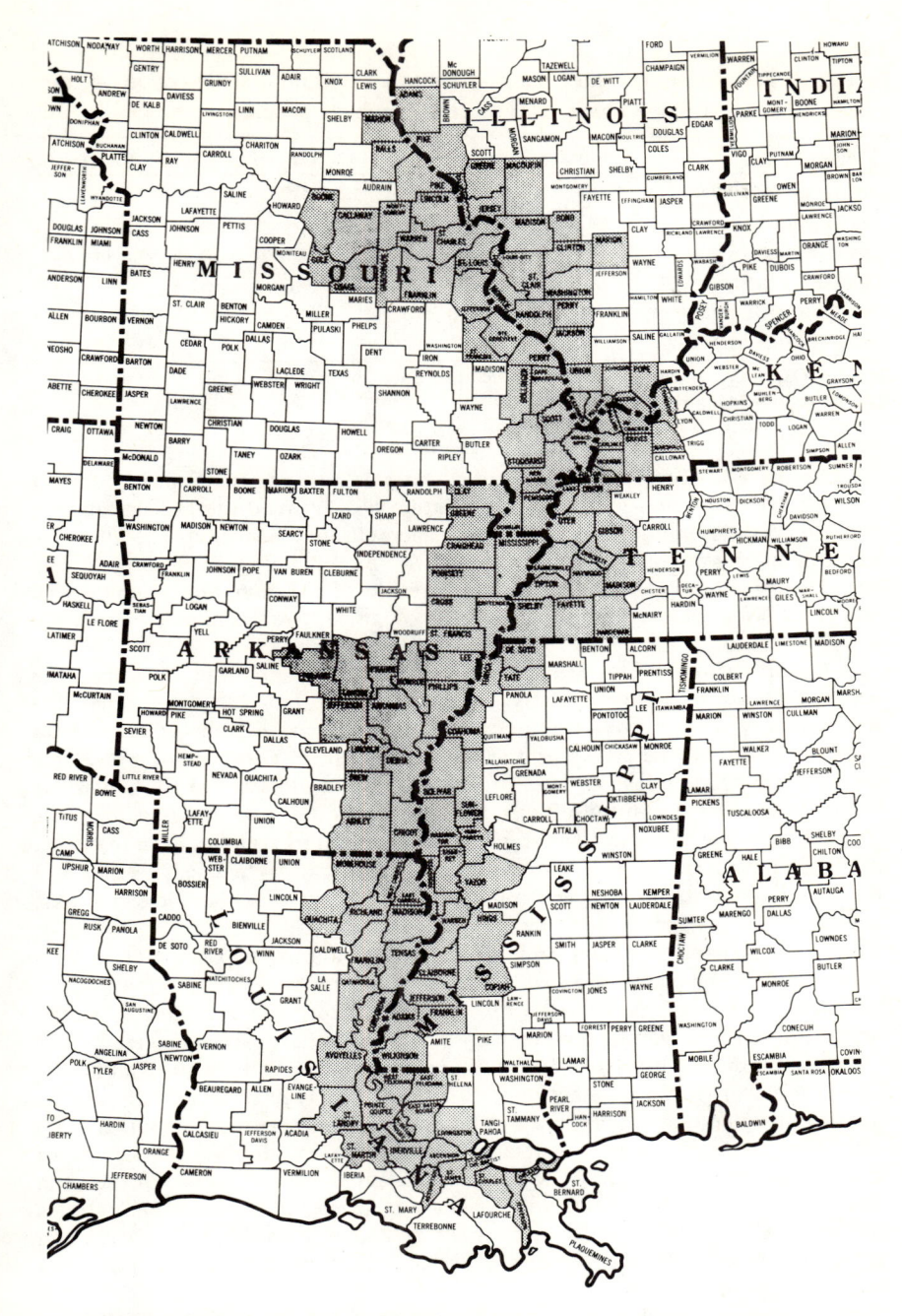

Figure 9.1. Regions Depositing Waste Directly Into the Lower Mississippi River.

Table 9-1
Projected Population for the Selected Counties Within Each Section of the Lower Mississippi River Through 1980 (thousands of people)

Section Ending At	1950	1960	1970 (est.)	1980 (Projected)
St. Louis	2,054	2,392	2,730	3,068
Cap Girardeau	243	279	315	351
West Memphis	1,381	1,469	1,557	1,645
Vicksburg	783	807	834	861
St. Francisville	615	710	805	900
New Orleans	1,000	1,271	1,542	1,813

Source: U.S. Department of Interior, Bureau of the Census 1950, 1960. 1970 and 1980 figures are based on a straight line projection of 1950 and 1960 actual population data.

A figure of .17 pounds of Biological Oxygen Demand (BOD) per capita per day [5:96] had been derived as the relationship between the population of a region and the nonindustrial waste load generated by that region. Therefore, by multiplying the population figures in table 9.1 by .17 pounds of BOD per capita per day, or 62.05 pounds per year, a measure of the potential nonindustrial waste load for each section of the river is obtained. Results of this calculation are shown in table 9.2.

While municipal wastes are a significant factor affecting the Mississippi, the waste load generated by industry is of equal importance in determining the water quality of the river. Hundreds of industries have established themselves and prospered on the banks of the Mississippi River. The type of processing varies greatly from industry to industry. Some do not use water at all. Others use water, but without harmful effects. Still others use the river waters and in the process harm the water in some way. Either dissolved oxygen has been removed, pH levels have been altered from the neutral 7.0, harmful chemicals are put into the water, unpleasant tastes and odors exist, the water is discolored, or toxic substances are present in the water.

In a well-balanced system none of these characteristics exists. But as such altered conditions occur, the availability of the water system for municipal and industrial water supplies, irrigation, transportation, animal drinking water, support of fish and wildlife, and recreational facilities becomes endangered.

In a moderately unbalanced system, wastes are entering the water that impair one or more of the above-mentioned functions. The type of waste entering the water will determine which uses will be impaired. Sewage wastes contain great quantities of organic matter that act as a fertilizer in speeding the growth of plants and algae in the water. The effect of this rapid growth can be a decrease in dissolved oxygen, which affects the balance of species in the water and affects the pH level. A natural waste such as silt and sand may affect navigation and perhaps some recreational activities. A waste rich in minerals may ruin the river as a source of drinking water.

In a severely unbalanced system the water is usually not fit for anything beyond navigation. The water is so toxic that animal life cannot exist, dissolved oxygen is very low, and strong odors and tastes are present.

Table 9-2
Projected Nonindustrial Waste Load Generated Within Each Section of the Lower Mississippi River Through 1980 (million lbs. of BOD per year)

Section Ending Act	1960	1970	1980
St. Louis	148.4	169.4	190.4
Cape Girardeau	17.3	19.5	21.8
West Memphis	91.2	96.6	102.1
Vicksburg	50.1	51.7	53.4
St. Francisville	44.1	50.0	55.9
New Orleans	78.9	95.7	112.5

Source: Population from Table 9-1 multiplied by 62.05 pounds BOD per year, WASTE MANAGEMENT (Regional Plan Association, N.Y.), p. 96.

Except for isolated cases, the Mississippi has not reached this extreme state of imbalance, but industrial pollution has substantially added to the imbalance that does exist.

Because of the different ways each industry uses water, the simple relationships to calculate municipal wastes cannot be used here. Instead, each industry must be analyzed separately to determine the extent of the waste it delivers into the river. As with municipal wastes, industrial wastes have been measured in terms of BODs.

In analyzing the industrial waste load for each section of the river, thirty industries were selected that are known to contribute to the BOD level.

Table 9.3 shows the derivation of the formula used to quantify the potential BOD contribution per industry per section of river. This procedure is essentially the same as that described in Chapter 2. In using this formula, trend lines were used to project the BOD level for 1970 and 1980 based on the data available for 1958, 1963, 1967, and 1969.

Combining the results of applying this formula and the nonindustrial waste loads calculated earlier gives tables 9.4-9.9. Table 9.10 summarizes the findings in these six tables.

In 1970 the potential BOD load on the Lower Mississippi passed one billion pounds, a staggering level for a river that has just recently been recognized as having a possible pollution problem.

The key factor that has allowed this gross neglect of the river is the dilution effect caused by the tremendous rate of flow down the river. Table 9.11 gives the flow rate at various points along the river.

Until recently this massive flow of water was sufficient to dilute the effect of industrial and municipal wastes poured into the river, as well as reduce organic concentrations and eutrophication caused by agricultural runoffs.

However, even the mighty Mississippi began to experience adverse effects from the massive waste load that was flowing into it. St. Louis and New Orleans have experienced periodic color and odor problems in their drinking water. There have been numerous fish kills and cases of inedible fish. Dissolved oxygen levels have dipped dangerously low at various times, especially in the St. Louis region.

Table 9-3
Derivation of Formula to Quantify Potential BOD Contribution for Each Industry Per Section of the River

LBS. of BOD Per Industry Per Section of the River	=	BOD per unit of output per industry	X	Total unit output per industry per section of the river	
	=	BOD per unit of output per industry	X	Total dollar output per industry per section of the river / Dollars per unit of output per industry	
	=	BOD per unit of output per industry	X	No. Employees per industry per section of the river	X Dollar output per employee per industry / Total U.S. dollar output per industry / Total U.S. units of output per industry
	=	BOD per unit of output per industry	X	No. employees per industry per section of the river	X Dollar output per industry per state bordering river / No. employees per industry per state bordering river / Total U.S. dollar output per industry / Total U.S. units of output per industry

Sources:

BOD per unit of output per industry	— Regional Plan Association, WASTE MANAGEMENT, New York, New York, March 1968, pp. 95-96.
	— The Cost of Clean Water—Industrial Waste Profile No. 1: Blast Furnaces and Steel Mills, Volume III, p. 79.
	— The Cost of Clean Water—Industrial Waste Profile No. 10: Plastics and Resins, Volume III, pp. 15-35.
No. of employees per industry per section of the river and per state bordering the section of the river	— Dr. Gould—Economic Information Systems Industrial Employment Statistics by Four-digit SIC codes for 1963, 1967, 1969.
	— U.S. Department of Commerce, Bureau of the Census, County Business Patterns for Illinois, Tennessee, Arkansas, Mississippi, Missouri, Louisiana, and Kentucky, 1958, 1963, 1967.
Dollar output per industry per state and per total U.S.; Unit output per total U.S.	— U.S. Department of Commerce, Bureau of the Census, Census of Manufactures, 1963, 1967.

Finally, about seven years ago, policies began to change concerning pollution in the river by the states involved. Capital expenditures for sewage treatment facilities were increased and laws were passed requiring certain treatment of industrial wastes before they could enter the river.

Table 9-4
Projected Potential BOD Load on the St. Louis Section of the Lower Mississippi River for 1970 and 1980 (Million Lbs. of BOD)

SIC Code	Description	1970	1980	% Increase
2011	Meat Packing	23.4	27.6	
2631	Paperboard Mills	20.2	22.5	
2818	Organic Chemicals	14.0	18.2	
2819	Inorganic Chemicals	43.8	53.6	
2911	Petroleum Refining	13.3	17.0	
3312	Blast Furnace and Steel Mills	34.3	36.6	
3321	Gray Iron Founderies	7.9	9.7	
3323	Steel Founderies	16.5	16.5	
	Other Industries	7.7	10.6	
	Total Industrial BOD Load	181.1(51.7%)	212.3(52.7%)	16.7%
	Total Nonindustrial BOD Load	169.4(48.3%)	190.4(47.3%)	12.4%
	Total BOD Load	350.5(100.0%)	402.7(100.0%)	14.9%

Sources: Industrial–see Table 9-3. Nonindustrial–see Table 9-2.

Table 9-5
Projected Potential BOD Load on the Cape Girardeau Section of the Lower Mississippi River for 1970 and 1980 (Million Lbs. BOD Per Year)

SIC Code	Description	1970	1980	% Increase
2819	Inorganic Chemicals	4.7	7.3	
	Other Industries	3.3	3.6	
	Total Industrial BOD Load	8.0(29.1%)	11.9(32.3%)	48.7%
	Total Nonindustrial BOD Load	19.5(70.9%)	21.8(67.7%)	11.7%
	Total BOD Load	27.5(100.0%)	33.7(100.0%)	22.5%

Sources: Industrial–see Table 9-3. Nonindustrial–see Table 9-2.

The effect of this change can be seen in table 9.12, which shows the percentage of waste water treated for the prime polluting industries in 1963 and 1967. It should be noted that the data in this table do not detail the type of treatment performed. It, therefore, can be used only to show trends in pollution control and not the quantitative reduction in the waste load as a result of the treatment.

Table 9-6
Projected Potential BOD Load on the West Memphis Section of the Lower Mississippi River for 1970 and 1980 (Million Lbs. of BOD Per Year)

SIC Code	Description	1970	1980	% Increase
2011	Meat Packing	4.7	6.5	
2621	Paper Mill Except Building	6.7	9.5	
2818	Organic Chemicals	22.6	29.0	
2819	Inorganic Chemicals	7.7	10.5	
2911	Petroleum Refining	5.2	7.0	
3111	Leather Tanning and Finishing	5.2	7.0	
3312	Blast Furnace and Steel Mills	3.6	4.9	
3321	Gray Iron Foundries	6.8	7.2	
	Other Industries	4.5	5.2	
Total Industrial BOD Load		67.0(40.9%)	86.8(45.9%)	29.5%
Total Nonindustrial BOD Load		96.6(59.1%)	102.1(54.1%)	5.7%
Total BOD Load		163.6(100.0%)	188.9(100.0%)	15.5%

Sources: Industrial—see Table 9-3. Nonindustrial—see Table 9-2.

Table 9-7
Projected Potential BOD Load on the Vicksburg Section of the Lower Mississippi River for 1970 and 1980 (Million Lbs. of BOD Per Year)

SIC Code	Description	1970	1980	% Increase
2621	Paper Mill Except Building	39.3	53.0	
2631	Paperboard Mills	26.2	28.0	
2654	Sanitary Food Containers	3.3	3.8	
2819	Inorganic Chemicals	9.2	11.0	
3323	Steel Foundries	4.1	5.0	
	Other Industries	5.0	7.8	
Total Industrial BOD Load		87.1(62.8%)	108.6(67.0%)	24.7%
Total Nonindustrial BOD Load		51.7(37.2%)	53.4(33.0%)	3.3%
Total BOD Load		138.8(100.0%)	162.0(100.0%)	16.7%

Sources: Industrial—see Table 9-3. Nonindustrial—see Table 9-2.

Table 9-8
Projected Potential BOD on the St. Francisville Section of the Lower Mississippi River for 1970 and 1980 (Million Lbs. of BOD Per Year)

SIC Code	Description	1970	1980	% Increase
2621	Paper Mill Except Building	7.7	12.4	
2631	Paperboard Mills	37.0	48.0	
	Other Industries	6.7	9.9	
	Total Industrial BOD Load	51.4(50.7%)	70.3(56.1%)	36.8%
	Total Nonindustrial BOD Load	50.0(49.3%)	55.9(43.9%)	11.8%
	Total BOD Load	101.4(100.0%)	125.2(100.0%)	23.5%

Sources: Industrial—see Table 9-4. Nonindustrial—see Table 9-2.

Table 9-9
Projected Potential BOD on the New Orleans Section of the Lower Mississippi River for 1970 and 1980 (Million Lbs. BOD Per Year)

SIC Code	Description	1970	1980	% Increase
2011	Meat Packing	3.7	5.5	
2061	Raw Cane Sugar	3.7	4.8	
2818	Organic Chemicals	47.9	59.0	
2819	Inorganic Chemicals	44.9	52.0	
2911	Petroleum Refining	28.5	34.0	
	Other Industries	9.1	11.6	
	Total Industrial BOD Load	137.8(59.0%)	166.9(59.7%)	21.1%
	Total Nonindustrial BOD Load	95.7(41.0%)	112.5(40.3%)	17.6%
	Total BOD Load	233.5(100.0%)	279.4(100.0%)	19.7%

Sources: Industrial—see Table 9-3. Nonindustrial—see Table 9-2.

Table 9-10
Summary of Tables 9-4—9-9

Section	Total Industrial BOD Load (millon lbs. BOD)		Total Nonindustrial BOD Load (million lbs. BOD)		Total BOD Load	
	1970	1980	1970	1980	1970	1980
St. Louis	181.1	212.3	169.4	190.4	350.5	402.7
Cape Girardeau	8.0	11.9	19.5	21.8	27.5	33.7
West Memphis	67.0	86.8	96.6	102.1	163.6	188.9
Vicksburg	87.1	108.6	51.7	53.4	138.8	162.0
St. Francisville	51.4	70.3	50.0	55.9	101.4	125.2
New Orleans	137.8	166.9	95.7	112.5	233.5	279.4

Source: Tables 9-4 through 9-9.

Table 9-11
Average Water Discharges in the Lower Mississippi River Basin Between 1958-1967

Measuring Point	Discharge (cubic feet per second)
St. Louis	174,000
Cape Girardeau	176,000
Memphis	410,000
Vicksburg	489,000
St. Francisville	406,000
New Orleans	406,000

Source: H.W. Poston and W.C. Galegar, "Pollution in the Mississippi River," Figure I.

Table 9-12
Percentage of Waste Water Treated by Major Polluting Industries Along the Mississippi River for 1963 and 1967

Section of the River and Industry	1963	1967
St. Louis		
Food and kindred products	3.0	9.9
Paper products	18.5	30.0
Chemical products	12.7	16.6
Petroleum products	19.4	20.3
Primary metals	25.4	39.2
West Memphis		
Food and kindred products	2.5	14.0
Paper products	40.0	60.0
Chemical products	18.0	25.0
Vicksburg		
Paper products	65.0	90.0
Chemical products	12.0	20.0
St. Francisville		
Paper products	14.0	65.0
New Orleans		
Food and kindred products	20.0	22.0
Chemical products	10.0	15.0
Petroleum products	50.0	75.0

Source: U.S. Department of Commerce, Bureau of the Census, CENSUS OF MANUFACTURES, 1963, 1967.

However, even with the improvements made, pollution control at the end of the 1960s was far from a complete success. In fact, in 1969 only 30 percent of industrial and municipal wastes received adequate treatment before going into the river.

The basic problem that persisted through the 1960s was the lack of an overall program to control pollution in the Lower Mississippi. This, coupled with a reluctance by some states to spend the money for municipal treatment facilities and force industries to perform treatment on their wastes, has effectively negated the concern over pollution that began in the 1960s.

The end of the decade, however, saw the federal government step into the picture in the form of the Federal Water Quality Administration. For the first time, there existed an organization that could make plans for the entire region, as well as require states and industries to make the expenditures necessary to control pollution.

Their job will be a substantial one. Well over one billion dollars will be needed in the next decade to provide adequate municipal and industrial waste treatment facilities [11: Vol. I, 27-59]. With the problems that have already occurred, it is obvious that the river is close to the saturation point; and this money will have to be spent if the Mississippi is to remain the life force of this region.

If the treatment facilities are not improved and the increase in the potential waste load projected here developes, severe problems, especially in the St. Louis and New Orleans regions, could result. The costs will be considerably higher if that is allowed to happen, and the services the river provides will be substantially reduced.

Notes

1. *Encyclopedia Britannica,* Volume 15, 1957.

2. H. W. Poston and W. C. Galegar, "Pollution in the Mississippi River," paper presented at winter meeting of the American Institute of Chemical Engineers, New Orleans, Louisiana, March 16-19, 1969.

3. U. S. Department of the Interior, Geological Survey, Water Resources Data for Illinois, Tennessee, Arkansas, Mississippi, Missouri, Louisiana, and Kentucky, 1964-1969.

4. U. S. Department of Commerce, Bureau of the Census, County Business Patterns for Illinois, Tennessee, Arkansas, Mississippi, Missouri, Louisiana, and Kentucky, 1958, 1963, 1967.

5. Regional Plan Association, *Waste Management* (New York, New York, March 1968).

6. U. S. Department of the Interior, Federal Water Pollution Control Administration, South Central Region, *Endrin Pollution in the Lower Mississippi River Basin,* June 1969.

7. U. S. Department of the Interior, Federal Water Quality Administration, "Clean Water for the 1970s" Status Report, June 1970.

8. U. S. Department of Interior Geological Survey, *Suspended Sediment Discharges at St. Louis,* 1948-1958.

9. U. S. Department of Agriculture, *Sediment Yields for U. S. River Basins,* 1965.

10. U. S. Department of the Interior, Bureau of the Census, 1950, 1960 Census.

11. U. S. Department of the Interior, Federal Water Pollution Control Administration, *The Economics of Clean Water.* Summary, Volume I — Detailed Analysis, Volume II — Animal Wastes Profile, March 1970.

12. U. S. Department of the Interior, Federal Water Pollution Control Administration, *The Cost of Clean Water,* Volume I — Summary Report January 1968; Volume III — Industrial Waste Profiles 1, 3-9, Sept. 1967.

13. Dr. Gould, Economic Information Systems, Industrial Employment Statistics by Four-digit SIC codes for 1963, 1967, 1969.

14. U. S. Department of the Interior, Bureau of the Census *Census of Manufacturers* 1963, 1967.

10 Savannah River

ROGERS R. FRENCH

The Savannah River is one of the principal waterways in the South. It begins at the confluence of the Tugaloo and Keowee Rivers at Hartwell Dam and empties into the Atlantic Ocean. For 310 miles it forms the boundary between Georgia and South Carolina.

The area covered by this chapter includes the lower 28 miles of the river. The river flows past the City of Savannah at river mile 14.5 and then branches off to form the North and South Channels at river mile 10.0 From there the channels proceed to the Atlantic Ocean.

The regions that feed the river receive an average annual rainfall of 46 inches. Only the Pacific Northwest has a higher annual rainfall [18:1-7]. The runoff from this rainfall creates an average daily flow in the river of 7.2 billion gallons of water, or about 10,000 gallons per day for each person living on the river's banks [19:9].

Today, in the study area, the main channels of the Savannah River are used for industrial water supply, navigation, and waste disposal. In past years, these and adjacent waters were also used for commercial fishing and shellfish harvesting.

Ocean-going vessels are able to navigate 20 miles up the river in order to reach the City of Savannah and the industrial area in Chatham County. This is made possible by the Savannah Corps of Engineers who must constantly dredge the river to maintain a width of 200 feet and a depth of 35 to 40 feet.

Over fifty industries with varying wastewater discharges are located in Chatham County from river mile 10 to 21. These industries include pulp and paper mills, marine terminals for asphalt and flammable petroleum products, a sugar refinery, plants producing naval stores, asphalt roofing and pressure creosoting plants, fertilizers, and chemical products.

The cities of Savannah, Savannah Beach, Port Wentworth, and Garden City, with a combined total population of over 350,000, are located in the study area. Of this population, 146,000 are served by sewers that discharge to the Savannah River.

In contrast, the South Carolina side of the river is mostly marshland and sparsely populated. The total population of Jasper County, South Carolina, was 12,730 in 1965 [17].

Sewage

One of the principal sources of pollution is sewage. From the time of its founding by James Oglethorpe in 1733, the City of Savannah has used the river for sewage disposal. In those days, the city's population was quite small so the human wastes were conveniently assimilated. However, as the population increased, the river's ability to absorb the wastes decreased. Today we suffer the result of over 200 years of abuse.

Sewage contains many constituents that have adverse effects on water quality in the Lower Savannah River. These effects can be described and measured in terms of bacterial and oxygen demand population equivalents. A population equivalent is the amount of oxygen-consuming pollutant that one person would produce as sewage in one day.

To understand the present problem, it is necessary to go back to 1965. All the known sewage discharges enter the river from the City of Savannah, Savannah Beach, and the surrounding industrial area.

The total sewered population is estimated to be 194,000. Approximately 25 percent of this total sewered population is located in areas that are drained toward the intrastate Vernon, Wilmington, and Little Ogeechee Rivers in the State of Georgia. Raw and treated sewage from the remaining 75 percent of the sewered population discharges to the Savannah River [13:22].

From the figures in table 10.1 it can be seen that less than one-fifth of the sewage receives primary treatment, which reduces the total amount of bacteria by nearly 7 percent. The other four-fifths of the sewage enters the river in its natural state. Thus, a population equivalent of 146,200 discharges its sewage into the lower portion of the Savannah River.

The main offender is the City of Savannah, which contributes almost 80 percent of the waste constituents entering the river. In population equivalents this is equal to approximately 108,000. The next most serious contributor is Savannah Beach, which is responsible for over 10 percent of the problem. Port Wentworth, Travis Field, Garden City, Chatham City, and Cloverdale account for the remainder of the untreated sewage.

Bacteria in Sewage

The bacteria from human sewage is another important aspect of pollution. A population of approximately 146,200 people discharges its sewage, most of it untreated, to the Savannah River. This sewage contains millions of pathogenic bacteria that can cause dysentery, hepatitis, diarrhea, or typhoid fever if water containing them is swallowed. For this reason many portions of the river and adjoining streams have been declared off limits for swimming and drinking.

In many cases aquatic animals have become infected with bacteria. If the diseased animals are eaten by humans the bacteria can be passed on to them. Oysters are notorious in their ability to harbor several forms of pathogenic

Table 10-1
Estimated Sewage Discharges Entering the Savannah River

Jurisdiction	Type of System*	Type of Treatment	Population Served	Population Equivalents				
				Oxygen Demand		Bacteria		
				Number	% of Total	Number	% of Total	
Port Wentworth	S	None	3,700	3,700	2.7	3,700	2.8
Garden City	S	None	5,200	5,200	3.8	5,200	4.0
Savannah Beach	C	Primary	25,000	16,700	12.3	12,500	9.5
City of Savannah	C	None	108,000	108,000	79.6	108,000	82.4
Other Areas	S	–	4,300	2,000	1.5	1,600	1.2
Total to Savannah River			146,200	135,600		131,000	

Source: U.S. Department of Health, Education and Welfare, CONFERENCE ON POLLUTION OF INTERSTATE WATERS OF THE LOWER SAVANNAH RIVER AND ITS TRIBUTARIES, SOUTH CAROLINA–GEORGIA (February, 1965), p. 22a.

*Type of System: S = Separate Sanitary Sewers
C = Combined Sanitary and Storm Sewers

bacteria. For this reason hundreds of acres of oyster beds in and near the river have been condemned for almost forty years. The areas where fishing was permitted have also faced similar restrictions in recent years.

Certain measures were taken prior to 1965 to reduce the bacterial content in the sewage. However, the measures only succeeded in reducing the bacteria by 10 percent. Thus, the sewage still contained the bacterial population equivalent to 131,000 people. The estimated bacterial content from the various sources of sewage, with estimated allowances for any treatment, are presented in table 10.1. A bacterial population equivalent is equal to the quantity of coliform bacteria added daily to a stream by the untreated sewage discharged by one person.

The raw sewage also has a detrimental effect on the biochemical oxygen demand (BOD) equal to a population equivalent of 146,200. The treatment provided reduces the oxygen demand by 7 percent with a resulting population equivalent of 135,600. The main offender is the city of Savannah, which accounts for almost 80 percent of the oxygen demand (see table 10.1).

Neither Georgia nor South Carolina has adopted total coliform standards for swimming, water skiing, boating, and fishing in the Savannah River. However, if the standards adopted for other bodies of water in the states are used, the seriousness of the problem becomes more evident.

The standards are expressed in terms of most probable numbers (MPN) per 100 milliliters of water. In everyday terms 100 milliliters is slightly more than one-half cup.

The commonly used limit for water contact sports is 1,000 per 100 milliliters (ml) of water. Where total coliform objectives for general recreational use have been adopted, an arithmetic average of 5,000 MPN per 100 milliliters (ml) is the commonly used upper limit. . . . Waters subject to sewage pollution are closed to shellfishing by Georgia and South Carolina State Health Departments, where sanitary surveys indicate that sewage may reach the waters, or where the median total coliform MPN value representative of the waters of the shellfish-growing areas exceeds 70/100 ml [13:37-8].

Figure 10.1 shows the average MPN values for total and fecal coliform bacteria in the Lower Savannah River during August 1963. From river mile 24 to the Atlantic Ocean, the total coliform density exceeded 5,000 MPN per 100 milliliters 10 percent of the time. It reached its peak of 62,000 at river mile 10 in the heart of the city of Savannah. At no point in the study area did the line meet the 70 MPN/100 ml median set forth by the Georgia and South Carolina Health Departments for shellfishing.

A properly functioning primary sewage treatment plant can remove nearly 35 percent of the oxygen-demanding substances and about 50 percent of the bacteria from the untreated sewage. If this process is carried another step, secondary treatment, it could remove approximately 90 percent of the oxygen-demanding substances and 95 percent of the remaining bacteria. No process is 100 percent effective, but if primary and secondary treatment were used the results would certainly have been better than the present reductions of 7 percent

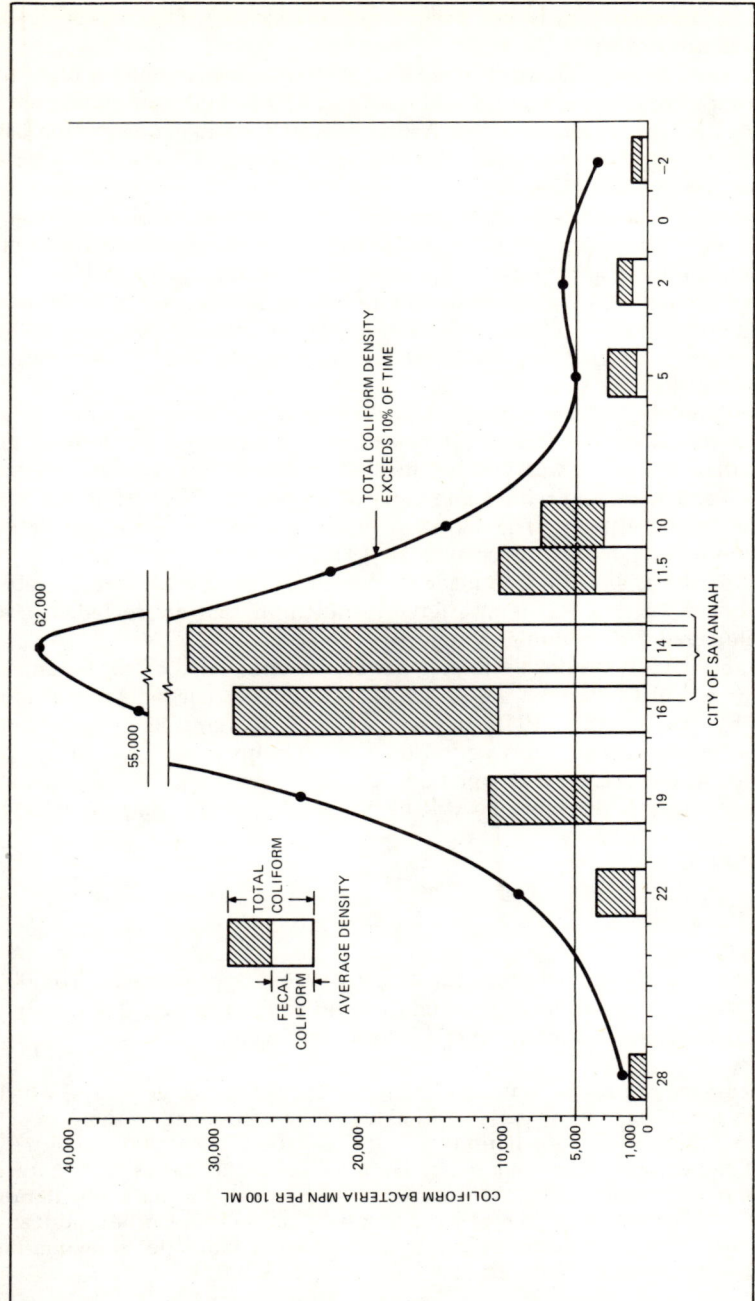

Figure 10.1. Coliform Bacteria, Savannah River Estuary (Average Densities—MPN per 100 ml in Main Channel). Source: U.S. Department of Health, Education and Welfare, *Conference on Pollution of the Interstate Waters of the Lower Savannah River and Its Tributaries, South Carolina-Georgia* (February 1965), p. 38a.

in oxygen-demanding substances and 10 percent in bacteria.

Why have these standards not even been approached? The answer again demands a return to 1965.

In 1965 the city of Savannah learned that it would have to build a plant to treat its sewage. Initially the plant would have cost $10,800,000 and should have been finished by the middle of 1968. At the present time the construction has not even begun, and the estimated cost has risen to over $21,000,000 with a completion date in late 1972.

The reasons for the delay can be summarized as economical, legal, technological and, most important of all, political. The Savannah politicians are by no means unique in this respect; many other areas suffer from the same malady.

A sewage system is not a beautiful thing to behold. A new civic center is, and that is why the sewage system was delayed. The politicians, seeking to be reelected, wanted to give the voters something tangible, and a new sewage system was not the answer.

The situation might have continued unchanged if the second session of the Lower Savannah River Pollution Conference had not been held on October 29, 1969. At that session it was decided that all municipalities were to provide secondary waste treatment with a minimum of 85 percent BOD reduction and disinfection of the effluent. The necessary facilities were also to be completed and in operation no later than December 31, 1972.

A good start has already been made in this direction, as evidenced by table 10.2. It should be noted that all of the jurisdictions that are included have already completed their preliminary reports.

From table 10.2 it is possible to see that all the cities cited by the Savannah River Enforcement Conference are on schedule to provide adequate wastewater treatment by December 31, 1972. In each case there is a great amount of work to be done in a relatively short period of time. For this reason, any delays of consequence would delay the completion. At the present time, it seems that all the evidence suggests the schedule will be followed and the completion dates attained.

Industrial Pollution

As previously mentioned, the untreated sewage from approximately 108,000 people in the city of Savannah is certainly significant. However, this is only a very small part of the Lower Savannah's pollution problem.

Industries discharge process wastes, cooling water, and chemical wastes, which include oxygen-demanding materials estimated to be equivalent to those in raw sewage from a sewered population of approximately 1,000,000. About 90 percent of the oxygen-consuming materials in the industrial wastes is discharged by the pulp and paper mills. An estimated 78 percent, having a population equivalent of 810,000, is discharged by Union Bag-Camp Paper Corporation; and the remaining 10 percent is discharged by the other six industries shown in the table (table 10.3) of industrial waste characteristics [13:12].

Table 10-2
Summary Status Municipal Waste Program Lower Savannah River

Jurisdiction	Final Plans	Financial Arrangements	Contract Awarded	Completion	Comments
Garden City	5/01/71	7/01/71	7/01/71	4/01/72	Will also serve Chatham City, Inc. City is presently exploring possibility of a combined venture with Port Wentworth and local industries.
Port Wentworth	5/01/71	5/01/71	7/01/71	12/31/72	City and local industries now plan a joint venture subject to approval by FWQA.*
Savannah Beach	4/12/71	4/12/71	7/11/71	4/08/72	
Chatham City Corp.					Chatham City Corp. sewerage is being combined with Garden City system.
Savannah-Travis Field	Completed	Completed	Completed	8/14/71	
Savannah-Savannah River	9/30/71	Completed	11/30/71	12/31/72	

Source: A letter from Mr. R.S. Howard, Jr., Executive Secretary of the Georgia Water Quality Control Board, to Mr. John R. Thoman, Regional Director of the Federal Water Quality Administration, Southeast Region, January 11, 1971.

*FWQA—Federal Water Quality Administration.

Table 10-3
Estimated Oxygen Demand of Industrial Wastes Discharged to the Savannah River, 1963

Plant	Classification	Type of Treatment	Oxygen Demand Population Equivalents Discharged	
			Number	% of Total
Continental Can Co., Inc.	Paper Mill	Save-all	130,200	12.7
Savannah Sugar Refinery Corp.	Sugar Refining	None	30,000	2.9
The Ruberoid Company	Asphalt Roofing & Felt	Save-alls	4,500	0.4
Hercules Powder Company	Naval Stores & Tall Oil	Skimmer	3,900	0.4
Certain-Teed Product Corp.	Asphalt Roofing & Felt	None	24,000	2.3
Union Bag-Camp Paper Corp.	Paper Mill	Save-all & Skimmer	810,000	78.6
Wesson Division, Hunt Food & Industry, Inc.	Food Products	Oil Separators	26,000	2.5
Meddin Packing Company	Packinghouse	None	1,700	0.2
Total to Savannah River			1,030,300	100.0

Source: U.S. Department of Health, Education and Welfare, CONFERENCE ON POLLUTION OF THE INTERSTATE WATERS OF THE LOWER SAVANNAH RIVER AND ITS TRIBUTARIES, SOUTH CAROLINA–GEORGIA (February, 1965), p. 27a.

Industrial wastes account for almost 89 percent of the pollution and sewage accounts for less than 13 percent. The resulting combination is responsible for the pollution of the Lower Savannah River.

With industrial wastes the main concerns are oxygen-demanding materials and acid wastes, rather than bacterial pollution of the water. Industrial wastes do not add significantly to the bacterial pollution problem.

It must be noted that the figures in the preceding quote were based on the findings of the Public Health Service in an August 1963 survey. In 1965, an engineering study revealed that those earlier figures were considerably understated.

At the first Savannah River conference in 1965, federal and state officials ordered all industries to decrease their pollution by 25 percent. From September 29 to October 1, 1969, the Georgia Water Quality Control Board conducted surveys to determine whether those polluters had complied with the order. Their results and findings from the 1965 study are included in table 10.4 as a basis for comparison and for reflecting the current situation as closely as possible.

Most of the major polluters have been quite successful in removing the solids from their discharges into the river, as shown by table 10.5.

Table 10-4
Lower Savannah River Biological Oxygen Demand per Day in Terms of Population Equivalents

Source	1965 Level	1969 Level	% Reduction
American Cyanamid	(1)	(1)	
Certain-Teed Products	50,000	14,000	72
Continental Can	130,000	80,000	38
GAF (formerly Ruberoid)	33,000	18,650	44
Hercules, Inc.	20,400	17,900	12
Hunt-Wesson Food, Inc.	25,000	13,000	48
Meddin Packing Co.	1,740	420	
Savannah Sugar Refinery	37,500	29,850	20
Union Camp Corp.	810,000	696,000	14
Total	1,107,840	869,820	21%

(1) American Cyanamid's effluent has been determined to exert a significant chemical oxygen demand due to the oxidation of iron in its waste. The 1969 Conference gave an admittedly very rough estimate of the oxygen demand in terms of a Population Equivalent of between 150,000 and 200,000. However, this is only an order of magnitude estimate as chemical oxygen demand and biological oxygen demand measure two separate reactions that occur in the receiving stream.*

Ralph Nader, et al., THE WATER LORDS (Washington, D.C.: Center for Study of Responsive Law, 1971), pp. 345-46.

Table 10-5
Industrial Solid Removal Efficiency, Savannah

Industry	Type of Treatment	Solid Removal Efficiency
Continental Can	Mechanical clarifier and vacuum filtration	48%
Savannah Sugar Refinery	Vacuum filtration	96%
G.A.F.	Disc filter	70%
Hercules		37%
Certain-Teed Products	Mechanical clarifier	79%
Union Camp	Mechanical clarifier	85%
Hunt-Wesson Foods		90%

Source: U.S. Department of the Interior, Federal Water Pollution Control Administration, CONFERENCE IN THE MATTER OF POLLUTION OF THE INTERSTATE WATERS OF THE LOWER SAVANNAH RIVER AND ITS ESTUARIES, TRIBUTARIES AND CONNECTING WATERS, GEORGIA—SOUTH CAROLINA (Second Session), (October, 1969), pp. 82-88.

Paper Mills

The paper industry, as a whole, is directly responsible for approximately one-fourth of this country's industrial water pollution. In order to understand the use and pollution of water in the production of paper it would be helpful to start at the beginning of the manufacturing process.

Water carries the logs down the sluices and to the mill. Once in the mill, the water continues to carry the wood and eventually washes it. Then the wood is reduced to a pulp. In the final step, a tremendous amount of water is used to dissolve the cellulose. This solution is then spread on a wire screen. The fiber that remains on top is dried and run through a series of rollers to remove as much water as possible. As the water expelled during this process becomes part of the paper mill's wastewater discharge. The finished rolls of paper contain 6 percent water.

"Kraft mills, like Union Camp and Continental Can, usually use 20,000 - 40,000 gallons per ton of paper" [4]. "All together the nation's paper mills use and contaminate so much water that they account for about 15 percent of all industrial wastewater and nearly 25 percent of biological pollutants" [15:17].

Large kraft mills such as Continental Can and Union Camp pollute a stream in three ways. First of the pollutants are the dirt and bits of wood washed off the mills' raw material. "A large mill like Union Camp puts out huge quantities of these settleable solids—up to half a million pounds on a bad day" [2:101]. Such sedimentation will exterminate bottom-dwelling organisms near the mill if discharged in sufficient quantities. Secondly, a kraft mill discharges toxic by-products, notably the reaction products of the sulfur compounds used to digest the raw wood. Finally, and most importantly, the effluent of kraft mills contains large amounts of oxygen-demanding substances that consume the stream's oxygen and renders it unfit for fish [2:101-2].

Most organic matter decomposes in water. The wood by-products of paper production are no exception to this rule. Decomposition occurs when bacteria break down the substance. During the process, the bacteria use up some of the dissolved oxygen in the water. The resulting oxygen extraction is measured as Biological Oxygen Demand (BOD).

In order to make one ton of kraft paper, thirty pounds of BOD are produced. This means that thirty pounds of the river's dissolved oxygen supply is used up for each ton of kraft paper produced [3:95]. In small quantities there is no problem because the river can regenerate itself quite easily. However, with more serious cases, like the Lower Savannah River, a condition could result in which there would not be enough dissolved oxygen left to support the usual forms of life in the river.

Union Camp is the main offender in this study area, dumping more than 240,000 pounds of BOD into the Savannah River on a day of heavy discharges. This BOD load is equivalent to the sewage from over one million people in oxygen demand.

Kraft mills are responsible for a tremendous amount of unnecessary pollution. If the wastes are permitted to settle and decompose before they enter the river, the amount of pollution can be greatly reduced.

One of the easiest and least expensive methods is to discharge the wastes into a clarifier. A clarifier is a large circular pond that is equipped with the necessary apparatus to skim the top and bottom layers of any solids. This purification process can reduce the amount of floating or settleable solids by as much as 90 percent.

However, this process does very little to accelerate decomposition. Secondary treatment is needed to reach the standards set by the Second Session of the Lower Savannah River Pollution Conference. At that conference it was decided that all industries, except American Cyanamid, were to provide treatment of their wastes commensurate with that recommended for municipal sources with a minimum of 85 percent BOD reduction, and disinfection of the effluent where appropriate.

The most common type of secondary treatment is the holding pond. Wastewater is allowed to sit for several weeks. During this period the wastes decompose naturally and then the water is allowed to return to the river with 90 percent of its BOD removed.

Until recently, paper mills had little incentive to try to reclaim their wastes. In recent years, it has been recognized that paper mill wastes contain valuable wood sugars and lignins, and paper mills are trying to save them to their benefit and that of the river. In the future, someone might find an important use for some of the other pollutants that are presently being discharged into the river.

As previously mentioned, Union Camp is the largest kraft paper plant in the world. It also is the most influential single enterprise in Chatham County. If Union Camp's top executives came out in favor of ending the pollution problem it is more than likely that the other industries in the area would follow suit.

However, this has not been the case in the past. By its bad example, Union Camp has encouraged others to neglect their antipollution duties.

After the 1965 conference, every industry was ordered to reduce the BOD of their effluent by 25 percent and to reduce the amount of settleable solids by 90 percent before the end of 1967. Union Camp built a 310-foot clarifier and made a few changes in the mill process, but this was not accomplished until July 1968. Solids removed were in the 85-90 percent range, but the BOD had only been reduced by 14 percent of the 1965 level.

Union Camp has not even managed to meet the 1965 standards, so they will certainly have a difficult time in meeting the 1969 standards. Their job will not be impossible, but they will have to build a secondary treatment system. This would probably involve the construction of a holding pond with aeration facilities. There is no reason why Union Camp should not be able to meet the new standards. However, they must rectify many of their old practices.

Acid Wastes

The main cause of acid wastes in the Lower Savannah River lies with American Cyanamid's titanium dioxide plant.

At the 1965 conference, American Cyanamid was charged with discharging 690,000 pounds of sulfuric acid per day into the Savannah River [14:89]. At that time it was believed that American Cyanamid did not contribute significantly to the oxygen demand of the river. Further tests refuted that idea. American Cyanamid's discharged wastes contain iron in a soluble ferrous state that reacts with dissolved oxygen in the river to form various iron oxides that deplete the oxygen in the water.

During the two-day period, May 19-20, 1969, a survey was conducted by personnel from the Corps of Engineers and Georgia Water Quality Control Board to determine the extent of American Cyanamid's pollution. It must be noted that a heavy rainfall during the first day of testing caused a greater than normal discharge of pollutants.

The survey indicated discharges of approximately one million pounds of sulfuric acid on the first day and 830,000 pounds the second. These acid discharges clearly violated established pH standards in the south channel of the river. The suspended solids discharged were 88 tons on the first day and 16 tons on the second. It was also determined that the discharged waste contained 132 tons of iron on the first day and 92 tons on the second. . . . it was estimated that the iron discharged would have exerted an oxygen demand of 34,000 pounds on the first day and 25,000 pounds on the second. The iron which is oxidized will precipitate and contribute significantly to the sediment load [14:90-91].

Most forms of aquatic life are quite sensitive to changes in the pH level of the water brought about by the addition of acid wastes. The pH is a measure of acidity or alkalinity. Pure water has a neutral pH of 7. Any substance above that figure is an alkaline and anything below it is an acid. For each decrease of 1pH, the acid concentration increases ten times.

In order to understand how delicate this balance actually is, it should be noted that a slight acid concentration of 6.5 can reduce the respiration rate of oysters by 90 percent. By the time the pH reaches 4.8 most of the fish in the area are dead.

The acid discharged by American Cyanamid is so concentrated that its pH often falls in the .9 to 1.0 range. As the acid waste flows down the river it is eventually neutralized. However, the company's own surveys reveal that some pockets of water contain a pH factor as dangerously low as 4.0 in the marshlands several miles downstream of the plant [2:63].

The previous figures do not tell the entire story. In 1968 a heavy rainfall caused a large section of an American Cyanamid holding pond to burst, spilling into the Savannah River over thirty million gallons of wastewater, containing approximately 3.5 million pounds of sulfuric acid. As a result, fish died all the way down to the Atlantic Ocean. The sad part is that the entire tragedy could probably have been averted if the necessary precautionary steps had been taken by the company. After the incident, they built a larger dike, but there is no guarantee that the same thing could not happen again.

American Cyanamid has been ordered by the Georgia Water Quality Control Board to make the following changes by the middle of 1972:

1. Reduce the quantity of their acid discharged into the river to a level that will never lower the pH of the river below 6.0 at any point more than 1,000 feet downstream of the plant's normal discharge area.
2. Reduce the oxygen demand of the effluent by 25 percent.
3. Reduce the suspended and settleable solids by 90 percent [21].

As of November 1970, American Cyanamid had only taken action to correct the portion of the first order—that is, to reduce the acid concentration. This was accomplished by installing a diffusion header on the end of their waste discharge pipe. The apparatus does not reduce the amount of acid entering the river, but spreads it out in the river. The only benefit is that there are fewer concentrated pockets of acid near the bank of the river.

It is obvious that American Cyanamid must take further steps in order to rid the Savannah River of pollution. They are not alone in this respect. There are many other industries in the area that have been ordered to comply with the standards set forth in the 1969 conference. They are on schedule as shown by table 10.6.

Groundwater

Up to this point, this chapter has only been concerned with pollution of the Savannah River. Pollution takes many forms, and one of them is related to groundwater.

Groundwater is becoming an increasingly important source of water. In 1960, it accounted for almost 20 percent of the water used in America [7:23]. If the most recent population projections are fairly accurate, the United States will be

Table 10-6
Summary Status Industrial Waste Program Lower Savannah River

Industry	Waste Materials	Quantity	Preliminary Plans	Final Plans	Award or Contract	Completion of Project
American Cyanamid Company	Sulfuric acid and ferric sulfate	26 MGD	9/70	Not available	Not available	Not available
Certain-Teed Corporation	Pulp & Paper	.43 MGD	9/70	3/71	6/71	6/72
Continental Can Company	Pulp & Paper	14.4 MGD	7/70	1/71	4/71	1972
G.A.F. Corporation	Roofing waste	.43 MGD	12/31/71	7/15/71	9/1/71	7/1/72
Union Camp Corporation	Pulp & Paper	33 MGD	1/31/71	6/16/71	8/31/71	11/30/72
Hercules Inc.	Chemical	.6 MGD	3/31/71	10/1/71	11/15/71	12/31/72
Hunt Wesson Foods	Cooking Oil	3.63 MGD	6/2/70	12/31/71	3/31/71	12/31/72
Savannah Sugar Refinery Company	Domestic & organic industrial waste	12.6 MGD				12/31/72
Meddin Packing Company	Rendering Waste	.029 MGD		Not available	Not available	Not available

Source: Adapted from a letter from Mr. R.S. Howard, Executive Secretary of the Georgia Water Quality Control Board, to Mr. John R. Thoman, Regional Director of the Federal Water Quality Administration, Southeast Region, January 11, 1971.

taking approximately 280 billion gallons per day from its groundwater sources in 1980. This amount will account for half of all the water used by then [7:85].

The Savannah area is richly endowed with groundwater, but this level has been drastically reduced in the last twenty years. If the level drops too much, salt water from the ocean could move in and permanently ruin the fresh groundwater.

The main reason for the reduction in the level lies with the industries that have sunk wells to pump water for their own purposes. The water level in regions as far away as fifteen miles from Savannah has dropped by as much as fifty feet [6:16]. The main offender has been Union Camp, whose one mill uses three times as much water as all the other industries in the city put together, as well as more than the area's 120,000 people require for all domestic purposes. Because the Union Camp load is so heavy and so concentrated, "the Savannah cone of depression (of the underground water table) has its absolute center under Union Camp's plant" [2:179].

The important fact to consider is that the groundwater is being used up at an alarming rate. Once it is gone or ruined by the intrusion of salt water, the polluted Savannah will be the only remaining source of useable water. The necessary steps must be taken before the situation gets completely out of control.

Pollution of the Marshes

The Savannah is not a rapidly flowing river. For this reason many of the pollutants it carries manage to find their way into the surrounding marshes. This was confirmed by a test conducted by the U. S. Army Waterways Experiment Station in 1964 on the lower portion of the river. Instead of moving into the ocean, some of the dye, simulating pollution, tended to hug the coastline and infiltrate the marsh areas [13:30-34].

The marshes are more important than most people realize. They are the breeding grounds for hundreds of different plants and animals. Many commercially important fish spend a portion of their life in the various marsh areas. If the delicate balance of the marsh is upset by the entrance of acid, sewage, or kraft mill wastes, many of the animals that live in the marsh will die. The consensus of several scientists interviewed in 1970 revealed they believed something harmful was happening to the marshland. However, they were not sure what it was or how bad it might become. The worst part was that they were afraid that its contamination might bring severe consequences to future generations [2:101-102].

The scientists were certainly not definite about the consequences, but that is no reason to treat the subject lightly. It would be ridiculous to wait for more signs to appear before taking corrective measures. By that time it might be too late. One look at a map of the condemned oyster beds will show that the process has already started.

Savannah Corps of Engineers

As previously noted, pollution can come in many shapes and sizes. One of the forms deals with the problems relating to the maintenance of Savannah Harbor.

The development and maintenance of the Harbor is the responsibility of the Savannah Corps of Engineers. The average annual cost associated with these operations exceeds $1 million. This includes dredging and associated costs, annual surveys, and special studies [12].

Certain areas along the river and harbor have a tendency to shoal quite rapidly. In order to maintain the proper dimensions these areas must be dredged two or three times a year.

The dredging operation is accomplished by the U. S. Dredge Henry Bacon. The vessel's overall length is 226 feet and it carries a crew of 64 men for its 24-hour day, 7 days per week operation. The dredging operation works on a suction principle with a suction pump sucking up the shoal material. The material is then carried by a pipeline to the disposal area. In some areas it is necessary to pump the dredged material over 10,000 feet because there are not enough disposal areas.

In order to prevent the dredged material from running back into the waterways it has been necessary to build dikes, which have been fairly effective in containing the material.

Actually, there are two major sources of material that cause shoaling. Firstly, there is the sediment that is brought down from more northern areas by the Savannah River. Secondly, there is the material that comes from the surrounding marshlands. The first source has been eliminated to some extent by the construction of the Clark Hill Dam and the second by dredging operations. The construction of the dam has also decreased the hazard of a flood and it has greatly increased low stream flows. Both measures have had the combined effect of improving the overall water quality of the river.

There are two projects still under construction that will improve the area and the water quality. Together they are part of a $17 million plan. One calls for widening the 30-mile channel by 100 feet and deepening it by 4 feet. The other project involves the construction of a sediment basin on Back River. The latter project will reduce the annual harbor maintenance cost of $1 million by 40 percent. A very important environmental safeguard provided in conjunction with the sediment basin is a fresh water canal. It would help to stop increased salinity which is a result of salt water intrusion [5:5].

Water Quality Monitoring Stations

In cooperation with the Georgia Water Quality Control Board and the Chatham County Health Department, the Savannah District of the Corps of Engineers has set up eight monitoring stations on the Savannah River and the Savannah Harbor.

The ninth monitoring station was set up in September 1969 in the harbor near Fort Jackson to measure anticiapted water quality improvements in the area as the pollution abatement program proceeds.

Water samples are taken by Corps of Engineers personnel around the first of the month at each of the eight water quality sampling stations established at concentrated industrial areas, spaced from 2 to 5 mile intervals along the harbor channel. The "grab" samples are taken on the incoming tide at depths of 3 feet and 20 feet. An analysis of the samples is accomplished by the Chatham County Health Department's laboratory which, in turn, forwards the data to the Georgia Water Quality Control Board's office in Atlanta, Ga [12].

The Georgia Water Quality Control Board has classified the Savannah River according to water quality. The best portions are classified as "fishing" or "recreational" and the worst as "industrial" or "navigational." In the study area from river mile 22 down to river mile 5 the water is classified as "industrial" or "navigational." In this area fishing and swimming are strictly prohibited [9].

The previous classification does not tell the whole story. Many stretches of the river have oxygen levels too low for most forms of aquatic life. This is especially true of the stretch from river mile 22 to river mile 5. Apart from the danger to local fish, this 17-mile stretch can hinder, stop, or even kill shad, striped bass, or other varieties of fish from swimming upstream to their spawning grounds.

Fish must breathe dissolved oxygen in the water to stay alive. In order to remain in good health, most varieties of fish need 4 to 8 parts per million (ppm) of dissolved oxygen in the water. Whenever the level drops below 4 ppm for a period of time the fish usually begin to slowly die. Their eggs will not hatch as often and if they do the young fish will be very weak. If the level drops below 2 ppm, death is almost a certainty. If the level dropped any further the river could lose its dissolved oxygen supply and become anerobic. That would be the ultimate catastrophe.

Over the past few years a series of government agencies have monitored the amount of dissolved oxygen in the Lower Savannah River. The data collected in August 1963 showed that half of the stations recorded average dissolved oxygen readings below 3.0 ppm and 37.5 percent recorded average readings below 2.5 ppm, which is below the minimum for water relegated to "industrial" status. The situation was actually much worse than it appeared. The data for minimum dissolved oxygen show that 75 percent of the stations recorded minimum dissolved oxygen levels below 2.5 ppm. On one day at station 11.5 the river came within .1 ppm of losing all its oxygen and becoming anerobic. The average for that day showed a dissolved oxygen content of 1.8 ppm. While that figure is considerably below the "safe" level it does not reflect that the minimum level for the day was .1 ppm. That is the shortcoming in using averages for water quality data [13:48 j-w].

The results of data collected from August 12-22, 1968, were similar to the 1963 survey in that half of the stations recorded average readings below 3.0 ppm and 37.5 percent recorded average readings below 2.5 ppm. The main improvement lies in the fact that 75 percent of the stations in the 1968 survey showed increased minimum dissolved oxygen levels over the 1963 survey. At first glance this should be cause for elation, but a closer look shows that this is not the case. The levels are still far below the 4.0 ppm that most fish need to remain healthy.

Since 1963 many of the industries have installed devices that reduce the oxygen demand on the river. However, this does not mean that the problem has been solved. New sewers and further industrial controls are needed to make the river as healthy as it once was.

It is distressing to note that the 1968 figures for fecal and total coliform bacteria reflect a rise over the 1963 level. On the other hand, the amount of suspended solids has appreciably decreased, although the reduction has not achieved the desired level yet.

The 1963 data revealed that the average pH level dropped below 6.5, 62.5 percent of the time. The more critical measure, the minimum pH level, dropped below 6.5, 100 percent of the occasions when the readings were taken. At one point it reached a low of 4.2 and an incalculable number of aquatic animals were killed.

The 1968 data reflected a definite decrease in the concentration of acid in the waters of the Lower Savannah. The average pH level never fell below 6.5 and the minimum pH level only fell below that point 50 percent of the time. Certain positive measures have been taken by the industrial offenders, but their job is far from complete. They must take the necessary precautions to maintain a minimum pH level of 6.7.

At this time it should be emphasized that one pollutant, like a high acid concentration, can make aquatic forms of life more vulnerable to other pollutants. This is the main reason why so many fish die even though each pollutant in the water is above the minimum for safety. There may be cumulative or perhaps synergistic effects from multiple pollutants present in concentrations approaching dangerous proportions.

Decline in Commercial Fishing

Ever since James Oglethorpe founded the City of Savannah, people have used the Savannah River as a source of food. During that time over a billion pounds of water animals have been harvested. Until recently the harvests have been quite large. However, all this has changed due to the pollution in the river. A study of the decline in fishing tends to parallel the decline in water quality.

One of the hardest hit were the oystermen. Not every kind of harvest has declined as much as the oyster catch, but the general trend has been remarkably downward. Table 10.7 shows that the harvests of every kind of commercial fish have fallen greatly since the start of the twentieth century in the Savannah River. Fishing statistics can be influenced by many factors, but it is hard to avoid the conclusion suggested by table 10.7 that catches have been adversely affected by pollution. In 67 percent of the cases the catch has fallen to less than 9 percent of its peak. The average total catch from 1918 to 1968 was 22,190,000 pounds. The total catch for 1968 was only 14,338,000 pounds. Therefore, the 1968 catch as a percentage of the fifty-year average was 64.8 percent [2:360-61].

It cannot be inferred that fish populations have declined in proportion to the catch, but it does seem certain that industrial pollution has played a large part in the decline that has occurred [2:362].

Table 10-7
Savannah Commercial Fishing Harvests in Pounds, Peak Year Since 1880 and 1968

Species	Maximum Harvest	Year	1968 Harvest	% of Maximum
Alewives	125,000 lbs.	1880	0	0
Bluefish	50,000	1928	0	0
Catfish	289,000	1902	66,000 lbs.	23
Croaker	46,000	1908	0	0
Flounder	307,000	1945	23,000	8
Grouper	160,000	1908	17,000	11
King Whiting	464,000	1955	123,000	27
Menhaden	34,102,000	1927	0	0
Mullet	194,000	1908	0	0
Sea Trout	144,000	1890	2,000	1
Shad	1,333,000	1908	569,000	43
Red Snapper	880,000	1908	17,000	2
Spot	103,000	1955	2,000	2
Sturgeon	354,000	1880	0	0
Hard Clams	43,000	1908	0	0
Blue Crabs	15,766,000	1960	3,669,000	23
Oysters	8,070,000	1908	191,000	2
Shrimp	16,392,000	1945	8,536,000	52
Total	47,458,000	1927	14,338,000	30

Source: South Atlantic Fisheries Historical Catch Statistics.

If one assumes that water quality deterioration has been the principal factor behind the decline in the fish catch and if the further assumption is made that the decline will continue at the same rate as in the past, we may project the catch for 1980. Using the data of table 10.7 and assuming a straight-line decline at the same rate as that which prevailed from year of maximum catch to 1968 continued to 1980, there would be a total disappearance from the catch of all species except catfish and shrimp. The 1980 harvest of the former would be 30,000 pounds or 10 percent of the maximum and the 1980 harvest of shrimp would be 4,600,000 pounds or 28 percent of maximum. The combined 1980 catch would be 4,630,000 pounds of the two species or 10 percent of the maximum catch of 47 million pounds of all species in 1927. The 1980 projection has no value as a forecast, but it does suggest the ruinous consequences for the fishing industry if the catch continues to decline at the same rate as in the past, and there is good reason to believe that this decline was associated, to a substantial degree, with a deterioration of water quality.

But if industry in particular and also the municipalities adhere to the decisions of the Second Session of the Lower Savannah River Pollution Conference [14], there should be a marked improvement in Savannah River water quality as suggested by tables 10.8 and 10.9. If this occurs, it seems most unlikely that the fish harvest will decline in 1980 to the level projected above. Rather some increase in the catch would appear more reasonable. Many factors enter into the fluctuations of the fishery harvest, and there is no satisfactory formula for predicting the results of an improvement of water quality. However, if we were to assume that as a consequence of improved water quality the 1980 catch would equal the 1968 catch or 20 percent of the maximum catch, whichever is larger, we might achieve some approximation. This assumes that the species that disappeared from the catch by 1968 were more sensitive to pollution. Following this formula, the total 1980 catch would be projected at 22 million pounds, or 150 percent of the 1968 harvest. The figure of 22 million pounds is not likely to be very close to the actual 1980 harvest, but if water quality improves it will no doubt be much closer to the mark than our earlier, more pessimistic projection. It also seems safe to predict happier days ahead for Savannah fishermen if water quality improves to the degree postulated by tables 10.8 and 10.9..

Table 10-8
Savannah River Estimated Industrial Waste Characteristics (Oxygen Demand of Wastes Discharged to the River in Population Equivalents)

Industrial Plant	1963[1]	1965[2]	1969[2]	1980[4]
Continental Can	130,200	130,000	80,000	24,400
Savannah Sugar Refinery	30,000	37,500	29,900	7,000
GAF (formerly Ruberoid)	4,500	33,000	18,600	6,200
Hercules Powder	3,900	20,400	17,900	3,800
Certain-Teed Products	24,000	50,000	14,000	9,400
Union Camp	810,000	810,000	696,000	151,900
Hunt-Wesson Foods	26,000	25,000	13,000	4,700
Meddin Packing	1,700	1,700	400	300
American Cyanamid	175,000[3]	175,000[3]	175,000[3]	32,800
Total	1,205,300	1,282,600	1,044,800	240,500

Sources:

[1] U.S. Department of Health, Education and Welfare, CONFERENCE ON POLLUTION OF THE INTERSTATE WATERS OF THE LOWER SAVANNAH RIVER AND ITS TRIBUTARIES. SOUTH CAROLINA–GEORGIA, 1965, p. 27a.

[2] Ralph Nader, et al., THE WATER LORDS, Task Force Report on the Savannah River (Washington, D.C.. Center for Study of Responsive Law), p. 345.

[3] This figure is based on the average of two estimates found in THE WATER LORDS, p. 346.

[4] Projected assuming a production increase of 25 percent over 1969 and a reduction in oxygen-demanding wastes of 85 percent after treatment on the 1965 base.

Table 10-9
Estimated Sewage Characteristics (Oxygen and Bacteria Demand of Wastes Discharged to the River in Population Equivalents)

Area	1963[1] Population	Oxygen[1] Demand After Treatment	Bacteria[1] Demand After Treatment	1970[2] Population	Oxygen Demand After Treatment	Bacteria Demand After Treatment	1980[4] Population	Oxygen[5] Demand After Treatment	Bacteria[5] Demand After Treatment
Port Wentworth	3,700	3,700	3,700	3,910	3,910	3,910	4,120	620	410
Garden City	5,200	5,200	5,200	5,740	5,740	5,740	6,050	910	610
Savannah Beach	25,000	16,700	12,500	26,000	15,000	10,000	27,000	4,050	2,700
City of Savannah	108,000	108,000	108,000	118,350	118,350	118,350	129,000	19,350	12,900
Other Areas	4,300	2,000	1,500	4,515[3]	1,950	1,500	4,700[3]	710	470
Total	146,200	135,600	130,900	158,490	144,950	139,500	170,870	25,640	17,090

Sources:
(1) U.S. Department of Health, Education and Welfare, CONFERENCE ON POLLUTION OF THE INTERSTATE WATERS OF THE LOWER SAVANNAH RIVER AND ITS TRIBUTARIES, SOUTH CAROLINA–GEORGIA, 1965, p. 22a.
(2) U.S. Department of Commerce, Bureau of the Census, 1970 CENSUS OF POPULATION, GEORGIA, Advance Report, 1970, p. 6.
(3) A projection based on an estimated 5 percent population increase.
(4) A projection based on 1960 and 1970 population figures from the 1970 CENSUS OF POPULATION, GEORGIA.
(5) The 1980 figures for oxygen and bacteria demand after treatment are based respectively on 85 and 90 percent reduction estimates.

Sewage and Industrial Waste Characteristics

Table 10.8 shows the estimated oxygen demand of industrial wastes discharged into the river in population equivalents. The figures for 1965 reflect an increase of nearly 6.5 percent over the 1963 level. Some action had to be taken to curb the rising pollution reflected in these statistics.

For this reason the *First Session of the Conference on Pollution of the Interstate Waters of the Lower Savannah River and its Tributaries, South Carolina–Georgia* was held. Even though the results did not reach the standards set down by the conference, the reductions were encouraging, as shown by the 1969 figures in table 10.8, which reflect a decrease in oxygen demand exceeding 20 percent. In order to motivate the industrial concerns to do their part for pollution reduction, a second conference was called in 1969. At that time the standards for pollution abatement were raised to 85 percent reduction, but they also show a 25 percent rise in production. Therefore, the 1980 figures show an 81 percent reduction over the 1965 base and a 77 percent reduction over the 1969 figures.

If the conference decisions are enforced, it appears that the Savannah River, which has been grossly polluted for many years by the industries in the area, will be starting to make a comeback. If so, by 1980 the water quality should return to a position it has not enjoyed for several decades.

Table 10.9 shows the estimated oxygen and bacteria demand of sewage wastes discharged into the river in population equivalents. The figures for 1963 reflect the small percentage of treatment. Oxygen demand was only reduced by 7.2 percent and the bacteria demand was only reduced by 10.4 percent before it entered the river. By 1970 the sewered population rose to 158,478, but the oxygen and bacteria demand were only reduced by 8.6 and 12 percent respectively. The reduction was very small and the municipalities were also ordered by the second conference to make the same reductions as the industrial polluters. The 1980 population was based on a projection based on 1960 and 1970 population figures from the *1970 Census of Population, Georgia*. Even though the population is projected to increase to 170,867, the oxygen demand is expected to decline by 85 percent and the bacteria demand by 90 percent.

Although the pollution from the sewered population is not as dynamic as industrial pollution in terms of population equivalents, it is still an important problem. The 1969 conference decided to correct the misuse of the Savannah that has existed for many years.

Conclusion

Today there is a great concern about the environmental crisis in this country. In this chapter only one problem in one small area has been discussed, but the author hopes that it has shown that other areas with pollution problems can find remedies with some hard work. This case study of the Lower Savannah River might also have the effect of showing other cities what could happen if pollution was allowed to run free. In their case, an ounce of prevention could have been worth more than a pound of cure. It might be an old saying, but the validity that it carries is exemplified in the case of the Lower Savannah River.

Over the past few years, Georgia has developed an effective program to abate and prevent water pollution in the Lower Savannah River. However the job is far from complete. In April 1970 the Georgia Water Quality Control Board's *Water Quality Data Lower Savannah River* indicated that while the water quality of the river had improved somewhat, the stretch between river miles 2 and 22 was still heavily polluted by large volumes of untreated or inadequately treated domestic and industrial wastes.

If the estimated sewage and industrial waste characteristics for 1980 (tables 10.8 and 10.9) are combined, the oxygen demand in population equivalents would be 257,590. This is still a fairly large number, but it reflects an 81 percent decrease over the combined figures for 1963, which totaled 1,351,000. This is probably the most significant fact with regard to the estimated quality of the water in 1980. How much such a reduction would help the river is a matter for conjecture, but it should definitely start it on the path to recover its former purity.

In order to reach the standards set down by the Georgia Water Quality Control Board by the end of 1972, everyone's cooperation is needed. The job will not be easy, but it is not impossible. If the 1972 standards are met, the job of pollution control in the future will be much easier. In that case it will be based on prevention rather than on cure. The author believes the water quality in the Lower Savannah River will continue to be within the prescribed limits for recovery by 1980.

Although the companies and the residents in the Savannah area have themselves to blame for the current situation, they will also have themselves to thank for the improvement in the future. Most important of all, they will be the ones to reap the benefits.

Notes

1. W. Wesley Eckenfelder, Jr., *Industrial Water Pollution Control* (New York: McGraw-Hill Book Company, Inc., 1966).

2. Ralph Nader, et al., *The Water Lords.* Task Force Report on the Savannah River (Washington, D.C.: Center for Study of Responsive Law, 1971).

3. Blair T. Bower, et al., *Waste Management* (New York: Regional Plan Association, March 1968).

4. *Southern Pulp and Paper Manufacturer,* September 1969.

5. Colonel John S. Egbert, *The Corps of Engineers Stewardship of Natural Resources* (Savannah, Georgia, June 16, 1970).

6. M. J. McCollum and Harland B. Counts, *Relation of Salt-Water Encroachment to the Major Aquifer Zones Savannah Area, Georgia and South Carolina.* U. S. Geological Survey Water Supply Paper 1613-D (Washington, D.C., 1964).

7. C. L. McGuinnes, *The Role of Ground Water in the National Water Situation,* U. S. Geological Survey Supply Paper 1800 (Washington, D.C., 1963).

8. Georgia Water Quality Control Board, *Fifth Annual Progress Report 1969.*

9. Georgia Water Quality Control Board, *Ogeechee, Little Ogeechee and Lower Savannah Rivers* (mimeographed).

10. Georgia Water Quality Control Board, *Water Pollution Control in Georgia*, 1970 (mimeographed).

11. Georgia Water Quality Control Board, *Water Quality Data Lower Savannah* (Atlanta, Georgia, 1970).

12. United States Army Corps of Engineers, Savannah Engineer District, *Savannah Harbor* (Savannah, Georgia, October 1970).

13. United States Department of Health, Education and Welfare, *Conference on Pollution of the Interstate Waters of the Lower Savannah River and its Tributaries – South Carolina–Georgia* (Washington, D.C.: Government Printing Office, 1965).

14. United States Department of the Interior. Federal Water Pollution Control Administration, *Conference in the Matter of Pollution of the Interstate Waters of the Lower Savannah River and its Estuaries, Tributaries and Connecting Waters – Georgia–South Carolina* (Washington, D.C.: Government Printing Office, 1970).

15. United States Department of the Interior, Federal Water Pollution Control Administration, *The Cost of Clean Water – Summary Report* Volume I (Washington, D.C.: Government Printing Office, 1968).

16. United States Department of the Interior, Federal Water Pollution Control Administration, *The Cost of Clean Water – Industrial Waste Profile Number 3: Paper Mills, except Building,* Volume III (Washington, D.C.: Government Printing Office, 1968).

17. United States Department of the Interior, Federal Water Pollution Control Administration, *Southeastern Comprehensive Water Pollution Control Project – Population Projection* (Atlanta, Georgia, 1966), (mimeographed).

18. United States Study Commission, "Plan for Development of the Land and Water Resources of the Southeast River Basins," Appendix 1, Savannah River Basin, 1963.

19. Report of the Flood Control Subcommittee of the House of Representatives, "Savannah River Basin Inspection," 1969. (mimeographed).

20. Letter from Mr. R. S. Howard, Executive Secretary of the Georgia Water Quality Control Board, to Mr. John R. Thoman, Regional Director, Federal Water Quality Administration Southeast Region, on January 11, 1971.

21. Letter from Mr. R. S. Howard, Executive Secretary of the Georgia Water Quality Control Board, to Mr. C. W. Sieber, on September 8, 1969.

11 Long Island Water

HENRY LANG

Long Island extends from the southeastern part of the mainland of New York State east-north-eastward about 120 miles into the Atlantic Ocean. Long Island consists of four counties; from east to west they are Kings, Queens, Nassau, and Suffolk Counties. Kings and Queens counties, which are also boroughs of New York City, occupy slightly less than 200 square miles of the western part of the island, while Nassau and Suffolk counties are about 290 and 920 square miles respectively.

Kings and Queens counties, the most densely populated areas, receive most of their fresh water supply through the New York City Public Water Supply system from surface-water sources in the Delaware and Hudson River basins mainly in upstate New York. Both Nassau and Suffolk counties derive their entire water supply from wells tapping the groundwater reservoir.

Before the intrusion of man the hydrological system of Long Island was in a long-term state of equilibrium. "That is, although inflow may have been larger or smaller than outflow for a short period of time, over periods of several decades or more the amount of water entering the system was balanced by, or was equal to, the amount of water discharging from the system" [5:56].

In the first stage of development of Long Island, which began with the arrival of the first European settlers, every house had a shallow well for its water supply. Since the water was being pumped from and returned to the same ground level, the upper glacial aquifer, the system remained in balance. However, this cycle of groundwater development and wastewater disposal resulted in the pollution of the shallow groundwater in the vicinity of the cesspools.

As a result of this pollution the shallow private wells had to be abandoned and were gradually replaced by deeper public supply wells. Many of these wells were required to draw fresh water from the deeper Magothy and Jameco aquifers. Since cesspools were still in use the net withdrawal was small, but cesspools returned the water to the upper glacial aquifer. As populations and fresh water usages increased, the downward movement of water from the upper glacial aquifer was not sufficient to compensate for the withdrawals of water from the Magothy and Jameco aquifers. "The imbalances resulted in local decreases in the amount of fresh groundwater in storage in the Magothy and Jameco aquifers and a concurrent landward movement of salty groundwater" [5:64].

In the next stage sewers were installed on the western portions of Long Island, notably Kings and Queens counties. Now the wastewater that had been discharged to the ground through cesspools was discharged to the seas. This meant a permanent loss of water from the groundwater system. This loss also resulted in a hydrological imbalance leading to a more rapid landward encroachment of salty water into the previously fresh groundwater reservoir.

The most striking example of this occurred in Kings and Queens counties during the 1930s. "Decreased natural recharge owing to urbanization and increased groundwater withdrawals... caused groundwater levels in Brooklyn locally to decline to as much as 35 feet below sea level. This local overdevelopment caused contamination of large parts of the groundwater reservoir in that area from sea water encroachment" [8:5].

Nassau County's water supply is derived from approximately thirty-five well installations located throughout the county. The water is delivered by forty-four distribution systems consisting of municipal agencies, special water districts, and privately-owned public utilities. There are also approximately 100 minor water supplies serving small communities and parks and about 500 private wells supplying homes and businesses where no public water is available [21:2].

Public water supply in Suffolk County is provided by the Suffolk County Water Authority, nineteen municipal water districts and village systems, and fifty-eight privately-owned water corporations. A large percentage of the population, 27.5 percent in 1965, is served by domestic wells serving individual households. In addition, the State of New York operates water supply systems at a number of state institutions, and other systems are operated by the Long Island State Park Commission, the county, and various towns [17:1, 8].

Kings and Queens Counties

Early in the twentieth century Kings and Queens received their fresh water primarily from groundwater supplies pumped by private companies. "The aquifers contained mostly fresh water in 1903, except in some places near the shore where salt water was present naturally or had moved toward wells near the shore" [15:136]. At this time most of the water used was discharged to the sea by public sewers, but in some areas cesspools and septic tanks were still in use.

In 1903 the water table in Brooklyn and Queens was well above sea level, "the highest altitudes of the water table were about 20 feet in northeastern Kings County and about 60 feet in northeastern Queens County" [15:136]. By 1933 the groundwater used for public supply was almost entirely being discharged through public sewers, and in addition, "natural groundwater recharge had decreased substantially because much of the permeable soil had been covered with impermeable surfaces such as streets, highways and buildings. These factors resulted in a marked imbalance in the groundwater system — groundwater discharge exceeded recharge, causing a net decrease in the amount of fresh groundwater in storage. The decrease in storage was evidenced by declining groundwater levels" [5:82].

As a result of the falling water table the depressions left were filled in by salty sea water, contaminating the fresh groundwater supply, and public supply wells in Kings County had to be abandoned. These conditions resulted in "the passage of legislation that required the approval of the New York State Water Resources Commission for the drilling of all wells, except agriculture, yielding more than 70 gpm (gallons per minute)" [15:138]. The legislation had come too late, however, and by 1936 "the water table in practically all of Kings County had declined below sea level; locally it had declined about 50 feet to a depth of more than 35 feet below sea level" [5:82].

Since the 1930s New York City has supplied most of Kings County's water needs from its upstate reservoirs. In 1947 "the wells of a private water company, which had supplied as much as 27 MGD in central Kings County were shut down as a result of condemnation proceedings by the City of New York" [15:138]. From this point on there was a marked rise in the water tables of this county, but much of this rise consisted of salt water that continued to fill in the depressions left by excessive pumping. "The water elevation has risen, but salt water pollution is still present in many areas" [10:14].

With the cessation of groundwater pumpage in Kings County, groundwater levels began to recover. By 1965 the water table in all but the northern part of the county had recovered to a position above sea level. But while groundwater pumpage was stopped in Kings County, it continued in Queens and Nassau counties. "In 1954 about 154 MGD (million gallons per day) of fresh water was pumped mainly for public supply, in southern Nassau and southeastern Queens Counties . . . Part of the water was returned to the ground through cesspools and septic tanks; part was lost to the sea through sewers; and part was exported from the area by New York City . . . Salty groundwater in the principal aquifer is believed to be encroaching very slowly, at rates of less than 100 feet per year" [13:A2].

Continued usage of the groundwater as a source of fresh water in Queens led to the fall of the water table, much the same as had already occurred in Kings County. "Calculations . . . indicate that about 51 billion gallons of fresh water was removed from storage in the water table aquifer in Queens County between 1903 and 1961," and this continued removal of fresh water led to the decline of the water table by as much as 20 feet along the central part of the Queens-Nassau boundary and "in 1961, water levels in the center of the cone in the Woodhaven-Jamaica area were about 15 feet below sea level" [15:138].

Figure 11.1 illustrates the rise and fall of the water table in Kings and Queens counties for the years 1903, 1936, and 1961. In illustration A, Kings County, the rise in the water table between 1936, when ground pumpage was ceased almost entirely, to 1961 is shown to have almost returned to its 1903 level. In Queens, however, where groundwater pumpage continued through 1961, the water level is shown to have continued to fall.

The history of the overdevelopment of fresh water supplies in Kings and Queens counties are graphic demonstrations of what can occur when this natural resource is tapped indiscriminately. The possibility of this overdevelopment taking place in Nassau and Suffolk counties is currently being debated by experts on both sides. According to one authority there is little likelihood excessive groundwater pumpage could lead to salt water intrusion on eastern Long Island. "It should be emphasized that the hydrogeologic setting in

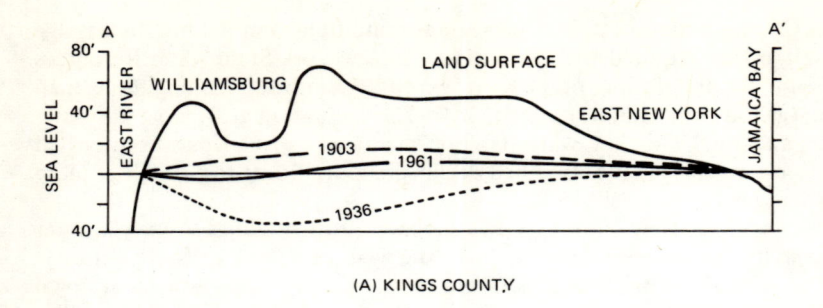

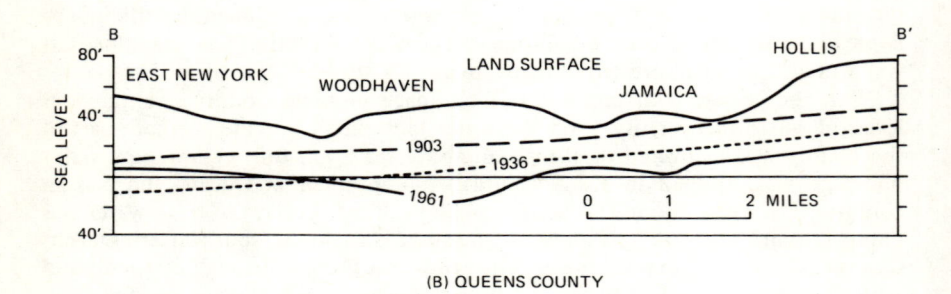

Figure 11.1. Profiles of the Water Table in Kings County (A) and Queens County (B). Source: N.M. Perlmutter and Julian Soren, *Effects of Major Water Table Changes in Kings and Queens Counties,* New York City, U.S. Geological Survey Professional Paper 450E, (Mineola, New York: 1963), p. 137.

Brooklyn and Queens bears little resemblance to Suffolk County. Salt water intrusion such as occurred there is not likely to take place here (Suffolk County) because of this difference" [18:II,107].

There is no denying, however, that Kings and Queens counties at one time contained independent fresh water supplies and are now dependent upon the New York City supply system to meet their fresh water needs. Developers of fresh water supplies in Nassau and Suffolk counties would be wise to review the case histories of Kings and Queens and plan accordingly.

Water Supply

While the entire land surface of Long Island receives precipitation, only a certain portion of the area is effective in transmitting the precipitation to the main groundwater body as recharge. This is defined as the water budget area. Areas subject to salt water flooding or where salt water intrusion may be possible must be excluded from the water budget area.

Estimates have been made that there is approximately 54 cubic miles, or about 60 trillion gallons of fresh groundwater beneath Long Island. The amount of groundwater recoverable, however, is considerably less and is assumed to be 5 to 10 percent of the total, or from 3 to 6 trillion gallons [5:26].

Precipitation is the main source of all of Long Island's underground fresh water supply. "Average annual precipitation on the water budget area for water years 1951-65 ... is about 44 inches. This is equivalent to an average of about 2.1 MGD (million gallons per day) per square mile, or about 1,600 MGD for the water-budget area" [5:30].

It is very difficult to estimate the amount of precipitation that actually percolates deep into the ground and recharges the groundwater supply. Evapotranspiration is the natural loss of water to the atmosphere through plants and from evaporation. There are many different variables that may influence evapotranspiration, but it has been estimated that approximately 46 percent of the precipitation is consumed in this manner [10:2].

Part of the precipitation results in direct runoff into streams. This direct runoff is estimated to be from 2 percent to 6 percent of the total [10:2]. The total amount of water available for natural recharge of the groundwater reservoir is precipitation less the losses of direct runoff and evapotranspiration, or approximately 50 percent.

Nassau and Suffolk counties have attempted to increase the amount of water percolating through the ground and replenishing the groundwater supply through the use of recharge basins (sumps). At the present time "more than 2,000 recharge basins on Long Island are used to dispose of storm runoff from residential and industrial areas and highways" [20:C1].

In Nassau and Suffolk counties storm sewers lead to recharge basins. "Recharge basins ... are unlined pits of various shapes and sizes which are excavated below land surfaces, mostly in moderately to highly permeable sand and gravel deposits of glacial outwash. The basins range in area from about 0.1 to 30.0 acres and average about 2 acres. Their average depth is about 10 feet, but some are 25 feet deep" [20:C3].

Large housing developers in Nassau and Suffolk counties are required to provide basins in relation to the area of ground developed. There is no way to measure whether these basins in fact make up for the fresh water infiltration lost through increased urbanization. Some experts are of the opinion that increased urbanization actually leads to an increased amount of fresh water replenishing the groundwater supply by catching water that otherwise may have been lost to stream runoff.

Kings and Queens counties do not have any recharge basins and quite possibly this was a compounding factor leading to the deterioration of the groundwater supply. Paved highways and buildings without the addition of recharge basins limit the surface area available for the natural filtration of fresh water into the ground.

The groundwater is contained in the voids in the geological sand deposits that form Long Island. "The geological formations below the water table which contain water sufficient for water supply development are called aquifers. ... The glacial aquifer, which is the uppermost, contains a large and readily accessible source of water. Directly below the glacial aquifer is the Magothy aquifer, which is the largest and most important, yielding 80 percent of the water supply. The Lloyd sand, the deepest aquifer, lies under an impervious formation called the Raritan clay and is used largely in such areas as Long Beach where the upper aquifers have been contaminated by salt water" [21:2].

The major hydrogeologic units of the groundwater reservoir of Long Island are illustrated in figure 11.2.

Water Quality

The groundwater on Long Island is generally of excellent quality and meets the standards for drinking water. Water quality is measured in terms of the concentration of a number of constituents. "The quality of the groundwater is influenced by many factors. Most basic is the physical and chemical makeup of the earth in the groundwater reservoir itself. Two of the common undesirable

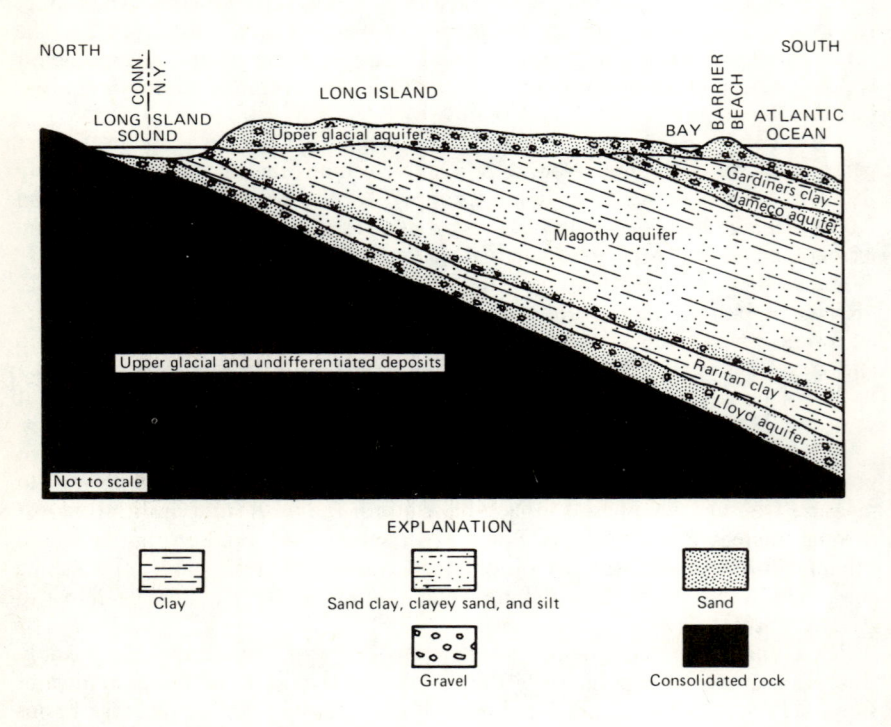

Figure 11.2. Major Hydrogeologic Units of the Groundwater Reservoir of Long Island, New York. Source: Philip Cohen, O.L. Franke, and B.L. Foxworthy, *An Atlas of Long Island's Water Resources,* New York Water Resources Commission Bulletin 62 (Albany, New York: State of New York, 1968), p. 19.

chemical constituents dissolved by groundwater as it flows through the aquifers are iron and manganese" [18:II,131].

Iron and manganese are by no means the only chemical constituents found in the groundwater of Long Island. "Practically all the elements and compounds above, on, and beneath the surface of Long Island are, at least to some extent, soluble in water. Therefore, all of Long Island's water contains at least some dissolved solids. In addition, the dissolved solids content of the water commonly increases as the water moves through the system" [5:48].

More serious are the number of constituents added to the groundwater by domestic, commercial, and industrial wastes. While Kings and Queens counties are completely sewered, as of December 1969 Nassau County was approximately 44 percent sewered. It is estimated district 3 of a four-district plan will be sewered by 1985, and while a schedule for district 4 has not yet been determined, it is expected this area will also be sewered by 1985 [22:16].

With the exception of some private sewage plants constructed by large housing developers, Suffolk County is virtually non-sewered. "In Suffolk County less than 4 percent of the population is presently sewered. A comprehensive plan was prepared for the five western towns, an area of 566 miles. It is impossible that this sewer construction can be completed even on a crash program in less than 30 years" [6:2-1].

The glacial aquifer of Nassau County is subject to contamination from various sources. It is estimated that approximately 146 million gallons per day of used water are discharged into this aquifer. ... The largest source of groundwater contamination is the effluents from the residential private disposal facilities primarily in the unsewered areas of the southeastern and northern watershed of the county [7:5].

The largest single source of contamination of Suffolk County groundwater is from the leaching of thousands of individual subsurface waste disposal systems. In 1965 these systems discharged an estimated 58 MGD of highly polluted waste water into the ground. The actual amount of treatment provided by individual disposal systems is quite limited, and depends upon many related factors such as the capacity of the system in relation to the loads imposed upon it, age of the system, depth to groundwater, characteristics of underlying soil, type of system, and adequacy of maintenance.

The constituents most commonly associated with domestic sewage are certain chemical compounds including the nitrogen compounds, sulfates, chlorides, phosphates, and synthetic detergents. In addition, suspended solids, bacteriological and biological organisms are normally present in high concentrations [18:II,133].

Detergents in groundwater are evidence of general pollution since they are associated with waste water. The presence of ABS (Alkyl Benzene Sulfonate) or MBAS (Methylene Blue Active Substances) is an indication that the water supply has been subjected to pollution from underground disposal of sewage. "Though not related to toxicity, ABS has served as the best and most convincing indicator of contamination or pollution. Concentrations of ABS in drinking water above

0.5 mg/1 (milligrams/per liter) can be expected to cause a number of consumers to experience slight tastes and cause foaming of the water. Moreover, at this concentration, approximately 5 percent of the water can be assumed to be of sewage origin [18:II,12].

The amount of nitrates found in the groundwater supplies are closely related to the amount of urbanization present in those areas that are not sewered. In Suffolk County, for example, the highest nitrate concentrations are found in the town of Babylon, decreasing in most instances in the eastern portion of the town of Islip and in the town of Brookhaven.

It has also been determined that nitrates are likely to be found in groundwater as a result of fertilization. Measuring the amount of nitrates resulting from fertilization is virtually impossible, but it is known that in predominantly agricultural areas in Suffolk County, the nitrates in the groundwater approach, and in some cases exceed, the allowable limits. The sparse population and the small number of private sewage disposal systems in these areas leaves agricultural fertilizer as the only significant source of this pollution. Since Long Island is largely made up of residential homes with lawns, it is likely the fertilization of the lawns may account for a large portion of the nitrates found in the groundwater.

Nitrates from fertilizers reach the groundwater supply principally through storm water runoff. "Storm water runoff contains appreciable quantities of fertilizer, animal wastes, and other street and land washings that contain nitrates. Landfill leaching can also contribute nitrogen to the groundwater. Specific contributions of nitrates by storm water and landfills in Nassau County are not known" [9:4].

Nitrate contamination is probably the most serious threat to the quality of Long Island's fresh water supply. "Serious and occasionally fatal disorders have occurred in infants following ingestion of water shown to contain nitrates. The condition, called 'nethemoglobinemia,' is caused by an interference in the mechanism whereby the blood carries oxygen to body tissues. Cases have been sufficiently frequent and widespread geographically to compel restrictions of nitrates in potable waters to 10 mg/l as nitrogen" [9:4].

Wells found to exceed these restrictions have been retired from service in Nassau and Suffolk counties. Investigations have been instituted into the feasibility of removing nitrates from the fresh water supplies, but the only real solution at the present time is the installation of an extensive public sewer system.

Suffolk County Detergent Ban

On November 10, 1970 the Suffolk County Legislature approved a law banning the sale of certain detergents in Suffolk County, effective March 1, 1971. Suffolk County thus became the first local government unit to prohibit the sale of detergents containing the commonly employed chemicals: alkylbenzene sulfonate, alcohol sulfates, or methylene blue active substances [25].

The aim of the legislation is to eliminate these substances from entering and contaminating the groundwater supply. Since Suffolk is almost entirely without sewers the substances enter the groundwater supply through septic tanks and cesspools. Once the substances are in the groundwater they fail to break down chemically (biodegrade) and dilution does not take place. The result is cloudy and bad-tasting water.

Contamination from the above-mentioned substances occurs primarily in the upper glacial aquifer. "Detergents are the most persistent and most commonly found pollutant in the Glacial aquifer and the most frequent cause for rejection of glacial wells as source of water supply. Synthetic detergent residues, even when other sewage constituents are minimal, have forced public water purveyors to abandon or curtail their use and development of the Glacial aquifer" [6:1-1].

Since the glacial aquifer contains a large supply of fresh water and is the most accessible, contamination of this aquifer has a serious effect on the availability of fresh water for Long Island.

Where the glacial aquifer has been contaminated, suppliers have placed wells in the deeper Magothy stratum. "This procedure has been followed to a large extent but poses the risk of overdevelopment of the Magothy aquifer. Overpumping of this aquifer will result in the increased transfer of contamination from the overlying Glacial formation to the underlying aquifer, and may also induce greater salt water intrusion from surrounding waters into the fresh water resources of Long Island" [6:1-2].

The Soap and Detergent Association has criticized the legislation, stating that lack of sewers is the real problem and not detergents. The Association has also noted that not all non-biodegradable substances have been banned. At this time it appears Nassau County does not plan to follow Suffolk County's lead in banning detergents.

As mentioned earlier, Nassau County has a more extensive sewer program than Suffolk County, and, additionally, most of Nassau's wells pump water from the Magothy stratum, which to date has not been seriously contaminated by detergents. "The large initial detergent concentration in residential waste, coupled with its limited degradation and adsorption, contribute to its presence in substantial quantities in the shallow groundwater, but at the same time, explains the slow penetration of deeper aquifers" [9:9].

Tests of Nassau County wells have indicated there is no immediate need for banning detergents. "Synthetic detergents are present in only seven public supply wells in Nassau, out of a total of 381 public wells. Only one of these seven, a shallow well in Wantagh, exceeds the standard of 0.5 mg/1 for detergents with a value of 0.54 mg/1. The other six wells have concentrations of between 0.04 to 0.12 mg/1. Both the number of wells with positive detergent values and the detergent concentrations within each well have been decreasing in recent years. The maximum number of public wells with positive detergents was 16 and occurred in 1967. . . . The surfacrant component of synthetic detergents does not pose an immediate nor significant threat to the quality of groundwater aquifers used for public supplies. Private supplies, which would be adversely affected by detergent contamination, are generally not used" [9:10-1].

Water Demand

As was mentioned earlier, there are vast amounts of fresh water below Long Island, approximately 60 trillion gallons. Not all of this water can be made available to supply to fresh-water needs of the public. Of the total groundwater supply, estimates have been made of the amounts of water that can be removed from the ground annually without causing any ecological imbalances or damaging the total supply of fresh water. This amount that can be withdrawn has been titled its "safe yield." Safe yield has been defined as "... the maximum quantity that can be withdrawn annually in perpetuity within reasonable economic limits while still maintaining the groundwater basin at a consistent level and at a reasonable productive size" [10:36]. The safe yield is probably not one specific amount at all points in time. "... the safe yield of the groundwater reservoir for all of Long Island or for subareas on the Island can range between large limits depending mainly upon future management decisions regarding (a) the undesired results (hydraulic, water quality, economic, and other) that are to be avoided, (b) the amount of natural groundwater that is salvaged, and (c) the amount of additional groundwater recharge (natural or artificial) that is induced" [5:92]. Since Long Island is surrounded by salt water, use of the fresh water supply beyond the safe yield could possibly lead to salt water intrusion, making the groundwater unusable, as has already occurred in Kings and Queens counties.

A first step in estimating the safe yield is to estimate the amount of rainfall replenishing the groundwater supply. It was mentioned earlier that the average annual precipitation is about 44 inches and this accounts for approximately 1600 MGD for the water budget area. Of this total rainfall, it is estimated approximately 50 percent of the precipitation actually replenishes the groundwater supply. This estimate is probably only accurate within plus or minus 25 percent. The reason for the large differences are the many simplifying assumptions used as well as disregarding the effects of man, which in some cases increase and in other cases decrease the aquifer recharge. By using the 50 percent estimate 800 MGD ± 200 MGD is the amount of precipitation expected to replenish the groundwater supply.

The second step involved with estimating the future safe yield is to estimate Long Island's population growth. There are many population projections available, and there is some great variance among them. Reasons for the differences are the basic assumptions used. Nassau County had its greatest growth between the years 1950 and 1960, increasing from 672,765 to 1,300,171. This growth slowed down substantially in the next decade, and the latest census figures show a population of 1,420,021. Nassau's growth has tapered off primarily because most available land has been developed to the extent possible.

Suffolk County's population during these periods increased from 276,000 in 1950 to 668,800 in 1960 to 1,114,164 in 1970. Suffolk's great growth is expected to continue for some time in the future as there are still large areas available for development.

The population estimates used in this chapter were drawn from Long Island Water Resources, which were "made in consultation with the Nassau-Suffolk Regional Planning Board and with the Office of Planning Co-ordination" [10:26]. It may be noted that Long Island's 1970 population was projected to 1970 as 2,605,000 while the announced census indicated a population of 2,534,000, a difference of 71,000 or 2.7 percent.

The third step in this analysis is to estimate the future water usage. Using the population and groundwater withdrawal figures for Nassau and Suffolk Counties for the years 1950 through 1969, correlation coefficients of .884 and .946 were computed, indicating close relationships.

The appendix develops regression lines to be used in projecting groundwater pumpages for Nassau and Suffolk Counties for the period 1970 to 1980. Due to the different natures of the two counties, Nassau is highly developed while Suffolk is not so highly developed, different equations were computed for each county. For Nassau County the equation developed was Estimated Groundwater pumpage (Y) equals 10 plus .1401 Projected population (X), (Y_{est} = 10 + .1401X), and for Suffolk County Y_{est} = 12 + .1135X.

Using the equations computed, the expected populations and corresponding groundwater withdrawals are shown in table 11.1.

It was mentioned earlier that 800 MGD ± 200 MGD is the amount of rainfall that can be expected to replenish the groundwater supply. When using the low figure of 600 MGD it appears obvious that at no time will the expected groundwater withdrawal exceed the amount replenished through rainfall during the next decade.

When using the Nassau County statistics a slightly different picture may appear. In the Greely and Hansen Study for Nassau County in 1963, it was estimated that 154 MGD was the safe yield for Nassau County [6:3-7]. This

Table 11-1

Nassau and Suffolk Counties, N.Y.: Estimated Population and Groundwater Withdrawal, 1970-1980

Year	Projected Nassau Population (thousands) (a)	Estimated Groundwater Withdrawal (MGD) (b)	Projected Suffolk Population (thousands) (c)	Estimated Groundwater Withdrawal (MGD) (d)	Total Groundwater Withdrawal (MGD) (e)
1970	1,420	189	1,114	128	317
1971	1,435	191	1,150	133	324
1972	1,451	193	1,190	139	332
1973	1,466	195	1,230	144	339
1974	1,482	198	1,270	150	348
1975	1,497	200	1,310	155	355
1976	1,513	202	1,350	161	163
1977	1,528	204	1,390	166	370
1978	1,544	206	1,430	171	377
1979	1,559	208	1,470	177	385
1980	1,575	211	1,500	182	393

Source: (a) and (b) LONG ISLAND WATER RESOURCES, Division of Water Resources, New York State Conservation Department, Albany, N.Y., 1970, p. 26; (b) and (a)—computed from data of columns (a) and (c) and regression equations (see text); column (e)—sum columns (b) and (d).

estimate is adjusted in Long Island Water Resources to 159 MGD in 1965 and was expected to increase to 170 MGD by the year 1990 "when increased urbanization is expected to affect the water budget" [10:37].

In 1969 Nassau County had sewers serving approximately 44 percent of the population, and it was expected to be completely sewered by the year 1985. Using this information in a straight-line fashion, table 11.2 attempts to project Nassau County's net groundwater withdrawal for the period 1970-1980, and to compare this with the estimated safe yield.

Inspection of table 11.2 indicates there is a possibility Nassau County on a countywide basis will be exceeding its safe yield quantity toward the end of the next decade, 1980. Withdrawals of water beyond this safe yield figure will be mined water, "... a one-time withdrawal whose minimum consequence would be a loss in aquifer size" [10:40]. Results of this mining of water cannot be predicted without actually measuring the effect upon the water table. The experiences of Kings (Brooklyn) and Queens counties indicate there is a possibility the mining of water may lead to salt water intrusion, but to what extent and at what point in time this may occur cannot be predicted with any certitude.

In Suffolk County the safe yield has been estimated at 501 MGD [6:3-7] and there is little likelihood of groundwater withdrawal exceeding this estimate before 1980. Even if Suffolk County were completely sewered by 1980 (a very

Table 11-2
Estimated Population and Estimated Groundwater Withdrawal, Nassau County 1970-1980

Year	Projected Groundwater Withdrawal (MGD) (a)	% of County Not Sewered (b)	Groundwater Recharge by Cesspools (MGD) (c)	Net Groundwater Withdrawal (MGD) (d)	Safe Yield* (MGD) (e)
1970	189	52.5	99	90	161
1971	191	49.0	94	97	162
1972	193	45.5	88	105	163
1973	196	42.0	82	114	164
1974	198	38.5	76	122	165
1975	200	35.0	70	130	166
1976	202	31.5	64	138	167
1977	204	28.0	57	147	168
1978	206	24.5	50	156	169
1979	208	21.0	44	164	170
1980	211	17.5	37	174	171

Source: (a) Table 11-1; (b) straight line projection assuming 44 percent sewered in 1969 and 100 percent in 1985; (c) column (a) times column (b); (d) column (a) minus column (c), (e) LONG ISLAND WATER RESOURCES, Division of Water Resources, New York State Conservation Department, Albany, N.Y., 1970, p. 37.

unlikely occurrence) the total groundwater withdrawal estimated for 1980 would only be 182 MGD (table 11.1).

While there appears to be little chance of an overdraft in either Nassau or Suffolk County before 1980, local problems could very likely arise. For example, in Nassau County, "The Town of Hempstead, accounting for 55 to 60 percent of the total consumptive water use in Nassau County but only 40 percent of the area from which water may be withdrawn at present, is one of the most critical areas in relation of consumptive use to safe yield" [10:36].

There are also local problems that may occur within Suffolk County, although not within the near future. "Certain towns will experience deficiencies, unless extensive upzoning is instituted to reduce projected water usage. Sometime after 2000 the area of East Hampton, east of Napeaque, will not have adequate local water" [18:III,9].

Current Water Supply Procedures

Kings and Queens counties receive their water primarily from upstate water supplies through New York City. Nassau and Suffolk counties receive their water primarily from groundwater distributed by public suppliers. There are still many private supply wells on Long Island, especially in eastern Suffolk County. In 1969 it is estimated that approximately 350,300 people were receiving their water from private sources [19:7]. Wherever possible this water is pumped from the upper glacial aquifer. Often this groundwater has been polluted, necessitating withdrawal from deeper aquifers.

In view of the continued growth of Long Island's population and water demand, continued use of the current water supply procedures will lead to a hydrological imbalance that may result in salt water intrusion and declining groundwater levels. In addition, the practice of going to deeper and deeper aquifers to supply water, while successful in the past, may not be the most practicable method for retrieving fresh water. The glacial (upper) aquifer is generally composed of coarse sand and gravel in which wells can easily be constructed. For these reasons, wells can be constructed in this aquifer at considerably less cost than in the deeper and less easily developed aquifers.

While the Magothy aquifer probably contains the greatest amount of fresh water, this water is not so easily retrieved as it is from the upper glacial aquifer. The greater drilling depths are an obvious reason for this, but there are also other reasons, "the individual beds in the formation (Magothy) generally do not have a wide lateral extent, and few of the beds have been correlated over distances of much more than a mile or two" [6:3-4]. As a result, data from existing wells reveal the nature of the geological formation of the wells at their specific sites but cannot be extrapolated elsewhere.

Accordingly, different water supply theories should be investigated. Since fresh water is more easily retrieved from the upper glacial aquifer, perhaps more emphasis should be put on cleaning up the waters of this aquifer rather than attempting to dig deeper wells. The following sections review various alternative water supply theories.

Importing Fresh Water

It is not uncommon for large metropolitan areas to import water from less populated areas. Since Long Island's Nassau County is very heavily populated, with almost all its land developed, and since it is projected that Suffolk County may very well one day reach this status, the importation of water is one method to be considered to satisfy Long Island's fresh water requirements. New York City's reservoirs and aqueducts are located in New York State. "New York City has no definite legal obligation to furnish water to Nassau County . . . any water it furnished would probably be surplus to the city system's demands and mutual advantages would be necessary" [10:40].

But there is no guarantee New York City will have this surplus to deliver to Long Island. According to one source "New York City must develop new sources of supply for about 500 MGD more by 1990 and another 500 MGD by 2020" [10:41]. Under drought conditions it has faced serious shortages in past years, and the satisfaction of its requirements was at the expense of Philadelphia, Pennsylvania, which also draws on the Delaware system.

As a result it seems unlikely New York City would be either able or desirous of supplying fresh water to Long Island. In view of the potential water shortages facing New York City, it is likely they may have to look outside of New York State for fresh water. This may prove to be very difficult since "in view of the rapidly increasing demand for water throughout the northeast, it is highly unlikely that any other state within hundreds of miles of the New York Metropolitan area would permit the export of water to Suffolk County" [18:III,66].

Sources other than the New York City water supply do not seem to hold much promise for Long Island either. Besides the obvious problems of the great expense and huge distances that must be covered, most neighboring states have water problems of their own. "Inclusion of a transmission main . . . from Long Island to Connecticut, presumably to bring water across from Connecticut, has been suggested; but Connecticut does not have excess quantities. Like New York, its sources were in short supply during the recent drought, and interstate problems would be difficult to surmount. Similarly, northeastern New Jersey suffered severely during the drought, and projected demands indicate water supply will be critical there. More distant areas offer no significant potential for Long Island's needs" [10:41].

Desalination

During the middle 1960s a severe drought occurred in the northeastern United States. This drought resulted in fresh water shortages in many areas. The fresh water shortages led to large-scale emphasis on the possibility of using sea water to supply our fresh water needs. Large-scale studies have been initiated in this area, and while it appears to be a very expensive solution at the present time, there is still hope desalination may one day be the solution for fresh water shortages.

Desalination consists of removing dissolved solids from water. There are many different methods of desalination. The following are the most common methods in use presently [10:41]: (1) distillation, (2) membrane processes, (3) freezing, (4) humidification, and (5) chemical.

The distillation process is, at present, the most advanced. "As of January 1, 1968, there were 627 desalting plants of 25,000 gallons per day capacity or greater that were either in operation or under construction throughout the world. The total capacity of these plants was approximately 222 million gallons per day. Distillation plants accounted for approximately 93 percent of this capacity" [18:III,70].

Desalination at the present time is a very expensive process. "Most desalination plants yield 1 MGD or less at an average cost of $1.00 per 1,000 gallons of fresh water, far above the present costs for fresh water supply by conventional means" [10:41]. There are indications, however, that present technology exists that could substantially lower these costs to more competitive levels. Table 11.3 details the projected costs for desalting water based upon economies of size as well as the sale of electric power generated in a joint process with desalination.

While these costs are projected to be substantially below current costs for desalination, they are still way above current supply costs. In addition, the costs quoted do not include any costs involved with the distribution of the water. Most populated areas of Long Island are completely mained, however, and distribution costs should consist only of tying in the main from the desalination plants to the already existing mains.

A more serious problem is the effect of desalted water upon the existing mains. "The aerated, distilled pure water, particularly if run in a way to reverse the flow, can cause catastrophic purging of existing accumulations in distribution mains, leading to delivery of water with high turbidity, color, and possibly unacceptable taste. The water will attack cast iron and soften the portland cement in asbestos-cement pipes" [10:42].

The costs quoted above are for the most part based upon the desalination of water termed moderately saline or only slightly saline. Table 11.4 gives the classifications of saline water by the United States Geological Survey.

Table 11-3
Projected Costs for Desalination of Water

Process	Projected (per thousand gallons)
Vacuum freezing-vapor compression	45¢-50¢
Distillation	15¢-45¢
Reverse osmosis process	35¢-40¢
Electrodialysis	30¢-35¢

Source: REPORT–COMPREHENSIVE PUBLIC WATER SUPPLY STUDY, SUFFOLK COUNTY, NEW YORK, CPW 5-24, Vol. III, Preliminary Copy Subject to Review, Holzmacher, McLendon and Murrell, Consulting Engineers (Melville, N.Y.: 1968), p. 75.

Table 11-4
Classification of Saline Water

Classification	Dissolved Solids Content Milligrams per Liter (mg/l)
Brine	35,000 plus
Slightly saline	1,000-3,000
Moderately saline	3,000-10,000
Very saline	10,000-35,000

Source: REPORT–COMPREHENSIVE PUBLIC WATER SUPPLY STUDY, SUFFOLK COUNTY, NEW YORK, CPW 5-24, Vol. III, Preliminary Copy Subject to Review, Holzmacher, McLendon and Murrell, Consulting Engineers (Melville, N.Y.: 1968), p. 69.

Sea water generally approximates 35,000 mg/l dissolved solids content. Desalination of highly salinated water is, of course, much more expensive than that of slightly desalinated water.

Water Recycling

It was indicated earlier that there is much water reuse on Long Island at the present time. The problem is this water is derived from cesspools that pollute the groundwater supplies. Eventually Long Island will be supplied with an extensive sewer system that will end this form of groundwater recharge. While this will stop the pollution of this fresh water supply, it will also result in a reduction of the amount of fresh water available to be pumped from the ground.

Many investigations have been made and serious consideration is being given to treating the sewage effluent and returning it to the ground in a purified form. There are two basic means to return the treated effluent to the ground. One is through the use of barrier injection wells, and the second is through inland recharge.

Barrier Injection Wells

The purpose of barrier injection wells is to inject the water directly to the aquifer along the shore lines. The injection of the fresh water will then prevent the intrusion of salt water at the shore and, to some extent, will replenish the groundwater supplies inland.

Barrier injection wells have the added advantage of being able to inject the water into the deeper aquifers. In Bay Park, Nassau County, a 400 gallons per minute (gpm) well was placed in operation on April 8, 1969. This is a research project designed to test the feasibility of injecting highly treated effluent into the Magothy formation. Tests have been made by the U. S. Geological Survey and the Nassau County Department of Public Works; "Preliminary data from both tests indicate that the aquifer has readily accepted the processed water" [1-1968:19].

A tertiary sewage treatment plant at Bay Park renovates 410 gpm of effluent from a secondary treatment plant to the drinking water standards of the U. S. Public Health Service [14:226]. Since the initial run in 1968 the Bay Park experiment has experienced difficulties in the injection phase of the operation. A problem has been the clogging of screens used in the injection process.

Once the water is successfully returned to the aquifer, some authorities claim, there has been no effect upon the quality of the groundwater. "Chemical data collected thus far have not indicated any major change in the injected water after passing through the aquifer" [1-1969:9].

The key phrase in the above quotation must be "major change," since there are evidences of bacterial growth in the water pumped from the injection well. Even though the injected water generally meets potable standards, high bacterial counts are observed in water pumped from the injection well after recharge. In fact, bacterial counts are higher than in injected water suggesting growth of bacteria in the acquifer around the injection well [27:1415].

Similar results of bacterial growth have been reported by the California State Water Pollution Control Board. The high bacteria counts are noted following the initial pumping of water from the injected well. "Continued pumping results in lower counts until virtually bacterial-free water is produced . . . Apparently the surging action of the water dislodges bacteria that otherwise would not enter the well" [27:1415].

Barrier injection wells are expected to prevent salt water intrusion by forming a pressure ridge of fresh water parallel to the coast. "This information ultimately will be used to help evaluate the feasibility, and possibly the design, of a proposed line of recharge wells along the south shore of Nassau County to return about 30 million gallons per day or more of treated wastewater to the Magothy aquifer. It has been conjectured that such recharge may permit increased withdrawals of fresh groundwater and retard the rate of salt-water encroachment in southern Nassau County by maintaining submarine outflow at or above its present rate" [14:226].

Since some of the injected water is expected to move inland to replenish the groundwater supply, it is necessary this water receive tertiary treatment. The injected water must meet the accepted standards for drinking water.

A second test program was conducted at the Riverhead Sewer District. The purpose of this project was to develop design and operating criteria for the injection of treated wastewater into the water table aquifer.

At Riverhead the water was injected under conditions designed to approximate natural elements. "The tank is filled with layers of soil to duplicate natural conditions around the injection well, as well as any other desired conditions" [10:42]. This project has recently been terminated and while progress was made, apparently much more experimental work has to be done.

While barrier injection wells are still in the experimental stage on Long Island, barrier injection wells are already in successful operation in Southern California. "In California, a seven and one-half mile barrier of 77 injection wells is used for injection 45,000 acre-feet of water per year" [10:42].

Inland Recharge

The use of inland water recharge is very similar to the use of storm water recharge. Treated effluent is transported inland and allowed to filter through the ground by use of recharge basins. There are a number of disadvantages present in the use of inland recharge that are not present when barrier injection wells are used. One is the cost of transporting the treated effluent inland, and the second is the availability of land for use as recharge basins. A third problem is that fresh water recharge through recharge basins results in the water directly recharging the upper glacial aquifer. If the fresh water is being pumped from the Magothy or a lower aquifer, hydrological imbalances could occur.

The transportation of treated effluent is only a problem assuming shoreline sewage treatment plants. If the plants were to be located inland then inland recharge is actually an advantage. In this instance there is no need to transport the sewage long distances to plants located on the shores. Inland treatment plants might be located in the most advantageous positions eliminating the need for long truck lines. The problem is the availability of land. With Nassau County highly developed and Suffolk County headed in that direction, the most advantageous locations for both inland sewage treatment plants as well as recharge basins are probably highly developed areas. In addition, inland recharge basins require fairly large areas of land. "Present design criteria of the Suffolk County Health Department for the recharge of treated effluent is for a maximum of five gallons per day per square foot. This is equivalent to 4½ acres of recharge surface required for each million gallons per day of recharge for average size basins of, say, 200 by 400 feet. This would necessitate a gross area on the order of 7 acres per million gallons of recharge capacity . . . on the basis of 100 percent recharge of an anticipated 86 mgd of sewage effluent from the Towns of Babylon and Islip in 2020, approximately 600 acres of land would be required for recharge basins" [18:III,57]. Not only would the acquisition costs for such large areas be impossible, it is unlikely the land is available, especially in the most effective locations.

The need for large areas of land for inland recharge would be partially offset through the use of existing storm water recharge basins. Those basins best suited for inland recharge are those in the mid island area that are normally dry. "If all the existing storm water recharge basins along the Long Island Expressway in the Babylon-Islip area were utilized for the recharge of treated sewage effluent during the substantial part of the time they are empty, approximately 23 MGD could be recharged in this manner" [18:III,58].

Cost Comparisons

When exploring various water supply procedures it is necessary to compare the costs associated with each procedure. The indications are that in the eastern United States the cost for developing new water supplies is in the neighborhood of 20 cents per 1000 gallons. This is what New York City charges the suburbs it supplies, while in nearby New Jersey the cost is 17 cents per 1000 gallons [16:638]. In addition to the 20 cents per 1,000 gallons New York City now charges Westchester County for water, New York City has proposed an increase in the rates it charges individual Westchester communities for water service [28].

By comparison, the Suffolk County Water Authority, the largest public supplier in Nassau or Suffolk Counties, supplied 21.7 million gallons of water during the fiscal year 1970 at a cost of $8.6 million, or an average cost of 4 cents per 1,000 gallons [2:3,5]. These are the direct costs involved with supplying fresh water and do not take into consideration any indirect costs that may be applicable. An example of an indirect cost that cannot really be measured might be the loss of autonomy suffered by the political subdivision that must rely upon other government localities for its fresh water supply.

Intimately related to the fresh water supply procedures are those costs associated with the treatment and disposal of sewage effluent. This is especially true when considering groundwater recharge through wastewater recycling. Since Suffolk County is virtually unsewered at the present time, the following section deals with the sewage treatment practices of Nassau County.

"There are at present twenty-five municipal sewage treatment plants in Nassau County with a combined capacity of over 100 million gallons per day. All of these plants have, for several years, provided secondary treatment, considerably in advance of a goal set by New York State Health Department for all treatment plants by 1972" [24:7].

At the present time all treated effluent is being discharged to the estuarine and harbor waters surrounding Long Island. Construction has begun on a Water Pollution Control Plant at Wantagh, Long Island, and it is anticipated this plant will be completed by the latter part of 1971 [1-1970:24]. This new sewage treatment plant at Wantagh will be equipped with an outfall extending more than two miles into the Atlantic Ocean [24:8]. The purpose of the ocean outfall is to protect the water quality of bays and harbors.

During 1970 Nassau County treatment plants discharged an average of 95.44 million gallons daily (MGD) with a treatment range of 62-93.2 percent removal of suspended solids and 57-94.7 percent removal of biological oxygen demand (BOD) [23]. The Bay Park Treatment Plant, the largest of the twenty-five treatment plants, treated a daily average of 65.0 MGD and "the Bay Park Plant had a removal efficiency of 87.6% of suspended solids and 88.6% BOD. This efficiency is down slightly from 1969 which can be attributed to the daily flow being 4.9 million gallons above design capacity" [1-1970:25].

Secondary treatment of sewage wastes produce effluents that in most cases can be assimilated by receiving waters without adversely affecting the water quality. With increased population and increased urbanization on Long Island this may no longer be true. To date three Nassau County beaches have been closed as a result of polluted waters. "In the South Shore bays at the western end of Nassau, the large quantities of treated sewage discharged, in spite of a high degree of treatment, have exceeded the assimilation capacity of the shallow bay waters with the result that large portions of the bay waters are presently closed to shellfishing. On the north shore, pollution originating in New York City is carried by tidal transport to Nassau waters. All of Manhasset Bay and Hempstead Harbor is closed to shellfishing; furthermore, Little Neck Bay is not suitable for bathing because of this situation" [24:8].

The proposed ocean outfall in Wantagh may alleviate the problems associated with discharging treated effluent in harbor and estuarine waters, but may create another problem, the "mining" of groundwater supplies. By discharging treated effluent into the ocean, a one-time use of the fresh water is created. As mentioned earlier, there is a possibility Nassau County will exceed its "safe yield" by 1980. The use of ocean outfalls will do nothing to conserve this fresh water supply. Additionally, recent estimates indicate costs of outfalls may increase cost of treatment and disposal as much as 25 cents/1,000 gal. [29].

Ocean outfall, then, involves costs of 25 cents/1,000 gallons, which are greater than the 20 cents/1,000 gallons cited earlier in this chapter for developing new fresh water supplies. Part of the reason for this high cost is the necessity of providing facilities that permit inspection of the lines and cleaning when needed. Sand and gravel may enter with sewage effluent and marine growth may occur. Accordingly, to permit inspection and cleaning and prevent biological growth, provision must be made for chlorination facilities and frequent access entries for inspection and cleaning [4:1805-6].

The costs associated with treatment of wastewater to a level suitable for groundwater recharge are extremely variable depending upon the size of the facility, operating labor rates, quality of the plant influent, and quality standards to be maintained in the product water. "To evaluate this economic feasibility, cost estimates of reclaimed water have been prepared based on the results of pilot-plant operations and conceptual design studies of water renovation plants" [16:638].

Table 11.5 represents water renovation charges, which are assumed to be a function of the size of the facility, particularly as it affects the cost of operating labor. The costs quoted are for treatment to suitable level for injection only, and do not include transmission or injection facilities. It is hoped the current experiments at Bay Park will yield information regarding both the feasibility of water injection as well as the potential costs involved.

Conclusion

During the next ten years there is a possibility Nassau County's consumptive water use may exceed its safe yield. The use of water beyond the safe yield will result in "mining" water from the underground supplies. This reduction in the groundwater supply will lead to a falling water table, which may result in salt water intrusion contaminating the fresh water supply. It is impossible to predict at which point this salt water intrusion may occur, but there is a precedent, since it has occurred in Kings and Queens counties. There appears little likelihood of consumptive use exceeding the safe yield in Suffolk County before 1980.

The use of cesspools in eastern Nassau and all of Suffolk County has led to contamination of the upper glacial aquifer. As a result public water suppliers have had to go deeper and deeper into the ground to obtain fresh water. By withdrawing water from the deeper aquifers hydrological imbalances have

Table 11-5
Estimated Unit Costs (Water Renovation)

Costs per 1000 Gal.	Plant Capacity		
	1 Million Gallons per Day	10 Million Gallons per Day	100 Million Gallons per Day
Process costs, less labor			
Coagulation	4.9	3.5	3.2
Filtration	1.8	1.1	1.0
Carbon absorption	6.3	4.5	4.0
	13.0	9.1	8.2
Operating labor	28.0	5.6	1.8
	41.0	14.7	10.0

Source: John H. Peters and John L. Rose, "Water Conservation by Reclamation and Recharge," JOURNAL OF THE SANITARY ENGINEERING DIVISION–PROCEEDINGS OF THE AMERICAN SOCIETY OF CIVIL ENGINEERS, (August, 1968), p. 638.

occurred, since the wastewaters are being returned to different aquifers. These hydrological imbalances may also result in salt water intrusion.

Before 1980 it may be necessary for Nassau County to seek new sources for its fresh water needs. This will be especially true in certain critical areas. Suffolk County will probably have a supply greater than its needs, but there are political factors to be considered before the importation of fresh water may be possible. This also is true as applies to transmitting water from water-rich areas of Nassau County to water-poor areas, since there are presently forty-four different major water suppliers in Nassau County.

Importation of fresh water from New York City or other neighboring states is another possible solution to satisfy Nassau's water needs. It has been projected, however, that New York City will have to find additional suppliers to meet its own increasing needs by 1990. Most neighboring states also are experiencing potential water shortages of their own.

Desalination has been discussed as a possible solution, but at the present time the costs involved with this process are very high. In order to reduce these costs it is necessary to build extremely large processing plants. Plants of such large capacity must, of necessity, serve very large areas. This results in high costs involved with transmitting the water. Due to these factors, "the net result is that the use of desalted water is totally unjustified from an economic standpoint whenever an adequate supply of groundwater is available" [18:III,81].

A third alternative solution is the recycling of treated sewage effluent. This method is considered by many to have the greatest potential and has been used in areas of southern California. Groundwater recharge through the use of barrier injection wells has been experimented with on Long Island with limited success.

When comparing the cost of water recharging with desalination, one expert has concluded recharge is the more economical. "In an economic comparison with the desalting of sea water, the recharge of a quaternary treated sewage effluent has an important advantage. The total dissolved solids in a typical sewage effluent is only a small fraction of the dissolved solids in brackish or sea water. The cost of removing dissolved solids tends to vary more or less directly with the concentration of dissolved solids at least in the membrane processes. As a result, the desalting of sewage effluent can be expected to result in substantial cost savings as compared to the desalting of brackish or sea water" [18:III,59].

There are a number of other advantages associated with water recycling. Unlike ocean outfalls, water renovation and its subsequent injection into the ground not only protects the quality of the surface waters surrounding Long Island, but at the same time it helps to replenish the underground fresh water supply. Additionally, by creating an underground wall of fresh water, it is expected water recycling through injection wells will prevent salt water contamination of the fresh water supply.

Information about the costs associated with and the feasibility of water recycling through injection wells is fairly speculative because of the lack of operating experience, but in light of the many advantages associated with this procedure I believe it should be the primary method of wastewater disposal as well as fresh water supply that should be investigated.

Appendix

Table A11-1
Population (X) as a Predictor of Groundwater Withdrawal (Y) Nassau County

Year	X Population	Public Water Supply	Industrial	Agriculture	Y Total	X^2	XY
1950	673	100.0	8.8	.9	110	452,929	74,030
51	745	69.5	12.0	.9	82	555,025	61,090
52	817	76.5	13.9	.6	91	667,489	74,347
53	889	100.0	17.7	.3	118	790,321	104,902
54	961	140.7	17.5	.5	159	923,521	152,799
55	1,032	106.2	20.1	.5	127	1,065,024	131,064
56	1,086	98.0	22.0	.3	120	1,179,396	130,320
57	1,140	136.8	24.3	.4	161	1,299,600	183,540
58	1,194	113.8	22.2	.2	136	1,425,636	162,384
59	1,248	124.8	27.1	.1	151	1,557,504	188,448
60	1,300	123.4	21.6	.1	145	1,690,000	188,500
61	1,323	127.2	26.0	.1	153	1,750,329	202,419
62	1,346	143.3	28.3	.1	172	1,811,716	231,512
63	1,369	149.9	30.0	.1	180	1,874,161	246,420
64	1,392	162.2	31.1	.1	193	1,937,664	268,656
65	1,401	175.4	33.8	.2	209	1,962,801	292,809
66	1,410	185.2	29.7	.2	215	1,988,100	303,150
67	1,422	149.9	31.0	.1	181	2,022,084	257,382
68	1,433	173.3	34.4	.2	208	2,053,489	298,064
69	1,446	168.0	33.3	.1	201	2,090,916	290,646
	23,627				3,112	29,097,705	38,424,82

$$a = \frac{(\Sigma y)(\Sigma x^2) - (\Sigma x)(\Sigma xy)}{n(\Sigma x^2) - (\Sigma x)^2} = \frac{3,112(29,097,705) - 23,627(3,842,482)}{20(29,097,705) - (23,627)^2}$$

$$= \frac{90,552,057,960 - 90,786,322,214}{581,954,100 - 558,235,129} = \frac{-235,262,254}{23,718,971} = -9.9$$

$$b = \frac{n(\Sigma xy) - (\Sigma x)(\Sigma y)}{n(\Sigma x^2) - (\Sigma x)^2} = \frac{20(3,842,482) - 23,627(3,112)}{23,718,971}$$

$$= \frac{76,849,640 - 73,527,224}{23,718,971} = \frac{3,322,416}{23,718,971} = .1401$$

$y_{estimate} = a + bx = -9.9 + .1401 x$

Table A11-2
Population (X) as a Predictor of Groundwater Withdrawal (Y) Suffolk County

Year	X Population (thousands)	Public Water Supply	Industrial	Agriculture	Y Total	X^2	XY
1950	276	24.3	12.4	3.6	42	76,176	11,592
51	303	24.5	13.4	9.4	47	91,809	14,241
52	330	25.6	16.0	16.0	58	108,900	19,140
53	357	27.8	17.6	16.0	61	127,449	21,777
54	384	27.5	20.0	16.9	64	147,456	24,576
55	412	27.1	21.9	17.6	67	169,744	27,604
56	463	28.9	17.4	8.2	55	214,369	25,465
57	514	34.7	20.6	17.6	73	264,196	37,522
58	365	33.0	20.5	7.8	61	319,225	34,465
59	616	37.2	24.8	6.1	68	379,456	41,888
60	667	40.3	26.6	8.2	75	444,889	50,025
61	703	43.3	27.4	9.2	80	494,209	56,240
62	739	48.8	28.9	11.0	89	546,121	65,771
63	784	55.0	32.4	11.1	98	614,656	76,832
64	830	64.4	34.6	20.9	120	688,900	99,600
65	885	70.0	35.2	18.0	123	783,225	108,855
66	939	75.1	36.3	18.7	130	881,721	122,070
67	977	73.3	35.5	2.1	111	954,529	108,447
68	1014	85.0	38.1	14.8	138	1,028,196	139,932
69	1059	87.2	35.8	8.3	131	1,121,481	138,729
	12,817				1,691	9,456,707	1,224,771

$$a = \frac{(\Sigma y)(\Sigma x^2) - (\Sigma x)(\Sigma xy)}{n(\Sigma x^2) - (\Sigma x)^2} = \frac{1,691(9,456,707) - 12,817(1,224,771)}{20(9,456,707) - (12,817)^2}$$

$$= \frac{15,991,291,537 - 15,697,889,907}{189,134,140 - 164,275,489} = \frac{293,401,630}{24,858,651} = 11.8$$

$$b = \frac{n(\Sigma y) - (\Sigma x)(\Sigma y)}{n(\Sigma x^2) - (\Sigma x)^2} = \frac{20(1,224,771) - 12,817(1,691)}{24,858,651} = \frac{24,595,420 - 21,673,547}{24,858,651}$$

$$= \frac{2,821,873}{24,858,651} = .1135$$

$$Y_{est} = a + bx = 12 + .1135x$$

Notes

1. *Annual Reports 1968-1970,* Nassau County Department of Public Works, Mineola, N.Y.

2. *Annual Report 1970,* Suffolk County Water Authority, Oakdale, N.Y.

3. J. J. Baffa and N. J. Bartilucci, "Waste Water Reclamation by Ground Water Recharge on Long Island," *Water Pollution Control Federation Journal,* March, 1967, pp. 431-35.

4. Wallace J. Beckman, "Engineering Considerations in the Design of an Ocean Outfall," *Journal of Water Pollution Control Federation,* October, 1970.

5. Phillip Cohen, O. L. Franke, and B. L. Foxworthy, *An Atlas of Long Island's Water Resources,* New York Water Resources Commission Bulletin 62, State of New York, 1968.

6. *Final Report of the Long Island Ground Water Pollution Study,* State of New York Department of Health, Albany, N.Y., April, 1969.

7. Francis J. J. Flood, "Water Pollution Control Program," *Programs in Environmental Health* (Mineola, N.Y.: Nassau County Department of Health, October 1969).

8. R. C. Heath, B. L. Foxworthy, and Phillip Cohen, *The Changing Pattern of Groundwater Development on Long Island,* U. S. Geological Survey Bulletin 524 (Washington, D.C.: Government Printing Office, 1966).

9. *Influence of Sewage Constituents on Quality of Nassau Waters* (Mineola, N.Y.: Nassau County Department of Health, March 11, 1971).

10. *Long Island Water Resources,* Division of Water Resources, New York State Conservation Department, Albany, N.Y.: 1970.

11. E. R. Lusczynski and W. V. Swarzenki, *Salt-Water Encroachment in Southern Nassau and Southeastern Queens Counties, Long Island, N. Y.,* U. S. Geological Survey Water Supply Paper 1613-F (Washington, D.C.: Government Printing Office, 1966).

12. Roscoe C. Martin, *Water for New York* (Syracuse, N.Y.: Syracuse University Press, 1960).

13. N. M. Perlmutter and J. J. Geraghty, *Geology and Ground-Water Condition in Southern Nassau and Southeastern Queens Counties, Long Island, N. Y.,* U. S. Geological Survey Water Supply Paper 1613-A (Washington, D.C.: Government Printing Office, 1963).

14. N. M. Perlmutter, F. J. Pearson, and G. D. Bennett, "Deep-Well Injection of Treated Waste Water — An Experiment in Re-use of Ground Water in Western Long Island, N. Y.," *Geological Association Guidebook, 40th Annual Meeting* (Mineola, N.Y.: U. S. Geological Survey, 1968).

15. N. M. Perlmutter and Julian Soren, *Effects of Major Water Table Changes in Kings and Queens Counties, New York City,* U. S. Geological Survey Professional Paper 450E, Mineola, N.Y., 1969.

16. John H. Peters and John L. Rose, "Water Conservation by Reclamation and Recharge," *Journal of the Sanitary Engineering Division — Proceedings of the American Society of Civil Engineers,* August 1963.

17. *Report-Comprehensive Public Water Supply Study, Suffolk County, New York*, CPW5-24, Vol. 1, Holzmacher, McLendon and Murrell, Consulting Engineers, Melville, New York, 1968.

18. *Report-Comprehensive Public Water Supply Study, Suffolk County, New York*, CPW 5-24, Vol. II and Vol. III. Preliminary copies subject to review, Holzmacher, McLendon and Murrell, Consulting Engineers, Melville, New York, 1968.

19. *Report of Long Island Groundwater Withdrawal During 1969*, New York State Water Commission, May 1, 1970.

20. G. E. Seaburn, *Preliminary Results of Hydrologic Studies at Two Recharge Basins on Long Island, New York*, U. S. Geological Survey Professional Paper 627-C, Mineola, N.Y., 1970.

21. Sheldon O. Smith, "Water Supply." *Programs in Environmental Health*, Nassau County Department of Health, October 13, 1967.

22. Sheldon O. Smith and Joseph H. Baier, *Report on Nitrate Pollution of Groundwater Nassau County, Long Island*, Nassau County Department of Health, December 1969.

23. *Wastewater Treatment Plants Flow Averages For Period January, 1970 - December, 1970*, Bureau of Water Pollution Control, Nassau County Department of Health, Mineola, N.Y.

24. *Water Pollution Controls in Nassau County*, Nassau County Department of Health, Mineola, N.Y.

25. *New York Post*, November 11, 1970, p. 5.

26. *Long Island Press*, U. S. census figures, December 12, 1970, p. 1.

27. John Vecchio, "A Note on Bacterial Growth Around a Recharge Well at Bay Park, Long Island, New York," *Water Resources Research*, Vol. 6, No. 5, U. S. Geological Survey (Mineola, N.Y.: October 1970).

28. Letter from Joseph F. Holt, Director, Water Division, State of New York Public Service Commission (April 27, 1971).

29. Letter from John L. Rose, Consulting Engineer for Burns and Roe, Inc. Consultant to Nassau County Department of Public Works (Oradell, New Jersey), April 20, 1971.

12 Water Recycling in Southern California

JAMES CARMICHAEL

The Southern California area is classified as semi-arid. Rainfall averages less than 15 inches a year. This ranges from 10 inches by the coast to approximately double that in the foothills. Rainfall in the higher mountains of the coastal plain drainage area is much heavier.

Until recently the area had been experiencing a long period of severe drought. One particular period of drought that lasted from 1893 to 1904 caused the leaders of the city of Los Angeles to look north to the Owens Valley in the high Sierra for an additional supply of water. Its 233-mile-long Los Angeles Aqueduct was completed in 1913, and it continues to supply the city with 320,000 acre-feet a year [6].

In the 1920s, the fast-growing city of Los Angeles realized that, due to the large increase in its population, an additional source of supplemental water supply must be developed. This time the Colorado River, which is 300 miles away, was considered to be the best source. The city therefore commenced studies into the feasibility of an aqueduct from the Colorado to the Southern California coastal plain.

At this juncture it was also evident that water shortages were even more likely to occur in other cities in the same area. Unlike Los Angeles, they did not have the benefit of importing water from the Owens Valley on the east side of the Sierra Nevada Mountains.

The cities of Long Beach and Pasadena, the next largest cities in Los Angeles County, were involved in litigation over rights to the surplus runoff of the San Gabriel River. Moreover, other cities experienced or were threatened with lower artesian well pressures or underground water levels, the start of seawater intrusion along the coast, and the decline in the quality of the water in their wells.

Thus, the Colorado River Aqueduct, which was planned as a Los Angeles project, became the undertaking of a number of Southern California cities. This was done to insure that the region as a whole would receive an adequate water supply. The Metropolitan Water District of Southern California (hereafter referred to as MWD) was the agency created to allow the interested cities to pool their resources to finance and build the aqueduct and distribution system.

Metropolitan Water District

In 1928 the MWD was organized as a public agency. Its purpose was to develop, store, transport, and sell water for municipal and industrial uses of its member agencies. At conception it consisted of eleven members and it immediately took over the task of investigating the feasibility of an aqueduct from the Colorado River. By 1931, thirteen cities in Los Angeles and Orange counties were members. These included Los Angeles, Pasadena, Long Beach, Beverly Hills, Burbank, Compton, Glendale, San Marino, Santa Monica, and Torrance. In Orange County the cities were Anaheim, Fullerton, and Santa Ana.

While the MWD only covers 12 percent of the Southern California geographic area, it services approximately 90 percent of the population of the area. The district covers an area of 4,800 square miles, has a total population of approximately 10 million people, and an assessed value of around $25 billion.

Consequently, the territory covered by the MWD plays a major role in the water problems of Southern California, and as such it will be used as a basis of study for this chapter. The study will consider the problems from the inception of the MWD through the year 1980.

The terminal reservoir from the Colorado River is Lake Mathews in Riverside County. The Colorado River Aqueduct crosses 242 miles of desert and mountains. Construction was started in January 1933, and the first water was delivered to a district member, the city of Pasadena, in June 1941.

Transportation of the Colorado water is carried out by the use of tunnels, conduits, canals, and five pumping plants. At full capacity the aqueduct can deliver more than one million acre-feet of water per year. It has been operating at full capacity since 1964. The power to bring this water along its journey is supplied by the Hoover and Parker Dams, and involves a total purchase of 2 billion kilowatt hours of electricity per year [6:15].

Population

The key to Southern California's water problem is its large population. The area's basic water shortage is compounded by the extremely heavy water consumption of the municipalities' residents. The MWD is no doubt one of the fastest-growing areas in the nation. In 1940 the six-county area had a population of 3.5 million. The 1960 census showed that the figure had grown to 8.8 million, and in 1969 the last estimate was that it had risen to 11 million. Furthermore, as is indicated in tables 12.1 and 12.2, the six counties of the MWD account for 94 percent of the population of Southern California, and 57 percent of the state as a whole. The ten counties known as Southern California represent 61 percent of California's total population [6:19].

Projections through 1980 indicate that the growth rate is tailing off. However, as is indicated in table 12.3, the MWD's population has been estimated to be 14,924,000 by 1980. The projected figure for the Southern California area is predicted to be 16,832,000. Consequently, the MWD will still be servicing approximately 90 percent of the Southern California area, and the water demands in this densely populated area will still be tremendously high.

Table 12-1
Area and Population in the Six Metropolitan Water District Counties, Southern California

County	Square Miles			Population		
	Total	In MWD	% In MWD	Total	In MWD	% In MWD
Los Angeles Co.	4,080	1,388	34	7,168,000	6,855,600	96
Orange Co.	786	654	83	1,384,000	1,370,000	99
Riverside Co.	7,249	1,004	14	444,000	313,000	70
San Bernardino Co.	20,154	241	1	705,000	225,000	32
San Diego Co.	4,314	1,170	26	1,352,000	1,275,000	94
Ventura Co.	1,865	336	18	365,000	200,000	55
Total Six Counties April 1969.	38,448	4,793	12	11,418,000	10,238,600	90

Source: Metropolitan Water District, Official Statement, Los Angeles, California, 1970.

Industry and Agriculture

Total employment in the six-county area has steadily risen over the last twenty years. In 1952 total employment for this area was 2.5 million, and by 1969 that figure had increased to 4.6 million. The employment distribution factor is the significant influence on water demand. Table 12.4 shows that the distribution of employment is varied. It is in the manufacturing area that industry makes its greatest demands on the water supply. A total of 25.1 percent of the employment distribution of the MWD is in manufacturing. There are 22,000 manufacturing establishments in this area employing more than 1,200,000 persons. The MWD's value output figures are stated at $17 billion annually.

Agriculture is of fundamental importance in the economy of Southern California. This six-county area still maintains 15,000 farms with 4.7 million acres of land. Farm output was valued at approximately 828 million in 1969.

Accordingly, it is found that industry in Southern California is not the major drain on the water supply. The type of manufacturing that is done in this area does not call for the use of large amounts of water.

However, in total the demand for the entire area is definitely on the way up, so that by 1980 water demand in this area will have reached approximately 4.08 million acre-feet per annum. This is directly due to the great size of the continually increasing population. The Southern California municipalities serve one of the most densely populated, affluent parts of the country, and this is exactly why so much water is needed.

Table 12-2
Population Distribution in Southern California

Six MWD Counties	Population	As a Percentage of: So. Calif.*	Calif.	Other So. Calif. Counties	Population	As a Percentage of: So. Calif.*	Calif.
Los Angeles Co.	7,168,000	58.8	36.2	Imperial Co.	82,000	0.7	0.4
Orange Co.	1,384,000	11.3	7.0	Kern Co.	339,000	2.8	1.7
Riverside Co.	444,000	3.6	2.2	San Luis Obispo Co.	97,000	0.8	0.5
San Bernardino Co.	705,000	5.8	3.6	Santa Barbara Co.	260,000	2.1	1.3
San Diego Co.	1,352,000	11.1	6.8	Total 10 So. Calif. Cos.	12,196,000	100.0	61.6
Ventura Co.	365,000	3.0	1.9				
Total Six Counties	11,418,000	93.6	57.7				

Source: Metropolitan Water District, Official Statement, Los Angeles, California, 1970.
*Southern California is defined as the ten counties listed in this table.

Table 12-3
Estimates of Population Growth, 10 Counties in Southern California

County	Estimated 1960 Jan. 1	Estimated 1969 Jan. 1	Projected 1970 Jan. 1	Projected 1980 Jan. 1
Los Angeles	6,002	7,180	7,260	8,400
Orange	684	1,370	1,440	2,275
Riverside	303	460	475	720
San Bernardino	501	705	725	1,000
San Diego	1,022	1,355	1,390	1,814
Ventura	195	367	390	725
6 County Area	8,707	11,437	11,680	14,934
Imperial	72	80	81	90
Kern	291	346	353	434
San Luis Obispo	80	109	111	141
Santa Barbara	162	261	266	350
10 County Area	9,312	12,233	12,491	15,949
43 Northern Counties	5,669	7,144	7,277	9,168
California Total	23,688	30,814	31,448	40,051

Source: Adopted from Preliminary Projection made by the Population Sub-Committee of the Research Committee, Los Angeles Area Chamber of Commerce.

Table 12-4
Distribution of Total Employment in the Six-County Metropolitan Water District Area

Manufacturing	25.7%
Trade	21.8%
Government	14.6%
Transportation & Communication	5.3%
Finance, Insurance & Real Estate	5.3%
Construction	4.4%
All Other	22.9%

Source: Metropolitan Water District, OFFICIAL STATEMENT 1970 (Los Angeles, California, April 1970), p. 21.

Water Requirements

Any forecast of future water requirements must consider probable increases in population. As previously stated, while increases in population have slowed down, the trend in Southern California is still upward towards a total of approximately 16 million people.

With the increases in population there follows a greater demand for water. Table 12.5 indicates that by 1980 water demand for industrial and municipal uses will be 3,290 thousands of acre-feet per annum.

While irrigated agriculture still plays a large role in the Southern California economy, the trend is a shift in the land from irrigated crop acreage to other uses. Consequently, water demand for this purpose is decreasing. A study of table 12.6 reveals that by 1980 agricultural demand will decrease from 970 to 790 thousand acre-feet per annum.

Water Supply

Local Surface and Ground Water

Southern California is classified as a semi-arid region. Rainfall averages less than 15 inches per year. In the last twenty-five years only eight years have had rainfalls above normal.

While there is a severe lack of rain in the area, what rain does fall is retained in reservoirs and underground basins, and only a small percentage is lost. This source of supply contributes approximately 1,523,000 acre-feet per year to the natural surface and groundwater reservoirs. Furthermore, as is indicated in table 12.7, this source of supply will actually increase. This is due to the fact that some of the imported water is being percolated into the ground to recharge the natural ground and surface reservoirs [6:29].

Table 12-5

Projected Municipal and Industrial Water Demand in Southern California (in thousands of acre-feet per annum)

Area	1970	1980
Coastal Los Angeles County	1580	1930
Orange County	320	490
Coastal San Bernardino County	170	260
Coastal Riverside County	100	180
Coastal San Diego County	210	290
Ventura County	80	140
	2460	3290

Source: Adapted from a report WATER DEMANDS AND SUPPLIES FOR SOUTHERN CALIFORNIA, Water and Power Committee, Los Angeles Chamger of Commerce (Los Angeles, California, 1968), p. 5.

Table 12-6
Projected Irrigated Agricultural Water Demand in Southern California (In thousands of acre-feet per annum)

Area	1970	1980
Coastal Los Angeles County	40	10
Orange County	90	40
Coastal San Bernardino County	200	140
Coastal Riverside County	310	300
Coastal San Diego County	90	100
Ventura County	240	200
	970	790

Source: Adapted from a report, WATER DEMANDS AND SUPPLIES FOR SOUTHERN CALIFORNIA, Water and Power Committee, Los Angeles Area Chamber of Commerce (Los Angeles, California, 1968), p. 23.

Table 12-7
Present and Future Water Supplies Available from all Sources for the Southern California Coastal Study Area (in Thousands of Acre-Feet)

Source of Supply	1970	1980
Local Surface & Ground Water	1,520	1,620
Los Angeles Aqueducts	472	472
Colorado River Aqueduct	1,200	1,200
California Aqueduct	0	562
Bolsa Island[a]	0	150
Reclaimed Water	20	76
Totals	3,212	4,080

Source: Adapted from a report, WATER DEMANDS AND SUPPLIES FOR SOUTHERN CALIFORNIA, Water and Power Committee, Los Angeles Area Chamber of Commerce.

[a]Bolsa Island. This refers to a man-made island off the Souther California coast. The island was made to accommodate a nuclear energy plant, that was to produce desalinised water as a by-product. The plan has been discontinued due to extremely high costs.

Imported Supplies. Local surface and groundwater is insufficient by far to cover total demand. Consequently, the greater portion of the supply is imported from the north. The Los Angeles Aqueduct now delivers 320,000 acre-feet per year. A second Los Angeles aqueduct has been constructed and it can deliver another 152,000 acre-feet per year. In addition to this the Colorado River Aqueduct delivers 1,200,000 acre-feet to MWD member agencies. In 1970 the total importation was therefore 1,672,000 acre-feet, as shown in table 12.7.

State Water Project. More than two-thirds of California's water supply originates in the northern part of the state, but the greatest need for water lies in the southern part of the state. Consequently, in order to distribute the water to areas of need, the California water plan, also known as the Feather River Project, was developed.

The key to this project is the Oroville Dam, which was built on the Feather River north of Sacramento. This dam is the largest in the world, and it is capable of handling 3,000,000 acre-feet. The facility to transport this water on its 450-mile journey to Southern California is known as the California Aqueduct [6:30].

Arizona vs. California. For many years the states of Arizona and Nevada have been fighting with Southern California for the rights to obtain a share of the mainstream Colorado River water. Arizona, in particular, has taken California to the U. S. Supreme Court, and in 1964 the Court decreed that Arizona was entitled to 2.8 million of the first 7.5 acre-feet of Colorado River water. However, the Central Arizona project, the facility to receive the Colorado River water, will not be completed before 1990. At that time California entitlement will be substantially reduced to approximately 550,000 acre-feet per year. The additional source of supply to replace this reduction will come from the California Aqueduct [1].

As is obvious from table 12.7, the imported water is Southern California's major source of supply. As of 1970, statistics indicate that aside from local surface and ground water, 0.6 percent of the total supply is other than imported water. To underline this, it should be noted that by 1980 the percentage of nonimported water will be reduced to 0.2 percent. The 1980 projections indicated in table 12.7 were based on the assumption that the Bolsa Island desalting plant would make a contribution to the water supply. However, plans for this plant have been scrapped due to high cost.

The MWD is therefore committed both philosophically and economically to the continued overland transportation of water to its member agencies. However, while Southern California does have the financial strength and tremendous political force to obtain the water it needs, it does not fully control its source of supply. The Supreme Court actually has the controlling hand.

The outlook through 1980 indicates that Southern California will not have any serious problems in obtaining the water it needs as long as the people of the area are willing and able to handle the substantial costs the district has to pay to import the bulk of its supply.

Total Cost Factors

The consideration at this point becomes one of economics. Is the importation of large quantities of water the most economical approach to supply the Southern California area the water it needs?

Costs. To make a true cost analysis, total cost figures must include bond redemption, interest on outstanding debt, operation and maintenance, power, and insurance. As stated by Dr. Jack E. McKee, true total cost figures are difficult to dig out, however necessary they may be to review the total prices charged by the MWD. However, the best estimates show that in 1965 the MWD's price for Colorado River water was as follows:

$13.80 per 1,000 cubic meters of untreated water
$32.40 per 1,000 cubic meters of softened filtered water [5].

To facilitate comparing the above prices with others in this chapter, a conversion factor of .000810708 was used to convert cubic meters to acre-feet. The above prices are the equivalent of:

$17.00 per acre-foot of untreated water
$40.00 per acre-foot of softened filtered water.

Debt Service. In 1966 the voters of the MWD authorized the district to incur an additional bonded indebtedness of $850 million. The purpose of this issue was to acquire funds to construct and improve the facilities for supplying water. It is the intention of the MWD to schedule principal repayment of the bonds to permit approximately overall equal annual debt service for the $850 million bond authorization. Assuming an interest rate of 6 percent, principal and interest payments will increase gradually and level off and remain at $53 million annually.

In addition to the above issue, there exist debt requirements that were issued in 1931 and 1956. However, combining debt service on all issues, the annual payouts will peak at approximately $65 million in the mid-1980s [6:60].

Delta Water Charge. The Delta Water Charge is levied by the MWD to return to the state the cost of developing and conserving of the water supply. This charge is uniformly passed on to the member agencies of the MWD. In 1970 estimated charges were $6.65 per acre-foot. However, total costs to the state have not been completely determined. The charge will rise as actual costs are accrued by the state, and could go as high as $10.00 in later years [6:60].

Transportation Charge. The Transportation Charge is to return to the state the cost of transporting the water to the service area of the MWD agencies. Each agency is charged according to location, so that charges are extremely varied.

However, since the MWD is located at the end of the California Aqueduct, which is used to transport state water, and it does in fact receive about one-half of the total available supply, it is allocated more than half of the cost of transportation facilities [6:60].

As was indicated above, the charges have been based on incomplete cost factors. The state has not as yet completed construction of its program, so that total costs are certain to rise. Escalation in construction costs must also be expected. The state water project increased in cost from $2 billion in 1960 to $2.8 billion in 1969 [6:62]. Consequently, the cost for developing, conserving, and transporting is based on the latest estimates of $2.8 billion.

The major charges are passed on to the member agencies, and it is at this level that understandable comparisons can be made. Table 12.8 shows the actual prices charged by the MWD over an approximate thirty-year period. These prices show an increase from $15 to $57 per acre-foot for softened and filtered water, a rise of 357 percent.

Each year's rates must be considered separately, as there are many variables that can alter the rates. However, estimates can be made. Table 12.9 shows the MWD's projected prices through 1980 and indicates that the price per acre-foot will rise to $75. However, with modifications and efficiencies, the price may be held at $68 per acre-foot. This price represents a total increase of 31.6 percent over the next ten years.

The charges to the member agencies have by no means leveled off and will continue to increase as true costs of construction are determined. However, the prices are still manageable and economically feasible over the next ten-year period.

While the MWD has made commitments to continue transporting water, the Southern California authorities are aware of the state's dependence on its out-of-state supply. Consequently, some consideration has been given to other possible sources of supply. The two main possibilities reviewed are sea water conversion and water recycling.

Sea Water Conversion. The technology to convert sea water to fresh is available today and presently used in more than 680 desalting plants throughout the world. Southern California water authorities, in conjunction with local electrical utilities, have considered such a plant. However, the plant, which was to have produced 150 million gallons per day, was never completed due to high cost.

In 1965, costs were estimated at 21.4 cents per 1,000 gallons [2:3]. Since in 1968 water could be imported from the north at 22 cents per 1,000 gallons ($72 per acre-foot), the conversion of sea water was not economically feasible, and it was therefore no longer considered as a possible source of supply. On the other hand, water recycling has been given more consideration as a practical solution to the Southern California water shortage problem.

Water Recycling

As early as 1949 California engineers were advocating water reclamation and reuse in a comprehensive report prepared by C. E. Arnold, H. E. Hedger, and A. M. Rawn, Los Angeles County sanitation engineers. The report was updated in 1958 and a proposal was submitted for a ten-MGD water reclamation plant. As a result of this, a plant was built at Whittier Narrows in 1962. This plant at Whittier Narrows functions solely to reclaim water for recharge purposes, and was made possible through a four-agency agreement, with the following division of labor:

1. Los Angeles County — supplied the capital.
2. The Sanitation Districts — designed and constructed the plant.
3. Central and West Basin Water Replenishment District — purchased the reclaimed water.
4. Los Angeles County Flood Control — spread the water [8:225].

Table 12-8
Metropolitan Water District's Water Rates per Acre-Foot

		For Municipal and Industrial Uses			For Agricultural and Groundwater Replenishment Uses	
From	To	Softened and Filtered	Filtered	Untreated	Softened and Filtered	Filtered	Untreated
8/1/41	6/30/48	$15.00		$8.00			
7/1/50	11/30/54	20.00		10.00			
7/1/60	12/31/60	23.00		15.00	20.00		12.00
7/1/70	6/30/71	53.00	49.00	44.00 Agric. Repl.	30.00	26.00	21.00
					31.00	27.00	22.00
7/1/72	6/30/73	57.00	53.00	48.00 Agric. Repl.	31.00	27.00	22.00
					33.00	29.00	24.00
				52.00 Agric. Repl.			23.50
							27.00

Source: Metropolitan Water District, Official Statement, Los Angeles, California, 1970.

Table 12-9
Southern California, Illustrative Water Rates (All values in dollars per acre-foot)

	Rate for Untreated Water for Domestic Purposes Based on[a]		
Fiscal Year	Present Construction Schedule[b]	Proposed Modified Construction Schedule[c]	Annual Difference
1966-67	31	31	0
1970-71	44	44	0
1975-76	64	58	−6
1980-81	75	68	−7
Totals	214	201	−13

Source: Adapted from a report of the Board of Directors of the Metropolitan Water District of Southern California, Los Angeles, California, 1970.

[a]Rates based on bond interest rate of 6 percent and maximum term of 50 years on all bonds sold in future years.

[b]Rates shown in brochure for Series C Bonds dated April 27, 1970. No additional bond authorization assumed.

[c]A second bond authorization of $370 million is assumed to be approved in June 1982.

Recharge Operations

Surface spreading as a method of recharging the water supply has been shown in practice to be the safest, most trusted, and accepted process of pumping reclaimed water back into the water supply system. The process of slowly filtering the water through sand has the advantage of mingling and diluting the waste water with the natural ground water. This accomplishes a complete loss of identity of the recycled water, and this process allows the use of the water for any purpose. It should be noted that groundwater recharging by the spreading process is and has been in operation in Southern California for many years. The treatment plants that exist in the San Gabriel Valley all percolate water through the soil back into the groundwater supply [7:3].

Public Health Standards

A major question that arises at the thought of reusing water is whether the sewage does lose its identity when processed through the ground, and whether or not it meets public health standards. To ascertain whether or not the Whittier Narrows Plant met the state health standards, a three-year independent investigation was made by the California Institute of Technology. The study concluded the following:

1. The secondary effluent meets the regional quality boards' standards.
2. It also meets U. S. Public Health Service drinking water standards.
3. Any suspended solids or oxygen-demanding matters contained in the effluent spread are filtered, absorbed, and biologically decomposed in the first few feet of soil the renovated water passes through [7:22].

Practical Applications of Recycling

As previously outlined, the major use of renovated water is for groundwater recharge. California laws demand that any wastewater to be used for domestic purposes be passed through many feet of soil. However, depending upon the level of quality, wastewater can be used for the purposes of irrigation, recreation, and also industrial applications. An example of this can be found at Santee, a small community that lies adjacent to San Diego. At Santee, renovated water is presently being used for all three of the above purposes. It should be noted, as well, that many industries are presently recycling renovated water back into their plant cooling systems, and thus creating considerable cost savings. The supply of high-quality wastewater that could be made available would not substitute for the need to develop additional sources of water in the Southern California area. However, if a supply of renovated water could be used to augment the water resources system, it could furnish a time buffer during interim periods of crises and prior to development of any additional sources that may be available [8:225-31].

Santee County Water District

The water management system of Santee County serves as a practical working example of how renovated water can be used to augment a water resources program.

The Santee County Water District is a public agency that provides water and sewage services for the unincorporated community of Santee, California, which is located close to the city of San Diego, and lies some 20 miles inland from the Pacific Ocean. In 1968 this community had a population of 14,500 in an area that is blessed with abundant sunshine and a warm climate. However, the district has no local water supply, and must depend on water from the Colorado River, which is located 300 miles away. The district presently uses 2,200 acre-feet per year. In the early 1970s the district will continue to import water from the Feather River as part of the California State Water Project. The wholesale cost to the district has doubled in the last ten years. The price increased from $17.00 an acre-foot in 1957 to $42.25 in 1968. In line with MWD prices it is expected that the price will rise to over $75.00 per acre-foot within the next decade [13:1].

The members of the board who run the district of Santee found that there was nothing they could do about the rising cost of imported water. However, by developing a water management system that includes the reclamation of sewage water and subsequent reuse of this water, the district can make multiple uses of the same water for recreational, agricultural, and industrial purposes. This thereby reduces the unit cost per use of the water. Thus by building a recycling plant, the district of Santee was able to keep its water costs at the same level, or at least to stop them from rising at the high rates that are prevalent in the Southern California area. In 1959 the members of the Santee Board were faced with two alternatives:

1. Join a metropolitan sewer system designed for the disposal of wastewater into the Pacific Ocean.
2. Design and build Santee's own wastewater disposal system.

The board chose to build its own water management system rather than throw water away, and developed the reclamation plant that exists today.

The reclamation plant was completed in 1967 and it reclaims 60 to 80 percent of the original imported water. The plant has proven to be a success, and to date the renovated water has been used for a variety of recreational purposes including boating, fishing, and swimming. The water is also used for irrigating crops and the local golf course. Industrial use has so far been limited to local sand and gravel pits which use less expensive reclaimed water. However, when a tertiary treatment facility is fully completed, more sophisticated uses will be accomplished. Santee's plans for the future include:

1. A park and lake system.
2. The attraction of agriculture and industry.
3. The sale of by-products of the treatment plants.
4. Underground water storage and recycling.

The main plan is, of course, to supply the people of Santee with a safe, high-quality water supply at the lowest possible price, and the luxury of water recreational facilities that could not be available had the district not chosen to build its own water reclamation plant [13].

Consequently, the process of recycling renovated sewage water is completely biologically and chemically suitable and acceptable for industrial and municipal reuse, and meets all U. S. Public Health standards. The only barrier that still hinders the use of renovated water is the psychological one. It is possible that some people will never be able to overcome this hurdle. However, the experience at Santee has shown that this community to a large extent has overcome this psychological problem and has totally accepted the reuse of renovated sewage water.

Sanitation

As Southern California has grown, so has its need for proper sanitation. Consequently, in this area today we find one of the largest sanitation districts in the world. This district includes seventy-one incorporated cities covering 729 square miles and housing 3.9 million people. This is a combination of twenty-six sanitation districts containing 937 miles of trunk sewers covering 7,388 miles of lateral sewers. The district operates eleven water pollution control, reclamation, or renovation plants [11:4].

This trunk sewer system flows into a large water pollution control plant. At this plant the sewage is given primary treatment and disposed of through two tunnels into the Pacific Ocean. The ocean outfalls disposing of the treated effluent have been successful so far, and consist of the following outlets:

a. One 60-inch diameter pipe extending 5,000 feet off shore, discharging into 110 feet of water.
b. One 72-inch diameter pipe extending 6,500 feet off shore, discharging into a depth of 165 feet.
c. One 90-inch diameter pipe extending 8,500 feet off shore, discharging at a depth of 195 feet.

These pipelines are designed to take advantage of favorable ocean currents and temperatures [11:5].

It is ironic that in an area like Southern California that needs water, so much wastewater that could be renovated is dumped into the ocean. Los Angeles County alone generates some two million gallons per day of wastewater that is now almost entirely wasted to the Pacific Ocean. In addition to this, Los Angeles dumps approximately 340 million gallons per day into the Santa Monica Bay. About one-third of this wastewater is given secondary treatment and the remainder primary treatment before being discharged. Furthermore, it is estimated that more than 55 percent of the total water imported into Southern California is discharged into the ocean [7:2].

The sanitation district is of course aware of the ever-increasing flow of waste that it must handle, and in 1965 a master plan was drawn up to help relieve the sewage problem. The plan calls for new effluent tunnels and ocean outfalls as well as additional trunk sewers. The master study also concluded that the construction of large secondary treatment plants in the upstream areas would be the most economical approach. These plants will make available a significant amount of highly treated water that eventually could be redirected into the water supply system. The plan is to discharge this water into the normally dry San Gabriel River, which will carry the flow to the ocean [7:5].

Glendale – Los Angeles Reclamation Plant. The cities of Glendale and Los Angeles contribute a large portion of the effluent that eventually flows into the ocean. Consequently, in 1968 the two cities entered into a joint powers agreement to conduct a study to determine the feasibility of a water reclamation plant. An analysis by the Los Angeles Bureau of Engineering has shown that it would be more economical to build a reclamation plant rather than construct additional trunk sewers. The overall conclusions of the study were as follows [3:3]:

1. Beginning in 1970 there was a need to divert 5 million gallons per day (MGD) of wastewater. This amount will increase to 10 MGD by 1980.
2. It is economically feasible to construct and operate a reclamation plant to accomplish the diversion of wastewater flows.
3. The proposed plant could be attractive in design and clean in operation.
4. The facility should have an optimum capacity of 20 MGD in 1970 and the capacity could be increased to 35 MGD by 1985 and to 50 MGD by 1995.
5. The additional available water would be suitable for industrial cooling water for the Glendale Municipal Steam Plant.

The reclamation plant design provides for staged construction in order to distribute cost according to need and to facilitate projected financing. The plan is to complete construction in three stages as follows:

Year	Capacity	Construction Cost (Per acre foot)
1970	20 MGD	$5.60
1987	35 MGD	4.23
1995	50 MGD	5.66

This will allow the facility to supply the cities with renovated wastewater at a total unit cost of $31.00 per acre-foot. However, as indicated in table 12.10, the cost will decrease over a thirty-year period.

Table 12-10
Total Annual Cost for the Plant, and Corresponding Unit Cost of Reclamation Water

Year	Total Annual Cost	Unit Cost
1970-1987	$ 694,000	$31 per acre-foot
1987-1995	$1,142,000	$34 per acre-foot
1995-2000	$1,681,000	$30 per acre-foot
2000-2045	$1,456,000	–
2000-2017	–	$26 per acre-foot
2017-2025	–	$23 per acre-foot
2025-2045	–	$19 per acre-foot

Source: Los Angeles Chamber of Commerce, WATER RECLAMATION PLANT (Los Angeles, California, March 1969), p. 4.

In furtherance of the theory that the reclamation of water is feasible, the sanitation districts of Los Angeles County propose to construct eleven water reclamation plants throughout the area. The Whittier Narrows Plant was constructed in 1962 and the facility has exceeded expectations. It is presently producing 12 MGD clear water for ground recharge at a cost close to the original estimates. Per unit costs are $11.05 per acre-foot. Furthermore, this plant can be expanded to a capacity of 24 MGD, reducing the cost to $9.81 per acre-foot; and second-stage plans for expansion would allow the plant to handle 50 MGD. It is estimated that the cost per acre-foot at this point would be $10.76 [12:38-40].

It is interesting to note that in six and a half years the Whittier Reclamation Facility has realized $1,500,000 from the sale of 97,000 acre-feet of high-quality recycled water, and operating costs have averaged approximately $9.00 per acre-foot. This cost is about 43 percent of the 1970-71 charge by the MWD for agricultural replenishment groundwater, with MWD relying on imported sources (see table 12.8). From this perspective, then, the Whittier Reclamation Facility must be regarded as an enormous success. It is a measure of how little that success is appreciated to recognize that the average annual output of the Whittier Facility cited above (about 15,000 acre-feet) represents only 0.2 percent of the 1970 consumption of water in Southern California (see table 12.7).

This is especially surprising in view of the success of the Santee recreation project near San Diego that employed reclaimed water (discussed earlier in this chapter). It would appear that the lesson of Santee for Southern California should be cyrstal clear. Los Angeles is expected to reach a population of 8,400,000 by 1980, an increase of 1,140,000 over 1970, or nearly 16 percent (see table 12.3). The projected increase for the six-county Southern California area is 3¼ million or nearly 28 percent (see table 12.3). This is in an affluent area where people generally have leisure and income to indulge their taste for the outdoor sports the climate invites. There is naturally a nearly insatiable demand for water-based recreation, not always easily met in a semi-arid region. Reclamation would appear to be one of the least expensive avenues to meet this very real social need.

Conclusion

As has been shown above, the water shortage in the southwest United States is a serious problem. Total surface and groundwater is completely inadequate to meet demand. Consequently, the greater portion of the supply has to be imported. This has been shown to be a costly approach, for great sums of money must be spent for transporting the water and building aqueducts to retain it. However, Southern California is fortunate in having an excellent water resources and management program that has taken advantage of the water that is available. In actual fact, very little local water escapes being used in this system.

Based on available evidence there is little indication to conclude that Southern California will have any difficulty in obtaining its water supply up to and through the year 1980. The demand for water will steadily increase until by 1980 the area will require 3,290,000 acre-feet per annum. However, based on contracts to import Colorado River and Northern California water, the supply in 1980 will be approximately 4,080,000 acre-feet, which should be ample to meet requirements.

Southern California competes with other southwest states for Colorado River water. In particular, Arizona has been awarded 2.8 million acre-feet of water that presently goes to Southern California. However, Arizona will not be ready to receive this water until after 1980.

The main problem in Southern California is the continually increasing high cost of importing water. In 1972 water for municipal and industrial use will cost as high as $57.00 per acre-foot and by 1980 that price will be around $75.00 per acre-foot, representing a total increase of 31.6 percent over 1972. The MWD is committed to import its major water supply, but with prices continually rising, the Southern California authorities have been forced to consider other possible sources of supplemental water. The choices considered are as follows:

1. Continue to import — $ 57.00 per acre-foot;
2. Construct a desalting plant — $197.50 per acre-foot;
3. Reclaim sewage water — $ 11.05 per acre-foot.

Serious consideration has been given to constructing a desalting plant, but extremely high cost makes it uneconomic at the present time.

Water recycling has been considered and found to be economically and practically feasible, and reclamation plants are functioning in the area at present, but they do not contribute any significant amount of water to the water resource system and the plans for the future do not indicate that this will change. Importation of the major source of supply will continue through 1980.

Another major problem related to this topic is the heavy sewage flows that continue to increase in volume in this area. The densely populated municipalities find that sanitation facilities must be increased to handle the flow. Plans are underway to build additional treatment plants to handle the load. However, this costs money and in many instances more than it does to renovate water for reclamation.

As has been shown above, the function of supplying water to the greater portion of this area is the responsibility of the MWD, while sewage is the job of the sanitation district. Both agencies serve the same area, but they seem to compete rather than augment their efforts. These are two separate bureaucracies functioning separate and apart. The MWD does not deny that recycling is feasible, but it does not consider recycling in its plans.

The MWD governs an extremely rich and powerful geographic area and it is able to afford the high price of importation at this time. However, it is the belief of the author that the feasibility and advantages of water recycling make it a primary consideration to help solve the sewage and also the water shortage problem. The efforts of the water and sanitation districts should be augmented to maximize the use of dollars that are presently spent to import water, clean it up, and throw it in the ocean.

As will be seen in Chapter 25, Los Angeles has been a pioneer in long-range planning to convert solid waste disposal sites to recreational uses. This intelligent effort to meet the burgeoning leisure-time demands of its residents might well be supplemented by a policy of making maximum use for recreation of its limited water resources, by recycling where that is indicated.

Notes

1. Arizona Interstate Stream Commission, *Water Resources* (Tucson, Arizona, October 1967).

2. Eel River Water Council, *Desalting Present and Future* (Santa Clara, California, November 1969).

3. Los Angeles Area Chamber of Commerce, *Water Reclamation Plant* (Los Angeles, California, March 1969).

4. Los Angeles Area Chamber of Commerce, *Water Demand and Supplies* (Los Angeles, California, June 1968).

5. Jack E. McKee, "Our Cheapest Source of Water," *Engineering and Science Magazine,* California, November 1966.

6. *Official Statement 1970,* Metrolpolitan Water District (Los Angeles, California, April 1970).

7. J. D. Parkhurst, *Reclamation and Reuse – The State of the Art* (Davis, California, June 1969).

8. J. D. Parkhurst, C. W. Carry, A. N. Masse, and J. N. English, "Practical Applications for Reuse of Wastewater" *Chemical Engineering Progress* 64, 90 (1968).

9. E. G. Paulson and B. Q. Welder, *Water Management – Water Reuse*, presented to International Congress on Industrial Waste Water, Stockholm, Sweden, November 1970.

10. *REPORT – 1970,* Metrolpolitan Water District (Los Angeles, California, September 1970).

11. Sanitation District of Los Angeles County, *Resume of Sewerage and Reuse Disposal Practices* (Los Angeles, California, June 1970).

12. Sanitation District of Los Angeles County, *A Plan for Water Reuse* (Los Angeles, California, July 1963).

13. Santee County Water District, *Total Water Reuse* (Santee, California, 1969).

13 The Effects of Increased Production on the Oceans

FRANK D. HUSSON, JR.

The world's oceans, which occupy over 70 percent of the globe's surface area, have acted as the ultimate sink for the natural wastes of erosion and decay for hundreds of thousands of years. The continents, covering the remaining 30 percent of the earth's area, are the natural habitat of man, who, at about the time of the industrial revolution, began dumping the by-products of production into the seas in amounts that were significant when compared with those materials being added by natural processes. This pollution has increased steadily over the past two centuries with a positive relationship to the population. Materials completely foreign to the natural environment such as DDT are added to the oceans in increasing amounts each year. Pollution due to naturally occuring materials such as oil, lead, and mercury has increased to quantities surpassing (in many cases by factors of ten) that introduced by nature.

It has just recently been realized that these same oceans, which supply mankind with one-quarter of its annual protein intake, in certain areas are polluted to the extent that the fish and shellfish existing within are not safe for human consumption. During a seven-year period, starting in 1953, over one hundred persons died or suffered serious neurological damage after eating (over a period of time) fish that contained abnormally high amounts of mercury. This disease is now referred to as Minamata Disease, obtaining its name from the bay where the fish were caught [2].

The pollution of the seas by oil has killed thousands of fish and birds, with similar effects being attributed to DDT and other chlorinated hydrocarbons. Numerous other pollutants are added to the oceans each year both through direct and indirect avenues, with effects to our environment that are not even fully realized. The quantities of naturally occuring pollutants such as mercury, lead, and petroleum are a good percentage of the total now existing in the seas, but it must be remembered that these quantities have been present for millions of years, and that the living creatures of this earth have, over this period of time, been able to adapt genetically to these levels. With man now adding significant quantities of certain pollutants to the oceans (basically all during the twentieth century), the time-consuming genetic changes to protect the species cannot be realized.

One of the most common assumptions with regard to pollutants is that the major percentage is introduced into the ocean system at the continent-ocean boundary, both through the rivers and direct dumping. Other unfavorable compounds enter the oceans through the atmosphere and are thus more widely distributed.

This chapter will attempt to review the source, quantities, and effects of a few of the most common or dangerous pollutants. Because of our limited knowledge with regard to ocean pollution, many of the coefficients used to compute amounts of pollution will be estimates.

The avenues by which pollution enters the oceans may be divided into two discernible groups, direct and indirect. Direct pollution will refer to pollution that is introduced directly into the seas as compared to indirect, which relies on other carriers such as rivers or the atmosphere to transport the pollutants. It should be noted that 100 percent of any pollutant is not introduced under either category, but, in general, a large part of any given pollutant will be predominantly from one avenue.

One major fact on which all parties are agreed is that the majority of this pollution occurs in coastal waters and the highest percentage of total pollution is found here, where the temperatures are relatively warm and the depths are relatively shallow. Depending on the continent, the shelf area will vary. Figure 13.1, though not drawn to scale, shows something about the nature of the cross-section of North America as it relates to continental shelves. It may be noted that the shelf on the west coast is much narrower than on the east, but in each case stretches for many miles. Until recently, it was assumed that the dumping of harmful wastes into the oceans was safe, because of the great dilution that was expected to occur when the pollutants mixed with the vast quantities of water over a period of time. This is not the case. Most of the direct pollution occurs on the continental shelves, where water with large quantities of dissolved oxygen, microorganisms, and plants support a great percentage of the bird and fish population. Pollution is significantly higher in these areas and a variety of the life found there will accumulate dangerous materials in their bodies which are then passed up the food chain. The spawning grounds of many of the larger deep-water fish are also in the relatively shallow water over the continental shelf. Because the water is warmer at shallower depths, with a high oxygen content, many forms of pollution are more easily dissolved and held in

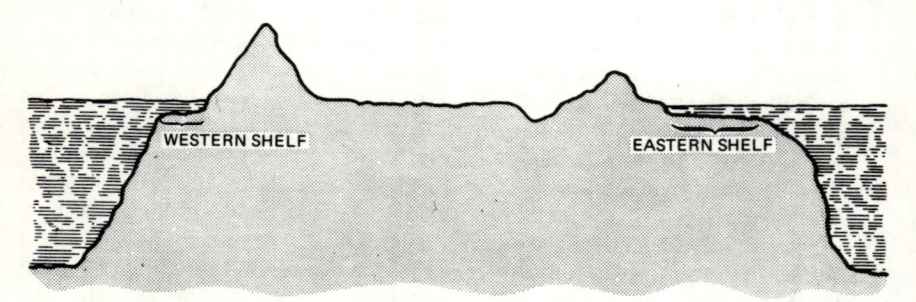

RELATIVE WIDTHS OF EASTERN AND WESTERN SHELVES, APPROXIMATE AND INDICATIVE ONLY.

Figure 13.1. Contours of Continental Shelves, Eastern and Western North America (not to scale).

solution. Depending on the tides, waves, currents, and winds, these pollutants may remain in one area for extended periods of time. Thus, pollution of these coastal waters poses a larger threat than in the open sea.

Direct Pollution

The most obvious direct pollutant is petroleum. It can be observed both on the high seas and continental ocean boundaries. Liquid oil pollution may amount to over 10 million metric tons per year according to some authorities [25:761]. Upon ignition, petroleum products produce a number of hazardous pollutants, such as heavy metals, aromatics, and carbon-based materials, etc., which enter the atmosphere, approximately 70 percent of which eventually find their way to the oceans. This section of the chapter is only concerned with the pollution by liquid petroleum, reserving the contamination by gaseous products for discussion in another section.

Liquid petroleum pollution may occur in basically four ways:

1. Shipping
2 Off-shore drilling
3. Refinery operations
4. Industrial and auto wastes.

Of the four avenues, only industrial and auto wastes are not predominantly a direct input.

Because of the lack of effective national and international legislation restricting oil pollution, there has been a reasonably close relationship between crude oil production and oil pollution. Figures 13.2 and 13.3 show the quantities of both crude and refined oil produced with projections to the year 1980. World production of petroleum products has greatly increased in recent years, as indicated by the doubling of the total tonnage produced in the ten-year period, 1953-1963. The production of crude oil has increased at a rate of approximately 1.1 times that of petroleum products [34:283]. Edward R. Corino of Esso has projected a faster growth in crude oil than shown in figure 13.3 [39: Table 1].

One of the significant contributors of total oil pollution and a major offender in concentrated spillage is the shipping industry. As of 1970, there were a total of 52,444 merchant ships in the world fleet [19], contributing over 1 million metric tons of oil per year directly to the world's oceans. This quantity of pollution can be conveniently divided into three basic classes:

1. Pollution by normal tanker operation.
2. Pollution by all other vessels, under normal operation.
3. Pollutions due to collisions, groundings, and sinkings.

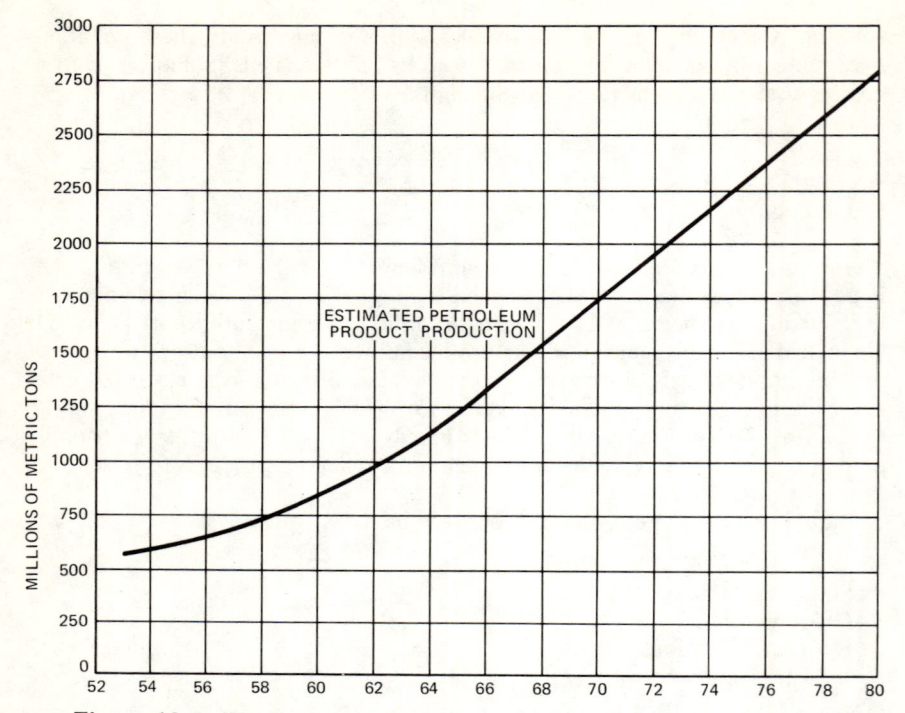

Figure 13.2. World Petroleum Products Production. Source: *United Nations Statistical Yearbook*, Table 123, p. 283.

Tanker Operation

At the present time, 12 percent of the world's fleet is comprised of tankers, which carry 38 percent of the total tonnage [19], and presently contribute approximately 50 percent of the oil pollution accounted for from normal operation of the shipping industry. This large percentage is not unreasonable since their cargo is the pollutant in question, with approximately 3 million metric tons moving on the seas each day [26]. Under normal operating conditions, after delivering their cargo, tankers fill oil storage tanks with sea water to insure stability. When arriving at the point where the oil is being processed or stored, they empty these tanks so that the pure oil may be loaded. The water being pumped out contains oil that had remained in the tanks (on the sides and bottom) when the previous shipment was unloaded. Conservative calculations indicate that approximately 0.1 percent oil remains in the tanks [35:68]; the loss is additive with additional trips annually.

Some authorities believe this avenue of pollution has accounted for approximately 2 million tons of oil pollution each year. By multiplying the amount of oil transported per year by the factor 0.2 percent, a value of this magnitude is achieved.

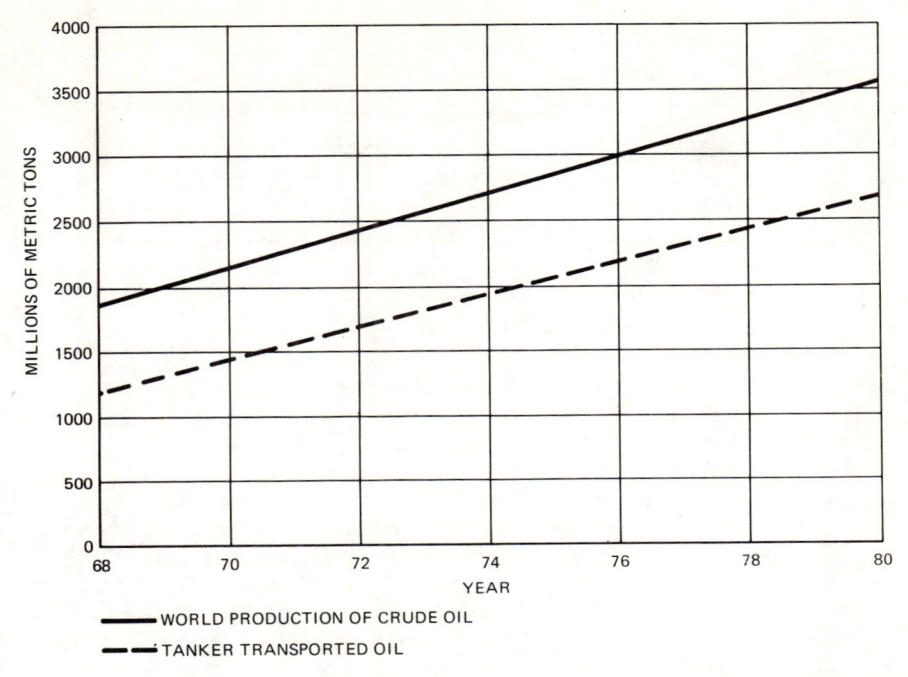

Figure 13.3. Tanker Transported Oil & World Production of Crude Oil. Source: 1968–*Oil and Gas Journal*, 1970–*U.S. Statistical Abstract*.

Recently (1962), a method of emptying ballast water, known as LOT (Load on Top), was introduced. The basics of this method are shown in Figure 13.4. The tanks can be either heated or soaked with detergent as sea water is being forcefully spread on the inner walls. This removes the oil from the sides and it forms a suspension in the water. Because of their different densities, the oil then rises to the surface. The top section (containing the oil) is then pumped into a holding tank and allowed to settle. After settling, all but three inches of water are pumped out to sea [24:15].

This procedure is believed to retain oil that would otherwise be lost to the extent of 0.37 percent of the ship's carrying capacity annually. For one British firm transporting 115 million tons a year, this represents a saving of 400,000 tons annually [24:14].

The prospect of such savings led to rapid adoption of the LOT system by major oil companies, and by 1967 it was estimated that 75 percent of the crude tonnage was moved in tankers equipped for LOT operation. Because of costs involved, which offset some of the savings, and other technical problems, smaller companies have been slow to adopt LOT.

Edward R. Corino of Esso Research estimates that LOT was employed by 80 percent of the world fleet in 1970 and that this portion of the fleet contributes only 30,000 tons of oil pollution compared to 500,000 tons by the 20 percent of the fleet who do not employ LOT. He suggests that by 1980 the portion of

*LOAD ON TOP

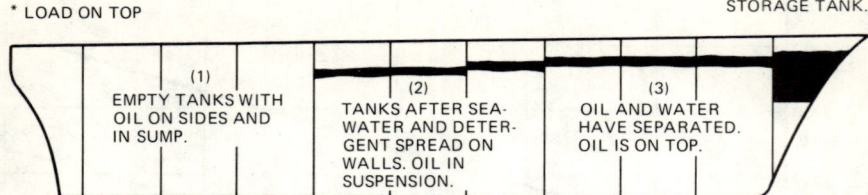

AFTER SEPARATION IN (3), OIL IS PUMPED TO STORAGE TANK.

(1) EMPTY TANKS WITH OIL ON SIDES AND IN SUMP.

(2) TANKS AFTER SEA-WATER AND DETERGENT SPREAD ON WALLS. OIL IN SUSPENSION.

(3) OIL AND WATER HAVE SEPARATED. OIL IS ON TOP.

Figure 13.4. Schematic Diagram of Operation of LOT* Tanker System.
*Load on Top

the fleet not employing LOT will be responsible for one million tons of oil pollution, which might be reduced to 13,000 tons if LOT were universally adopted [39: Table 4].

The larger firms feel that by efficiently using the LOT system, the money obtained from salvaging the oil more than pays for the extra man-hours spent. The fact that newer tankers can be inexpensively fitted for the LOT system (the average age of tankers is eighteen years [33:89]) and that LOT has been available for ten years should begin to have an influence on this avenue of pollution. Organizations such as NATO appear ready to restrict spillage on the open seas, and the United States has proposed the elimination of tanker flushing at sea by 1975 [23].

It is expected that LOT will be virtually globally accepted by 1980. Thus, the amount of pollution occuring from normal tanker operation by 1980 and 2000 is expected to be negligible when compared to present-day standards.

Merchant Ships

Other ships are no different from tankers when it comes to ballast water. As the nontanker ships use fuel in their engines, they refill the empty tanks with water. As in the case of the tanker, these fuel tanks contain residual petroleum which upon adding ballast water mixes, so when the ballast water is emptied, the oil pollution occurs. At the present time, there are not any systems for removing this fuel from the ballast water on nontankers. Though a quantitative value for this form of pollution is not available, most published estimates are around 500,000 metric tons per year [39: Table 4]. This figure may be arrived at by assuming that 0.2 percent of the oil used as fuel is eventually pumped into the seas. Without controls and assuming that LOT is widely accepted on tankers but not on others, merchant ships may soon surpass tankers in oil pollution and would contribute over ten times the pollution due to tankers by 1980 [39: Table 4].

Tanker Accidents

At the present time, tanker accidents, in which large quantities of oil are released in concentrated amounts into the sea, appear to cause greater harm than extensive dilute pollution. Estimates have suggested that approximately 100,000 metric tons per year of oil is lost in this fashion [39].

Examples of tanker accidents involving spills are extensive. In recent years, there have been approximately 100 ships lost annually [19]. Since approximately 20 percent of the merchant fleet is composed of tankers, we shall assume that approximately twenty tankers are lost each year. Until recently, 1950, tanker capacity averaged approximately 20,000 metric tons. The average age of tankers sunk recently has been twenty years. If it is assumed that one-half of these ships were carrying oil upon sinking, a value of 200,000 metric tons of oil pollution would be obtained. Other accidents, where the ship is only damaged and not lost, have contributed to oil pollution in significant quantities. The value of 200,000 metric tons/year will be thus used as a conservative estimate of this form of pollution at the present time.

Because of various worldwide problems, the most significant of which was the closing of the Suez Canal, tankers capable of carrying 325,000 metric tons of oil are on the seas. Mitsubishi Heavy Industries are preparing to build tankers of up to 720,000 metric tons capacity and both Japan and Great Britain are talking about building one million ton ships [26].

Even with advanced technical equipment, the possibility of damage to, or the sinking of one of these ships is possible. Thus, in the future, the oil lost in the ocean from one ship accident could cause more pollution than all the tanker accidents have for the past five years.

Off Shore Wells

Another significant contributor to ocean oil pollution directly related to the oil industry and the second-largest contributor to direct and concentrated oil spills is off shore oil wells. Pollution from these rigs may occur during both normal operation and drilling. Though empirical data are lacking, the drilling operation appears to contribute the largest percentage of concentrated spills. At the present time, Standard Oil of New Jersey alone has 5,600 wells in operation.

Before delving into the actual oil pollution, we might consider the technique that is used for locating oil wells, and until recently was responsible for a large quantity of dead fish each year. The general term of "seismograph surveying" refers to the practice of bouncing shock waves (traditionally supplied by dynamite) off the ocean floor and by interpreting the results, attempting to locate the potential oil deposits. It has been determined that the negative pressure resulting from dynamite blasts killed large quantities of fish. In 1966, 80 percent of the energy pulses generated were from dynamite, with the proportional amount of fish casualties. Recently, other less harmful systems have been developed, such as the aquapulse system; and as of 1968, 80 percent of the energy pulses were supplied by other than dynamite sources [3:8].

Off shore oil production began in 1948 off the coast of Louisiana. As of 1968, the United States off shore production was responsible for one million barrels of crude oil per day. This is more than 10 percent of the total daily United States production and is expected to increase 15 percent per year until 1975.

The drilling operation is carried on by movable rigs. From 1950-1964, 100 of these rigs were assembled, with approximately another 100 being added between

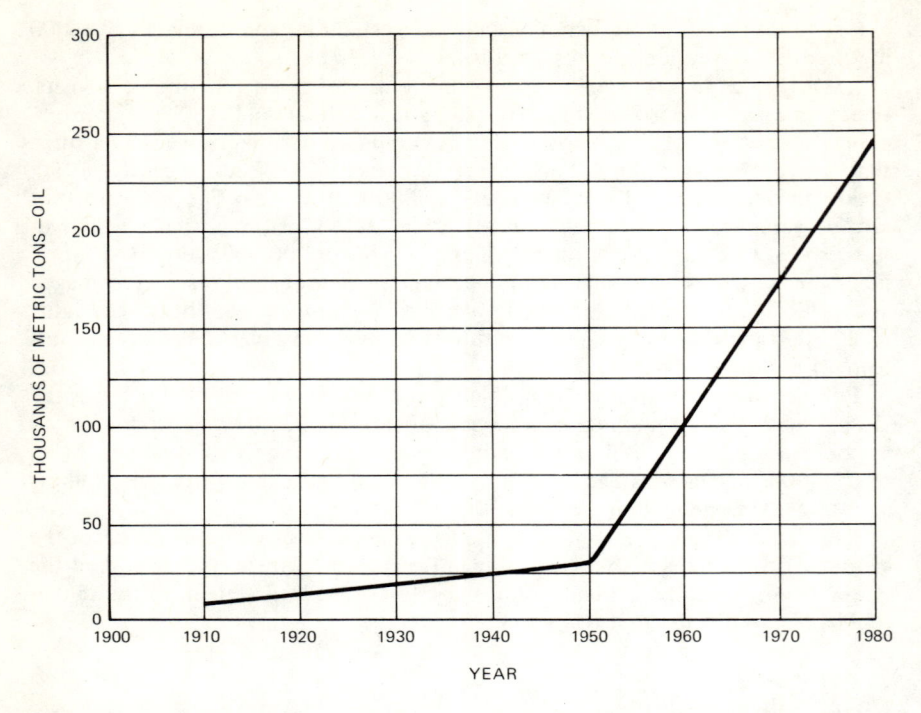

Figure 13.5. Projected Average Tanker Capacity. Source: Statistics from *Life Science Library*, Volume on SHIPS, 1965, p. 105.

1964 and 1968 [3:11]. By 1965, oil had been found off the coast of seventeen nations [37:9].

As previously stated, the greatest percentage of concentrated spills occurs during the drilling phase. An example of a large-scale drilling accident occured in early 1969 when in the Santa Barbara Channel, off California, a hole blew. Initially, the hole was capped, but back pressure caused a fissure in the earth's crust and 21,000 gallons of oil per day poured out. It required eleven days before the fissure was sealed and allowed approximately 200,000 gallons of oil to pollute the sea. An oil blanket nearly one inch thick covered 800 square miles of ocean because of this tragedy [35:64].

Estimates by Standard Oil of New Jersey and the American Petroleum Institute indicate that approximately 50,000 metric tons of concentrated oil per year may be lost in this fashion [20:267].

The other form of oil pollution that may occur from off shore rigs is through normal operations. When oil is collected on the surface, it contains quantities of natural gas and salt water, which are circulated through separators, in an attempt to retain only wanted material. The water is removed from the crude oil and then returned to the sea.

A variable amount of residual oil remains in the water after separation and, though in low concentrations, it contributes large overall quantities of oil pollution to the oceans. An estimate of this pollution has been set at 100,000 metric tons per year [20:267]. At the present time, the large oil firms are working on perfecting the method of separation to minimize this pollution.

Refinery Operations

Refining, though not a contributor of concentrated spills, is responsible for a substantial amount of total pollution. Both through normal operations and accidents, refineries are estimated to be responsible for approximately 300,000 metric tons of oil pollution per year [20:267].

Considering the quantities of petroleum products produced each year (see figure 13.2), this quantity is not excessive. Continuing research with separators and distilling apparatus by the oil firms should considerably reduce the quantity in the near future.

Industrial and Automotive Wastes

Though quantitative information is again absent, this source of indirect oil pollution appears to be quite significant. Many studies of discarded oil have been attempted in various parts of the world. A rather concise study done by Arthur D. Little Inc., of waste oil in Massachusetts, indicates that 3.55 percent of all lubricating oil both industrial and automotive is disposed of into sewer systems that eventually reach rivers and the ocean [29:3].

UN statistics indicate that lubricating oils constitute between 1 and 2 percent of total petroleum products. Between 1953 and 1966, the percentage decreased from 1.83 percent to 1.4 percent [34:283]. Thus, we will adopt the conservative value of 1.2 percent total petroleum products which are lubricating oils. Thus, in 1970 approximately 21 million metric tons of lubricating oils were produced on a global basis. By multiplying the pollution factor 3.55 percent times this quantity, we would receive 745,000 metric tons.

Another method of estimating possible indirect oil pollution that uses rivers to transport the organic material was used by the group working on oil pollution at MIT. Using the values of 0.85 part per billion average hydrocarbon content in U.S. rivers, plus 1,750 billion tons of water per year runoff in the United States, they estimated that 150,000 tons of hydrocarbons per year are carried to the oceans each year by rivers in the United States. Then assuming that the river runoff of the world is three times that of the United States, they arrived at the value of 450,000 metric tons/year [20:267]. The estimate of 500,000 metric tons per year will thus be adopted as a value of oil pollution occuring from industrial and automotive wastes.

It must be remembered that this quantity has been arrived at by considering only United States' statistics and thus may vary significantly from absolute global values. Some authorities have estimated the pollution from this source may amount to more than 3 million metric tons per year.

Indirect Pollution

The largest percentage of pollution added to the oceans is through indirect means (i.e., atmosphere and rivers). Great quantities of heavy metals, chlorinated hydrocarbons, and other toxic materials are introduced into the oceans each year, in amounts that are presently unknown with possible effects not yet conceived.

Mercury is one form of indirect pollution that mankind has only recently recognized as critical. Incidents such as the deaths resulting from Minamata Disease and high levels of mercury found in tuna and swordfish have initiated widespread studies of this pollutant, but quantitative data quoting actual amounts being released into the oceans each year on a worldwide basis are not yet available. The quantities of pollution and coefficients used for this material (and in fact for both of the pollutants discussed in this section) will therefore be estimates, since empirical data do not exist.

Mercury Pollution

Metallic mercury has become a commonly used element in many industries in recent years. At the present time, approximately 25 percent of all mercury being produced is used in the production of chlorine and caustic soda, with roughly the same amount being used for the production of electrical units. Table 13.1 indicates these and other basic uses of mercury.

Table 13-1
Percentage Distribution of Mercury Usage—U.S., 1968

Group A	
Chlorine and caustic soda preparation	24%
Catalysts	3%
Others (including chlorine plant start up)	15%
Group B	
Electrical	28%
Mechanical	11%
Paints	15%
Agriculture	4%
	100%

Source: "The Realities of Mercury Pollution," METALS WEEK, Vol. 42, January 18, 1971, p. 6.

It is assumed that the major portion of mercury pollution released into the oceans is from those industries where mercury is in general not part of the finished product (Group A in table 13.1).

The 1968 edition of Minerals Year Book reports that 46 percent of all mercury was re-used, inferring that 54 percent was discarded. The quantities of this material deposited in the oceans versus on-land masses was not estimated. Some authorities estimate that one-quarter of all mercury produced is deposited in the oceans, while others place this proportion as high as one-half [2:2], [38:178].

Figure 13.6 plots the production of mercury on a worldwide basis from 1953 to 1966 with the dotted lines indicating the probable range of mercury pollution each year. These pollution estimates were obtained by using the previously discussed pollution coefficients of 25 and 50 percent.

By extrapolating the line in figure 13.6, a value of 16,000 metric tons of mercury produced for the year 2000 results. A forecast by the U. S. Bureau of Mines indicated that mercury production would be twice that produced in 1968 or 20,000 metric tons (i.e., 1968 = 10,000 tons) by the year 2000 with over one-third being used for the preparation of chlorine and caustic soda, which appears to be a major contributor of mercury pollution [31:6].

It is estimated that 5,000 tons of mercury per year are washed from the continents, and thus at the present time, man's contribution to this form of pollution is about equal to that caused by natural avenues [38:178]. Of importance is the concentration and placement of mercury pollution with regard to other pollutants. This topic will be discussed in detail in the section entitled, the Effects of Mercury Pollution.

DDT Pollution

Since the early 1940s when man discovered DDT's insecticide potential, he has been polluting the oceans with this highly toxic material, which is completely foreign to the natural environment. Data on the quantities of DDT produced each year on a global basis are totally lacking, but we do know that since 1944, the United States has produced 1,220,000 metric tons and it now appears that production in this country is slightly declining [15:132].

Production in the United States has averaged 67,000 metric tons per year, while consumption has been steadily falling since 1960 (see figure 13.7). This indicates the widespread use of DDT in countries not capable of producing their own insecticide. As the technology of these countries advances, DDT production will increase. At this point in time, 1971, the extent of this production is not known.

Because of the lack of data, it will be assumed that 10^5 metric tons of DDT are produced each year, which is less than twice the U. S. production.

As with the methyl mercury problem, the DDT that enters our environment in any one year does not all degrade immediately. The half life of DDT in the ocean is estimated at ten years and thus, the pollution in any given year is much less significant than the cumulative effects (see figure 13.8) [11] [38:186].

———— MERCURY PRODUCTION
━ ━ ━ ━ MERCURY POLLUTION–HIGH PROJECTION
━ ━ ━ MERCURY POLLUTION–LOW PROJECTION

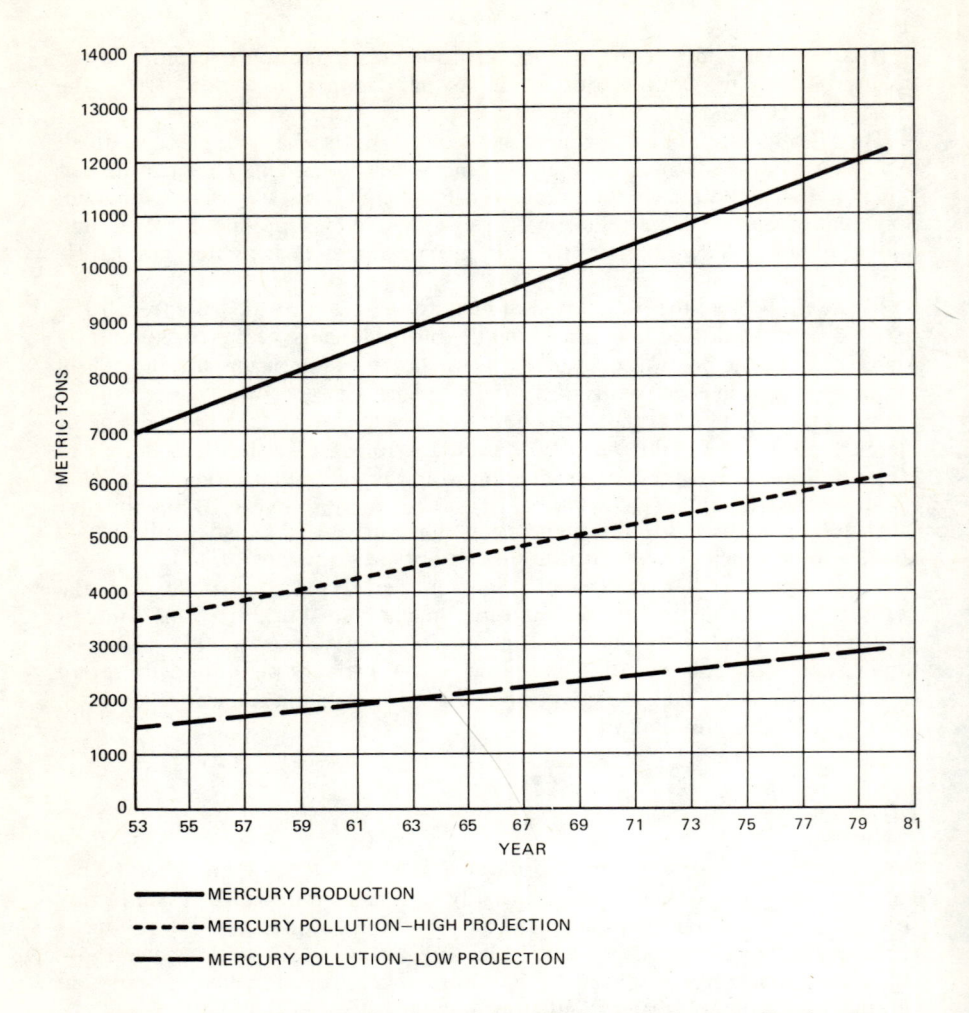

Figure 13.6. Mercury Production and Pollution. Source: Production data for years 1953-1966 from *United Nations Statistical Yearbook*, 1968, increasing world totals shown by estimates of output in U.S.S.R., mainland China, and Czechoslovakia made by U.S. Bureau of Mines. Low projection–based on assumption that one-fourth of production is deposited in the oceans. Estimate of B. Birnbaum, *CBS News Special*, January 12, 1971, p. 2. High projection–based on assumption that one-half of production is deposited in the oceans, S.F. Singer, *Global Effects of Environmental Pollution*, (New York, 1968: Springer-Verlag), p. 178.

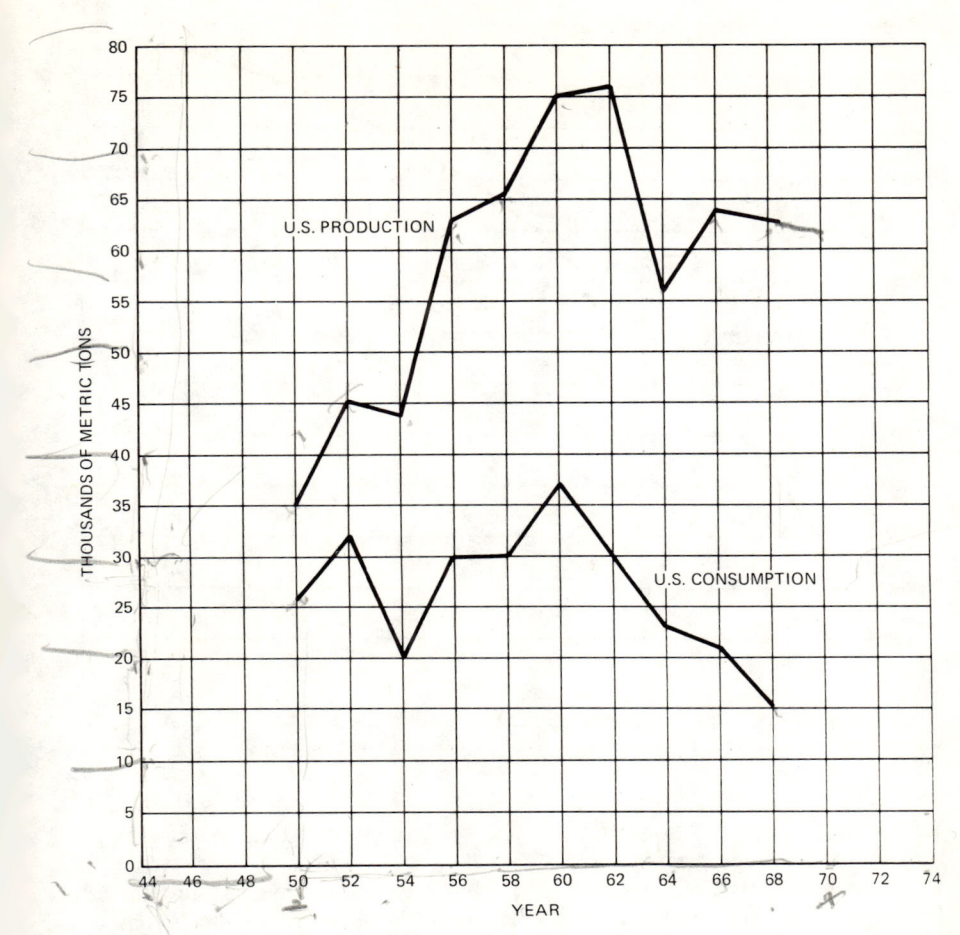

Figure 13.7. U.S. Production and Consumption of DDT. Source: *Man's Impact on the Environment.* Report to the 1972 United Nations Conference on the Human Environment (Cambridge, Mass.: M.I.T. Press, 1970).

DDT has its major use in agriculture and there are three ways that it might be introduced into the oceans. Runoff is one possibility, but this quantity appears negligible. The other two possibilities are both through the atmosphere, either by addition directly during spraying or volatilization after deposit on the soil.

The inability to measure the amount of DDT that enters the oceans in a given period of time is a result of two major factors:
a. DDT's rapid distillation and codistillation rates;
b. DDT's varying solubilities in water and organic compounds.

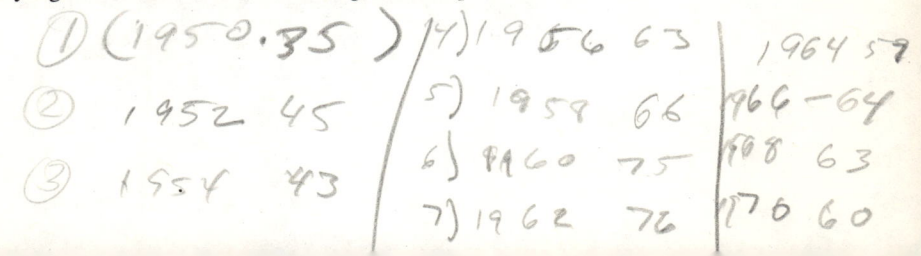

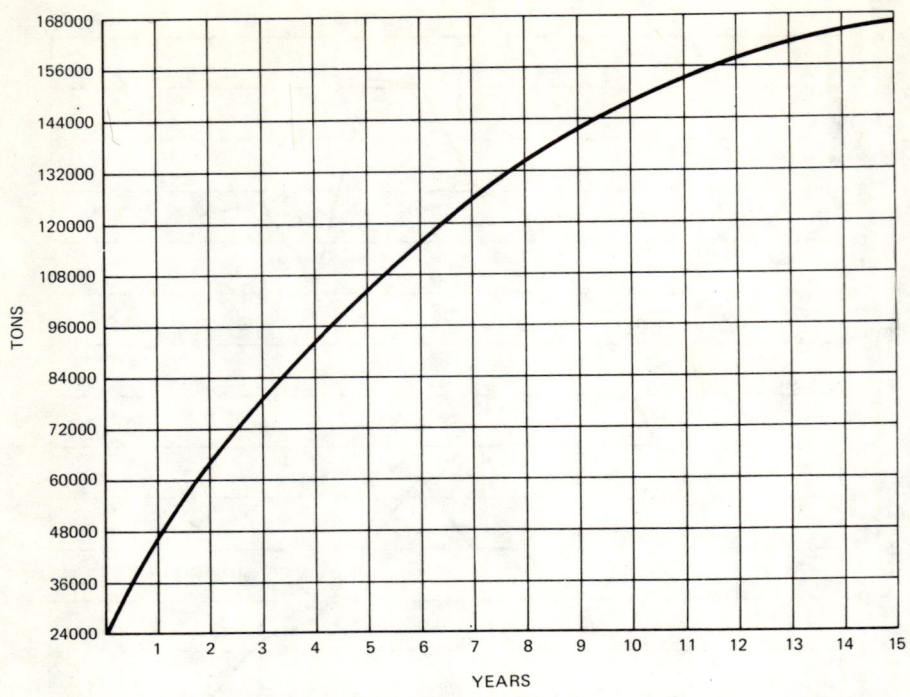

Figure 13.8. DDT Accumulation. (Assuming 24,000 tons DDT added each year and half life of 5 years, a conservative estimate.)

Taking these two conditions into account, we will attempt to arrive at a value for DDT pollution per year.

A number of studies have been done that measure the average amount of DDT in rainfall. The most extensive tests, run in Great Britain, indicate that an average of 80 ppt (parts per trillion) DDT is found in rainwater [8]. Tests run in Florida indicate 1,000 ppt DDT in rainwater [13].

The total annual precipitation over the oceans has been estimated to be 2.97×10^{14} metric tons [6]. By multiplying the precipitation by the observed data (both 80 and 1,000 ppt), we may obtain the theoretical contamination by rainfall alone. The relevant formula is:

metric tons water as rainfall per year annually $\times \frac{DDT}{Water}$ = metric tons DDT entering ocean

Thus, at 80 ppt:

$$3 \times 10^{14} \times \frac{80}{10^{12}} = 240 \times 10^2 = 2.4 \times 10^4 = 24,000 \text{ metric tons}$$

and, at 1,000 ppt:

$$3 \times 10^{14} \times \frac{1,000}{10^{12}} = 3000 \times 10^2 = 3 \times 10^5 = 300,000 \text{ metric tons}$$

The second value of 300,000 metric tons exceeds the estimated production value by three times. This seemingly unrealistic value is explained by the fact that it is more likely for DDT to vaporize on the warmer surface water nearer the equator and thus change from ocean to atmosphere more readily. The codistillation rate of DDT with water has been recorded at rates of up to 100 ppb (parts per billion) in 24 hours [17:278].

The codistillation rate varied directly with the ppt concentration and temperature. This information helps to explain both the exaggerated rainfall values received in Florida, and the system by which DDT moves from land to atmosphere to ocean. The fact that DDT can vaporize without water vaporizing is of importance when considering this phenomenon. Also, because of the much larger requirements for DDT in areas near the equator, we should expect an increased quantity to be found in rainfall in this area.

Data have also been collected on the Antarctic Continent that indicate that substantial quantities of DDT may be building up on the world's ice caps. In the Southern Hemisphere's summer, winds blow from the Pacific that apparently carry DDT in suspension or as a vapor. Upon reaching the colder air mass, the DDT is removed from the air by falling snow. Because of the relatively small quantity of snow that evaporates or sublimes in this region, these data should give a truly representative indication of DDT pollution in this section of the world.

The data available gives a value of 5×10^{-12} grams per gram of DDT found in snow samples. With the average snowfall of 19.8 cm/year, new snow, an estimate of the amount of DDT deposited over the entire continent for the past 22 years would be 2.4×10^6 kg = 2,400 metric tons of DDT.

Thus, the values obtained from Antarctica indicate that only 0.05 percent of the expected global DDT pollution during the year occurs over this area, but considering the relatively small quantities of contaminated air forming its immediate atmosphere, it seems to be in accord with the rainfall values.

Data on the quantities of DDT existing in the ocean at the present time are not available. This is due partly to the fact that the solubility of DDT increases rapidly with increasing water temperatures, salinity, and greatly depends on the quantity of organic material existing on and near the surface of the water.

Thus, it appears that from an analytical standpoint the values of DDT pollution arrived at by the rainfall experiments are the only ones available. For this reason, the value of DDT pollution as obtained by the British experiments will be used. It is the author's opinion, when considering the half life and codistillation rates of DDT, that as much as 70 percent of the total DDT used will eventually reach the oceans. Thus, the value that has been adopted, of 25,000 metric tons per year of DDT pollution, is probably quite conservative.

Effects of Oil Pollution

The effects attributed to oil pollution can be divided into two groups:

1. Effects of concentrated spills;
2. Effects of dilute spills.

Because of recent investigations of shipping and off shore production accidents, a quantity of information is available about the effects of concentrated spills. Because of the lack of evidence linking human illness to increased hydrocarbon concentration in sea water, little research has been done in the area of dilute concentrations. Both the oil companies and independent institutions are presently beginning studies of the overall effects of carbon content in sea water. Thus, we will discuss only effects attributed to concentrated spills in this section.

It is a generally accepted fact that most spills occur in coastal waters where marine productivity is concentrated. In September 1969 a spill of 160,000-175,000 gallons of #2 fuel oil occurred off West Falmouth, Massachusetts.

Because of the various densities of both crude and refined oil, an absolute weight/volume relationship is not available. Generally, oil has a density from .6 to 1.0 and thus, one ton could be anywhere in the range of 240-400 gallons [25:762].

Members of the Woods Hole Oceanographic Institution trawled the area three days after the spill and found that 95 percent of their catch was dead, with the remaining 5 percent dying. This finding was not unexpected and was explained by two different possibilities.

1. Death occured when the organisms were exposed to toxic materials in the oil.
2. Death occured through asphyxiation from direct coating.

A year after the accident occurred, bottom life (shellfish) was still being poisoned from the toxic material that had been absorbed by the sediment. Other effects, such as the fouling of beaches, are also evident on a global basis.

Because of large-scale accidents such as the Torrey Canyon and Santa Barbara incidents, great effort is being expended by the oil companies to both prevent spills and remove oil from the water after spills have occured. During recent spills, detergents were used in an attempt to disperse the spill. It has been found that these detergents were more toxic and caused more damage to ocean life than the oil itself.

Research has now developed new water soluble dispersants (an example is Corexit 7664) that have no deleterious effects on ocean life [30:99]. Other methods of collecting oil such as huge rubber tanks, flexible booms, air barriers, and urethane chips are being developed to cope with the problem [36:49-59].

Lengthy studies are in progress that will help determine if oil or its components which are absorbed by ocean life either in concentrated or dilute doses affect future generations of that species or man himself.

Effects of Mercury

Until recently, the dumping of inorganic mercury (Hg) was not considered a hazard because of two major reasons:

1. Metallic mercury is a heavy metal with a negligible solubility in water.
2. The poisoning effect of metallic mercury is reversible and only affects the alimentary tract and kidneys when taken in large quantities.

Thus, the assumption was that metallic mercury dumped into the oceans would lay inert on the bottom, with no effect on the environment or living creatures.

As early as 1965, Dr. Alf Johnels and his associate, Mr. M. Olsson, announced their opinion that microorganisms (later identified as methano-basterium omelanski) in anaerobic areas on lake and ocean bottoms could digest metallic mercury and form methyl mercury, a highly poisonous material. Methyl mercury attacks the central nervous system and has crippled and killed on many occasions. These facts were later confirmed at the University of Illinois by Dr. J. M. Wood, Dr. Carl Rosen, and Mr. F. S. Kennedy in 1968 [21:50].

In 1969, the U. S. government placed a maximum allowable value of 0.5 ppm (parts per million) mercury on all foods, which provided a relative safety factor of 10x, as compared to the normal 100x factor, traditionally used when considering highly poisonous materials. On December 4, 1970, Professor Bruce McDuffie of the State University of New York at Binghampton analyzed tuna from a can and found it to contain 0.75 ppm mercury. Immediately, the FDA began testing other tuna and fish species. A report issued January 7, 1971, by the FDA indicated that less than 2 percent of canned tuna contained more than the arbitrarily selected value of .5 ppm mercury.

Simultaneously, tests were run which showed that 89 percent of the swordfish contained more than 0.5 ppm mercury. An explanation for the high mercury content found in these fish was by now readily at hand. After the formation of methyl mercury, methano-basterium omelanski, by the process of organic complexing, spread the mercury through the ocean environment where algae absorbed the mercury-containing bacteria, and fish ate the algae. As the fish grew in size or were eaten by larger fish, the concentration of mercury increased. It has been determined that the half life of mercury in fish is 200 days, and thus, the poison was concentrated thousands of times in sections of the larger fish [21]. The half life of mercury in man is 70 days.

Tests by the FDA on tuna show that the larger fish have more than 0.5 ppm Hg, 23 percent of the time, compared to smaller tuna, which only exceed the limit approximately 2 percent of the time [2:3]. But the major problem was just becoming apparent. That is, we do not know the rate at which methyl mercury is formed from metallic mercury (Hg°). It is assumed by most authorities that great quantities of metallic mercury still lie dormant on the ocean bottom. The addition of other pollutants (particularly those which decompose and therefore remove oxygen from the environment) to these same waters help contribute to the condition necessary for the formation of an oxygen-free environment and increase the concentration of microorganisms that transform metallic mercury to methyl mercury. Thus, a large percentage of the estimated 60,000 metric tons poured into the oceans since 1950 will affect the marine ecosystem for many years to come, even if the addition of new mercury is completely halted.

Recently, on May 20, 1971, the first United States case of mercury poisoning from fish was reported. A Long Island woman who had consumed regular portions of swordfish for a prolonged time became mentally and neurologically disoriented [14:28].

Warnings to pregnant women to stop eating fish with possible mercury contamination have resulted from findings made in Japan after the Minamata Bay experience. It was found that women who had eaten mercury-contaminated fish had given birth to an abnormally high percentage of brain-damaged children without themselves being affected.

Tests run off the California coast indicate levels of up to 25 ppm mercury found in parts of California sea lions, and up to 170 ppm in seals, from the Bering Sea [1].

It thus appears that the mercury problem is one of enormous concern. The apparent concentrations required to cause severe neurological damage seems to be about a milligram per day, but little is known about the quantities needed to do other, less detectable damage to man or his offspring [2:4].

The amount of metallic mercury, in certain parts of the underwater world, has been greatly increasing. Depending on many other pollutants and conditions, these inert deposits may some day be transformed into methyl mercury, and lead to widespread Minamata disasters.

Thus, the question of mercury pollution is a complex one. Not only must the quantity of new mercury pollution be decreased, but ways to prevent or delay the already existing mercury deposits from forming methyl mercury must be studied.

The Effects of DDT

One of the undesirable attributes of DDT is that it is not selective. That is, not only is it poisonous to insects, but it has deleterious effects on most living creatures. Sharp reductions in the population of birds and mammals have been attributed to DDT poisoning [25:758].

One example concerns many species of birds where the existence of this pesticide has, in general, accounted for abnormally thin-shelled eggs. A case in point is the Mallard. Experimental tests indicate that on the average, only one egg out of twelve hatched when the parent had been fed a diet that contained DDT and its derivative, DDE. This is opposed to the normal nine to eleven hatchings normally obtained from twelve eggs.

Shrimp, crabs, and sea plankton are killed when exposed to DDT in concentrations in the ppb range. Blubber from whales that exist in the open seas has had as much as 6 ppm DDT [10:8], and California mackerel are no longer commercially available because greater than .5 ppm of DDT is consistently found in their meat.

Experiments indicate that oysters, exposed to DDT in concentrations of 0.1 ppb, have concentrated this material up to 7 ppm in their tissue over a month's time. This is a ratio of 70,000 to 1, and with a solubility of DDT at 1.2 ppb, 100 ppm of DDT would have built up during this period [5] [28]. Other studies on Sheepshead Minnows indicate that fish whose parents were exposed to high concentrations of DDT were more sensitive to pesticides than fish whose parents had not been exposed [27]. It has been determined that DDT inhibits photosynthesis in single-celled marine plants [38:178].

The upper 100 meters of the ocean is known as the mixed layer, and at the surface a film exists that contains oil, fatty acid, and alcohol. It is expected that the DDT in solution would have its highest concentrations in this film. Here, phytoplankton absorb the DDT and it begins its way up the food chain, increasing in concentration as the size of the fish increases.

Let us now review the cumulative effects of DDT. If DDT only has a half life of five years, then after fifteen years there will still be 10 percent of the original amount in the system. This, coupled with its high codistillation and vaporization rates, and apparent effects on many of this earth's species, should suggest an existing crisis.

Pest control in the United States has in recent years begun the move away from the use of DDT and other pesticides and to more use of herbicides and other systems for the control of insect pests. Since 1964, more herbicides that are generally less toxic to marine and related organisms have been sold than insecticides [4].

Also, the use of other than chemical control measures, such as the addition of exotic species to the environment, is being practiced by the United States and other advanced nations to eliminate unwanted insects. Because of its relatively cheap cost and tremendous success to this time, DDT is becoming more widely used by less advanced nations which, in general, support large agricultural sectors. Thus, the United States and other advanced countries will have to supply means for the control of insect pests for these developing nations, if we are to avoid a crisis in the area of DDT pollution.

General Effects of Pollution

In the previous sections, the quantities and effects of oil, mercury, and DDT pollution were reviewed. In this section, we will consider some of the interactions between these and other pollutants with regard to the cumulative effects, which in many cases constitute problems of a much more severe nature than either of the individual pollutants would.

An explanation of the food chain phenomenon will conclude this discussion for a better understanding of the avenue by which pollution reaches man.

Food Chain

The pollutants studied in this chapter each have a certain affinity for parts of living organisms; examples are the concentration of DDT built up in fatty tissue and lipids; methyl mercury build-ups in livers, flesh, etc.

As these organisms that ingest the pollutants are absorbed by larger species, these pollutants tend to concentrate. Depending on the stability of the pollutant and the digestive system of the species, this concentration will vary. Figure 13.9 illustrates the quantities of absorbed food required to add the indicated weights to the predators.

Basically, for every 10 parts material absorbed, 1 part is retained. In the case of DDT, methyl mercury, etc., the methyl mercury and DDT retention is enormously greater than for most materials. The ratio is in the region of 100 parts absorbed, 99 parts retained. In this way, small insignificant percentages of pollutants found in vegetable plankton can concentrate and eventually yield dangerous levels in man's food supply [25].

Mercury and Decaying Material

As previously discussed, metallic mercury is transformed into methyl mercury when digested by methano-bacterium omelanskii, a microorganism. These organisms are only found in anaerobic areas. Of the possible ways an oxygen-free environment might be formed on ocean and lake bottoms, the decomposition of dead and decaying material is the most likely. Because of phosphorous and other nutrient pollution, entrophication occurs on lake, river, and shallow ocean waters. As the algae population increases, quantities die and drop to the bottom where they decompose. This decaying action absorbs oxygen from the water and replaces it with carbon dioxide. As the layer of dead plant life builds up on the bottom, an oxygen-free environment develops, which can then support methano-basterium omelanskii. When this decaying process (which is magnified by eutrophication) takes place over deposits of metallic mercury, the formation of methyl mercury is enhanced.

Another type of pollution that can greatly increase the formation of this lethal poison is untreated sewage. As with the algae, sewage decomposes and in doing so, absorbs oxygen from the water.

Because of the fact that most mercury, nutrient (phosphorous), and sewage pollution occurs in shallow waters, frequently at mouths of rivers, this problem of methyl mercury formation is a formidable one. The large quantities of

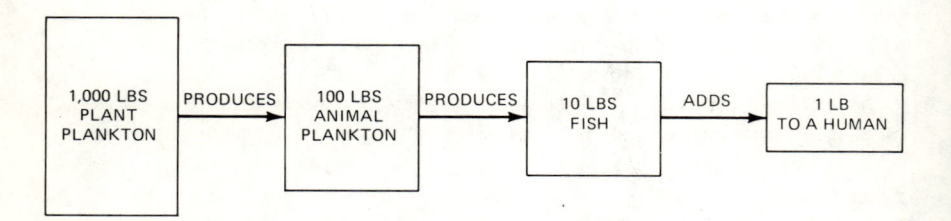

1,000 LBS PLANT PLANKTON → PRODUCES → 100 LBS ANIMAL PLANKTON → PRODUCES → 10 LBS FISH → ADDS → 1 LB TO A HUMAN

Figure 13.9. Food Chain. Source: *National Geographic*, Centerfold "Pollution, Threat to Man's Only Home," December 1970.

metallic mercury that have built up over the years, and to this time have been inert on the ocean floors, are becoming more subject to transformation into the dangerous and stable compound methyl mercury.

DDT And Oil

Each of the pollutants (DDT and oil) to be discussed here has previously been reviewed individually with the effects already noted. Of importance in this section is the effect of a thin film of oil that exists on the surface of water for periods of time after an oil spill has occured. The fact that DDT has a limited solubility in water (e.g., 1.2 mg/liter) but a much greater solubility in lipids (e.g., 100 gm/liter) was previously discussed [18:506]. It has been found that DDT has a high solubility in oil (very soluble); thus, relatively large quantities of DDT may be absorbed in the film of oil that remains on the surface after a spill. Many water life forms, especially microscopic plants, exist, at least for a period of time each day, in the surface waters of the oceans to take advantage of sunlight. Because of the higher concentrations of DDT in the surface film, it is expected that significant quantities of DDT and oil may be absorbed by those creatures and thus introduced into the food chain.

Here the types of pollution are not as localized as in the previous example, but because of the extensive evaporation and precipitation of DDT from the ocean-atmosphere system, it may be a problem that contributes significant contamination to man's food supply.

Conclusion

It has been the aim of this chapter to review some of the pollutants that are causing current problems to the inhabitants of the world's oceans, and thus, to man himself.

The quantities of production and pollution considered were meant to be orders of magnitude, since empirical data in these areas are insufficient, ambiguous, and contradictory. The major and significant points on each pollutant (oil, mercury, and DDT) will now be briefly reviewed in an attempt to establish the magnitude of the existing problem and outlook for the future.

Oil

The petroleum industry in recent years has maintained a policy of eliminating oil pollution. The use of LOT in tankers and precautionary techniques used on underwater wells should significantly reduce the pollution from this sector. Great emphasis will be required in the area of both merchant ships' oil spills and industrial and automotive wastes which, though not observed in concentrations proportional to tanker and oil well spills, are presently of substantial significance.

Little light has been shed on the long-term effects of oil pollution and increased carbon content in the oceans. Direct short-term effects observed have included oil spills that kill birds, fish, and other less developed sea creatures. To date, no human deaths have been attributed to long-term oil pollution.

Mercury

When a form of pollution becomes so severe that it is responsible for the deaths or serious crippling of hundreds of human beings, immediate and gross action should be taken. In the case of methyl mercury poisoning the immediate action was present, but the gross action is lacking. The quantities of mercution and pollution have not been grossly curtailed, except for a few nations of the world, a small group to which the United States does not belong. Unless drastic restrictions are placed on the amount of mercury entering our water systems, the yearly amount of metallic mercury pollution will nearly double by the year 2000. This, coupled with the increasing amount of sewage disposal and man-initiated eutrophication, could lead to the formation of large quantities of methyl mercury and widespread Minamata disasters. Not only should we consider the virtual elimination of mercury pollution, but also the problem of preventing the large quantities of metallic mercury now existing on the ocean bottom from forming the lethal compound methyl mercury.

DDT

The problems with DDT are in part quite different from those of either oil or mercury. We are not dealing with concentrated spills or potentially dangerous quantities of pollutant requiring only a catalyst to form a deadly poison. We are faced with a pollutant whose half life is measured in years and toxicity in ppm. To date, no reports of a human death has been directly linked with DDT pollution, but more subtle effects of this material on man are only now being discovered. Because of its long life and rapid codistillation with water, DDT pollution is a genuine global problem. Its usage in the more technically advanced countries is declining, but it appears that this decline is more than matched by increases in new technically emerging countries. The deaths of both birds and ocean creatures that have been attributed to DDT to date may have far-reaching ecological effects in the future.

Considering the absence of international laws, and weak national laws, the ocean pollution problem has no end in sight. One of the heralded United States laws, the Federal Water Pollution Act of 1965, is in reality nearly useless on a short-term basis in that the lawful minimum time required to study a case before reference to the Attorney General is 58 weeks [16:84 col. 1].

This chapter has only covered three of an ever-growing list of potentially dangerous pollutants in our global waters.

Without greatly improved monitoring and increased study of this problem, mankind may soon discover that irreversible damage has been done to the oceans, and the creatures that live in its realms, which will consequently have tremendous effects on man himself.

Notes

1. Lawrence K. Altman, "Mercury Is Found in Coast Sea Lion," *New York Times*, November 7, 1970.

2. B. Birnbaum, "Is Mercury a Menace," *CBS News Special*, January 12, 1971.

3. T. D. Barrow, *Oil and Gas Extraction Technology*.

4. T. W. Duke, "Estuarine Pesticide Research — Bureau of Commercial Fisheries," *Proceedings of the Gulf and Caribbean Fisheries Inst.*, November 1969.

5. David J. Hansen and Alfred J. Wilson, Jr., "Residue in Fish, Wildlife and Estuaries," *Pesticide Monitoring Journal* 4, 2 (September 1970).

6. Roger Revelle, "Water," *The Oceans* (New York: Prentice Hall, 1942).

7. Norman Ritter, "Oil and the Undersea World," *Lamp*, Summer 1966.

8. K. R. Tarrant and J. O'G. Tatton, "Organochlorine Pesticides in Rainwater in the British Isles," *Nature* 219 (August 17, 1968).

9. Maretta Tubb, "Control of Oil Spills — Recent Technical Developments," *Ocean Industry* 5 (June 1970).

10. A. A. Wolman and A. J. Wilson, Jr., "Occurrence of Pesticides in Whales," *Pesticides Monitoring Journal*.

11. G. M. Woodwell, "Changes in the Chemistry of the Oceans: The Pattern of Effects," *Global Effects of Environmental Pollution*.

12. C. F. Wuster, "DDT Reduces Photosynthesis by Marine Phytoplankton," *Science*, Vol. 159.

13. Yates, Holswade, and Higer, "Pesticide Residues in Hydrobiological Environments," *American Chemical Society*, 1970.

14. "Bare Swordfish Poisoning Case," *Daily News*, May 21, 1971.

15. *Chemical Economics Handbook*, 1969.

16. "Clean Water Act Held Insufficient," *New York Times*, May 26, 1971.

17. "Codistillation of DDT With Water," *Agricultural and Food Chemistry Journal* 11, 4 (July-August 1963).

18. *Handbook of Analytical Toxicology*, Chemical Rubber Company.

19. *Lloyd's Register of Shipping, Casualty Reports*, 1965-1970.

20. *Man's Impact on the Global Environment. Report to the 1972 United Nations Conference on the. Human Environment* (Cambridge, Mass.: M.I.T. Press, 1970).

21. "Mercury: How Much Are We Eating?" *Saturday Review*, February 6, 1971.

22. *Nature* (November 8, 1969).

23. *New York Times,* November 3, 1970.

24. *Ocean Industry,* November 1967.

25. "Pollution, Threat to Man's Only Home," *National Geographic* (December 1970).

26. *Sea Frontiers* (Jan.-Feb. 1971).

27. "Sensitivity to Pesticides in Three Generations of Sheepshead Minnows," *Bulletin of Environmental Contamination & Toxicology* 5 (1970).

28. "Solubility of Carbon-14 DDT in Water," *Agriculture and Food Chemistry* 8, 5 (Sept.-Oct. 1960).

29. *Study of Waste Oil Disposal Practices in Massachusetts,* Arthur D. Little, Inc.

30. *The Lamp—Anthology on Conservation.*

31. "The Realities of Mercury Pollution," *Metals Week* 42 (January 18, 1971).

32. "Tuna Safe; Taint Bars Swordfish," *Daily News,* January 8, 1971.

33. *U. N. Statistical Institute Abstracts.*

34. *U. N. Statistical Year Book,* 1968.

35. U. S. Naval Institute Proceedings, May 1969.

35. *U. S. Naval Institute Proceedings,* May 1969

36. *Ocean Industry,* June 1970.

37. *The Lamp,* Summer 1966.

38. S. F. Singer, *Global Effects of Environmental Pollution* (N.Y.: Springer-Verlag, 1968).

39. Edward R. Corino, Eng. Assoc., Esso Research and Engineering Co., *Industrial Wastes: Oils of Petroleum Origin* (mimeographed, 1970).

14 Summary of Findings Concerning Water

ALFRED J. VAN TASSEL

What will be the state of the environment in the United States of 1980? Is there any way in which we attempt a quantified answer to that question? On the face of it, the attempt to answer these questions would appear presumptuous. Surely, the economy of the most powerful nation in the world is too complex to permit a meaningful forecast as far ahead as 1980. And certainly any forecast that might be essayed should be cast in qualitative terms.

Nevertheless, we have made an attempt to answer the first question with respect to water pollution and the effort was made easier, not more difficult, by the quantitative approach. The quantitative methodology employed in almost all of the first chapters on water (3 through 11) and described in some detail in Chapter 2, gave a common framework for observation that was useful. In many ways, the quantitative method gave a basis for judging whether a given sector of a specific river was likely, over a period of time, to have improved or deteriorated in quality. The method also permitted comparisons on an objective basis between bodies of water. In each, the sum of the potential BOD associated with various levels of input provided a basis that was always objective even if at times approximate.

Uniting all the chapters was the conviction that at any given state of the art, the problems of combating water pollution increased in some rough proportion to population and the scale of industrial output. This approach made it possible to relate the egregiously polluted condition of both the lower Hudson River in the prime example of an urban environment and the almost equally polluted Savannah River in the sleepy cradle of the old South. The essential condition of these two strikingly different waterways could be expressed, at least roughly, in terms of the oxygen demand of the industrial and municipal sewage effluents entering its waters. This oxygen demand was compounded, in both cases, of two types of demand on the streams' dissolved oxygen. One was designated as a chemical oxygen demand and was exerted mostly by industrial effluents, expressing the chemical affinity of constituents of the effluent for oxygen. The other was known as the "biological oxygen demand" and measured the amount of oxygen required to "burn up," figuratively speaking, the biological pathogens that might exist in the waters as well as other more harmless organic (that is, carbon-containing) matter. Since the disease-bearing bacteria are the more dramatic manifestation of this situation, pollution is usually thought of in terms of their continued presence, but actually the same amount of a stream's dissolved oxygen may be consumed in decomposing a piece of wood pulp as in destroying a colony of cholera bacilus—the point is that it cannot be available in the same place at the same time for both jobs which, in nature's scheme of priorities, stand on the same level.

Actually, it is possible to achieve a rough addition of the polluting effects of the discharge from a sewage treatment plant that may exert a relatively small demand on a stream's oxygen supply and the effluent of a pulp mill that is likely to have a voracious appetite for oxygen. The effluent of a single pulp mill on the Coosa River, for example, was estimated to exert the equivalent BOD (biological oxygen demand) as that of the untreated sewage of a city of 200,000 people [1]. The same dissolved oxygen in a waterway will not be available for both jobs. If consumed by the effluent of a pulp mill, it cannot also destroy the remaining pathogens from the sewage effluent. In Chapter 2, technical justification and explanation of method was set forth supporting the use of the BOD of entering effluents as the fundamental measure of the potential pollution of a waterway. The limitations and shortcomings of this measure were discussed as well in Chapter 2. Despite these, it is felt that this fundamental approach gives the common standard needed for significant judgments concerning the state of pollution in this nationwide sample of waterways.

What did we learn about the individual bodies of water studied? There are risks in capsule summaries and the best way to learn what was found is to read the individual chapters themselves. What follows is a serious effort to summarize in a short space some of the principal conclusions.

Lake Michigan

Jay Parker found his BOD calculations pointed clearly to the trouble spots on Lake Michigan and spotlighted as well the areas likely to be scenes of the most serious future problems and the nature of the difficulties likely to prove most troublesome.

For example, some of the most severely polluted Lake Michigan waters are in the Green Bay area where the paper industry releases large quantities of oxygen-demanding effluents. Paper and related industries accounted for two-thirds of the potential BOD in the Green Bay area in 1969 compared to only 31 percent from municipal sources. Since paper and allied products production was projected to grow faster than population, the potential industrial BOD was expected to expand by 45 percent to 1980 compared to a trivial 2.8 percent increase in potential BOD from Green Bay municipal sources. These comparisons are of BOD of industrial and sewage effluents before treatment. Plans call for greatly improved treatment in both sectors raising effectiveness of treatment of municipal wastes from the present 60 percent level to 80 to 90 percent by 1980, and of industrial wastes from present unknown but extremely low levels to the recommended 90 percent BOD reduction by 1980. Clearly the fate of Green Bay waters depends mainly on how close the paper industry comes to accomplishing the goals laid out for it.

The situation is the opposite in another trouble spot—the Milwaukee area—where BOD from municipal sources accounted for three-fourths of the potential BOD in 1969 and was still expected to represent 70 percent in 1980. The overall potential BOD in the Milwaukee area was expected to be 176 million pounds in 1980 compared to 150 million pounds in 1969. The waters around Milwaukee Harbor were regarded as highly polluted in the 1960s despite an 86 percent BOD removal level in sewage treatment. Improved treatment is unlikely to achieve more than a 95 percent removal and this may not be sufficient to make a significant improvement in water quality in view of population growth and a 32 percent increase in potential industrial BOD from 1969 to 1980. More encouraging was the scheduled completion by the end of 1972 of facilities designed to remove 80 percent of the phosphate nutrient otherwise discharged.

Some phases of the battle were not easily measured in Lake Michigan. Generation of power employing nuclear fuel necessitates the discharge of heated waters to the sources from which cooling water is withdrawn—a process known as thermal pollution. Such nuclear generation of power on the shores of Lake Michigan is expected to increase 100-fold by 1980.

Unlike Milwaukee, Chicago suffers few of the consequences of water pollution because it exports its sewage effluent southward via the Chicago River which ultimately empties into the Mississippi. The most important industrial operations in nearby northwestern Indiana are in the steel industry, and the method employed elsewhere in this chapter to measure pollution potential in terms of BOD of treatment plant effluents was not applicable here.

Despite some problem areas along the shore, one can say in summary that the open waters of Lake Michigan are uniformly satisfactory and few areas betray more than the subtlest hint of the overenrichment that has plagued Lake Erie.

Lake Erie

The Lake Erie scene is full of paradox. A heavily populated, highly industrialized U. S. shore is matched by a rural Canadian side that contributes little to the lake's pollution. Despite the lake's reputation as a prime example of pollution, most of the municipalities along its shores draw their drinking water from it and most of its area is suitable for swimming. To be sure, Lake Erie water is sometimes potable only after heroic measures to remove objectionable tastes and odors, and swimming and other water contact sports may be unsafe at beaches closest to major centers of population. Nevertheless, it is true that the open waters of Lake Erie are usually of satisfactory bacteriological quality.

The pollutional load on Lake Erie is very unevenly distributed, with the heaviest inflow affecting the relatively shallow western basin west of Sandusky, Ohio. This portion receives the effluent of sewage treatment plants in Detroit and Toledo as well as extensive industrial discharges from the same cities. As much as two-thirds of the BOD of municipal and industrial discharges in the Lake Erie drainage basin arose in the western basin. Much of the municipal and industrial waste entering the Lake Erie drainage basin was inadequately treated, but whatever the degree of treatment the nitrate and phosphate nutrients in the effluents remained unaffected. In consequence, nutrient levels were high in many parts of the lake and especially in the western basin. The latter being relatively shallow readily admitted the light necessary to support the growth of vegetation encouraged by the presence of phosphates and nitrates. The growth and subsequent decay of vegetation not only adds to the burden on Lake Erie's supply of dissolved oxygen, but has also in bottom deposits created, according to Professor Barry Commoner, a future possible claim on oxygen supplies that may be collected at some time with disastrous consequences.

Puget Sound

Donald Niddrie decided that "the State of Washington is probably the most advanced state in the Union when it comes to water pollution control." This, together with the enormous assimilative capacity of Puget Sound with its powerful tidal movements, argues the likelihood of overcoming even the enormous polluting potential of a substantial pulp and paper industry. The paper and allied products group accounted for about 78 percent of the potential BOD in the five counties of the Puget Sound area compared to only 16 percent attributable to municipal sewage. This ratio held despite the fact that one of the five counties, King, contained Seattle and no paper industry, and another, Pierce, seat of Tacoma, had only a relatively small paper industry.

In the other three counties—Snohomish, Whatcom, and Skagit—the role of the paper industries was overwhelming, accounting for 92 percent of the potential BOD in 1969 compared to a trivial 3.4 percent of the total for municipal wastes. This was the highest proportion of BOD attributable to industrial sources of any area studied with respect to water pollution. The importance of the paper industries as a source of potential pollution in the three counties was not expected to be changed substantially in the 1980 projections.

In view of the importance of the paper industry for Puget Sound, there was reason for encouragement in the 1972 report by the Council on Priorities that the pulp and paper industry had made outstanding progress in overcoming pollution problems [6]. Niddrie found specific reason for hope in the fact that the State of Washington had exact knowledge of all potential sources of pollution and a plan for dealing with them.

Much of the stimulus for the creation of a state water pollution control program came from the earlier threat to the beauty and usefulness of Lake Washington, Seattle's greatest recreational asset. Excessive growth of algae, attributable to nutrient concentrations, marred the appearance of the lake in the middle 1960s, and led as well to unpleasant odors. The installation of a sewer system that diverted all sewage effluent from the lake had effected a 400 percent improvement in transparency of the water by 1969. A full-scale sewage collection and treatment program has saved Lake Washington in one of the most dramatic environmental success stories.

San Francisco Bay

The citizens of the San Francisco Bay area first became fully aware of the threat to their beautiful bay in the early 1960s. Landfill operations were reducing the area of the bay especially in the southern sections, and little had been done to clean up the many sources of pollution. The first step was a halt to further landfill, a measure that created something of a crisis in solid waste disposal (see Chapter 25). In addition, the very considerable scientific resources of the area were mobilized to study the problem and recommend solutions. The result was an ambitious and far-reaching plan that called for an expenditure of a billion dollars in the construction and operation of sewage collection systems and treatment plants to 1980. The final phases of the program even envision some measure of water reclamation, but in the first phase to 1980 the aim is to deal with the most pressing problems by increased treatment and export of the treated waste waters from the extremities of the bay to the central regions where tidal flushing is more powerful.

The effluent from sewage treatment plants serving 3.7 million people is either discharged to San Francisco Bay or into streams flowing into the bay. In addition, there is a large and growing BOD load from industrial establishments. Problems center in four counties—Contra Costa, Alameda, San Joaquin, and Santa Clara—where population was projected to increase by 22 percent to 1980. However, the projected rise in industrial BOD load for these counties to 1980 is 107 percent. Treatment of this greatly expanded industrial load is clearly a key to progress. Presently, water quality as judged by concentrations of coliform bacteria is unacceptable in all portions of the bay, but poorest in the southern sections. Raymond Matern found that, despite the billion dollar expenditure on pollution control, likelihood is for only a slight improvement or maintenance of present quality levels by 1980. The BOD of wastewaters is projected to increase by about 40 percent over 1965 levels by 1980 in the southern sections. Looming as a possible further threat requiring study is the sharp increase in nutrients in both the southern sections, the shallow part of the bay, and also in San Pablo and Suisun Bays.

Long Island Sound

Robert Caponi's findings concerning Long Island Sound were somewhat inconclusive since he was unable to locate data sources that permitted systematic analysis. Nevertheless, his evidence did point to increasing nutrient concentrations, particularly in the western part of the sound. Thus, despite the fact that most sewage treatment plants discharging effluent to the sound offer secondary treatment, the quality of waters near discharge points declines.

M. Grant Gross, research director of the Marine Sciences Research Center at the State University at Stony Brook, N.Y., completed a six-month study of North Shore Long Island harbors and found three of six unsatisfactory. "The more severe (pollution problems) tend to be those with more population and more sewage," Gross said [7:4]. He warned in an interim report that nitrate levels were so high in Manhasset Bay that plans for construction of further sewage treatment facilities should be halted. The reason a high concentration of nitrogen compounds is associated with lower water quality, according to Dr. Gross, is that the nitrates stimulate such a rapid growth of marine organisms that they cannot survive on the food available. The death of the organisms in large quantities and their decomposition deprives fish of the oxygen required for survival [29:19]. This potentially unstable situation appeared to be the most sinister threat to the sound's future. Open waters of the sound are generally suitable for high level usage such as shellfish cultivation except for those areas closest to New York City. Harbors and embayments are frequently unsatisfactory, often as a consequence of heavy use by boats.

Hudson River

Richard Martin found that the Hudson River epitomizes many of the contradictions involved in the study of water pollution. Although one of the most heavily polluted waterways in the nation as it flows through New York City, it is suitable for swimming some 25 miles north of the city line. Industrial wastes are a relatively small part of the Hudson's problem, the potential BOD of municipal wastes is over four times that of industrial wastes considering the entire length from Albany to New York City and in the mid-Hudson region where population is increasing rapidly and industry is light, the ratio is almost 9 to 1. Moreover, industry in the Hudson Valley is growing at a moderate rate and much of the growth is in such nonpolluting areas as machinery production and research and development. Thus, there is little likelihood that the growth of industrial waste will overpower efforts to clean up sources of pollution. The river from 30 miles north of the New York City line should show a steady improvement in quality as new treatment works come on stream between now and 1980.

Nevertheless, there is little prospect for substantial improvement in the water quality of the lower Hudson in the vicinity of New York City. Heavy inputs of raw or inadequately treated sewage from both sides of the river plus oxygen-demanding wastes from New Jersey industries appear to be impossible to clean up by 1980.

Actually, there is a possible danger that the heavy concentrations of nutrients and pollutants in the lower Hudson combined with sluggish net flow during summer months of the Hudson in the stretch affected by tidal action might reverse some of the improvement upstream.

Some 5 million cubic yards of solid waste from New York City sewage treatment plants are dumped annually in the Atlantic 15 miles offshore together with 6 million yards annually of dredging spoils and 2 million yards of industrial acid wastes. A four-year study by the National Marine Fisheries Service's Sandy Hook Marine Laboratory found that: "Bottom-dwelling animals and organisms are virtually nonexistent in the dumping grounds, smothered under layers of bacteria-laden, oxygen-depleting wastes [5]," but the same study "revealed little or no damage to fish life." Concentrations of heavy metal in the dumping area were reported to be "100 times greater than normal and bacteria counts were high" [5]. It would appear that such irresponsible "export" of pollution could hardly continue for long with impunity.

The Lower Mississippi River

Robert Schwartz points out that the Mississippi River has long been the carrier of a vast cargo of pollution—the suspended solids that make it the "big muddy" and which ultimately form the ever-extending delta at the river's mouth. For generations, the communities along the river have been accustomed to make careless use of its facilities, drawing water for drinking and other purposes at the same time using it as a depository for sewage and other wastes. There is no reason to expect any sudden change in this relationship.

The projected population growth from 1969 to 1980 is low (under 6 percent) for two communities, moderate (12 percent or under) for three others, and vigorous (at 17.5 percent) only in New Orleans. The growth in the pollutional load attributable to industry is substantially greater. Thus, the industrial BOD load at St. Louis was projected to increase 17 percent to 1980 compared to 12 percent in the same period for BOD of municipal waste. Chemicals and petroleum refining increased most rapidly as was also the case at Memphis and New Orleans. The growth in BOD load from paper products was the important factor at Vicksburg and St. Francisville.

Overwhelming the effects of these discharges was the diluting effect of the Mississippi's enormous rate of flow from 174,000 cubic feet per second at St. Louis to 489,000 feet at Vicksburg and 406,000 at the river's mouth. As the area through which the river flows rouses from its industrial somnolence, dilution will no longer suffice. Over $1 billion will be needed in the next decade to provide adequate municipal and industrial waste treatment.

The Savannah River

The lower Savannah River presents the most clear-cut case of any of those studied in which industrial pollution is basically responsible for the present deplorable state of water quality. It also suggests that such a situation may be subject to a dramatic improvement by 1980. Drawing upon reports of federal conferences on pollution of the river as well as a study of the Center for Study of Responsive Law, Rogers French concludes that the outlook is for approximately an 80 percent reduction in the pollutional load on the river between 1969 and 1980. If this occurs, 76 percent of the BOD reduction will be attributable to improvement in the industrial waste discharges. If it comes about as planned, it will no doubt be a first in the history of industrial pollution in America.

Support for the view that such progress is possible may be had from the knowledge that only nine industrial firms are responsible for 78 percent of the 1970 BOD load on the river. A minimum of surveillance should be necessary to determine whether the decisions of the second conference on the pollution of the lower Savannah River are being carried out. A more sizable threat to this happy prospect of a dramatic clean-up might come from a relaxation of administrative vigilance which would make compliance with conference decisions a matter of choice.

Long Island Water Supply

In a careful and thorough study, Henry Lang has done a skillful job of tracing the extraordinary history of the Long Island water supply. The first settlers on Long Island were blessed with an immense underground supply of fresh water variously estimated at 60 trillion gallons, of which 3 to 6 trillion gallons was recoverable. It was usually only necessary to sink a shallow well and pure water was available, and this was the typical source of supply all over the Island through the 1920s. As the New York City portion of the Island became sewered and paved over, an increasing proportion of the precipitation *and* water drawn from the ground was lost in runoff to the ocean. By the 1930s, it was necessary to supply Kings County (Brooklyn) with imported water and, by the 1950s, all of Queens County. Only the non-New York City counties still draw their water from the underground supply. Nassau County is now about half sewered and this proportion is expected to approximate 80 percent by 1980. By that date, Nassau County will be discharging enough to the ocean so that it will be beginning to exceed its safe yield and will be "mining" water. Suffolk County, which has only begun to install sewers, is not likely to exceed its safe yield for the county as a whole in the foreseeable future. Both counties have benefited from a policy instituted many years ago requiring developers to install recharge basins ("sumps") to permit the percolation into the ground of storm runoff from a separate storm sewer system.

Forward-looking officials in both counties have been considering the future water supply and weighing costs of alternative sources. Importation from the New York City system is a possibility, but New York may face shortages itself in the future. Desalination is another possible solution, but the indications are that it is too costly. A third alternative is the recycling of sewage effluent, which is attractive from a cost standpoint, but presents problems in removing dissolved solids, notably nitrates. Presently, one of the Nassau villages is seeking funding for an experimental plant that will remove 85 percent of the nitrates [8]; nitrate removal is believed essential for safe drinking water.

A full-scale study of the problem would have to take account not only of the cost of water supply, but of getting rid of sewage effluent, including construction costs for ocean outfall, plus costs of impairment of water recreational assets.

Southern California Water Supply

James Carmichael traces the history of the effort to assure a water supply for the populous and semi-arid Southern California counties. The struggle began in 1913 with the construction of the Los Angeles Aqueduct to bring water from the Sierra Nevadas and continued to the 1930s, which saw the construction of the Colorado Aqueduct and the beginnings of bruising contests with Arizona for rights to that river's waters. Most recent chapter is completion of the California Aqueduct which adds waters brought down from Northern California to meet the ever-burgeoning demands. During much of this period, this has been the responsibility of the Metropolitan Water District of Southern California, which has used its control of water to assure the hegemony of Los Angeles in Southern

California. That the delivered cost of softened and filtered water rose from $15 per acre-foot in 1941 to $53 per acre-foot in 1971 was not regarded as surprising in view of other cost increases.

Meanwhile, as the population of the area boomed, the initial problem of water supply was matched by a mounting problem of sewage disposal. Organized in an entirely separate bureaucracy from the MWD, the Los Angeles Sanitation districts have begun to consider water reclamation in an effort to reduce disposal costs. One plant constructed in 1962 produces reclaimed water at a cost of about $11 per acre-foot. Ten more plants are contemplated, but today reclaimed water accounts for only 0.2 percent of the consumption of water in Southern California.

Meanwhile, south of Los Angeles in Santee, a suburb of San Diego, it has been shown that water can be recycled and used for water sports. In an affluent area with an almost insatiable demand for water sports, this suggests a useful line of development.

What's the Source of the Problem?

The shortcoming of the sort of case by case summary just completed is that it may obscure important general conclusions concerning the nature of the pollution problem. For example, what do we conclude concerning the question: is the main source of difficulty industrial pollution or the inadequacy of municipal sewage treatment? The only situations in which the bulk of the BOD was attributable to industry were in the Savannah River and portions of Puget Sound, though the BOD of industrial effluents was an outstanding factor in the pollution of the central basin of Lake Erie, the Green Bay area of Lake Michigan, and the central portion of San Francisco Bay. On the other hand, in the worst cases of pollution, municipal sewage seemed to account for most of the BOD; these included: the western basin of Lake Erie, the lower Hudson River and extreme western portion of Long Island Sound, and the Milwaukee area of Lake Michigan. Such a variety of situations and degrees of definiteness do not lend themselves to firm generalizations. However, two broad conclusions do emerge with crystalline clarity: (1) The rate of growth in the input of industrial pollutants in almost all of the waterways was greater than that attributable to population; (2) Nevertheless, the effluent of municipal sewage plants and specifically that portion beginning as household sewage was far and away the most important source of the nutrients responsible for the phenomenon of overenrichment, or eutrophication. It will be useful to enlarge on these two clear-cut general conclusions.

*Consequences for Waterways of Rapid Expansion
of Industrial Output*

It is, of course, a truism that in the United States industrial output has grown faster than population. There is some limit to the size of family we want, but human desires for goods and services appear to be nearly boundless. If the amount of potential pollution tends to increase in some rough proportion to industrial output, and it has at least since World War II, then it follows that an increasing portion of the pollutants entering our waterways is attributable to industrial discharges. In any event, it appears to have been true that the rate of increase in potential BOD attributable to industry in almost all sectors of all waterways studied was greater than that traceable to population. The few situations where this was not the case—the mid-Hudson area and the easterly portions of Long Island Sound—were also the places where the prospects seemed best for decisively cleaner water by 1980. It is necessary to qualify this conclusion by emphasizing that this refers to "potential" pollution rather than actual. This formulation—this reference to "potential" BOD—was essential because adequate data on extent of treatment and especially on the change in extent of treatment were not available for either industrial or municipal effluents. While some information was available on projected expansion or improvement of municipal sewage treatment plants, almost nothing systematic was available on changes underway or planned in extent of industrial treatment.

Compounding the difficulty in allocating the share of blame between municipal and industrial sources for the observed increase in pollution is the fact that most industrial discharges pass through municipal sewage treatment plants on their way to waterways. Again, precise information on this is not available, but there is good reason to believe that the principal sources of industrial pollution, such as pulp mills and iron and steel plants, usually discharge their effluents directly to the waterway involved.

While solid fact is scanty, there seems to be good reason to assume that thus far the progress in extent and thoroughness of treatment of municipal sewage effluents—with the exception of removal of phosphates and nitrates—has been greater than with industrial discharges. This view found support in a study by the General Accounting Office, referred to in Chapter 2, which found that the increase in industrial pollution had overwhelmed and negated the very considerable amount of federally aided construction and upgrading of municipal sewage treatment plants.

Presumably, more rapid progress in controlling the pollution from industrial effluents may be anticipated in the future. New York State has enacted a Pure Water Program and while most of the funding is to aid municipal sewage treatment facilities there are provisions for assistance to industry as well. A comprehensive state and federal program to clean up San Francisco Bay has been instituted using federal, state, and municipal resources.

The Federal Water Quality Administration has scheduled hearings to establish programs for pollution control affecting several of the waterways included in the Hofstra study. Lake Erie was one of the first covered by hearings in 1967 and a report in 1968. Others have dealt with Lake Michigan, New York Harbor, the Hudson, and the Savannah River with a plan in the offing for study of Long Island Sound.

These efforts and pressures should begin to bear fruit by the middle of the 1970s and have a substantial impact on sources of industrial water pollution by 1980. Meanwhile, it is safe to say that the disclosures of new sources and aspects of industrial water pollution have outpaced progress in control of such sources. Among the most dramatic of these findings have been the evidences of toxic substances in food attributable to industrial discharges. Frank Husson cites a number of such incidents.

The Perils of Overenrichment – Eutrophication

More insidious, more subtle, but more pervasive manifestations of pollution than the toxic consequences of industrial pollution such as mercury poisoning are the results of eutrophication–the distinctively American expression of water pollution that arises from overenrichment with plant nutrients. When Jewish peasants migrated from Middle Europe to Israel, they introduced the cultivation of carp in ponds to provide the basis for *gefilte* fish. To speed the growth of carp, these Israelites introduced fertilizers containing such plant nutrients as phosphates and nitrates to speed the growth of the vegetation on which the carp fed and prospered. It is precisely an excess of these nutrients which lies at the base of the malaise that afflicts such bodies of water as Lake Erie. Such overenrichment, or eutrophication, is responsible for the excessive growth of vegetation which, when it dies, places a crushing burden on the supplies of dissolved oxygen.

Unlike the carp ponds of Israel, the enrichment of Lake Erie with plant nutrients was unintended, a consequence of discharge of sewage treatment effluents rich in phosphates and nitrates reflecting partly the diets of an affluent America, but also and in the case of phosphates, the detergents widely employed in household cleansers. Today eutrophication is held chiefly responsible for the persistent and predictable oxygen deficiencies in the western and central basins of Lake Erie during the summer months.

Will eutrophication create more Lake Eries by 1980? If the question is limited to major bodies of water such as those covered in the Hofstra study, the answer is no, though the blight of excessive vegetation afflicting another body of water is a daily fact of life at smaller lakes and ponds across the country. But the overenrichment of a body of water as massive as Lake Michigan is a long-term process not likely to be completed by 1980, although Jay Parker in his chapter dealing with Lake Michigan warns of the need for monitoring. Estuaries flushed by tidal action are less likely than lakes to be afflicted by eutrophication, but Long Island Sound in its western portions shows unmistakable signs of overenrichment. The particular manifestation is large populations of mossbunker, a source of bait and commercial fish meal, which has led to excellent catches of bluefish and striped bass, but also on calm summer days to massive mossbunker kills.

Raymond Matern cited data showing that concentrations of both nitrates and phosphates were high enough to support excessive algae growths in Delta area of San Francisco Bay, and projections for 1980 suggested high nitrogen concentrations in the Suisun Bay section.

Richard Martin found that in summer months the sluggish flow of the Hudson River was substantially offset by tidal action so that the preconditions were created for a nutrient build-up. Thus far, however, the annual flushing by spring runoff has prevented eutrophication, though he urges monitoring to be forewarned.

Dilution – The Traditional American
Solution to Pollution

Where eutrophication has not adversely affected the waterways, it is not because of what has been done to offset it, but simply a matter of sufficient dilution–

which has, incidentally, always been the traditional American solution to pollution problems. It is not that the citizens, industry, and agriculture on the shores of the Mississippi are more careful and discharge less nutrients to its waters than do the same sources on Lake Erie; rather, the powerful flow of the mighty river drowns the effects of the added chemicals. For the same reason, no problem of eutrophication exists or looms as a threat in the Savannah River despite the heavy pollution present in its lower stretches. Much of the dramatic improvement in sewage treatment in the Seattle area was inspired by the threat of eutrophication to the beauty and recreational values of Lake Washington. No such threat exists in the powerful tidal currents to which the treated sewage effluent was directed in part from Lake Washington.

As noted earlier, potential problems of eutrophication have been held in check thus far by dilution, in such areas studied as Lake Michigan, Long Island Sound, the Hudson River, and San Francisco Bay. Whether this will be sufficient for the future remains to be seen.

Exporting the Effects of Pollution

Where dilution has not sufficed to overcome the effects of pollution by powerful cities, the favorite way out has been to export the pollution. Bathers swim at Lake Michigan beaches in the city of Chicago where they dare not do so at many beaches in the Cleveland area and certainly not in the lower Hudson at New York, solely because the city of Chicago exports its sewage effluent via the Chicago River to the Mississippi. Actually, New York City exports much of its water pollution by transporting the sludge generated by that portion of its sewage that is treated out into the Atlantic Ocean some thirty miles offshore [5]. This practice, together with the dumping of industrial wastes in the same area, has given rise to something described dramatically, if not entirely accurately, as a "Dead Sea." At the same time, New York City enjoys good drinking water and to that extent escapes the consequences of its careless handling of water resources by importing its water some 200 miles, from the upper reaches of the Delaware. Similarly, Los Angeles uses its municipal and political power to draw upon the water supplies of Northern California and to compete with parched Arizona for the relatively scanty supply from the Colorado River.

Consequences of Water Pollution

What have been the consequences of a set of policies that have depended on such evasions as the hope of "dilution" or of "export" as a solution to water pollution? The answer is that, by and large, the evasions haven't worked to effect a satisfactory answer, though they may have appeared to do so in some cases.

Most striking, perhaps, has been the impact on such water contact sports as swimming and water skiing. Such recreational opportunities are likely to be most impaired where they are most needed. Thus, the millions from New York City and urban New Jersey across the river are forced to jam the highways to Atlantic beaches rather than dare the waters of the Hudson River. The northern suburbs of New York for at least 25 miles above the city line are denied use of the river. Similarly, most of the beaches in the vicinity of Cleveland, many near Toledo, and some of the most important near Buffalo are unsafe for swimming because of pollution of Lake Erie.

It is, perhaps, fortunate that waterways may be used for boating long after they have been rendered unfit for other recreational uses. Thus, the Chicago River, which serves as the conduit southward of most of the sewage effluent of the Chicago metropolitan area, is home for a considerable flotilla of pleasure craft. The same is true of the lower Hudson River in the vicinity of New York City, as well as Lake Erie. This state of affairs is fortunate in the sense that an affluent segment of the middle class, those able to afford boats, a group of considerable potential political power, thus become conscious of the polluted condition of the waterways.

It may sound contradictory at first to suggest that one of the consequences of water pollution is to interfere with the supply of drinking water. How is it possible to use water that is unquestionably afflicted with pollution as a source for public water supply? It is an important fact and lesson that exactly this is done by many communities. It is true that such municipalities as New York and San Francisco drink pure, imported water while also enjoying the doubtful pleasure of fouling their own nest, but others such as the cities on the shores of Lake Erie and the Mississippi draw their water from the nearest available source. This is not possible for Lake Erie and Mississippi communities without some difficulty. Thus, many of the major communities drawing their public water supply from Lake Erie in 1966 reported that special treatment was required to correct for odor and taste [2] arising, in this instance, principally from eutrophication. The important fact is that this can be done with safety, modern techniques of treatment and disinfection being what they are. The important lesson is that these Lake Erie communities that use the lake as both a depository of sewage effluent and a source of drinking water are practicing the very modern technique of water recycling. So, of course, are numerous communities along the Hudson and Mississippi, to name two covered in this study, as well as countless other towns on American waterways. The capacity of the streams to renew their dissolved oxygen and to purify themselves has made this a satisfactory solution for a large sector of the nation's communities throughout our history. It is not certain that such usually haphazard recycling will suffice for a future in which increasing population and especially rising industrial output impose a heavy BOD on our waterways.

Prof. Barry Commoner suggests that Lake Erie may have passed a sort of point of no return. By accumulating wastes for a century it has contracted "a huge and growing oxygen debt" that may someday come due "in a biological cataclysm that could possibly exhaust, for a time, most of the oxygen in the greater part of the lake waters" [4:105-106]. Even if all further pollution of Lake Erie were halted tomorrow, "there would still remain the problem of the accumulated wastes on the lake bottom" [4:110].

If this view is even partially correct, Lake Erie must be the nearest thing in our list of waterways to being a hopeless case, because the encouraging thing about the remainder is that none can be so regarded. Even the lower Hudson could soon be rehabilitated with proper waste treatment. As these conclusions were being written, a congressional conference committee gave its approval to the largest authorization for water pollution in history [3]. If that sort of support survives the remaining steps in the legislative process—and they are many—the outlook for the water pollution fight will be bright indeed.

The Oceans — Where the Buck Stops

All pollutants must find some final resting place, and for those that enter waters this ultimate destination must be the ocean. Frank Husson was assigned the task of surveying such phenomena in the full realization that no one person could cover the topic fully.

He chose to investigate the mechanism by which certain pollutants enter the life cycle, and one that he came across is very close to the subject of Lake Erie and eutrophication. One of the mysteries has been how mercury enters the chain of life and eventually is ingested by man. It had been supposed that if mercury, an extremely heavy and generally chemically inert element in its metallic form, were to enter a body of water, it would sink to the bottom and remain there in splendid isolation. This hypothesis reckoned without the occurrence of situations in which in the absence of oxygen—a circumstance that might come about in the decaying matter on the bottom of a eutrophic lake—mercury might react with organic compounds to form methyl mercury. As such it could be absorbed by microorganisms and enter the life chain.

Among other mechanisms investigated by Mr. Husson is the way in which DDT finds its way into the chain of life, a process in which petroleum pollutants may play an important part. The question of oceanic pollution, then, is rarely a simple one, but rather the beginning of a process whose end cannot be readily foretold with our present knowledge.

Notes

1. Donald E. Carr, *Death of the Sweet Waters* (New York: W. W. Norton, 1966), p. 150.

2. U. S. Department of Interior, Federal Water Pollution Control Administration, *Lake Erie Report*, August 1968.

3. *New York Times*, Sept. 15, 1972, p. 1.

4. Barry Commoner, *The Closing Circle* (New York: Alfred A. Knopf, 1971), p. 104-105.

5. *Newsday* (Garden City, N.Y.), Sept. 28, 1972, p. 109.

6. David Bird, "Paper Industry Gains in Cleanup," *New York Times*, Aug. 28, 1972, p. 1.

7. Sylvia Carter, "Pollution Scores on Six Harbors," *Newsday* (Garden City, N.Y.), Feb. 7, 1972, pp. 4, 19.

8. Dan Hertzberg, "A Pure But Not So Simple Success," *Newsday* (Garden City, N.Y.), May 4, 1972, p. 21.

**Part 2
Our Environment: Air**

15 A Methodology for the Study of Air Pollution in the United States

ALFRED J. VAN TASSEL

Editor's Note: The initiative in developing the practical application of the general methodology outlined below was taken by Frank Caulson who was largely responsible for its detailed elaboration. This methodology is not relevant to the problems of Chapters 19 and 20, which required independent approaches.

The central thrust of this entire volume is to seek to discover and put to work techniques for the quantitative evaluation of the extent of various forms of pollution in specific localities. It may be dramatic to say 27 million tons of dangerous sulfur oxides are emitted to the atmosphere in the nation as a whole each year [1:199], but it is not especially meaningful. Pollution occurs in specific places subject to their own unique atmospheric conditions that powerfully affect the consequences. Whatever statistics we derive to express the magnitude of the air pollution problem must be in some sense "small area" statistics applicable to a particular and definable place and time.

Any effort at localized projection of the air pollution consequences of a given level of combustion of a particular fuel is bound to be a very rough approximation. However, it will almost certainly have a more precise meaning than general totals of pollutant emissions for the country as a whole. Accordingly, the attempt is believed to be worthwhile.

Sample for Study of Air Pollution

This chapter sets forth the methodology followed in pursuit of local statistics and projections on the volume of air pollutants in fifteen urban areas, distributed across the nation. The sample consisted of five eastern cities (Chapter 16), five midwestern cities (Chapter 17), and five western cities (Chapter 18). The procedure was to examine the statistics relating to three main sources of air pollution: (1) automobiles, (2) thermal electric generating plants, and (3) space heating furnaces—residential, commercial, and industrial. This list omits two of the five major sources of air pollution: (1) industrial processes, and (2) refuse disposal. The latter is a relatively minor source, accounting for only about 3.5 percent of the national total, though, as usual, it may be hard to persuade someone who lives near a municipal incinerator or, worse, a dump subject to open burning. On the other hand, emissions from industrial processes (other than power generation, which is covered) are a very important source of particulates and sulfur oxides. Accordingly, the failure to account statistically for emissions from industrial processes is unquestionably a very serious shortcoming of the study. However, there appeared to be no feasible procedure for estimating statistically the present and projected emissions from a host of industrial processes, each with its own set of emission coefficients, and a quantitative approach to industrial emissions was, accordingly, reluctantly abandoned. However, in the treatment of individual cities, qualitative information on the contribution of industrial emissions was included where available.

Federal data from 1966 appearing in an earlier Hofstra study [1] serve to place the question in perspective. The three sources chosen for statistical study in the next three chapters account for roughly: 80 percent of all pollutants, 96 percent of carbon monoxide, 62 percent of sulfur oxides, 77 percent of nitrogen oxides, 74 percent of hydrocarbons, and 42 percent of particulates, the main constituents of smoke. The smoke of mills and factories, which account for about half the total particulates, is a highly visible indicator of the importance of industrial emissions. Industrial processes are also important as a source of sulfur oxides, representing about a third of the problem, and nitrogen oxides, where about a fourth are from industry. But electric power generation is responsible for much more of both the last two pollutants than other industrial processes [1:199]. It seemed wise, accordingly, to confine the quantitative estimates of pollution to the bulk of the problem for which reasonably good local estimates could be derived.

Estimating Automobile Pollution

The automobile is responsible for over 90 percent of carbon monoxide emitted, nearly half the nitrogen oxides, and nearly two-thirds of the unburned hydrocarbons [1:199]. The latter two pollutants in the presence of sunlight enter into a complex set of reactions that forms the eye-irritating constituents of the haze known as smog that has become a familiar feature of the air of many U. S. cities and a sort of trademark for Los Angeles. The first step in estimating present and projected pollution for a city was to determine the number of automobiles in the city contributing to the problem. For a given city let $A_{1959}, A_{1969}, A_{1980}$ represent the number of automobiles for the city in question for the indicated year. (Number of passenger vehicles was used as the proxy for all motor vehicles in this exercise.) A_{1959} and A_{1969} were obtained from *Standard Rate and Data Service* market statistics for the year in question [2] and

$$\frac{A_{1969} - A_{1959}}{10}$$

was used to establish the average annual increase. The assumption was made that this linear trend would prevail until 1980 and A_{1980} was projected accordingly. The formula used was:

$$A_{1980} = A_{1969} + (1980 - 1969)\frac{A_{1969} - A_{1959}}{10}$$

Pollution, however, is not directly proportional to the number of vehicles, but rather to the amount of fuel consumed, and there is evidence that fuel consumption per vehicle year has increased with the passage of time. It seemed unnecessarily complicated to compute and project the change in gallons per car per year for each city studied. In addition, a projected figure based on a broad aggregate is likely to be more reliable, in the sense that random factors are more likely to cancel out than one based on a narrow statistical base. Factors decisive for the trend in gallons per vehicle, such as size of car and design of engine, are determined mainly by national trends. Accordingly, data for the nation as a whole were collected for the years 1959, 1964, and 1969, and projected to 1980. Data for the nation as a whole for these years were obtained on population of cars from Bureau of Public Roads [3,4,5] and on population of people from Bureau of the Census [6], thereby establishing a series on a persons per vehicle ratio which was then projected on a straight line basis to 1980. From the same sources, data on gallons of fuel consumed and the number of people were used to compute a gallons per person ratio for each year, and this ratio was projected in the same way as before to 1980. Combining the two series of ratios yielded a series on gallons per car as shown in table 15.1, which reflects the trend that more people are using more cars more intensively and thus burning more gasoline per car. This series in table 15.1 was used in estimating automobile emissions.

The central purpose of this chapter is to provide a basis for estimating the volume of pollutants emitted by automobiles in each city under study for the years 1969 and 1980. Table 15.1 is an important step in fulfilling our central purpose. If we designate the ratio of gallons per car for 1969 as G_{1969} and that for 1980 similarly as G_{1980}, we can combine these with the number cars symbol employed earlier to provide an expression for gasoline consumed in automobiles in each city in each year. Thus: Gallons of gasoline consumed in City A in 1969 = $A_{1969} G_{1969}$; same in 1980 = $A_{1980} G_{1980}$.

Now it remains to determine a suitable set of emission factors expressing the amount of each pollutant formed per gallon of gasoline consumed. Unhappily for simplicity, such emission factors are not static multipliers either. Rather, as a consequence of increased environmental awareness and under the prodding of governmental agencies and legislation, automobile manufacturers have perforce succeeded in designing automobiles that greatly reduce the pollutants emitted per gallon of gasoline consumed and legislation on the books calls for still more drastic cuts in emissions by 1976. Table 15.2 reflects the progressive reductions that occurred or were called for to 1971 and that are mandated for 1976. Table 15.2, together with our estimates and projections of gasoline consumption by city, provides a basis for estimating and projecting the volume of automobile pollutants by city.

Factors Affecting Emission Coefficients

How reliable are the coefficients of emission given in table 15.2? That depends, in part, on which column is under consideration. The column headed "Uncontrolled Vehicles" is the result of federal studies during the 1960s of actual emissions on presumably typical vehicles of the period before any efforts at controls had been undertaken. The "National Standards" for 1968 and 1971 set forth the emission ratios specified by federal agencies for new automobiles sold beginning in the indicated year. That is, federal regulations specified that all new cars sold in 1968, 1969, and 1970 meet or exceed the 1968 standards and that all automobiles sold in 1971 and subsequent years fulfill the 1971 standards. Research on the measures necessary to reduce emissions to the extent needed to meet 1968 and 1971 standards has been under way for some time and the problems are well understood [1:224-25] [8:I-H5-9]. There is no evidence to suggest that manufacturers have experienced insuperable difficulties in meeting these standards for prototype cars though much less is known about how ordinary assemblyline automobiles have performed, especially under actual driving conditions and extended usage. Certainly, there are major problems associated with testing and inspection of devices for the control of emission under actual operating conditions. (See a report of a committee of the Office of Science and Technology for a discussion of these matters [8:35-37, I-H2-3].)

Aside from such practical problems, major difficulties have been experienced even with prototype models in simultaneously reducing carbon monoxide and hydrocarbons while also cutting down on nitrogen oxides.

Various approaches to control the amounts of these unwanted pollutants in automotive exhaust depend generally on exploiting known principles of combustion and their applications to engineering solutions. For example, in internal-combustion engines the air-fuel mixture can be modified to influence the amounts of the various pollutants emitted. By making the mixture "leaner" (less fuel and more air), more complete combustion is favored and there is a reduction in the emissions of carbon monoxide and unburned hydrocarbons. However, this increased combustion efficiency tends to increase combustion temperatures and thereby favors formation and emission of nitrogen oxides. Conversely, the use of a "rich" mixture (more fuel and less air) tends to keep combustion temperatures down and restrict the oxygen supply, thus diminishing the amount of nitrogen oxides formed; but the same conditions tend to increase the emissions of carbon monoxide and unburned hydrocarbons. One is faced with the dilemma that measures which are effective for control of HC and CO tend to produce more NO_x, and vice versa [8:13-14].

Table 15.2 reflects this dilemma by showing a higher value for nitrogen oxide emissions for 1968 and 1971 standards than for pre-1968 cars; i.e., a lowering of carbon monoxide and hydrocarbon was achieved at the expense of higher nitrogen oxide values.

The 1968 and 1971 standards were contained in regulations issued by federal environmental agencies after extensive study and hearings. Meanwhile, congressional hearings were also held that led to the enactment of the Clean Air Act of 1970, Public Law 91-604, 91st Congress, which contained the emission standards for 1976 shown in the right-hand column of table 15.2. At the legislative hearings, scientists and engineers were given ample opportunity to testify, but it is not unfair to the Congress to say that the reductions specified from 1971 to 1976 were a function of what it was felt was necessary to clean up the air rather than a study of what was technically feasible. It will be noted that

Table 15-1
Average Consumption Gasoline per Vehicle in the United States, Selected Years (gallons)

	1959	1964	1969	1980
Persons/Vehicle	2.9	2.6	2.4	1.8*
Gallons/Person	225.8	247.6	306.7	418*
Gallons/Car	666	652	718	773

Source: Years 1959-1969—U.S. Department of Transportation, Federal Highway Administration Bureau of Public Roads. HIGHWAY STATISTICS, "Estimated Motor Vehicle Travel in the U.S. and Related Data" (Washington, D.C.: Government Printing Office, 1959), pp. 65, 70. 1980—Estimated. The Statistical Abstract of the United States. 1970 edition states that the population of the United States in 1959, 1964, 1969 was 177,830,000, 192,120,000 and 203,213,000 for the respective years. Series P-25 of the Bureau of Census current population reports "Population Estimates and Projections" estimates that the U.S. population in 1980 will be 232,412,000. Dividing total automobile population by total population establishes a persons per vehicle ratio. Dividing total consumption of gasoline by population establishes a gallons/person ratio. Multiplying persons/vehicle by gallons/person establishes gallons/car.

Table 15-2
Estimated Automobile Emissions—Selected Years (Pounds/Car/Year)

	Uncontrolled Vehicles Pre-1968	National Standards 1968	National Standards 1971	National Standards[a] 1976
Hydrocarbons	450[b]	170	75	14
Carbon Monoxide	1370	630	410	41
Nitrogen Oxide (NO_2)	160	200	200	20

Source: U.S. Department of Health, Education and Welfare, Public Health Service Environmental Health Service, National Air Pollution Control Administration, CONTROL TECHNIQUES FOR CARBON MONOXIDE, NITROGEN OXIDE, AND HYDROCARBON EMISSIONS FROM MOBILE SOURCES (Washington, D.C.: Government Printing Office, 1970), Section 3, p. 8.

[a]Estimated on basis of public law 91-604 91st Congress, December 1970, which requires that hydrocarbon and carbon monoxide emissions be reduced 90 percent from the standards of 1970. The act also requires that nitric oxide emissions be reduced 90 percent below the standards of 1971.

[b]Applies only to pre-1963 cars. Arbitrarily reduced to 350 for 1969 estimates.

the values in table 15.2 for 1976 for carbon monoxide and nitrogen oxides are precisely one-tenth of the corresponding values for 1971, and this is because this is exactly how the 1976 values were determined, by a reduction to 10 percent of 1971 values. From the standpoint of the automobile industry this looked like a fiat as ill-founded technically as King Canute's command for the waves to stand still. From the congressional viewpoint the 1976 standards expressed the urgent necessity to clean up the air and an abiding conviction that a nation capable of putting a man on the moon could solve all the technical problems involved. The reaction of industry has been, yes, but at what cost?

Assumptions Respecting 1980 Automobile Emissions

The industry received powerful support from an ad hoc committee of experts, mostly from federal agencies, which was created by the White House Office of Science and Technology. The committee concluded that meeting the 1976 standards, especially for nitrogen oxide emissions, would add enormously to the cost of a car without any benefit of corresponding value [9] [8]. Several auto manufacturers had already made formal application for delay in enforcement of the standards and Henry Ford II, chairman of Ford Motor Co., forecast an industry shutdown if the standards were applied on schedule. However, on May 12, 1972, the Environmental Protection Agency rejected the industry application for delay [10].

This dispute raised a very serious question of procedure respecting our projections of automobile emissions. It was decided to make two sets of projections based on the following assumptions:

Assumption 1: The national standards set forth in table 15.2 would be met approximately on schedule in accordance with table 15.2. (Actually this allows for the Environmental Protection administrator's optional delay of one year in meeting 1975 standards.)

Assumption 2: The 1976 standards set forth in table 15.2 would not, in fact, be achieved until after 1980 and therefore would not enter into our projections.

The justification for Assumption 1 is simple: it reflects the "law of the land" and in the absence of proof to the contrary may be assumed to govern. The justification for the second assumption is that law or no law, there is substantial evidence that the full reduction in emissions postulated by table 15.2 will not be achieved by 1976. The Ad Hoc Committee points out, for example, that:

Regulation should not be based on blind faith in technology. Establishing standards beyond the known state of the art on the theory that industry can do anything if enough pressure is put on it is not likely to result in wise governmental decision making or to provide the greatest net benefits to society [8:x]. ... The ultimate feasibility of meeting the 1975-1976 emission standards has not yet been demonstrated except on a limited laboratory basis. There is a difference between the ability to synthesize prototype automobiles which can be certified by the EPA and the ability to build them on a mass production basis within the time frame prescribed by recent legislation [8:xix].

There is clearly no way of determining with certainty which was preferable, so each of the above assumptions was employed as a basis for an independent set of projections of automobile emissions. Table 15.2 and one of the above assumptions provided a necessary but not sufficient basis for projection. It is obvious that a new standard for emissions would not become effective for the entire stock of automobiles on the first date on which it was legally required of new cars; some rate of replacement of automobiles in the stock is required to complete our model. The basis employed was the age distribution of the U. S. automobile population in 1969, which was known to be: 52.4 percent under five years old; 35.6 percent more than five but under ten years old; and 12 percent ten years old or older [7].

Accordingly, there was no distinction employed in calculating 1969 emissions under either Assumptions 1 or 2 outlined above inasmuch as that portion of the stock purchased new in 1968 and 1969 was assumed to be governed by standards of Column 2, table 15.2 and all others by Column 1 standards for "Uncontrolled Vehicles."

For 1980 emissions:

Under Assumption 1 (national standards prevailing in all years):
a. 52.4 percent under 1976 standards;
b. 35.6 percent under 1971 standards
c. 12 percent under 1968 standards
   (No effort made to divide final category)

Under Assumption 2 (national standards adhered to through 1971, no improvement thereafter):
a. 88 percent under 1971 standards (i.e., 52.4 under 5 years old plus 35.6 more than five but under ten years old)
b. 12 percent under 1968 standards.

It is believed that the procedure outlined above will yield an approximately accurate picture of the contribution of automobile emissions to the air pollution of the fifteen cities selected for study.

Emissions From Power Plants

The data respecting fuel consumption by electric generating stations are both abundant and authoritative. This fact made it possible to devise a relatively simple procedure for estimating present and projected emissions from power plants in the fifteen cities selected for study. In addition, the responsible federal agencies have compiled emission factors believed to be applicable to electric generating stations for coal, gas, and oil. These are shown in table 15.3.

The primary source of data on fuel consumption in 1959 and 1969 by electric utilities was an annual publication of the National Coal Association entitled *Steam-Electric Plant Factors*. This source provided detailed information on the type and amount of fuel consumed by each utility. From this source, total fuel consumption by type was computed for each year studied and for each city covered. Then for each city and each year, fuel consumption was multiplied by the appropriate emission factor from table 15.3 to obtain an estimate of emissions from electric generating stations for each city and each year. Clearly, the weak link in this procedure is the information on emission factors. These were compiled on the basis of careful studies in the late 1960s, but it has not proven feasible to adjust them to cover changes from one city to another or over time in the efficiency of control equipment. This is probably a more serious weakness in estimating projected emissions for 1980 than with respect to earlier years. By 1980, under public and governmental pressure, there is likely to have been substantial progress in the technology of control and a wider variance in practice.

Projecting Future Fuel Comsumption in Electricity Generation

While *Steam-Electric Plant Factors* provided an excellent basis for estimating fuel consumption by type and city for 1969 and earlier, it did not contain 1980

**Table 15-3
Emission Factors for Electrical Generation**

Pollutant	Pounds/Ton of Coal Burned	Pounds/Million Cu ft. of Gas Burned	Pounds/1000 Gals. of Oil Burned
Aldehydes	0.005	1	0.6
Carbon Monoxide	0.5	Neg.	0.04
Hydrocarbons	0.2	Neg.	3.2
Oxides of Nitrogen	20.	390	104.
Sulfur Dioxide	38S[a]	0.4	157S[a]
Sulfur Trioxide	Neg.	Neg.	2.4S
Other Organics	Neg.	3	Neg.
Particulate	16A[b]	15	10

Source: U.S. Department of Health, Education and Welfare Public Health Service, Bureau of Disease Prevention and Environmental Control. National Center for Air Pollution Control. COMPILATION OF AIR POLLUTANT EMISSION FACTORS (Durham: Government Printing Office, 1968), pp. 4-7.

[a]S equals percent sulfur in coal (e.g., if sulfur content of coal is 2 percent there will be 76 pounds of sulfur oxides formed per ton of coal burned). Estimated on basis of correspondence with company official or estimated on basis of information in Regional Power Studies by Federal Power Commission.

[b]A equals ash content in coal estimated at 8.55 percent.

projections of these values.

For the basis of his projections of fuel consumption by electric utilities in five midwestern cities, Frank Caulson used a series of authoritative studies conducted for the Federal Power Commission and known under the collective title of *The 1970 National Power Survey* [12]. Mr. Caulson was able to locate four regional studies in this series covering his five cities and to determine from these an authoritative projected annual percentage increase to 1980 by type of fuel. Applying the appropriate annual increase rate to 1969 *Steam Plant Factor* data for each utility yielded 1980 projections.

The Federal Power Commission projections were less readily applicable to the eastern and western cities covered in Chapters 16 and 18, mainly because the regions into which the studies were divided were much broader than in the midwest. For these cities, 1959 and 1969 *Steam Plant Factors* data were employed to determine a straight-line rate of change in fuel use by city and the resultant trends were extrapolated to 1980. The individual city projections were later checked against the broad projections of the Federal Power Commission for general compatibility. As a consequence of such checks it was decided to lower the rate of increase in coal usage in Atlanta from 1969 to 1980, to hold Los Angeles natural gas consumption to the 1969 level, and to take account of the use of low sulfur oil in Los Angeles electric utilities by 1980. In all cities, the final step involved multiplication of 1980 projected fuel consumption by the appropriate emission factor from table 15.3.

Space Heating

Space heating represents a relatively unimportant source of air pollutants, contributing fewer nitrogen oxides than the automobile and less sulfur oxides than electric generating stations. Computing space heating pollutants proved a complex project, however, because of the multiplicity of sources and some ambiguities of classification.

The first step in computation, a table of emission factors, was not hard to come by, being available from the same federal source employed elsewhere (see table 15.4).

Only two fuels were considered in the study, fuel oil and natural gas; coal and LP gas were of negligible importance for space heating, and electricity, another minor source, was already counted in pollution from power stations. The procedure for both natural gas and fuel oil was to construct series representing national consumption, project these series to 1980 and allocate to each city that portion of national consumption representing its share of space to be heated, adjusted to take account of differences in ambient temperatures as measured by degree days. Actually, two sets of series were constructed, projected, and allocated, the first corresponding to heating of individual residences (hereinafter called residential space heating), the second for space heating of commercial and industrial establishments, plus multi-unit housing (this series hereinafter called commercial-industrial). Neither of these series was intended to include fuels used for industrial processing such as smelting, baking, sterilizing, etc.; while no

Table 15-4
Emission Factors for Fuel Oil and Natural Gas Consumed in Space Heating

Pollutant	Fuel Oil (Pounds per 1,000 gallons of oil burned)	Natural Gas (Pounds per million cubic feet of gas burned)
Carbon monoxide	2	0.4
Hydrocarbons[a]	5	neg.
Nitrogen oxides	12	116
Sulfur oxides[b]	159S[c]	0.4
Particulates	8	19

Source: U.S. Dept. of Health, Education and Welfare, COMPILATION OF AIR POLLUTANT EMISSION FACTORS (Durham, N.C.: Government Printing Office), p. 5-7.
[a]Includes aldehydes.
[b]Includes sulfur dioxide and trioxide.
[c]S equals percentage of sulfur in oil in round numbers.

certain method can be devised to assure this exclusion, roughly satisfactory estimates were developed that will be described presently.

The rough division of fuel oil consumption by function is rather easily accomplished. Federal statistics on fuel oil sales published annually [13] distinguish "heating oils" and within that category, five types by number. It is known that individual household heaters use only the lighter oils, and the assumption was made that all of the No. 1 oil and four-ninths of the No. 2 oil was consumed in residential heating. Commercial-industrial space heating was assumed to account for five-ninths of No. 2 oil. This division is in line with the findings of Landsberg and Schurr that the ratio of overall consumption of energy by the commercial and residential sectors was about 5 to 4 [14:15]. The American Petroleum Institute's description of uses for Grades 4, 5, and 6 oil makes it evident that all of these grades may be classified under commercial-industrial [15].

Gas consumption statistics have no such handy labels as fuel oil grades to aid in classification. The American Gas Association publishes data on natural gas consumption by customer divided between residential, commercial, industrial, and other [16]. Using very rough approximations, we have assumed that 55 percent of gas classified as commercial is used for space heating and 20 percent of the gas classified as industrial.

The result of these classifications of national fuel oil and natural gas consumption for selected years 1959 to 1969 will be found in tables 15.5 and 15.6. The projection to 1980 of each series was made assuming continuation of rates of increase in number of units and type of fuel usage that prevailed prior to 1969.

Given the national oil and gas space heating consumption figures contained in tables 15.5 and 15.6, the next step was to allocate the fuel to the individual cities covered by the chapter. For allocating residential gas consumption, there were two steps involved: (1) To determine the preliminary allocation to a city of

Table 15-5
Residential Consumption of Natural Gas and Heating Oil 1959-1980

	Gas thousands of therms	Oil thousands of barrels
1959	30,511,938	195,568
1964	39,886,340	204,051
1969	49,751,297	242,764
1980	62,457,680	309,651

Source:

Gas 1959-1969—American Gas Association, AMERICAN GAS ASSOCIATION MONTHLY, Vols. 43 p. 5; 48, p. 5; 52, p. 4; New York 1961, 1966, 1970. Adjusted as explained above.

1980—Estimated assuming continuation in next decade of rates of increase which prevailed prior to 1969.

Oil 1959-1964—National Petroleum News ANNUAL FACTBOOK ISSUE, Vols. 53, p. 180; 57, p. 180. New York 1961, 1966. Adjusted as explained above.

1969—U.S. Dept. of the Interior, Bureau of Mines, MINERAL INDUSTRIAL SURVEY: FUEL OIL SALES, ANNUAL (Washington: Government Printing Office). Adjusted as explained above.

1980—Estimated—assuming continuation in next decade of rates of increase which prevailed prior to 1969.

Table 15-6
Estimated Commercial/Industrial Consumption of Natural Gas and Heating Oil in Space Heating 1959-1980

	Gas thousands of therms	Oil thousands of barrels
1959	13,252,134	304,900
1964	18,174,063	336,302
1969	25,942,138	419,424
	46,789,061	606,771

Source:

Gas 1959-1969—American Gas Association, AMERICAN GAS ASSOCIATION MONTHLY Vols. 43, p. 5; 48, p. 5; 52, p. 4. New York, 1961, 1966, 1970. Adjusted as explained above.

1980—Estimated assuming continuation in next decade of rates of increase which prevailed prior to 1969.

Oil 1959-1964—National Petroleum News ANNUAL FACTBOOK ISSUE, Vols. 53, p. 180; 57, p. 180, New York, 1961, 1966. Adjusted as explained above.

1969—U.S. Department of the Interior, Bureau of Mines, MINERAL INDUSTRIAL SURVEYS: FUEL OIL SALES ANNUAL (Washington: Government Printing Office). Adjusted as explained above.

1980—Estimated assuming continuation in next decade of rates of increase which prevailed prior to 1969.

its share of national residential gas consumption by calculating its proportion of single-unit houses employing gas as fuel as revealed by Census of Housing, (2) To adjust this share of national consumption up or down to account for any difference for the city in question in ambient temperatures from the national average as measured in degree days. For purposes of this calculation, it was assumed that the term "owner-occupied" houses could be treated as identical with single-unit houses. A precisely analogous method was employed to allocate fuel oil from the national totals to individual cities.

On the assumption that commercial and industrial space heating was mostly for places of employment, the initial allocation of each city's share was made on the basis of its share of national employment. As before, this was adjusted for the city's annual degree days relative to the national average to take account of temperature differences.

Given historical and projected fuel consumption series for each city under study, it only remained to multiply by the appropriate emission factor from table 15.4 to determine space heating contribution to air pollution in the city in question.

Notes

1. Alfred J. Van Tassel (Ed.), *Environmental Side Effects of Rising Industrial Output* (Lexington, Mass., D.C. Heath and Company, 1970), 548 pp.

2. *Spot Television Rates and Data* (Consumer Markets Section) (Skokie, Illinois: Standard Rate and Data Service, relevant years).

3. U. S. Department of Transportation, Federal Highway Administration, Bureau of Public Roads, *Highway Statistics, 1959* (Washington, D.C.: Government Printing Office, 1959), pp. 1-84.

4. U. S. Department of Transportation, Federal Highway Administration, Bureau of Public Roads, *Highway Statistics, 1964* (Washington, D.C.: Government Printing Office, 1965), pp. 1-84.

5. U. S. Department of Transportation, Federal Highway Administration, Bureau of Public Roads, *Highway Statistics, 1969* (Washington, D.C.: Government Printing Office, 1970), pp. 1-84.

6. U. S. Department of Commerce, Bureau of the Census, *Statistical Abstract of the United States* (Washington, D.C.: Government Printing Office, 1970).

7. Automobile Manufacturers Association, *1970 Automobile Facts and Figures*, "Passenger Cars in Use by Age Groups," Detroit: 1970.

8. Office of Science and Technology, *Cumulative Regulatory Effects on the Cost of Automotive Transportation* (Final Report of the Ad Hoc Committee) Washington, February 1972.

9. *New York Times,* March 20, 1972, p. 1.

10. *Newsday,* (Garden City, New York) May 13, 1972, p. 11.

11. *New York Times,* May 2, 1972, p. 5.

12. Federal Power Commission, *The 1970 National Power Survey* in 4 parts (Washington, D.C.: Government Printing Office, 1970).

13. U. S. Department of Interior, Bureau of Mines, *Mineral Industry Surveys,* "Sales of Fuel Oil and Kerosine in 1969," (Washington, D.C.: Government Printing Office, Sept. 30, 1970).

14. Hans H. Landsberg and Sam H. Schurr, *Energy in the United States* (New York: Random House, 1960), p. 15.

15. American Petroleum Institute, *Petroleum Facts and Figures* (New York: 1967), pp. 184-89.

16. American Gas Association *Monthly,* Volume 52, p. 4.

16 Air Pollution in Five Eastern Cities

Peter L. Goffin

Air pollution, the contamination of the air we breathe, recently has become an important issue in political, economic, and environmental circles.

Polluted air is a mixture of gases, vapors, solids, mists, and fumes, composed of particulate material, carbon monoxide, hydrocarbons, nitrogen oxides, and sulfur compounds.

Particulate matter exists in a finely divided form as a liquid or solid. Particulates result from industrial plants, power stations, and stacks of many other buildings, fugitive dusts of all kinds from the ground, as well as other sources.

Carbon monoxide is a product of combustion, mostly from motor vehicles, but from other sources as well. Hydrocarbons are mostly emitted from the normal operation of motor vehicles, but also are emitted in smaller quantities from some industrial processes. Nitrogen oxides are the result of combustion of all kinds, while sulfur effluents usually result from the burning of coal and heavy fuel oil.

The effects of these pollutants vary from locale to locale and these effluents can even vary within the same area, depending on industrial patterns, traffic patterns, waste disposal, heating, normal living activities, and especially the weather.

This analysis of five SMSAs (Standard Metropolitan Statistical Areas) along the eastern coast of the United States will concentrate on covering at least 80 percent of the sources of air pollution problems in each SMSA. Each major category of polluters (e.g., automobiles, power plants, space heating, weather, and topographical conditions) contributing to the problem will be dealt with separately, reviewing conditions in each area in 1959, 1964, 1969, and after carefully computing all possible factors, finally arriving at the probable air pollutant quantities and problems in the year 1980.

General Assumptions

Several assumptions will be made in this analysis: (1) it is assumed that there will not be any major change in the internal combustion engine for at least the next ten years; (2) there will not be any drastic shifts to new types of fuels in powering automobiles, in heating residential and commercial buildings, and in providing the materials necessary in producing electrical power in each SMSA. Additional assumptions, if needed, will be introduced and labeled as the author progresses throughout the study.

New York SMSA

The Standard Metropolitan Statistical Area of New York is composed of New York City, Westchester, Rockland, Nassau, and Suffolk Counties. New York City has the most persistent and severe air pollution problem because of the population density of the region, the traffic patterns of motor vehicles, the smoke and pollutants blown by the winds from New Jersey and deposited in New York, and the electrical power generation activity needed to satisfy the demands of the inhabitants of the city. Few of the other SMSAs in the United States have the massive problem of New York.

The use of the inefficient internal combustion engine may have to be restricted in the future, especially in densely populated areas such as New York. "New York has hired one of the foremost auto engineers in Detroit to build a car powered by a gas turbine engine that minimized pollution. The action was taken under a program aimed at prodding the major auto makers to adopt an alternative engine" [1].

The car is built by the Williams Research Corporation. The action of New York City comes at a time when the auto industry has said that it has doubts that it can meet the strict standards set by the Clean Air Bill sponsored by Senator Edmund S. Muskie. New York City is also planning to experiment with more effective controls such as catalytic converters and nonleaded gasoline for present automobiles already on the road.

Former Commissioner of Air Resources Austin Heller recently states in an article in the *New York Times* that more than 60 percent of New York City's air pollution problem based on estimates of weight of emissions into the air now comes from motor vehicles. Mr. Heller feels that it would take until 1979 to have all cars operating with control devices effectively in the New York area. Mr. Heller also mentioned that a study showed that congestion and stop-and-go driving patterns in New York City have actually increased gasoline consumption 10 percent for taxi fleets [2]. This has added more nitrogen oxides and lead to pollute the air.

Congressional determination to control emissions from automobiles is seen in the National Air Quality Standards Act of 1970 where national emissions standards based on protection of public health and welfare provide that the 1975 model cars must achieve at least a 90 percent reduction from the 1970 standards. Table 16.1 shows the tons of carbon monoxides, hydrocarbons, nitrogen oxides, sulfur oxides, and particulates that were thrown into the atmosphere in New York by automobiles in 1959, 1964, and 1969. There was a 39 percent increase in the number of automobiles from 1959 to 1969 in the New York SMSA and the projection for 1980 assumes a continuation of this rate of increase.

If the 1971 standards and the standards prescribed in the law drafted by Senator Muskie are strictly adhered to, then it is projected that in 1980 the total emissions from automobiles in the New York SMSA will have decreased 88 percent. This dramatic improvement is reflected in table 16.1 in column (a) under 1980. However, it may not prove possible to effect a simultaneous reduction of this magnitude in all automobile pollutants for reasons discussed in Chapter 15. If only the 1971 standards are achieved, the reduction will be less

Table 16-1
Emissions from Automobiles—New York (In Thousands of Tons)

Emissions	1959	1964	1969	1980 (a)	1980 (b)
Carbon Monoxide	1,500.2	1,880.3	2,084.5	475.9	857.4
Hydrocarbons	492.8	617.6	532.5	108.0	171.7
Nitrogen Oxides	175.2	219.6	243.4	209.3	396.5
Sulfur Oxides[a]	7.2	9.0	10.0	13.9	13.9
Particulates[b]	9.6	12.0	13.3	18.5	18.5

[a] 9 lbs. per 1000 gallons of gasoline burned
[b] 12 lbs. per 1000 gallons of gasoline burned

Source: Number of automobiles from Standard Rate and Data Service projected to 1980 in accordance with procedures outlined in Chapter 15. Emissions estimated by multiplying by appropriate emission factor in Table 15-2 in accordance with procedure of Chapter 15.
For 1980:
   1980 Column (a) based on Assumption 1:
     National standards for 1976 and 1971 met on schedule
   1980 Column (b) based on Assumption 2:
     National standards for 1971 highest achieved before 1980.
Distribution of car population for 1980 computations following Automobile Manufacturers Association data: 52.4 percent under 5 years; 35.6, 5 to 10 years; 12 percent over 10 years old.

substantial, as shown in 1980 column (b). In this event, the increase in the number of automobiles by 1980 would result in a more than 60 percent increase in emissions of nitrogen oxides—a principal constituent of smog.

Electrical Power Generation

Although the United States has only 6 percent of the world's population, production of electricity in the United States represented 35 percent of the world's total in 1967 [4]. The American public consumes 2 1/3 times the electricity generated by our closest rival, Russia. Our continued trend towards rapid urban development in several regions throughout the country reinforces our need for a continuing growth of electrical power in satisfying our needs.

Although emissions from generating stations are released through stacks designed to introduce them into the upper atmosphere so that their effect at ground level is negligible, this is not always the case.

In the five SMSAs currently being analyzed, the prime electric power utilities in each SMSA were selected in order to obtain specific data relevant to the quantities of fuels used for electric power generation in each region. Fuel oil, coal, and natural gas combustion coefficients in conjunction with the quantities of each type of fuel utilized in each electric power utility were obtained to enable us to arrive at the total tons of pollutants emitted in the years 1959, 1964, 1969, and to project emissions in 1980 for each of the five SMSAs.

Table 16-2
Gaseous Emissions for Electric Power Utilities[a]
New York (In Thousands of Tons)

Emissions	1959	1964	1969	1980
Carbon Monoxide	1.6	1.5	1.1	.67
Hydrocarbons[b]	1.5	2.1	4.0	15.2
Nitrogen Oxides	97.8	114.2	192.0	723.1
Sulfur Oxides[c]	376.8	379.4	263.2	342.9
Particulates	70.7	68.7	52.1	40.7

[a]Includes coal, fuel oil, and natural gas combustion
[b]Includes aldehydes.
[c]Includes SO_3

Source: Based on coefficients taken from the COMPILATION OF AIR POLLUTION EMISSION FACTORS, 1968, pages 5, 6, and 7, and on quantities of natural gas, fuel oil, and coal used for each of the prime power plants in New York's SMSA. Quantities of fuels obtained from the NATIONAL COAL ASSOCIATION'S STEAM ELECTRIC PLANT FACTORS, 1959, 1964, and 1969–Fuel consumption and costs, pages 5-28. Projections based on percentage increase from 1959 to 1969. Sulfur content of fuel differs with each year of electric power production. Average sulfur content used in compilation of emission factors. Computations in accordance procedures outlined in Chapter 15.

The two prime suppliers of electricity in New York are Consolidated Edison and the Long Island Lighting Company. Table 16.2 shows the total tons of pollutants discharged into the atmosphere by both Con Edison and LILCO. Both companies have invested heavily in air pollution control equipment. Consolidated Edison and the LILCO have switched to burning low sulfur fuels well in advance of legislative mandates, and they have increased the use of natural gas while reducing the amount of coal burned. Older coal-burning equipment was either converted to natural gas or oil, or was retired from service. Since April 1968, some three years ahead of the existing legal requirements, all coal and oil burned by the companies were down to sulfur content of 1 percent or less. Con Edison is required by New York State law to use very low sulfur oil (.37 percent sulfur) in all new fossil fuel plants constructed by the company in the future. One such plant, a 1.6 million kilowatt plant expansion at Astoria, Queens, is scheduled for completion in 1974-1975. Natural gas, which is desirable as a boiler fuel because it contributes practically nothing to the atmosphere in the way of sulfur dioxide or particulate matter is currently being utilized by both companies. The increased availability of natural gas will enable Con Edison and LILCO to make further reductions in sulfur dioxide and particulate emissions.

Both companies have also invested in electrostatic precipitators to control particulate emissions. The electrostatic precipitators presently in operation all have guaranteed collection efficiencies of 99 percent or greater, though not all generating stations are so equipped. Con Edison has the world's largest electrostatic precipitator, which serves the company's one million kilowatt unit at the Ravenswood station. These precipitators combined with the reduced use of coal and the increased use of natural gas and low sulfur oils have reduced the emissions considerably for both companies. Both Con Edison and LILCO have reviewed all new sulfur removal techniques. These techniques remove the sulfur from the coal and fuel oil before combustion or from the stack effluent after combustion, although a technologically feasible process has not been developed that can be applied on a large scale.

Since there is not any combustion in nuclear power generation, there are not any products of combustion to pollute the air. Nuclear plants emit none of the oxides of sulfur, or nitrogen, or particulate matter that are generated in the burning of fossil fuels. While both companies utilize some nuclear energy, at present it is only 3 percent of the total energy output. There definitely will be a reduction in tons of effluents emitted in the New York SMSA by the electrical power utilities in coal combustion; however, projected emissions in 1980 are greater than those in previous years. The rapid increase in fuel consumption based on the percentage increases from 1959 to 1969 accounts for the projected increase of pollutants.

Space Heating

An additional contributor to air pollution difficulties in the use of fuel, both natural gas and fuel oil, is seen in space heating, both for commercial and industrial uses, and in owner-occupied houses. Table 16.3 shows the total tons of effluents emitted from this source in 1959, 1964, and 1969, along with projected emissions for the year 1980.

For space heating units in the New York SMSA, the Clean Air Act of 1970, administered by the Environmental Protection Agency, introduced tough national air quality standards for six of the principal pollutants [5].

William D. Ruckelshaus, administrator of the Environmental Protection Agency, has announced standards for sulfur oxides, particulates, carbon monoxides, hydrocarbons, nitrogen oxides, and photochemical oxidants, and expects individual states to have in effect, by July 1, 1975, plans for meeting the standards established. Although Mr. Ruckelshaus claims that most regions will be able to meet the standards by switching to low sulfur fuels and by requiring plants to install electrostatic precipitators to capture the soot, some cities might have a difficult time meeting the standards by the 1975 deadline. However, the use of low sulfur fuel oils for space heating is now required by New York City and this is expected to effect a reduction in sulfur oxides from space heating to about a fourth of the 1969 level in 1980. Increased use of natural gas also reduces sulfur oxide emissions, but at the expense of heightened nitrogen oxide release.

Philadelphia SMSA

The SMSA of Philadelphia consists of Bucks, Chester, Delaware, Montgomery, Philadelphia, Burlington, Camden, and Gloucester counties. It is the fourth largest such area in the United States.

Table 16-3
Space Heating: Commercial-Industrial and Owner Occupied Houses[a]
New York (In Thousands of Tons)

Pollutant	1959	1964	1969	1980
Carbon Monoxide	2.9	3.8	4.4	7.1
Hydrocarbons[b]	7.3	9.5	10.8	17.7
Nitrogen Oxides	20.0	25.8	27.9	47.3
Sulfur Oxides[c]	463.4	603.1	697.8	170.6[c]
Particulates	12.1	15.4	18.2	29.0

Source: Based on coefficients taken from the COMPILATION OF AIR POLLUTION EMISSION FACTORS, 1968, pages 5, 6, and 7, and on quantities of fuel oil consumed in SMSA with respect to percentages of residences using oil. Adjustments were made with respect to the number of degree days in SMSA. Projections to 1980 based on percentage increases from 1959 to 1969.
Computations in accordance procedures outlined in Chapter 15.
[a]includes fuel oil and natural gas.
[b]Includes aldehydes.
[c]Includes SO_3.
[d]Reduced to reflect of new ordinance limiting sulfur in fuel to no more than 0.3 percent.

The Philadelphia SMSA region is within an eleven-county Air Quality Control Region. "Most of the area lies in the gently rolling piedmont or flat coastal plain of the Mid-Atlantic states. The terrain level increases gradually to the west and northwest where the extremities of the region are located in outlying ridges of the Appalachian Mountains. There are no serious complications of wind flow caused by topography" [6]. The majority of emissions occur in the Delaware Valley in a limited area along the Delaware River.

Philadelphia is not far behind New York with regard to the air pollution problems caused by automobile emissions. Philadelphia had a 45 percent increase in the number of automobiles from 1959 to 1969. Table 16.4 shows the tons of carbon monoxide, hydrocarbons, nitrogen oxides, sulfur oxides, and particulates that were emitted by automobiles in Philadelphia in 1959, 1964, and 1969. Projected emissions for the year 1980 were obtained by the same methods used in table 16.1. A corresponding decrease similar to that previously described for the New York SMSA for the year 1980 will occur in Philadelphia if the 1971 and 1975 standards are supported and enforced. This is shown in 1980 column (a) of table 16.4. However, if for reasons discussed in Chapter 15 such a reduction does not prove feasible, there is likely to be a large increase in nitrogen oxide emissions, although improvement with respect to carbon monoxide and hydrocarbons, as shown in 1980 column (b).

Electrical Power Generation

The prime supplier of electricity in this region is the Philadelphia Electric Company, which has long been a pioneer in the fight against air pollution. Long

Table 16-4
Emissions from Automobiles—Philadelphia (In Thousands of Tons)

Emissions	1959	1964	1969	1980 (a)	1980 (b)
Carbon Monoxide	804.2	957.0	1,167.2	293.3	523.7
Hydrocarbons	264.2	413.3	298.2	65.3	103.7
Nitrogen Oxides	94.0	111.8	136.3	126.6	239.7
Sulfur Oxides[a]	3.9	4.6	6.0	8.2	8.2
Particulates[b]	5.1	6.2	7.5	10.8	10.8

[a]9 lbs. per 1000 gallons of gasoline burned
[b]12 lbs. per 1000 gallons of gasoline burned
Source: Number of automobiles from Standard Rate and Data Service projected to 1980 in accordance with procedures outlined in Chapter 15. Emissions estimated by multiplying by appropriate emission factor in Table 15-2 in accordance with procedure of Chapter 15.
For 1980:
   1980 Column (a) based on Assumption 1:
   National standards for 1976 and 1971 met on schedule
   1980 Column (b) based on Assumption 2:
   National standards for 1971 highest achieved before 1980.
Distribution of car population for 1980 computations following Automobile Manufacturers Association data: 52.4 percent under 5 years; 35.6, 5 to 10 years; 12 percent over 10 years old.

before any ordinances were put into effect, this company installed equipment in reducing particulate emissions from its generating stations.

Installation of double collection systems—mechanical and electrostatic—was begun in 1948. These systems, used extensively at the generating stations, have an efficiency as high as 99.2 percent.

Philadelphia Electric has converted most of their coal-fired boilers to oil. Table 16.5 shows the total amount of pollutants discharged by Philadelphia Electric. For several years this company has been purchasing increasing quantities of low sulfur oil, but with much larger quantities of fuel consumed, the total emissions projected for 1980 has increased for both fuel oil combustion and natural oil combustion. Philadelphia Electric hopes to have pollution-free nuclear generation expanded materially, so that by the mid-1970s nuclear generation will hopefully provide between 25 and 30 percent of the company's electric output. This will further reduce the total tons of effluents emitted in the Philadelphia SMSA area by Philadelphia Electric Company.

Space Heating

Philadelphia was included as one of the seven cities mentioned by the Environmental Protection Agency as having a possible problem meeting the air quality standards by the 1975 deadline. Table 16.6 shows pollutants emitted from space heating sources in the Philadelphia SMSA in 1959, 1964, 1969, and projected

Table 16-5
Gaseous Emissions for Electric Power Utilities[a]—Philadelphia (In Thousands of Tons)

Emissions	1959	1964	1969	1980
Carbon Monoxide	.97	1.2	.8	.69
Hydrocarbons[b]	.7	.8	2.4	.3
Nitrogen Oxides	46.5	57.1	75.2	155.4
Sulfur Oxides[c]	204.7	237.9	214.2	218.2
Particulates	41.3	49.2	37.0	23.3

[a]Includes coal, fuel oil, and natural gas combustion.
[b]Includes aldehydes.
[c]Includes SO_3.
Source: Based on coefficients taken from the COMPILATION OF AIR POLLUTION EMISSION FACTORS, 1968, pages 5, 6, and 7, and on quantities of natural gas, fuel oil, and coal used for each of the prime power plants in New York's SMSA. Quantities of fuels obtained from the NATIONAL COAL ASSOCIATION'S STEAM ELECTRIC PLANT FACTORS, 1959, 1964, and 1969—Fuel consumption and costs, pages 5-28. Projections based on percentage increase from 1959 to 1969. Sulfur content of fuel differs with each year of electric power production. Average sulfur content used in compilation of emission factors. Computations in accordance procedures outlined in Chapter 15.

Table 16-6
Space Heating, Commercial-Industrial and Owner Occupied Houses[a]—Philadelphia (In Thousands of Tons)

Pollutant	1959	1964	1969	1980
Carbon Monoxide	1.0	1.5	1.8	2.5
Hydrocarbons[b]	2.5	3.7	4.5	9.4
Nitrogen Oxides	8.7	12.6	15.2	28.7
Sulfur Oxides[c]	105.2	233.0	286.8	595.2
Particulates	4.4	6.5	7.9	16.0

[a]Includes fuel oil and natural gas.
[b]Includes aldehydes.
[c]Includes SO_3.
Source: Based on coefficients taken from the COMPILATION OF AIR POLLUTION EMISSION FACTORS, 1968, pages 5, 6, and 7, and on quantities of fuel oil consumed in SMSA with respect to percentages of residences using oil. Adjustments were made with respect to the number of degree days in SMSA. Projections to 1980 based on percentage increases from 1959 to 1969. Computations in accordance procedures outlined in Chapter 15.

emissions for 1980. It is hoped that the use of gas as a heating fuel will be increased from the 30 percent presently used, while fuel oil as a heating fuel will decrease from the present 55 percent of units in the Philadelphia SMSA. The increased use of natural gas as a heating fuel, while reducing sulfur oxides, will have to be carefully watched since large quantities of nitrogen oxides are emitted as a pollutant. Pollutants from space heating units utilizing fuel oil as a heating fuel will substantially increase as the quantity of fuel oil consumed increases.

Pittsburgh SMSA

The SMSA of Pittsburgh consists of Allegheny, Beaver, Westmoreland, and Washington counties. The Pittsburgh SMSA region is within an area comprised of seven counties. The population and industrial center of the area is the City of Pittsburgh in Allegheny County. Urban and industrial development of this area historically took place along the valleys of the rivers that traverse the area.

The SMSA of Pittsburgh mostly consists of "a plateau dissected by alternating ridges of rugged, flat-topped hills and steep-sided stream valleys" [7:2]. Within this area, the Allegheny and Monongahela Rivers merge to form the Ohio River. The flatlands of these river valleys have provided ideal sites for steel mills and other heavy industry, in addition to the fact that the rivers are used for transport of raw materials and finished products.

"In many areas the effects of emissions of pollutants into the air is aggravated by geographic and meterological conditions. Severe inversion conditions are far more frequent than is generally appreciated" [7:63]. Weather and climatology will be dealt with later on in the analysis of this region's air pollution problems.

Because of its strategic location at the point where the Allegheny and Monongahela Rivers merge, Pittsburgh was and is the center of growth and development for the region. The early dependence upon waterways promoted the establishment of many river-oriented communities that grew into significant industrial centers. A number of smaller communities have been founded adjacent to coal fields and other mineral deposits. The main source of employment in the region are manufacturing industries, and in manufacturing this area is highly specialized in the production of primary metals.

Although Pittsburgh only has approximately one-third the number of automobiles compared to New York, the percentage rate of increase from 1959 to 1969 is the same as that of New York's SMSA, 39 percent. There is likely, nevertheless, to be a sizable overall reduction in automobile pollutants in the region by 1980. Table 16.7 shows the tons of effluents discharged by the operation of automobiles in the Pittsburgh SMSA in 1959, 1964, and 1969. Projected emissions for the year 1980 are included and were arrived at based on calculations already mentioned earlier in this chapter.

Table 16-7
Emissions from Automobiles—Pittsburgh (In Thousands of Tons)

Emissions	1959	1964	1969	1980 (a)	1980 (b)
Carbon Monoxide	464.4	548.7	646.0	150.1	269.8
Hydrocarbons	153.0	180.2	165.0	33.3	52.8
Nitrogen Oxides	54.2	64.1	75.4	64.2	121.9
Sulfur Oxides[a]	2.2	2.6	3.1	4.3	4.3
Particulates[b]	3.0	3.5	4.1	5.7	5.7

[a] 9 lbs. per 1000 gallons of gasoline burned
[b] 12 lbs. per 1000 gallons of gasoline burned

Source: Number of automobiles from Standard Rate and Data Service projected to 1980 in accordance with procedures outlined in Chapter 15. Emissions estimated by multiplying by appropriate emission factor in Table 15-2 in accordance with procedure of Chapter 15.
For 1980:
  1980 Column (a) based on Assumption 1:
    National standards for 1976 and 1971 met on schedule
  1980 Column (b) based on Assumption 2:
    National standards for 1971 highest achieved before 1980.
Distribution of car population for 1980 computations following Automobile Manufacturers Association data: 52.4 percent under 5 years; 35.6, 5 to 10 years; 12 percent over 10 years old.

Electrical Power Generation

The prime supplier of electricity in this region is the Duquesne Light Company. Table 16.8 shows the total tons of pollutants discharged into the atmosphere by Duquesne Light. It should be noted that this company does not burn any natural gas in providing electricity for residents of this SMSA region, but relies primarily on coal and fuel oil combustion in operating their plants. Although Duquesne Light has started to equip its generating stations with modern air quality control devices (i.e., in their Cheswick Power Station, a 750-foot stack was erected, along with the installation of a 99.5 percent efficient electrostatic precipitator as well as special equipment for the removal of sulfur from coal before it is burned), the total quantities of fuel consumed means projected emissions, in 1980, which will increase from the previous level emitted in 1969, 1964, and 1959. Duquesne Light Company has plans to convert their fuel burning stations to low sulfur oil burning plants in the near future.

Duquesne Light has applied to the Public Utility Commission for permission to close down, effective January 1, 1971, part of its oldest power utilities, the Colfax Power Station in Springdale. Pending approval, ash emissions at Colfax will be reduced to less than 50 percent of the present level as a result of shutting down several units and generally reducing overall plant operations.

This company is also installing a wet scrubber system for the removal of both particulate (fly ash) matter and sulfur dioxide at several of their power stations. This is a new development in pollution control for fossil fuel power plants and it will be the first major application of this type in the nation.

Table 16-8
Gaseous Emissions for Electric Power Utilities[a]—Pittsburgh (In Thousands of Tons)

Emissions	1959	1964	1969	1980
Carbon Monoxide	.86	1.1	1.1	1.3
Hydrocarbons[b]	.35	.44	.43	.57
Nitrogen Oxides	34.0	42.9	43.0	53.4
Sulfur Oxides[c]	160.3	203.3	201.1	253.3
Particulates	35.4	44.9	44.4	3.5

[a]Includes only coal and fuel oil combustion as there is no natural gas utilized.
[b]Includes aldehydes.
[c]Includes SO_3.
Source: Based on coefficients taken from the COMPILATION OF AIR POLLUTION EMISSION FACTORS, 1968, pages 5, 6, and 7, and on quantities of natural gas, fuel oil, and coal used for each of the prime power plants in New York's SMSA. Quantities of fuels obtained from the NATIONAL COAL ASSOCIATION'S STEAM ELECTRIC PLANT FACTORS, 1959, 1964, and 1969—Fuel consumption and costs, pages 5-28. Projections based on percentage increase from 1959 and 1969. Sulfur content of fuel differs with each year of electric power production. Average sulfur content used in compilation of emission factors. Computations in accordance procedures outlined in Chapter 15.

Space Heating

Pittsburgh does not have the problem from space heating units that New York has, primarily due to the fact that 77 percent of owner-occupied houses in the Pittsburgh SMSA use gas as their heating fuel while only 4 percent consume fuel oil. Table 16.9 indicates what additional quantities of fuel oil consumed mean in relation to the total tonnage of effluents emitted from space heating units in this SMSA. Pittsburgh is not one of the seven cities that the Environmental Protection Agency mentioned as having a difficult time in meeting the air quality standards established, and although Pittsburgh's air pollution problems are not as severe as those of New York, they are present, nevertheless.

Atlanta SMSA

The SMSA of Atlanta, Georgia, consists of Clayton, Cobb, DeKalb, Fulton, and Gwinnett counties. Atlanta had an 87 percent increase in the number of automobiles in its SMSA from 1959 to 1969. This area has been one of the fastest growing regions in the country and its massive and continuing highway building program is one of the best in the United States.

Table 16.10 illustrates automobile emissions for the Atlanta SMSA in 1959, 1964, and 1969, along with projected emissions in 1980. While Atlanta does not have as serious a problem regarding automobile air pollution as does New York, Philadelphia, or Pittsburgh in terms of total tonnage of effluents emitted, the

Table 16-9
Space Heating, Commercial-Industrial and Owner Occupied Houses[a]—Pittsburgh
(In Thousands of Tons)

Pollutant	1959	1964	1969	1980
Carbon Monoxide	.3	.3	.4	.5
Hydrocarbons[b]	.8	.9	1.0	1.4
Nitrogen Oxides	5.5	6.2	7.0	9.1
Sulfur Oxides[c]	47.2	53.1	62.4	87.1
Particulates	1.8	2.1	2.4	3.2

[a]Includes fuel oil and natural gas.
[b]Includes aldehydes.
[c]Includes SO_3.
Source: Based on coefficients taken from the COMPILATION OF AIR POLLUTION EMISSION FACTORS, 1968, pages 5, 6, and 7, and on quantitites of fuel oil consumed in SMSA with respect to percentages of residences using oil. Adjustments were made with respect to the number of degree days in SMSA. Projections to 1980 based on percentage increases from 1959 to 1969. Computations in accordance procedures outlined in Chapter 15.

Table 16-10
Emissions from Automobiles—Atlanta (In Thousands of Tons)

Emissions	1959	1964	1969	1980 (a)	1980 (b)
Carbon Monoxide	219.0	307.0	409.0	110.3	197.3
Hydrocarbons	72.0	101.0	134.3	24.5	39.0
Nitrogen Oxides	26.0	36.0	48.0	47.6	90.0
Sulfur Oxides[a]	1.1	1.5	2.0	3.7	3.7
Particulates[b]	1.4	2.0	3.0	5.0	5.0

[a]9 lbs. per 1000 gallons of gasoline burned.
[b]12 lbs. per 1000 gallons of gasoline burned.
Source: Number of automobiles from Standard Rate and Data Service projected to 1980 in accordance with procedures outlined in Chapter 15. Emissions estimated by multiplying by appropriate emission factor in Table 15-2 in accordance with procedure of Chapter 15.
For 1980:
    1980 Column (a) based on Assumption 1:
      National standards for 1976 and 1972 met on schedule
    1980 Column (b) based on Assumption 2:
      National standards for 1971 highest achieved before 1980.
Distribution of car population for 1980 computations following Automobile Manufacturers Association data: 52.4 percent under 5 years; 35.6 percent, 5 to 10 years, 12 percent over 10 years old.

rapidity of the urbanized growth in this region has been spectacular.

Considering that the number of automobiles in this region will grow from only 264,920 in 1959 to approximately 1,116,000 in 1980, based on previous increases from 1959 to 1969, it can be seen that without any reduction in automobile emissions, Atlanta would be heading towards a severe automobile air pollution problem. The situation in Atlanta is considerably better than that of any of the previously mentioned SMSAs being investigated for two reasons: (1) Although Atlanta has seen an unusual growth pattern for a decade, this growth rate will most probably taper off at the end of the next decade. The density of the area even at that time will not even come close to the situation presently encountered in New York City; (2) Atlanta's highway system is far superior than that of New York, Philadelphia, or Pittsburgh's SMSA.

Electrical Power Generation

The prime supplier of electricity in this region is the Georgia Power Company, one of four operating companies of the Southern Company System, which serves over 57,000 of the 59,000 square miles in Georgia [8]. Table 16.11 shows the total tonnage of effluents discharged into the atmosphere by the Georgia Power Company.

Georgia Power began to control air pollution effects from its coal-fired generating plants in 1936. Installation of electrostatic and mechanical ash precipitators began in that year.

Between 1970 and 1972 Georgia Power plans on spending $12 million in installing more efficient precipitators and other clean air equipment at existing plants and plants now under construction [9:3]. With this new equipment, Georgia Power will hope to remove over 98 percent of the particulate matter from the gases caused by fuel combustion. Additionally, the new plant that Georgia Power is constructing will use very tall stacks to disperse combustion gases. This will virtually eliminate gases at ground level.

One of Georgia Power Company's biggest problems with respect to emitting pollutants into the atmosphere stems from the fact that relatively insignificant quantities of oil and gas are consumed in providing power for the generating plants. Large quantities of coal are used in providing electrical power. A 800,000 kilowatt plant is planned in Baxley, Georgia. This is to be the first nuclear generating station in the Southern Company System, and is expected to begin producing electricity in 1973. Plans for a second 800,000 kilowatt unit were announced early in 1970, with a completion date set for 1976 [9:9].

It is hoped that the addition of the nuclear plants, coupled with decreasing quantities of coal used, and increasing quantities of natural gas, will help to alleviate some of Atlanta's air pollution difficulties caused by Georgia Power Company's electrical generating plants operating in that region.

Table 16-11
Gaseous Emissions for Electric Power Utilities[a]—Atlanta (In Thousands of Tons)

Emissions	1959	1964	1969	1980
Carbon Monoxide	.4	1.0	1.4	1.96
Hydrocarbons[b]	.2	.4	.6	.84
Nitrogen Oxides	20.7	38.3	61.3	85.8
Sulfur Oxides[c]	79.8	181.5	270.7	378.9
Particulates	26.0	40.0	78.8	15.2

[a]Includes coal, fuel oil and natural gas.
[b]Includes aldehydes.
[c]Includes SO_3.

Source: Based on coefficients taken from the COMPILATION OF AIR POLLUTION EMISSION FACTORS, 1968, pages 5, 6, and 7, and on quantities of natural gas, fuel oil, and coal used for each of the prime power plants in New York's SMSA. Quantities of fuels obtained from the NATIONAL COAL ASSOCIATION'S STEAM ELECTRIC PLANT FACTORS, 1959, 1964, and 1969—Fuel consumption and costs, pages 5-28. Projections based on percentage increase from 1959 to 1969. Sulfur content of fuel differs with each year of electric power production. Average sulfur content used in compilation of emission factors. Computations in accordance procedures outlined in Chapter 15.

Space Heating

Atlanta's air pollution problem with respect to pollutants from space heating units using fuel oil and natural gas are minimal compared to New York, Philadelphia, and Pittsburgh. Table 16.12 shows the total effluents emitted from space heating units in Atlanta. It should be noted that 84 percent of space heating units in Atlanta use natural gas. Atlanta should not have a serious air pollution problem in comparison to the other SMSAs being reviewed in this chapter.

Savannah SMSA

The Standard Metropolitan Statistical Area of Savannah, Georgia, consists entirely of Chatham County. Savannah's SMSA had a 34 percent increase in the number of automobiles in the area from 1959 to 1969. The area is not large, nor is it as densely populated as are the other SMSAs being reviewed in the study.

Table 16.13 shows automobile emissions for the Savannah SMSA in 1959, 1964, 1969, and projected emissions from automobile traffic in 1980. Savannah has the least difficulties of the five SMSAs with respect to air pollution problems in terms of total tonnage of pollutants emitted, since there is a definite correlation between population density, or urbanization of an area, industrial activity in the region, automobile traffic patterns developed within the region, and the quality of the air in an area.

Table 16-12
Space Heating, Commercial-Industrial and Owner Occupied Houses[a]—Atlanta (In Thousands of Tons)

Pollutant	1959	1964	1969	1980
Carbon Monoxide	.3	.3	.4	.7
Hydrocarbons[b]	.6	.6	.9	1.7
Nitrogen Oxides	3.5	6.1	7.9	13.5
Sulfur Oxides[c]	41.4	99.1	62.6	117.6
Particulates	1.3	1.8	2.5	4.4

[a]Includes fuel oil and natural gas.
[b]Includes aldehydes.
[c]Includes SO_3.

Source: Based on coefficients taken from the COMPILATION OF AIR POLLUTION EMISSION FACTORS, 1968, pages 5, 6, and 7, and on quantities of fuel oil consumed in SMSA with respect to percentages of residences using oil. Adjustments were made with respect to the number of degree days in SMSA. Projections to 1980 based on percentage increases from 1959 to 1969. Computations in accordance procedures outlined in Chapter 15.

Table 16-13
Emissions from Automobiles—Savannah (In Thousands of Tons)

Emissions	1959	1964	1969	1980 (a)	1980 (b)
Carbon Monoxide	36.2	43.3	48.6	11.2	20.0
Hydrocarbons	11.9	14.2	12.4	2.5	3.9
Nitrogen Oxides	4.2	5.1	5.7	4.8	9.1
Sulfur Oxides[a]	.2	.2	.2	.3	.3
Particulates[b]	.2	.3	.3	.4	.4

[a]9 lbs. per 1,000 gallons of gasoline burned.
[b]12 lbs. per 1,000 gallons of gasoline burned.

Source: Number of automobiles from Standard Rate and Data Service projected to 1980 in accordance with procedures outlined in Chapter 15. Emissions estimated by multiplying by appropriate emission factor in Table 15-2 in accordance with procedure of Chapter 15.
For 1980:
    1980 Column (A) based on Assumption 1:
      National standards for 1976 and 1971 met on schedule
    1980 Column (b) based on Assumption 2:
      National standards for 1971 highest achieved before 1980.
Distribution of car population for 1980 computations following Automobile Manufacturers Association data: 52.4 percent under 5 years; 35.6 percent 5 to 10 years; 12 percent over 10 years old.

Considering that the number of automobiles in this SMSA will grow from 52,830 to approximately 94,500 in 1980, based on previous percentage increases from 1959 to 1969, and taking into consideration that the increase in the growth of the population was only 27 percent from 1959 to 1969, this SMSA does not have any appreciable air pollution problems directly attributable to automobile emissions.

Electrical Power Generation

The major contributor of electrical power in this region is the Savannah Electric and Power Company. Gaseous emissions for electrical power generating stations in Savannah are illustrated in table 16.14 in total tons emitted in 1959, 1964, 1969, and projected emissions in 1980. It should be noted that until 1969, more coal was consumed in providing electrical power than either fuel oil or natural gas, but by 1969 there wasn't any coal used in any of Savannah's generating stations. Instead, 87 percent of the fuel was oil and 13 percent was natural gas. Presently Savannah Electric has two steam electric generating stations with an additional unit being made available in the Spring of 1971. A 2,700-acre site in adjacent Effingham was purchased for a third steam plant. The initial unit at this location is scheduled to be in service in 1974. The Riverside station plant, which is located in the downtown area of Savannah can burn either No. 6 oil or natural gas. Fort Wentworth station, approximately twelve miles up the Savannah River, is capable of burning coal, No. 6 oil, or natural gas. Natural gas has been the primary fuel used at the Riverside station with No. 6 oil being used as a secondary fuel at both plants, and coal being burned as a last resort. The No. 6 oil presently being used is imported from South America and has a fairly high sulfur content. Savannah Electric and Power Company claims that as soon as low sulfur oil becomes available in their region, they will switch all their stations and not consume any gas [10].

Space Heating

Air pollution problems from space heating units using natural gas in the Savannah SMSA are minimal. Table 16.15 includes quantities of effluents emitted from space heating units using both natural gas and fuel oil. If one were to choose which of the five SMSAs currently being reviewed has the least air pollution problems, Savannah would have to be this author's choice; and if past trends continue (i.e., if utility gas continues to increase in usage while fuel oil decreases), then Savannah's air pollution problems with respect to space heating units will decrease as well.

Conclusion[a]

[a]This section assumes Clean Air Act Standards are met on schedule. —ED.

Table 16-14
Gaseous Emissions for Electric Power Utilities[a]—Savannah (In Thousands of Tons) Tons)

Emissions	1959	1964	1969	1980
Carbon Monoxide	.03	.07	–	–
Hydrocarbons[b]	.01	.03	.03	.09
Nitrogen Oxides	2.0	3.4	3.4	9.5
Sulfur Oxides[c]	5.2	13.6	3.2	9.6
Particulates	1.5	3.0	.2	1.5

[a]Includes coal, fuel oil and natural gas.
[b]Includes aldehydes.
[c]Includes SO_3.
Source: Based on coefficients taken from the COMPILATION OF AIR POLLUTION EMISSION FACTORS, 1968, pages 5, 6, and 7, and on quantities of natural gas, fuel oil, and coal used for each of the prime power plants in New York's SMSA. Quantities of fuels obtained from the NATIONAL COAL ASSOCIATION'S STEAM ELECTRIC PLANT FACTORS, 1959, 1964, and 1969–Fuel consumption and costs, pages 5-28. Projections based on percentage increase from 1959 to 1969. Sulfur content of fuel differs with each year of electric power production. Average sulfur content used in compilation of emission factors. Computations in accordance procedures outlined in Chapter 15.
Note: By 1969, 87 percent oil and 13 percent gas was used; (therefore a total withdrawl of using coal).

Table 16-15
Space Heating, Commercial-Industrial and Owner Occupied Houses[a]—Savannah (In Thousands of Tons)

Pollutant	1959	1964	1969	1980
Carbon Monoxide	.2	.2	.1	.2
Hydrocarbons[b]	.2	.3	.3	.5
Nitrogen Oxides	1.0	1.2	1.5	2.3
Sulfur Oxides[c]	14.8	16.9	21.3	35.1
Particulates	.4	.5	.7	1.1

[a]Includes fuel oil and natural gas combustion.
[b]Includes aldehydes.
[c]Includes SO_3.
Source: Based on coefficients taken from the COMPILATION OF AIR POLLUTION EMISSION FACTORS, 1968, pages 5, 6, and 7, and on quantities of fuel oil consumed in SMSA with respect to percentages of residences using oil. Adjustments were made with respect to the number of degree days in SMSA. Projections to 1980 based on percentage increases from 1959 to 1969. Computations in accordance procedures outlined in Chapter 15.

New York

As exemplified in table 16.16, total emission of pollutants per square mile from major contributors in the New York SMSA was projected to be nearly the same in 1980 as it had been in 1959. Substantial decreases in carbon monoxide and hydrogen emissions were mostly due to a drastic reduction in the carbon monoxide emitted from automobiles in this area, but strict adherence to the Clean Air Act of 1970, coupled with efficient and effective means of inspecting automobiles, will be necessary in realizing these projected emissions by 1980. Any further reduction for minimization of carbon monoxide and hydrocarbons would require changes in the internal combustion engine design.

Additionally proposals for subsidizing unleaded gasoline are currently being reviewed to help to alleviate some of the air pollution problems attributed to the operation of automobiles. Whether the automobile manufacturers can meet the standards established for 1975 remains to be seen. Henry Ford II declared that the "Federal Government had set an impossible task for automobile manufacturers in meeting the established criteria levels, and major changes to public transportation facilities in most cities would necessitate massive rebuilding of metropolitan areas that would take generations to complete" [11].

Alternative propulsion systems for New York seem rather unlikely by 1980, therefore some limitation on vehicle use would be needed in helping to combat the air pollution problem.

Nitrogen oxides, sulfur dioxides, and sulfur trioxide emissions have noticeably increased from 1959 levels. Space heating and electric power utility emissions predominantly contributed to this increase, although most of the increase can be attributed to fuel usage in space heating units rather than for combustion in supplying electrical power for residents of the New York SMSA.

Table 16-16
Total Emissions from Major Contributors in New York (In Thousands of Tons)

Pollutant	1959	1964	1969	1980 (a)	1980 (b)
Carbon Monoxide	1504.7	1885.7	2089.9	483.7	871.2
Hydrocarbons[a]	501.6	629.2	547.3	140.9	204.6
Nitrogen Oxides	291.5	357.7	465.5	979.7	1166.9
Sulfur Oxides[b]	849.5	991.5	2051.1	527.4	527.4
Particulates	92.4	96.4	83.2	88.2	88.2
Total Emissions	3236.0	3957.0	5234.0	2219.9	2858.3
Total Pollutants Per Square Mile	1.51	1.85	2.45	1.04	1.34

[a]Includes aldehydes.
[b]Includes SO_3.
Source: Tables 16-1, 16-2, and 16-3.

Local restrictions previously mentioned for Con Ed and LILCO in utilizing low sulfur fuels have helped reduce the total tons of sulfur dioxide, and sulfur trioxide effluents emitted into the air, but increased usage of these fuels have overwhelmed reductions resulting in increased quantities of SO_2, SO_3, and nitrogen oxides for 1980.

A total of 1,040 tons of pollutants per square mile by 1980 is by far the most severe air pollution problem of the five SMSAs being reviewed. This is actually an understatement of New York's problem since most residents of this SMSA reside within thirty miles of the metropolitan center. The severity of the air pollution problem therefore can be amplified many times. Fortunately, New York is only ten feet above sea level and has favorable geographic conditions that tend to alleviate some of the problem.

Philadelphia

Philadelphia's projected total emissions from major contributors in 1980 shows increases in nitrogen oxides, sulfur dioxides, sulfur trioxides, and decreases in carbon monoxide and hydrocarbons. Conditions similar to New York with respect to air pollution problems also prevail for Philadelphia (table 16-17).

Carbon monoxide will decrease 83 percent from 1959 levels; particulate emissions will decrease less than 1 percent, but nitrogen oxide will increase 110 percent, and sulfur dioxides 122 percent from the 1959 levels.

The severity of Philadelphia's air pollution problem compared to New York, in terms of total pollutants per square mile from major contributors, is two-thirds less than New York's (1,040 tons vs. 440 tons).

Considerably fewer household, commercial, and industrial heating units, fewer automobiles, and electric utility that hopes to expand nuclear generation to 25-30 percent before 1980, coupled with increased use of low sulfur fuels, makes Philadelphia's outlook much brighter than that of New York's. Also assisting in minimizing Philadelphia's air pollution burdens is a favorable topographic condition representing an elevation of seven feet above sea level [12:82].

Temperatures for this region, coupled with other beneficial meteorological aspects, also play an important role in reducing the gravity of the air pollution problem, so that by 1980 total emissions of pollutants per square mile is actually below 1959 levels.

Pittsburgh

Table 16.18 also illustrates total emissions from major contributors for the Pittsburgh SMSA, along with total pollutants per square mile in 1959, 1964, 1969, and projected emissions for 1980.

With only one-third the number of automobiles that New York has, with the Duquesne Light and Power Company having erected 750-foot stacks for their generating stations, 99.5 percent efficient electrostatic precipitators, utilization

Table 16-17
Total Emissions from Major Contributors in Philadelphia (In Thousands of Tons)

Pollutant	1959	1964	1969	1980	
				(a)	(b)
Carbon Monoxide	806.2	959.6	1169.9	139.9	370.3
Hydrocarbons[a]	267.3	318.8	304.6	246.4	284.8
Nitrogen Oxides	149.1	181.4	226.7	314.6	427.7
Sulfur Oxides[b]	365.8	475.5	506.7	821.6	821.6
Particulates	50.8	61.8	52.3	50.0	50.0
Total Emissions	1639.0	1997.0	2260.0	1573.0	1954.4
Total Pollutants Per Square Mile	.46	.56	.63	.44	.55

[a]Includes aldehydes.
[b]Includes SO_3.
Source: Tables 16-4, 16-5, and 16-6.

Table 16-18
Total Emissions from Major Contributors in Pittsburgh (In Thousands of Tons)

Pollutant	1959	1964	1969	1980	
				(a)	(b)
Carbon Monoxide	465.6	550.1	647.4	73.7	193.4
Hydrocarbons[a]	156.4	184.7	170.0	130.0	149.5
Nitrogen Oxides	93.5	113.1	124.8	131.7	189.4
Sulfur Oxides[b]	209.7	259.0	266.5	344.6	344.6
Particulates	40.2	50.4	50.9	12.4	12.4
Total Emissions	965.0	1157.0	1260.0	692.0	889.3
Total Pollutants Per Square Mile	.31	.37	.41	.22	.28

[a]Includes aldehydes.
[b]Includes SO_3.
Source: Tables 16-7, 16-8, and 16-9.

of low sulfur oil fuel, and a new wet scrubber system for removing both particulate matter and sulfur dioxide, Pittsburgh does not seem to encounter severe air pollution difficulties at first glance.

In 1980 pollution levels will probably be below 1959 levels in terms of total emissions, and in terms of the total pollutants per square mile. Nitrogen oxides will show a 40 percent increase while sulfur dioxides will surge ahead by 64 percent.

Carbon monoxide and particulate emissions will have decreased substantially from 1959 levels, representing an 84 percent and a 70 percent decrease respectively, while hydrocarbons will have decreased only 18 percent from 1959.

In 1980, 220 tons of pollutants per square mile will be realized, but remembering that Pittsburgh's SMSA has 3,049 square miles illustrates that this does not necessarily portray a clear picture of the air pollution problem in the concentrated urban areas, and in areas along the valleys of the rivers.

Pittsburgh's high elevation level of 1,151 feet above sea level [12:83], linked with more rigorous meteorological conditions, amplifies the problem of air pollution for this region rather than diminishing it, but Pittsburgh's difficulties in combating its problem can be minimal if previously established standards are adhered to, and if use of low pollutant fuels in larger quantities is an actuality by the mid-1970s.

Atlanta

Table 16.19 shows the total emissions from major contributors for Atlanta's SMSA, and total pollutants per square mile for 1959, 1964, and 1969, along with projected emissions for 1980.

Atlanta will have a 48 percent increase in hydrocarbons from 1959 levels by 1980; a 545 percent increase in nitrogen oxides, and a 423 percent increase in sulfur oxides. Decreases of 70 percent for carbon monoxide and 14 percent for particulates will also occur by 1980.

While the total tonnage of effluents emitted seems small compared to New York, Philadelphia, and Pittsburgh, Atlanta could be heading toward a serious air pollution problem by 1980.

Rapid urban development over the past decade has enabled a "multiplier effect" to compound Atlanta's air pollution problems. Two of the critical areas are: (1) automobile air pollution; (2) power plant pollution. Even with well-designed highways, sheer numbers of automobiles have added to the air pollution difficulties of the region (264,920 cars in 1959 vs. 1,116,000 cars projected for 1980—if past growth trends continue).

Georgia Power Company utilizes considerably more coal than either fuel oil or natural gas, both of which are used "sparingly." Emissions from space heating units are not considered a major problem, especially since 84 percent of the space heating units use natural gas as a fuel.

The major problems lie in the areas of rapid urban development, and large emissions of sulfur oxides from power plants; even though a gradual reduction of sulfur oxide emissions will be accomplished by reducing the sulfur content of the fuels used, substantial usage nullifies any reduction of the sulfur content. Meteorological and topographical features do not help to alleviate Atlanta's air pollution issues either, and with an elevation of 975 feet above sea level, the "pockets of pollution" do not have an opportunity to difuse and dilute [12:33] into the outlying areas. Atlanta's SMSA, with a total of 1,728 square miles, shows an increase from 280 tons of pollutants per square mile in 1959 to 870 forecasted tons of effluents in 1980.

Table 16-19
Total Emissions from Major Contributors in Atlanta (In Thousands of Tons)

Pollutant	1959	1964	1969	1980 (a)	1980 (b)
Carbon Monoxide	219.2	308.1	410.6	66.8	153.8
Hydrocarbons[a]	72.6	101.9	135.8	108.7	123.2
Nitrogen Oxides	49.7	80.1	117.3	271.1	313.5
Sulfur Oxides[b]	118.8	228.0	335.2	500.2	500.2
Particulates	28.7	43.8	83.9	24.5	24.5
Total Emissions	489.0	762.0	1083.0	971.3	115.2
Total pollutants per square mile	.28	.44	.62	.87	.99

[a]Includes aldehydes
[b]Includes SO_3
Source: Tables 16-10, 16-11, 16-12.

Hopefully, considerable quantities of low sulfur oil and natural gas will be used in the future in providing electrical power for the residents of this SMSA, which will help in reducing the amount of sulfur dioxides being emitted into the atmosphere.

Savannah

Table 16.20 is a representation of the total emissions from major contributors, linked together with the total pollutants per square mile in 1959, 1964, 1969, and projections for 1980.

Savannah has the slightest air pollution problem of the five SMSAs reviewed. This SMSA is not only the smallest in size (445 square miles), has the least number of residents (65,430 in 1969), but also has the fewest tons of effluents emitted into the air from the major sources of pollution. It is estimated that the total pollutants will have increased from 160 tons to 170 per square mile in 1980.

Statistically, it may seem that Savannah has an enormously complex air pollution problem considering the fact that nitrogen oxides will increase 446 percent, sulfur dioxides 120 percent, and particulates 49 percent from 1959 levels. This is not the case, however, since Savannah is dealing in much smaller numbers than any of the other four SMSAs.

Carbon monoxide emissions will decrease 90 percent, and hydrocarbons will decrease 25 percent in 1980 from 1959 levels. Considering the diminutive amount of effluents emitted, and the fact that Savannah's meteorological and topographical situation (48 feet above sea level) benefits the air pollution

Table 16-20
Total Emissions from Major Contributors in Savannah (In Thousands of Tons)

Pollutant	1959	1964	1969	1980 (a)	1980 (b)
Carbon Monoxide	36.3	43.4	48.8	2.6	11.4
Hydrocarbons[a]	12.9	14.5	12.9	9.6	11.0
Nitrogen Oxides	3.0	9.6	10.6	16.8	21.1
Sulfur Oxides[b]	20.1	30.8	25.5	45.4	45.4
Particulates	2.1	3.8	1.2	3.0	3.0
Total Emissions	74.0	102.0	99.0	77.0	91.9
Total Pollutants Per Square Mile	.16	.23	.22	.17	.20

[a]Includes aldehydes.
[b]Includes SO_3.
Source: Table 16-13, 16-14 and 16-15.

problems, the future for Savannah looks quite bright, especially when considering that Savannah Electric and Power Company is using 87 percent oil and 13 percent natural gas in providing electrical power for this area, and has not used coal since 1964.

The four other SMSAs analyzed should only have as slight a problem with air pollution as does Savannah!

Notes

1. *New York Times,* January 9, 1971; special dispatch to the *New York Times* by Agis Salpukas.

2. *New York Times,* "City Gains in Controlling Air Pollution," March 8, 1970, p. 23.

3. *Restoring the Quality of Our Environment,* Report of the Environmental Pollution Panel, President's Science Advisory Committee, November 1965, p. 67.

4. *Edison Electric Institute Pamphlet,* 1969-70 edition (New York, New York: Edison Electric Institute), p. 3.

5. E. W. Kenworthy, "Air Quality Rules for 6 Pollutants Given for Nation," *New York Times,* May 5, 1971, p. 1.

6. Commonwealth of Pennsylvania, Department of Health, *Proposed Implementation Plan for Philadelphia Air Quality Control Region* (Harrisburg, Pennsylvania, 1970), p. 1.

7. Commonwealth of Pennsylvania, Department of Health, *Southwestern Regional Planning Board,* 1970, p. 2.

8. Georgia Power Company, *Annual Report 1969,* Generating and Primary Transmission System, p. 53.

9. Georgia Power Company, *Supplementary Report on the Environment, Air Pollution and the Georgia Power Company,* 1969.

10. Special report prepared for author by W. C. Smith, Vice-President of Operations, Savannah Electric and Power Company, December 20, 1970.

11. Jerry M. Flint, "Ford Calls Clean-Air Goal Impossible Task in Cities," *New York Times,* May 27, 1971, p. 66.

12. H. M. Conway (Ed.), *The Weather Handbook* (Atlanta, Georgia: Conway Publications, 1963).

17 Air Pollution in Five Midwestern Cities

FRANK CAULSON

Introduction

Until the recent past, the air we breathe was regarded as a free and unlimited commodity that was to be used, even exploited, indiscriminately. The consequences of this policy have created a crisis that requires immediate remedial action and a rethinking and reevaluation of our traditional priorities.

Man has only himself to blame for the immensity of the present problem and the future consequences of air pollution. Although a completely pollution-free atmosphere never existed, man's increasing production and concomitant consumption of the world's resources has magnified a natural occurence to the point where it now threatens to become the ultimate man-made disaster.

During 1968 air pollution emissions in the United States amounted to 214 million tons [27:3] — more than one million tons of airborne debris for each member of the population. These emissions can be classified into the two broad categories of primary and secondary emissions.

Primary emissions include fine and coarse particulates, carbon compounds, halogen compounds, sulfur compounds, organic compounds, nitrogen compounds, and radioactive compounds.

Secondary pollutants may be described as an atmospheric reaction between two or more chemical compounds. The dissociation of NO_2, providing NO and O radicals, causing photochemical smog, is an example. The breakdown of sulfur trioxide (SO_3) when combined with water in the atmosphere to form sulfuric acid mist (H_2SO_4) is another. These secondary pollutants produced by such atmospheric combinations are among the most troublesome and hazardous types of air pollution.

Another determinant of the severity of a particular area's pollution problem is the meteorological conditions. Some cities, such as Los Angeles and Denver, are cursed with an air envelope that magnifies their problems. Others, notably New York and Boston, are blessed with strong winds that mitigate the effects of air pollution. This may be a mixed blessing since an incipient problem may be either ignored or unrecognized until it reaches such proportions that clean air is an impossibility.

This study will attempt to investigate some of the causes of air pollution in five American cities and project the levels of pollution that will exist in these cities in 1980. The automobile, electric power generation, and space heating are the three general categories that will be investigated.

Automobile Air Pollution

The conventional automobile engine emits air pollutants from four sources: engine exhaust, crankcase blowby, carburetor, and the fuel tank. Emissions are mainly hydrocarbons, carbon monoxide, and nitric oxides. About 60 percent of the unburned hydrocarbons come from engine exhaust, 20 percent escape from crankcase blowby, and 20 percent result from evaporation in the carburetor and the fuel tank [15, Sec. 2:12]. Nearly 100 percent of carbon monoxide and nitric oxide emissions are vented to the atmosphere from the automobile's exhaust.

Carbon monoxide pollution from automobile sources amounted to 59.2 million tons in 1968 and 70 percent of these emissions occurred in urbanized areas. Oxides of nitrogen pollution result from the reaction of the nitrogen and oxygen contained in the combustion air at the high temperature prevailing during combustion. Both nitrogen oxides and carbon monoxide pollution are closely related to the distribution of population in the country since motor vehicles contribute significantly to total emissions.

It is these components of automobile pollution that have been virtually the exclusive subject of legislative restrictions on automobile pollution. The others, including particulates and sulfur oxides, have not been important as far as automobile emissions are concerned. It is likely that they will continue to be emitted by cars at a constant or slightly increasing rate.

Tables 17.1 and 17.2 are the coefficients of automobile pollution for this study. From these bases it becomes a relatively simple exercise in multiplication and division to determine the amount of emissions in any given year.

Table 17-1
Estimated Automobile Emissions—Selected Years (Pounds/Car/Year)

	Uncontrolled Vehicles Pre-1968	1968	National Standards[a] 1971	1976
Hydrocarbons	450[b]	170	75	14
Carbon Monoxide	1370	630	410	41
Nitrogen Oxide (NO_2)	160	200	200	20

[a]Estimated on basis of public law 91-604, 91st Congress, December 1970, which requires that hydrocarbon and carbon monoxide emissions be reduced 90 percent from the standards of 1970. The act also requires that nitric oxide emissions be reduced 90 percent below the standards of 1971.

[b]Applies only to pre-1963 cars. Arbitrarily reduced to 350 for 1969 estimates.

Source: U.S. Department of Health, Education and Welfare. Public Health Service Environmental Health Service, National Air Pollution Control Administration, CONTROL TECHNIQUES FOR CARBON MONOXIDE, NITROGEN OXIDE, AND HYDROCARBON EMISSIONS FROM MOBILE SOURCES (Washington, D.C.: Government Printing Office, 1970), Section 3, p. 8.

Table 17-2
Emission Factors for Automobile Exhaust

Type of Emission	Emissions (pounds per 1,000 gals. of gas)
Sulfur Oxides (SO_2)[a]	9
Particulates[a]	12

Average Consumption of Gasoline Per Vehicle[b]			
1959	1964	1969	1980
666	652	718	773

Source:
(a) U.S. Department of Health, Education and Welfare, Public Health Service, Bureau of Disease Prevention and Environmental Control, National Center for Air Pollution Control, COMPILATION OF AIR POLLUTANT FACTORS (Durham: Government Printing Office, 1968), p. 50.
(b) Years 1959-1969 U.S. Department of Transportation, Federal Highway Administration Bureau of Public Roads, HIGHWAY STATISTICS, "Estimated Motor Vehicle Travel in the U.S. and Related Data" (Washington: Government Printing Office, 1959), pp. 65, 70.
1980–Estimated. The Statistical Abstract of the United States, 1970 edition states that the population of the United States in 1959, 1964, and 1969 was 177,830,000, 192,120,000 and 203,213,000 for the respective years. Series P-25 of the Bureau of Census current population reports "Population Estimates and Projections" estimates that the U.S. population in 1980 will be 232,412,000. Dividing total automobile population by total population establishes a persons per vehicle ratio. Dividing total consumption of gasoline by population establishes a gallons/person ratio. Multiplying persons/vehicle by gallons/person establishes gallon/car.

	1959	1964	1969	1980
Persons/Vehicle	2.9	2.6	2.4	1.8*
Gallons/Person	225.8	247.6	306.7	418*
Gallons/Car	666	652	718	773

*Between 1959-69 P/V decreased by .7910. Assuming that decrease in the next ten years we arrive at 1980 estimate. Between 1959-69 gallons/per person increased by 1.850. Assuming that increase in the next ten years we arrive at the 1980 estimate of Gallons/Person. Dividing 1980 population by Persons/Vehicle establishes total automobile population in 1980.

In 1980 the automobile population will obviously consist of both old and new cars. To make a reliable estimate of automobile air pollution, a means of determining the makeup of the automobile population is required. In 1969 52.4 percent of U. S. automobiles were under five years old, 35.6 percent were more than five but less than ten years old, and 12 percent were ten years old or older [46:22]. Assuming the population mix in 1980 is to be the same as 1969, we can apply the national standards of 1976 to 52.4 percent of the automobiles, the national standards of 1971 to 35.6 percent of the automobiles, and the national standards of 1968 to the remaining 12 percent of the automobile population.

As the protest against air pollution has become more intense, alternative sources of powering the automobile have been increasingly investigated. The steam engine, the Wankel engine, the gas turbine engine, and the electric-powered car are the best known. Undoubtedly, these and other experimental engines are possible alternatives to the internal combustion engine. However, it is highly unlikely that there will be any substantial replacements of the gasoline-powered engine by 1980. According to a recent article in the *New York Times*, Detroit predicts that Americans will still be driving cars powered by the internal combustion engine well into the 1980s [40]. This study will therefore discount the effects on pollution that these prototype engines might make.

Electricity

Traditionally, the energy inputs in electrical generation have been the fossil fuels—coal, oil, and natural gas. An alternative source of energy is nuclear power, which has been inconsequential in the past but will become more significant in the future. A concomitant to energy usage is the resultant pollution that occurs. Power plants using coal or oil to drive the generators will cause significant sulfur oxide pollution while natural gas combustion results in high emissions of nitric oxides. With the exception of those generators being powered by nuclear energy, which is free of conventional pollution, all cause particulate pollution. Although utilities might use multiple energy sources in generating electricity, the preponderant source of energy is coal.

Approximately 74 percent of the 33 million tons of sulfur oxides emitted into the atmosphere of the United States in 1968 resulted from the combustion of sulfur-bearing fuels [27:9]. Coal combustion accounted for the largest part of this total. Power plants emitted 16.8 million tons or 50.6 percent of this total. These emissions are unevenly distributed throughout the country. Over 60 percent of the sulfur oxides from the burning of coal by utilities is emitted in eight states of the East North Central and East South Central regions [64:2]. In the Middle Atlantic and South Atlantic states utility fuel used inland is generally high-sulfur coal; in a shallow region along the coast, relatively large quantities of residual oil that is also high in sulfur content are used.

Throughout most of the West and Southwest the principal utility fuels are natural gas and low-sulfur coal; and the sulfur oxides problem is thus far less acute but there is an increased nitric oxides problem. Tables 17.3 through 17.6 indicate the total historical and projected electric generation—by type of fuel—for the areas investigated in this chapter. Data in these tables are the basis for projecting the increase in electrical generation that will occur between 1969 and 1980.

Another step in building a model of air pollution from power generation is determining the amount and type of fuel consumed by a particular utility. The

Table 17-3
Electric Generation by Type of Fuel 1966-1980 S East Region

Thermal Generation	1966 Billion Kwh	%	1970 Billion Kwh	%	1975 Billion Kwh	%	1980 Billion Kwh	%
Coal	161.1	73.6	218.1	74.7	245.8	57.0	259.4	41.8
Oil	18.0	8.2	14.7	5.0	12.4	2.9	12.4	2.0
Gas	14.8	6.8	23.2	7.9	22.8	5.3	24.6	4.0
Nuclear	—	—	7.2	2.5	114.5	26.6	286.7	46.3
Int. Comb.	—	—	.3	.1	.1	—	.1	—
Total	193.9	88.6	263.5	90.2	˙395.6	91.8	583.2	94.1

Source: Federal Power Commission. SOUTH EAST POWER SURVEY 1970-1990: A REPORT PREPARED BY THE SOUTHEAST REGIONAL ADVISORY COMMISSION (Washington: Government Printing Office 1969), Section 6, p. 15. (Mobile, Alabama is located in SE Region.)

Table 17-4
Electric Generation by Type of Fuel 1966-1980 West Central Region

	1966 Billions Kwh	%	1970 Billions Kwh	%	1975 Billions Kwh	%	1980 Billions Kwh	%
Thermal Generation								
Coal	99.9	72.2	124.8	67.2	146.3	55.2	142.4	38.4
Gas	23.9	17.3	30.0	16.2	35.0	13.2	34.5	9.3
Oil	.9	.7	1.1	.6	1.3	.5	1.5	.4
Nuclear	1.4	1.0	17.5	9.4	69.5	26.2	178.4	48.2
Total	126.1	91.2	173.4	93.4	252.1	95.1	356.8	96.3

Source: Federal Power Commission, WEST CENTRAL REGIONAL POWER SURVEY 1970-1990: A REPORT TO THE FEDERAL POWER COMMISSION. PREPARED BY THE WEST CENTRAL REGIONAL ADVISORY COMMITTEE (Washington, D.C.: Government Printing Office, 1969), Section 2, p. 5. (Chicago, Illinois and St. Louis, Missouri are located in the West Central Region.)

Table 17-5
Electric Generation by Type of Fuel 1966-1980 South Central Region

	1966	1970	1975	1980
Thermal Generation (Millions of kwhr)	116,033	172,118	264,824	405,662
% Coal	4.56	4.28	9.18	10.61
% Oil	.03	.07	.32	.20
% Gas	95.22	95.47	85.15	71.14
% Nuclear	–	–	5.24	17.98
% Internal Comb.	.19	.18	.11	.07

Source: Federal Power Commission, Electric Power in the South Central Region 1970-1980-1990: A REPORT TO THE FEDERAL POWER COMMISSION PREPARED BY THE SOUTH CENTRAL REGIONAL ADVISORY COMMITTEE (Washington, D.C. Government Printing Office, 1969), p. 14. (New Orleans, Louisiana is located in the South Central Region.)

Table 17-6
Electric Generation by Type of Fuel 1966-1980 East Central Region

Thermal Generation	1966 Billion Kwh	%	1970 Billion Kwh	%	1975 Billion Kwh	%	1980 Billion Kwh	%
Coal	199.7	98.2	244.2	96.9	279.9	81.4	348.9	74.8
Oil	.2	.1	.3	.1	.1	–	.1	–
Gas	.2	.1	.4	.1	.2	–	.3	.1
Nuclear	.9	.4	4.3	1.8	57.5	16.8	110.0	23.6
Int. Comb.	.1	.1	.4	.1	.6	.1	.5	.1
Total	201.1	98.9	249.6	99.0	338.3	98.3	459.8	98.6

Source: Federal Power Commission, ELECTRIC POWER IN THE EAST CENTRAL REGION 1970-1980-1990: A REPORT TO THE FEDERAL POWER COMMISSION, Prepared by the East Central Regional Advisory Committee (Washington: Government Printing Office, 1969), Section 3, p. 4. (Cleveland, Ohio is located in the East Central Region.)

National Coal Association publishes an annual edition of *Steam-Electric Plant Factors* which provides a detailed breakdown of fuel consumed by most of the private utilities in the United States. This data established the historical base for determining the amount of pollution emitted by a particular utility. Table 17.7 gives the coefficients of pollution used in this study.

Table 17-7
Emission Factors for Electrical Generation

Pollutant	Pounds/Ton of Coal Burned	Pounds/Million Cu ft. of Gas Burned	Pounds/1000 Gals. of Oil Burned
Aldehydes	0.005	1	0.6
Carbon Monoxide	0.5	Neg.	0.04
Hydrocarbons	0.2	Neg.	3.2
Oxides of Nitrogen	20.	390	104.
Sulfur Dioxide	38S[a]	0.4	157S[a]
Sulfur Trioxide	Neg.	Neg.	2.4S
Other Organics	Neg.	3	Neg.
Particulate	16A[b]	15	10

Source: U.S. Department of Health, Education and Welfare Public Health Service, Bureau of Disease Prevention and Environmental Control. National Center for Air Pollution Control. COMPILATION OF AIR POLLUTANT EMISSION FACTORS (Durham: Government Printing Office, 1968), pp. 4-7.

[a]S equals percent sulfur in coal (e.g., if sulfur content of coal is 2 percent there will be 76 pounds of sulfur oxides formed per ton of coal burned). Estimated on basis of correspondence with company official or estimated on basis of information in Regional Power Studies by Federal Power Commission.

[b]A equals ash content in coal estimated at 8.55 percent.

Space Heating

This chapter will attempt to assess air pollution associated with space heating in two categories: that caused by residential space heating (single-unit houses) and commercial-industrial space heating. Two sources of energy, oil and natural gas, will be investigated. Of the other sources of energy, two—coal and LP gas—are insignificant factors in space heating. The other, electricity, is considered as a nonpollutant in space heating because its impact has been investigated at the source of contamination—the power plant.

There are multiple difficulties in attempting to determine the contribution of space heating to air pollution. As a rule, no source of energy enters the home, office, or plant earmarked for one use only. Statistics are compiled at the point of wholesale, or at best retail, distribution without distinction as to use. Where a source comes nearest to meeting a single use, as is the case for oil in space heating, the quantities used in homes and those used in commerce or industry are not separately accounted for. The problem is not insurmountable, but its solution requires multi-step exercises in deductive reasoning and qualified assumptions.

The minimal requirements for building a model of space heating air pollution are establishing the amount of fuel consumed in a given year and determining the gross national consumption of fuel on some per capita basis.

Once gross national consumption has been established, a procedure is necessary to adjust consumption to the climates of a localized area. Failure to adjust consumption would result in an unwarranted assumption that the same amount of fuel is consumed in a house or plant located in Bismarck, North Dakota, and one of the same proportions located in Miami, Florida.

Of the two fuels, oil lends itself more readily to refinement in the statistical determination of use. This study assumed that all of grade 1 and four-ninths of grade 2 distillate heating oil was consumed in residential space heating. The basis for this assumption was Landsberg and Schurr's finding that the commercial-residential consumption of energy approaches a ratio of 1.25/1 [49:15]. (Since this study brings apartments and other multi-unit housing under the category of commercial establishments, we could not assume that all of grade 2 distillate was consumed in single-unit houses.) The remainder of grade 2 distillate heating oil and all of grades 4, 5, and 6 heating oil were assumed to be consumed in commercial-residential space heating.

Distinguishing between the proportionate amounts of natural gas consumed in the household sector and that consumed in commerce and industry is more difficult. Not only is there no distinction as to use, but the American Gas Association's customer classification overlaps — service supplied to five or more households served as a single customer on one meter is considered as commercial. To resolve the latter, we relied on Census of Housing data. In 1960 total housing units numbered 58.3 million. Of these 9.34 percent were households of five units or more [65]. Adjusting the AGA's commercial category by this amount and adding the resultant figures to the household category gives us a more reliable figure of household consumption of gas.

Much of the gas energy consumed in commerce and industry furnishes furnace heat or uses steam heat as a direct element in processing, for example to cook, bake, dry, sterilize, etc. The remainder is used in space heating. We are thus forced to make two broad assumptions regarding the AGA's classification of commercial and industrial customers. We assume that 55 percent of energy classified by the AGA as commercial is used for space heating. The second assumption is that only 20 percent of gas classified as industrial by the AGA is burned for space heating purposes. These assumptions are little more than rough guesses and to the extent that they are inaccurate, to that degree will our estimation of commercial industrial gas space heating air pollution be erroneous.

The fuel entering the home is utilized for purposes other than space heating. However, the process that converts the fuel to a specific functional use (space heating, washers, dryers, stoves, etc.) is usually attached to a single unit or is vented to the atmosphere through a common exhaust system. Consequently, all gas and oil usage in the house is classified under the broad category of space heating.

As stated earlier, our estimation of residential space heating is limited to single-unit houses; multiple residences are lumped under the category of commercial-industrial space heating air pollution. Numerous household statistics are compiled, all sharing a common trait—a failure to distinguish between single-unit (houses) and multi-unit structures (apartments). In this chapter it is assumed that the classification of owner-occupied households by the Department of Commerce represented single-family houses and nonowner-occupied households represented apartments.

The Census of Housing supplied the means of estimating the number of houses using oil or gas as heating fuel in 1959. For subsequent years the figure was adjusted by the percentage increase or decrease in fuel usage as determined by national sales during the decade of the sixties.

It is assumed here that owner-occupied households represents single-unit houses, and that this figure multiplied by the percentage of oil or gas heated houses produces the number of houses using gas or oil. Dividing the fuel consumption by gas- or oil-using houses results in gross consumption of fuel consumed per house on a nationwide basis.

After gross consumption had been determined, it was only necessary to devise some means of adjusting gross national consumption to a localized area. The author felt that the degree day concept was a good determinant. Thirty year normals for one hundred cities were averaged, resulting in an average number of degree days for the nation. The degree days for a particular city were then measured against this average and the gross consumption of fuel adjusted accordingly.

Table 17-8
Residential Consumption of Natural Gas and Heating Oil 1959-1980

	Gas (thousands of therms)	Oil (thousands of barrels)
1959	30,511,938	195,568
1964	39,886,340	204,051
1969	49,751,297	242,764
1980	62,457,680	309,651

Source:

Gas 1959-1969—American Gas Association. AMERICAN GAS ASSOCIATION MONTHLY, Vols. 43 p. 5; 48, p. 5; 52, p. 4, New York 1961, 1966, 1970. Adjusted as explained above.

1980—Estimated assuming continuation in next decade of rates of increase which prevailed prior to 1969.

Oil 1959-1964—National Petroleum News ANNUAL FACTBOOK ISSUE, Vols. 53, p. 180, 57, p. 180, New York 1961, 1966. Adjusted as explained above.

1969—U.S. Dept. of the Interior, Bureau of Mines, MINERAL INDUSTRIAL SURVEY: FUEL OIL SALES, ANNUAL (Washington: Government Printing Office). Adjusted as explained above.

1980—Estimated—assuming continuation in next decade of rates of increase which prevailed prior to 1969.

Table 17-9
Heating Degree Days (65° F. Base) for 15 Selected U.S. Cities, Thirty Year Normals, 1931-1960

State	City	Normal Annual Degree Days
Alabama	Mobile	1,560
California	Los Angeles	1,799
	San Francisco	3,012
Colorado	Denver	6,283
Georgia	Atlanta	2,983
	Savannah	1,819
Illinois	Chicago	6,155
Louisiana	New Orleans	1,385
Missouri	St. Louis	,4900
New York	New York	4,871
Ohio	Cleveland	6,196
Pennsylvania	Philadelphia	5,101
	Pittsburgh	5,987
Washington	Seattle	5,145

Source: American Gas Association, 1969 GAS FACTS: A STATISTICAL RECORD OF THE GAS UTILITY INDUSTRY, New York, 1970, p. 128.

In summary, we established the amount of fuel consumed in the household sector, determined the number of single-unit houses in the nation, established national gross consumption of fuel per house, and adjusted the gross consumption in each localized area to allow for temperature differences.

Commercial-Industrial Space Heating

The method of establishing a commercial-industrial index of space heating pollution parallels that established for residential space heating. The procedure is to establish total national consumption and allocate it to local areas. Rather than attempt to ascertain consumption per building we will establish a per capita consumption based on the number of the employees in the United States for the specified years.

To determine gross national consumption per employee one need only: (a) establish the amount of fuel consumed in the commercial-industrial sector; (b) determine the number of employees in the United States in this sector—for this study *County Business Patterns* was the data source; and (c) devise a method of determining the division of fuel usage between gas and oil on a national basis.

Table 17.12 is the basis of determining the percentage consumption of gas and oil on a national basis.

To adjust gross consumption per employee, multiply by data in table 17.11 and multiply the result by number of employees in the appropriate area. As

Table 17-10
Proportionate Degree Days Selected Cities

Mobile	.318
Chicago	1.256
New Orleans	.283
St. Louis	1.000
Cleveland	1.264

Source: Table 17-9, American Gas Association, 1969 GAS FACTS: A STATISTICAL RECORD OF THE GAS UTILITY INDUSTRY, New York, 1970.

Table 17-11
Estimated Commercial-Industrial Consumption of Natural Gas and Heating Oil in Space Heating 1959-1980

	Gas (thousands of therms)	Oil (thousands of barrels)
1959	13,252,134	304,900
1964	18,174,063	336,302
1969	25,942,138	419,424
1980	46,789,061	606,771

Source:
Gas 1959-1969–American Gas Association, AMERICAN GAS ASSOCIATION MONTHLY, Vols. 43, p. 5; 48, p. 5; 52, p. 4, New York 1961, 1966, 1970. Adjusted as explained above.
1980–Estimated assuming continuation in next decade of rates of increase which prevailed prior to 1969.
Oil 1959-1964–National Petroleum News, ANNUAL FACTBOOK ISSUE, Vols. 53, p. 180; 57, p. 180, New York 1961, 1966. Adjusted as explained above.
1969–U.S. Department of the Interior, Bureau of Mines, MINERAL INDUSTRIAL SURVEYS: FUEL OIL SALES ANNUAL (Washington: Government Printing Office). Adjusted as explained above.
1980–Estimated assuming continuation in next decade of rates of increase which prevailed prior to 1969.

above, the employee data were obtained from *County Business Patterns*. We assumed that the proportion of gas/oil users in a *localized* area was the same as that developed for residential heating.

Table 17-12
Gross Consumption of Energy Resources 1959-1980* Trillions of BTU

	1959	%	1964	%	1968	%	1980	%
Gas	9713.6	54.3	12,711	56.6	15,709	58.7	25,401	62.9
Petroleum	8176.7	45.7	9,354	43.4	11,055	41.3	14,924	37.1
Total	17,890.3	100.0	22,065	100.0	26,764	100.0	40,325	100.0

Source:
1959-1964—American Petroleum Institute, PETROLEUM FACTS AND FIGURES, 1967 edition, New York, p. 250.
1968—U.S. Bureau of Mines, MINERAL YEARBOOK 1968, Vol. II (Washington: Government Printing Office, 1970), p. 24.
1980—Estimated, assuming continuation in next decade of rates of increase which prevailed in the past.
*Household, Commercial-Industrial Segments Only—excluding electricity and coal.

Chicago, Illinois

For the Chicago area, both daily and seasonal wind direction patterns are related to differences in air quality levels. Many days of high suspended particulate levels have been correlated to sustained wind directions conducive to pollutant

Table 17-13
Automobile Air Pollution Chicago, 1959-1980 (Thousands of Tons/Yr.)

	1959	1964	1969	1980 (a)	1980 (b)
Hydrocarbons	323	370	339	93	138
Carbon Monoxide	982	1,126	1,323	417	748
Nitric Oxides	115	131	155	181	343
Sulfur Oxides	4	5	6	12	12
Particulate	6	6	8	16	16
Total	1,430	1,638	1,831	719	1,257

Source: Number of automobiles from Standard Rate and Data Service projected to 1980 on basis of projected population increase and persons/per vehicle ratio as explained in footnote to Table 17-2.
Emissions estimated by multiplying by appropriate emission factor in Table 15-2 in accordance with procedure of Chapter 15.
For 1980:
    1980 Column (a) based on Assumption 1:
      National standards for 1976 and 1971 met on schedule
    1980 Column (b) based on Assumption 2:
      National standards for 1971 highest achieved before 1980.
Distribution of car population for 1980 computations following Automobile Manufacturers Association data: 52.4 percent under 5 years; 35.6 percent 5 to 10 years; 12 percent over 10 years old.

exposure from specific point sources and source areas. On a seasonal basis northwest winds, which are predominant during the heating season, have the effect of lowering particulate emissions on the northern and western fringes of the city while raising the levels of the inner city.

Yearly wind speed in Chicago averages 10.2 miles per hour. The effect of wind speed on pollution levels varies. When winds are below 6 miles per hour conditions favor buildup of pollutants in the air. When wind speed is excessive downwash of emissions from elevated stacks cause increased pollution downwind.

A stable atmosphere is often created in the late spring by easterly winds moving in across Lake Michigan. As air passes over the cold water surface it becomes stable in the lower layer and any air pollutants released into the atmosphere undergo little mixing. As the air passes over the warm land, thermal mixing breaks up the stability near the surface which results in low suspended particulate levels at shore line but higher levels at outer city and peripheral suburban areas.

Another contribution to the area's pollution levels is the industrial complex that stretches eastward along the Lake Michigan shorefront to Gary, Indiana.

Table 17-14
Electric Power Air Pollution Chicago 1959-1980 (Thousands of Tons/Yr.)

	1959	1964	1969	1980
Hydrocarbons[a]	1	1	1	1
Carbon Monoxide	2	3	3	3
Nitric Oxides[b]	77	114	137	151
Sulfur Oxides	393	619	737	785
Particulate[c]	76[d]	118[d]	143[d]	10[e]
Total	549	855	1021	950

Source: Table 7 p. 12. Fuel Consumption 1959-1969 for Fisk, Ridgeland, Crawford, Northwest, Calumet, Waukegan, Joliet, plants of Commonwealth Edison Company determined from STEAM PLANT FACTORS, National Coal Association, Washington 1959, 1965, 1970 editions. 1980 estimated on basis of projected increase of 205.76 percent in generation per table 4 assuming that 48.3% of this generation will result from nuclear power.
[a]Includes aldehydes and other organics.
[b]Includes NO_2 and NO_4.
[c]Includes SO_2, SO_3.
[d]Assumes Precipitators operating at 84% efficiency.
[e]Assumes Precipitators operating at 99% efficiency.

Table 17-15
Space Heating Air Pollution Chicago 1959-1980: Residential/Commercial/Industrial (Thousands of Tons Per Yr.)

	1959	1964	1969	1980
Hydrocarbons[a]	1	1	1	1
Carbon Monoxide	1	d	d	d
Nitric Oxides[b]	22	21	25	32
Sulfur Oxides[c]	68	57	63	63
Particulate	5	5	5	6
Total	97	84	94	102

Source: Household Data Standard Rate and Data Service. SPOT TELEVISION RATES AND DATA, Skokie, Illinois 1959, 1964, 1969. Employee Data U.S. Department of Commerce, County Business Patterns, 1980 estimated assuming rate of increase in housing and employee population of previous decade, procedure described in Chapter 15.
[a]Includes Aldehydes.
[b]Includes NO_2, NO_4.
[c]Includes SO_2, SO_3.
[d]Less than one-half.

Table 17-16
Total Air Pollution Chicago, Illinois 1959-1980* (Thousands of Tons/Yr.)

	1959	1964	1969	1980 (a)	1980 (b)
Hydrocarbons[a]	325	372	341	95	140
Carbon Monoxide	985	1,129	1,327	420	751
Nitric Oxides[b]	214	266	317	364	526
Sulfur Oxides[c]	465	681	806	860	860
Particulate	87	129	156	32	32
Total	2,077	2,577	2,947	1,771	2,309

Source: Tables 17-13, 17-14, 17-15. Totals made not add due to rounding.
*Space Heating, Electric Generation, Automobile Pollution-only.
[a]Includes aldehydes, other organics.
[b]Includes NO_2, NO_3.
[c]Includes SO_2, SO_3.

Table 17-17
Pollution Tons/Square Mile Chicago, Illinois

1959	1964	1969	1980 (a)	1980 (b)
9,258	11,496	13,142	7,901	10,350

Source: Table 17-16 divided by area of city.

St. Louis

Located at an altitude of 560 feet above sea level and encompassing an area of 61 square miles, St. Louis, Missouri has a climate that is classified as humid continental. Springs and autumns are moderate, winters cold, and summers hot. The significance of these meteorological conditions is the pronounced effect that the hot summers exert on air quality.

Maximum temperatures of 90° or above are registered on the average of forty days each year. Due to the heating characteristics of the streets, buildings, and other urban activities, a "heat island" is created that prevents the air adjacent to the ground from cooling. A resultant concentration of pollutants occurs and there is an increased health hazard. Fortunately, St. Louis is located in "open" country and these "inversions" quickly dissipate.

Diversified manufacturing and wholesale trade activities account for a large measure of St. Louis' continued economic viability. Space capsules, jet aircraft, and automobiles are all manufactured in St. Louis. Other major industries include food and kindred products, primary metal products, nonelectrical machinery, and chemicals.

Table 17-18
Automobile Air Pollution St. Louis, Missouri 1959-1980 (Thousands of Tons/Yr.)

	1959	1964	1969	1980 (a)	1980 (b)
Hydrocarbons	97	101	117	27	43
Carbon Monoxide	295	308	442	121	218
Nitric Oxides	34	36	52	53	100
Sulfur Oxides	1	1	2	3	3
Particulate	2	2	2	5	5
Total	429	448	615	209	369

Source: Number of automobiles from Standard Rate and Data Service projected to 1980 on basis of projected population increase and persons/per vehicle ratio as explained in footnote to Table 17-2.

Emissions estimated by multiplying by appropriate emission factor in Table 15-2 in accordance with procedure of Chapter 15.

For 1980:
    1980 Column (a) based on Assumption 1:
        National standards for 1976 and 1971 met on schedule
    1980 Column (b) based on Assumption 2:
        National standards for 1971 highest achieved before 1980.

Distribution of car population for 1980 computations following Automobile Manufacturers Association data: 52.4 percent under 5 years, 35.6 percent 5 to 10 years, 12 percent over 10 years old.

Table 17-19
Electric Power Air Pollution St. Louis, Missouri 1959-1980
Thousands of Ton/Yr.

	1959	1964	1969	1980
Hydrocarbons[a]	N	N	N	N
Carbon Monoxide	N	N	N	N
Nitric Oxides[b]	14	25	46	49
Sulfur Oxides[c]	70	139	259	276
Particulate	15[d]	26[d]	49	3[e]
Total	110	191	336	330

Source: Table 7, p. 12, Fuel Consumption 1959-1969 for Ashley Mound and Meramec plants of Union Electric Company determined from STEAM PLANT FACTORS National Coal Association. Washington: 1959, 1965, 1970 editions. 1980 estimated on basis of projected increase of 205.76 percent in generation per table 4 assuming that 48.2 percent electrical generation will result from nuclear power.
[a]Includes aldehydes, other organics.
[b]Includes NO_2, NO_4.
[c]Includes SO_2, SO_3.
[d]Assumes precipitators operating at 84% efficiency.
[e]Assumes precipitators operating at 99% efficiency.
May not add to totals due to rounding.

Table 17-20
Space Heating Air Pollution St. Louis, Missouri 1959-1980
Residential/Commercial/Industrial Thousands of Tons/Yr.

	1959	1964	1969	1980
Hydrocarbons[a]	N	N	N	N
Carbon Monoxide	N	N	N	N
Nitric Oxides[b]	3	4	4	6
Sulfur Oxides[c]	6	6	7	9
Particulate	1	1	1	1
Total	11	11	12	17

Source: Houshold Data—Standard Rate and Data Service. SPOT TELEVISION RATES AND DATA—U.S. Department of Commerce. COUNTY BUSINESS PATTERNS 1959, 1964, 1969. 1980 estimated. Assuming rate of increase in housing and employee populations in previous decades will prevail in future. May not add to totals due to rounding.
[a]Includes Aldehydes.
[b]Includes NO_2, NO_4.4
[c]Includes SO_2, SO_3.
May not add to totals due to rounding.

Table 17-21
Total Air Pollution St. Louis, Missouri 1959-1980*
Thousands of Tons/Yr.

	1959	1964	1969	1980 (a)	1980 (b)
Hydrocarbons[a]	97	101	118	28	44
Carbon Monoxide	294	308	443	122	219
Nitric Oxides[b]	52	55	102	108	155
Sulfur Oxides[c]	78	146	267	286	286
Particulate	18	29	53	9	9
Total	539	639	983	553	713

Source: Tables 17-18, 17-19, 17-20.
*Space Heating, Electric Generation Automobile Pollution-only.
[a]Includes aldehydes, other organics.
[b]Includes NO_2, NO_3.
[c]Includes SO_2, SO_3.

Table 17-22
Air Pollution Tons/Square Mile St. Louis, Missouri

1959	1964	1969	1980 (a)	1980 (b)
8,826	10,639	16,134	9,109	11,751

Source: Table 17-21 divided by area of city.

Cleveland, Ohio

Like Chicago, Cleveland is situated alongside a great lake, and like Chicago the presence or absence of winds blowing off the lake affects the levels of air pollution. When there is little or no wind off Lake Erie, air pollution is serious, even hazardous. With the exception of Lake Erie there are no meteorological conditions that improve or impede air quality.

Of far greater importance is the concentration of industry that is located in or around Cleveland. Gigantic steel mills and foundries, automobile assembly plants, petroleum refining and storage are the principal sources of employment and direct contributors to the area's pollution problems.

Table 17-23
Automobile Pollution Cleveland, Ohio 1959-1980 (Thousands of Tons/Yr.)

	1959	1964	1969	1980 (a)	1980 (b)
Hydrocarbons	132	142	135	30	47
Carbon Monoxide	401	433	526	132	238
Nitric Oxides	47	50	62	57	109
Sulfur Oxides	2	2	2	4	4
Particulate	2	2	3	5	5
Total	584	629	728	228	403

Source: Number of automobiles from Standard Rate and Data Service projected to 1980 on basis of projected population increase and persons/per vehicle ratio as explained in footnote to Table 17-2.

Emissions estimated by multiplying by appropriate emission factor in Table 15-2 in accordance with procedure of Chapter 15.

For 1980:

    1980 Column (a) based on Assumption 1:
      National standards for 1976 and 1971 met on schedule
    1980 Column (b) based on Assumption 2:
      National standards for 1971 highest achieved before 1980.

Distribution of car population for 1980 computations following Automobile Manufacturers Association data: 52.4 percent under 5 years, 35.6 percent 5 to 10 years, 12 percent over 10 years old.

Table 17-24
Electric Power Air Pollution Cleveland, Ohio 1959-1980 (Thousands of Tons/Tr.)

	1959	1964	1969	1980
Hydrocarbons[a]	N	N	N	1
Carbon Monoxide	N	1	1	2
Nitric Oxides[b]	27	34	46	64
Sulfur Oxides[c]	143	176	238	367
Particulate	30[d]	37	50[d]	4[e]
Total	200	248	335	438

Source: Table 7 p. 12. Fuel Consumption 1959-1969 for Avon, Eastlake, Lakeshore plants of Cleveland Electric Illuminating Company determined from STEAM PLANT FACTORS. National Coal Association. Washington: 1959, 1965, 1970 editions. 1980 estimated on basis of projected increase of 184.21 percent in generation per table 6 assuming 23.6 percent of this generation will result from nuclear power.

[a]Includes aldehydes and other organics.
[b]Includes NO_2, NO_4.
[c]Includes SO_2, SO_3.
[d]Assumes precipitators operating at 84% efficiency.
[e]Assumes precipitators operating at 99% efficiency.

Table 17-25
Space Heating Air Pollution Cleveland, Ohio 1959-1980—Residential, Commercial, Industrial (Thousands of Tons/Yr.)

	1959	1964	1969	1980
Hydrocarbons[a]	N	N	N	N
Carbon Monoxide	N	N	N	N
Nitric Oxides[b]	5	6	7	10
Sulfur Oxides[c]	1	1	2	2
Particulate	1	1	1	1
Total	7	8	10	13

Source: Household Data-Standard Rate and Data Service. SPOT TELEVISION RATES AND DATA. Employee data, U.S. Dept. of Commerce. COUNTY BUSINESS PATTERNS 1959, 1964, 1969. 1980 estimated assuming continuation of rate of increase in household, employee populations which prevailed in preceding decade.
[a]Includes Aldehydes.
[b]Includes NO_2, NO_4.
[c]Includes SO_2, SO_3.

Table 17-26
Total Air Pollution Cleveland, Ohio 1959-1980* (Thousands of Tons/Yr.)

	1959	1964	1969	1980 (a)	1980 (b)
Hydrocarbons[a]	132	142	135	31	48
Carbon Monoxide	402	434	527	134	240
Nitric Oxides[b]	79	90	115	131	183
Sulfur Oxides[c]	146	179	242	373	373
Particulate	33	40	54	10	10
Total	792	885	1,073	679	854

*Space Heating, Electric Generation, Automobile Pollution—only.
[a]Includes aldehydes, other organics
[b]Includes NO_2, NO_3
[c]Includes SO_2, SO_3.
Source: Tables 17-23, 17-24, 17-25—totals may not add due to rounding.

Table 17-27
Total Air Pollution Tons/Square Mile Cleveland, Ohio

1959	1964	1969	1980 (a)	1980 (b)
9,747	10,895	13,213	8,360	10,534

Source: Table 17-26 divided by area of city.

New Orleans

The semi-tropical climate of New Orleans combined with high humidity and low wind speeds are conditions that increase the hazards of air pollution. However, the relatively low levels of industrialization and the heavy rainfall, averaging 53.9 inches yearly, are factors that discount the inherently poor meteorological conditions.

In and around the city are important natural resources: petroleum, natural gas, and huge facilities for aluminum production. Shipbuilding and ship repairing, in extensive yards along the Mississippi in and adjacent to the city, constitute a growing field. By far the most important industry in terms of value added by manufacture, however, is the processing of food and kindred products. Fish, shrimp, and seafood are caught in abundance, and neighboring regions produce sugar, rice, strawberries, pecans, oranges, and vegetables.

Table 17-28
Automobile Air Pollution New Orleans 1959-1980 (Thousands of Tons/Yr.)

	1959	1964	1969	1980 (a)	1980 (b)
Hydrocarbons	33	48	49	11	18
Carbon Monoxide	99	147	193	51	92
Nitric Oxides	11	17	28	22	42
Sulfur Oxides	a	1	1	1	1
Particulate	1	1	1	2	2
Total	144	213	272	87	155

Source: Number of automobiles from Standard Rate and Data Service projected to 1980 on basis of projected population increase and persons/per vehicle ratio as explained in footnote to Table 17-2.

Emissions estimated by multiplying by appropriate emission factor in Table 15-2 in accordance with procedure of Chapter 15.

For 1980:
    1980 Column (a) based on Assumption 1:
       National standards for 1976 and 1971 met on schedule
    1980 Column (b) based on Assumption 2:
       National standards for 1971 highest achieved before 1980.

Distribution of car population for 1980 computations following Automobile Manufacturers Association data: 52.4 percent under 5 years, 35.6 percent 5 to 10 years, 12 percent over 10 years old.

aLess than one-half.

May not add to totals due to rounding.

Table 17-29
Electric Power Air Pollution New Orleans 1959-1980 (Thousand Tons/Yr.)

	1959	1964	1969	1980
Aldehydes	N	N	N	N
Nitric Oxides	4	5	12	22
Sulfur Dioxide	N	N	N	N
Other Organics	N	N	N	N
Particulate	N	N	N	1
Total	4	5	12	23

Source: Table 5, p. 12. Fuel Consumption 1959-1969 for Market Street. A.B. Paterson, Michoud plants of New Orleans Public Service determined from STEAM PLANT FACTORS. National Coal Association, Washington: 1959, 1965, 1970 editions. 1980 estimated on basis of projected increase of 235.18 percent in generation per table 5 assuming that 17.9 percent of this generation will result from nuclear power in 1980.
N=Negligible.

Table 17-30
Space Heating Air Pollution New Orleans, Louisiana 1959-1980—Residential, Commercial/Industrial (Thousands of Tons/Yr.)

	1959	1964	1969	1980
Hydrocarbons[a]	N	N	N	N
Carbon Monoxide	N	N	N	N
Nitric Oxides[b]	N	N	N	1
Sulfur Oxides[c]	N	N	N	N
Particulate	N	N	N	N
Total	&	&	&	1

Source: Household Data 1959-1969. Standard Rate and Data Service. SPOT TELEVISION RATE AND DATA, Employee Data 1959-1969. U.S. Department of Commerce. COUNTY BUSINESS PATTERNS 1959, 1964, 1969. 1980 estimated assuming rate of increase in housing and employee population which prevailed in previous decade.
[a]Includes aldehydes;
[b]Includes NO_2, NO_4.
[c]Includes SO_2, SO_3.
N=Negligible
&=less than one-half.

Table 17-31
Total Air Pollution New Orleans 1959-1980* (Thousands of Tons/Yr.)

	1959	1964	1969	1980 (a)	1980 (b)
Hudrocarbons[a]	33	48	49	11	18
Carbon Monoxide	99	147	193	51	92
Nitric Oxides[b]	15	23	41	45	65
Sulfur Oxides[c]	N	1	1	1	1
Particulate	1	1	2	3	3
Total	148	220	286	111	179

Source: Table 17-28, 17-29, 17-30.
*Space Heatings, Electric Generation Automobile Pollution—only.
[a]Includes aldehydes, other organics,
[b]Includes NO_2, NO_4,
[c]Includes SO_2, SO_3.,.

Table 17-32
Air Pollution Tons/Square Mile, New Orleans

1959	1964	1969	1980 (a)	1980 (b)
745	1,104	1,436	562	916

Source: Table 17-31 divided by area of city.

Mobile, Alabama

A mild climate with high rainfall and a location near water are the primary climatological influences that distinguish Mobile, Alabama's air envelope.

In recent years Mobile has experienced a marked increase in industrial development. The leading industries include shipbuilding and repairing, paper and pulp manufacture, bauxite processing, oil refining, food processing, paint manufacture, steel fabrication, and the manufacturing of asphalt and asbestos roofing.

Table 17-33
Automobile Air Pollution Mobile, Alabama 1959-1980 (Thousands of Tons/Yr.)

	1959	1964	1969	1980 (a)	1980 (b)
Hydrocarbons	20	24	23	6	10
Carbon Monoxide	60	75	89	28	51
Nitric Oxides	7	9	10	12	24
Sulfur Oxides	a	a	a	1	1
Particulate	a	a	1	1	1
Total	88	109	123	48	87

Source: Number of automobiles from Standard Rate and Data Service projected to 1980 on basis of projected population increase and persons/per vehicle ratio as explained in footnote to Table 17-2.
Emissions estimated by multiplying by appropriate emission factor in Table 15-2 in accordance with procedure of Chapter 15.
For 1980:
    1980 Column (a) based on Assumption 1:
      National standards for 1976 and 1971 met on schedule
    1980 Column (b) based on Assumption 2:
      National standards for 1971 highest achieved before 1980.
Distribution of car population for 1980 computations following American Manufacturers Association data: 52.4 percent under 5 years, 35.6 percent 5 to 10 years, 12 percent over 10 years old.
[a]Less than one-half.
May not add to totals due to rounding.

Table 17-34
Electric Power Air Pollution Mobile, Alabama 1959-1980 (Thousands of Tons/Yr.)

	1959	1964	1969	1980
Hydrocarbons[a]	N	N	N	N
Carbon Monoxide	N	N	N	N
Nitric Oxides[b]	12	16	15	18
Sulfur Oxides[c]	44	70	58	69
Particulate	11[d]	17[d]	15[d]	1[e]
Total	67	103	88	88

Source: Tables 7, p. 12. Fuel Consumption 1959, 1969 for Barry and Chickasaw plants of Alabama Power Company determed from STEAM PLANT FACTORS, National Coal Association. Washington: 1959, 1965, 1970 editions. 1980 estimated on basis of projected increase of 231.32 percent in generation per table 3 assuming that 46.3 percent of this generation will result from Nuclear Power.
[a]Includes aldehydes, other organics.
[b]Includes NO_2, NO_4.
[c]Includes SO_2, SO_3.
[d]Assumes precipitators operating at 84% efficiency.
[e]Assumes precipitators operating at 99% efficiency.
N=Negligible.

Table 17-35
Space Heating Air Pollution Mobile, Alabama 1959-1969—Residential, Commercial/Industrial (Thousands of Tons/Yr.)

	1959	1964	1969	1980
Hydrocarbons[a]	N	N	N	N
Carbon Monoxide	N	N	N	N
Nitric Oxides[b]	N	N	N	N
Sulfur Oxides[c]	N	N	N	N
Particulate	N	N	N	N
Total	&	&	&	&

Source: 1959-1969 Household Data. Standard Rate and Data Service. SPOT TELEVISION RATES AND DATA. Employee Data. U.S. Department of Commerce, COUNTY BUSINESS PATTERNS 1959, 1964, 1969. 1980 estimated assuming rate of increase in housing and employee population during previous decade will prevail.
[a]Includes Aldehydes.
[b]Includes NO_2, NO_4.
[c]Includes SO_2, SO_3.
N=Negligible.
&=Less than one-half.

Table 17-36
Total Air Pollution Mobile, Alabama 1959-1980* (Thousands of Tons/Yr.)

	1959	1964	1969	1980 (a)	1980 (b)
Hydrocarbons[a]	20	24	23	6	10
Carbon Monoxide	60	75	89	28	51
Nitric Oxides[b]	19	25	25	30	42
Sulfur Oxides[c]	44	70	58	70	70
Particulate	11	18	15	1	1
Total	154	212	210	135	174

Source: Table 17-33, 17-34, 17-35. Totals may not add due to rounding.
*Space Heating, Electric Generation, Automobile Pollution-only.
[a]Includes Aldehydes, other organics.
[b]Includes NO_2, NO_3.
[c]Includes SO_2, SO_3.

Table 17-37
Air Pollution Tons/Square Mile Mobile, Alabama

1959	1964	1969	1980 (a)	1980 (b)
357	1,334	1,326	854	1,102

Source: Table 17-36 divided by area of city.

Conclusion[a]

[a]This section assumes Clean Air Act Standards are met on schedule —**ED.**

Air pollution will continue to be a significant problem in 1980, but the light at the end of the tunnel can be glimpsed. The two major sources of pollution, the automobile and power plants, will emit less pollution in 1980 than they did in 1969. As new nuclear power plants are built and old cars are phased out carbon monoxide, sulfur oxide, nitrogen oxide, hydrocarbon, and particulate emissions will decrease whether measured absolutely or relatively.

National, state, and local ordinances are directly responsible for these advances in air pollution abatement. Historically, business corporations have been reluctant to undertake policies aimed at elimination of pollution until they were prodded by directives from the legislatures. Pollution abatement is expensive and corporations were disinclined to part with tangible assets in order to achieve an image of concerned environmentalists.

But the past is not always a prologue to the future. It is now both politically expedient and economically rewarding to "do something" about pollution. Because the law of parsimony operates in business as well as scientific circles, and since man is by nature pragmatic, the focus of attention in air pollution has been directed to those problems that can be resolved quickly or to those currently regarded as technologically feasible. Thus power plants have concentrated on reducing particulate emissions and building nuclear power plants as an answer to the problem of air pollution. Detroit has dramatically lowered hydrocarbon and carbon monoxide emissions by improving and modifying the conventional internal combustion engine.

Both Detroit and the electrical utilities share a common disinclination to commit adequate resources to resolve the more difficult problems of air pollution. Detroit has adamantly maintained that nitric oxides could not be controlled while the utilities have been remiss in attempting to conquer the problems of sulfur oxide and nitric oxide emissions.

But as a result of the stimulus of legislation these problems and their resolution have been forced upon the auto-makers and the utilities. Nitric oxides are required to be reduced in all automobiles manufactured by model year 1976 and, in the case of utilities, legislation and public opinion require that sulfur oxides and nitric oxides emissions be reduced.

When measured against the volume of effluents discharged by the power companies and the automobiles, space heating pollution is inconsequential. Unlike the electrical utilities and the automobile manufacturers, the individual polluter has not been subjected to legislative restrictions to curb pollution. The problem is thus likely to increase in significance unless it is resolved by the manufacturers of heating equipment or is coincidentally resolved by legislation, aimed at the utilities, which requires the burning of low sulfur fuel.

The latter is the likely course of events. The cities of Chicago, St. Louis, and Cleveland have enacted ordinances that restrict the sulfur content of the fuel burned in these cities. It is not likely that Mobile and New Orleans will soon follow suit. With the recent enactment of federal standards defining maximum allowable concentrations of pollutants in a city's air, it is likely that all cities will soon restrict the sulfur content of fuel.

The Cities

If our projections are correct the air in Chicago will be nearly 40 percent cleaner in 1980 than it was in 1969. Automobile pollution will cause only 41 percent of the problem decreased from 62 percent of total pollution in 1969. The improvements will be especially dramatic in the reduction of unburned hydrocarbons and carbon monoxide, which will be reduced by 246,000 tons and 906,000 tons respectively.

However, those citizens whose respiratory tracts are sensitive to the fumes of sulfur oxides will find the air of Chicago about as irritating in 1980 as in 1969. Our projections show that sulfur oxide pollution will increase by 48,000 tons in 1980. Yet this represents considerable progress because total electrical generation will more than double by 1980.

When calculated in terms of pollution per square mile, St. Louis, Missouri, was the dirtiest city investigated by this chapter. But by 1980 pollution will be only 57 percent of the 1969 level—a rate of improvement that is better than all other cities excepting New Orleans. (It is not quite fair to compare the other cities to New Orleans because the southern city has a distinct advantage. Electrical power generation uses natural gas as the sole fuel and virtually all of space heating is accomplished by natural gas.)

Cleveland, Ohio, best demonstrates the reversal of roles by the two major sources of pollution that will occur in most cities by 1980. In 1969 the automobile accounted for 67 percent of total pollution, but in 1980 it will account for only 34 percent of the problem. Power plant emissions, however, accounted for only 31 percent of the problem in 1969, but will account for 64 percent of the problem in 1980. Total pollution is expected to decrease by 37 percent from the 1969 level of 1,072,000 tons.

Mobile, Alabama, illustrates the advances that are possible in controlling particulate emissions from power plants. While demand for electricity is expected to increase by 231 percent in 1980 particulate emissions will amount to only 8 percent of the levels of 1969. Total pollution from all sources will decrease 36 percent from the 211,000 tons emitted in 1969.

The significance of natural gas in the reduction of air pollution is amply demonstrated in the study of New Orleans, Louisiana. Because its power plants use natural gas as an energy source, this generally major contributor to an area's pollution problem caused only 4.3 percent of total emissions in 1969 and will result in only 21 percent of total emissions in 1980.

As a result of the area's reliance on natural gas in power generation and space heating, automobile pollution accounted for a disproportionate amount of total emissions. Ninety-five percent of the total air pollution problem was caused by the automobile in 1969. By 1980 the automobile will still cause 78 percent of all emissions in New Orleans. However, total pollution will be only 39 percent of the 1969 levels.

Each of the cities studied shows a dramatic decrease in pollution levels by 1980. Two critical assumptions account for the conclusion: (a) that the federal standards for automobiles will be met, (b) that although the requirements for electrical usage will increase dramatically, utilities will respond by building more nuclear power plants and will employ electrostatic precipitators of 99 percent efficiency in conventional power plants.

It can be argued that reliance on nuclear power as the means of decreasing power plant pollution is unwarranted, that all past studies by any committee or governmental body that predicted X amount of nuclear generation always fell far short of the mark. The author realizes this. However, it is a fact that federal standards regarding the levels of sulfur oxide and other emissions have been established, that they are likely to be enforced, and that nuclear power is one way of meeting these standards.

There are two other alternatives, control techniques that will reduce sulfur oxide and nitric oxide emissions. Undoubtedly, the devices employed in these techniques will eventually become durable enough to withstand the high heat loads and great tonnage volume with which they must cope. They are not yet ready and it is unlikely that they will be installed in all plants by 1980.

Building power plants in outlying, sparsely settled areas is another way of reducing urban pollution. This is not a solution; rather, it transfers a metropolitan area's problem to the suburbs.

Whatever is done will require tradeoffs. The author believes that nuclear power is the most feasible and the one method that has the greatest short- and long-range potential. It is costly. But the alternatives are expensive, unproven, and unworkable.

Notes

1. T. G. Aylesworth, *This Vital Air, This Vital Water* (Chicago, New York, San Francisco: Rand McNally and Company, 1968), pp. 1-112.

2. M. A. Benarde, *Our Precarious Habitat* (New York: W. W. Norton and Company, Inc., 1970), pp. 1-305.

3. D. E. Carr, *The Breath of Life: The Problem of Poisoned Air* (New York: W. W. Norton and Company, Inc., 1965), pp. 1-175.

4. K. W. Kapp, *The Social Costs of Private Enterprise* (Cambridge, Massachusetts: Harvard University Press, 1959), pp. 1-244.

5. R. M. Linton, *Terracide: America's Destruction of Her Living Environment* (Boston-Toronto: Little Brown and Company, 1970), pp. 3-349.

6. A. R. Meetham, *Atmospheric Pollution: Its Origin and Prevention* (New York: Pergamon Press, 1961), pp. 5-73.

7. R. G. Ridker, *Economic Costs of Air Pollution: Studies in Measurement* (New York: Frederick A. Praeger, 1967), pp. 6-47.

8. E. Robinson and R. C. Robbins, *Sources, Abundance and Fate of Gaseous Atmospheric Pollutants* (Huntsville, Alabama: Stanford Research Institute, 1968), pp. 1-110.

9. A. C. Stern (Ed.), *Air Pollution Vol. II* (New York: Academic Press, 1968), pp. 1-654.

10. A. C. Stern (Ed.), *Air Pollution Vol. III. Sources of Air Pollution and Their Control* (New York: Academic Press, 1968), pp. 3-866.

U. S. Government Publications on Air Pollution

11. U. S. Congress, House, Committee on Interstate and Foreign Commerce, *Report on the Air quality Act of 1967*, H. Rept. to Accompany S. 780, 90th Congress, 1st Session, 1967, p. 1-97.

12. U. S. Congress, House, Committee on Government Operations, *Federal Air Pollution R. & D. On Sulfur Oxides Pollution Abatement*, Hearing before a subcommittee of the Committee on Government Operations 90th Congress, 2nd. Session, 1968.

13. U. S. Congress, House, Committee on Interstate and Foreign Commerce, *Air Pollution Control Research Into Fuels and Motor Vehicles*, Hearing before a subcommittee on Public Health and Welfare of the Committee on Interstate and Foreign Commerce, 91st Congress, 1st Session, June 1969, p. 7-37.

14. U. S. Department of Health, Education and Welfare, Public Health Service, Division of Air Pollution, *Air Pollution: A National Sample* (Washington: Government Printing Office, 1970).

15. U. S. Department of Health, Education and Welfare, Public Health Service, Environmental Health Service, National Air Pollution Control Administration, *Air Quality Criteria For Hydrocarbons* (Washington: Government Printing Office, 1970).

16. U. S. Department of Health, Education and Welfare, Public Health Service, National Air Pollution Control Administration, *Air Quality Criteria For Carbon Monoxide* (Washington, D.C.: U. S. Government Printing Office, 1970), pp. 2-1 to 9-19.

17. U. S. Department of Health, Education and Welfare, Public Health Service, Consumer Protection and Environmental Health Service, National Air Pollution Control Administration, *Air Quality Criteria for Particulate Matter* (Washington: Government Printing Office, 1969), p. 34-73.

18. U. S. Department of Health, Education and Welfare, Public Health Service, Consumer Protection and Environmental Health Services, National Air Pollution Control administration, *Air Quality Criteria For Photochemical Oxidants* (Washington: Government Printing Office, 1969), pp. 2-1 to 10-7.

19. U. S. Department of Health, Education and Welfare, Public Health Service, Consumer Protection and Environmental Service, National Air Pollution Control Administration, *Air Quality Criteria For Sulfur Oxides* (Washington: Government Printing Office, 1969), p. 25-158.

20. U. S. Department of Health, Education and Welfare, Public Health Service, Bureau of Disease Prevention and Environmental Control, National Center for Air Pollution Control, *Compilation of Air Pollutant Emission Factors* (Durham: Government Printing Office, 1968), p. 1-52.

21. U. S. Department of Health, Education and Welfare, Public Health Service, Environmental Health Service, National Air Pollution Control Administration, *Control Techniques For Carbon Monoxide Emissions From Stationary Sources* (Washington: U. S. Government Printing Office, 1970), pp. 2-1 to 5-7.

22. U. S. Department of Health, Education and Welfare, Public Health Service, Environmental Health Service, National Air Pollution Control Administration, *Control Techniques For Carbon Monoxide, Nitrogen Oxides, and Hydrocarbon Emissions From Mobile Sources* (Washington: Government Printing Office, 1970), pp. 1-1 to 9-3.

23. U. S. Department of Health, Education and Welfare, Public Health Service, Consumer Protection and Environmental Health Service, *Control Techniques for Sulfur Oxide Air Pollutants* (Washington: Government Printing Office, 1969), pp. 2-1 to 7-7.

24. U. S. Department of Health, Education and Welfare, Public Health Service, Consumer Protection and Environmental Health Service, National Air Pollution Control Administration, *Emissions From Coal Fired Power Plants: A Comprehensive Summary* (Durham: Government Printing Office, 1967), p. 1-24.

25. U. S. Department of Health, Education and Welfare, Public Health Service, Consumer Protection and Environmental Health Service, National Air Pollution Control Administration, *Report for Consultation On The Greater Metropolitan Cleveland Intrastate Air Quality Control Region* (Washington: Government Printing Office, 1969), p. 1-31.

26. U. S. Department of Health, Education and Welfare, Public Health Service, Consumer Protection and Environmental Health Service, National Air Pollution Control Administration, *Report For Consultation On the Metropolitan Chicago Interstate Air Quality Control Region* (Washington: Government Printing Office, 1968), p. 3-26.

27. U. S. Department of Health, Education and Welfare, Public Health Service, Consumer Protection and Environmental Health Service, National Air Pollution Control Administration, *Nationwide Inventory Of Air Pollutant Emissions 1968* (Raleigh: Government Printing Office, 1970), p. 1-36.

28. U. S. Department of Health, Education and Welfare, Public Health Service, Consumer Protection and Environmental Health Service, National Air Pollution Control Administration, *Workbook Of Atmospheric Dispension Estimates* (Cincinnati: Government Printing Office, 1970), p. 1-84.

Other Government Publications

29. Federal Power Commission, *1959 Statistics of Privately Owned Electric Utilities in the U. S.* (Washington: Government Printing Office, 1964), p. 1-618.

30. Federal Power Commission, *1968 Statistics of Privately Owned Electric Utilities in the U. S.* (Washington: Government Printing Office, 1969), p. 1-618.

31. Federal Power Commission, A Report to the Federal Power Commission Prepared by the South Central Regional Advisory Committee, *Electric Power In the South Central Region 1970, 1980, 1990* (Washington: Government Printing Office, 1969), p. 1-242.

32. Federal Power Commission, *West Central Region Power Survey 1970-1990*, A Report to the Federal Power Commission. Prepared by the West

Central Regional Advisory Committee (Washington: Government Printing Office, 1969), Section II, p. 1-21.

33. Federal Power Commission, *East Central Power Survey 1970-1990,* A Report to the Federal Power Commission, Prepared by the East Central Regional Advisory Committee (Washington: Government Printing Office, 1969), p. 1-210.

34. Federal Power Commission, *South East Power Survey 1970-1990,* A Report to the Federal Power Commission, Prepared by the South East Regional Advisory Committee (Washington: Government Printing Office, 1969), p. 1-215.

35. "Industry Urged to Begin Control of Pollution On Voluntary Basis," *New York Times,* January 13, 1971, p. 77.

36. "Mercury in Air Stirring Concern," *New York Times,* February 16, 1971, p. 18.

37. "City Reports Air Is Still Sooty, But Sulfur Dioxide Is Lower," *New York Times,* January 7, 1971, p. 69.

38. "Polluters Set on Sit on Antipollution Bands," *New York Times,* December 7, 1970, p. 1.

39. "Back to the Engine for Evaporative Emissions," *SAE Journal* 77 (October 1969), pp. 46-48.

40. "Whither Internal Combustion?" *New York Times,* April 4, 1971, p. 10 of section 5.

41. R. E. Harrington, R. H. Borgnardt, and H. E. Potter, "Reactivity of Selected Limestones and Dolomites with Sulfur Dioxide," *American Industrial Hygiene Association Journal* Vol. 29, March-April 1968, p. 156.

42. A. L. Plumley, O. D. Whiddon, F. N. Shutko, and J. Jorakin, "Removal of SO_2 and Dust From Stack Gases," *Combustion* 40 (July 1968), pp. 16-23.

43. "Antipollution Demonstration Unit Slated," *Oil and Gas Journal* 68, (Sept. 7, 1970), p. 103.

44. S. Kattel and K. D. Plants, "Here's What SO_2 Removal Costs," *Hydrocarbon Processing* 46 (July 1967), pp. 161-164.

*Publications Used for Gathering Supporting
Data on Population, Weather, Passenger Cars,
and Fuel*

45. *Automobile Facts and Figures 1967* (Detroit: Automobile Manufacturers Association 1966), pp. 1-70.

46. *Automobile Facts and Figures 1970* (Detroit: Automobile Manufacturers Association 1969), pp. 1-70.

47. *American Gas Association Monthly* (New York: American Gas Association), Vol. 48, pp. 1-5.

American Gas Association Monthly (New York: American Gas Association), Vol. 52, pp. 1-5.

48. H. M. Conway (Ed.), *The Weather Handbook* (Atlanta, Georgia: Conway Publications, 1963), pp. 3-255.

49. H. H. Lansberg and S. H. Schurr, *Energy In The United States: Sources, Uses and Policy Issues* (New York: Random House, 1968), pp. 15-75.

50. *Petroleum Facts and Figures 1967* (New York: The American Petroleum Institute, 1967), pp. 6-344.

51. *Spot Television Rates and Data* (Skokie, Illinois: Standard Rate and Data Service, December 1959).

52. *Spot Television Rates and Data* (Skokie, Illinois: Standard Rate and Data Service, December 1964).

53. *Spot Television Rates and Data* (Skokie, Illinois: Standard Rate and Data Service, December 1969).

54. *Steam-Electric Plant Factors 1959 Edition* (Washington: National Coal Association, 1959), pp. 3-30.

55. *Steam-Electric Plant Factors 1964 Edition* (Washington: National Coal Association, 1965), pp. 3-30.

56. *Steam-Electric Plant Factors 1969 Edition* (Washington: National Coal Association, 1970), pp. 3-30.

57. U. S. Department of Commerce, Bureau of Census, Census of Housing 1960: *States and Small Areas* "Type of Fuel and Selected Equipment" (Washington: Government Printing Office, 1963).

58. U. S. Department of Commerce, Bureau of Census, Current Population Reports, *Population Estimates and Projection,* Series P-25, #448 (Washington: Government Printing Office, August 6, 1970), p. 1-47.

59. U. S. Department of the Interior, Bureau of Mines, Mineral Industry Surveys, *Burner Fuel Oils, 1969* (Bartlesville: Government Printing Office, 1969), p. 2-30.

60. U. S. Department of Transportation, Federal Highway Administration, Bureau of Public Roads, *Highway Statistics, 1959* (Government Printing Office, 1959), p. 1-84.

61. U. S. Department of Transportation, Federal Highway Administration, Bureau of Public Roads, *Highway Statistics, 1964* (Government Printing Office, 1965), p. 1-84.

62. U. S. Department of Transportation, Federal Highway Administration, Bureau of Public Roads, *Highway Statistics, 1969* (Government Printing Office, 1970), p. 1-84.

63. Legislation. *An Act to Amend the Clean Air Act to Provide for a More Effective Program to Improve the Quality of the Nation's Air,* Public Law 91-604, 91st Congress, H. R. 17255 (Washington: December 1970).

64. U. S. Department of Health, Education and Welfare, Public Health Service Consumer Protection and Environmental Health Services, National Air Pollution Central Administration, *Sulfur Oxides Pollution Control* (Washington: Government Printing Office, 1968).

65. U. S. Department of Commerce, Bureau of the Census. Census of Housing 1960, U. S. Summary.

18 Air Pollution in Five Western Cities

GEORGE FENNIMORE

Los Angeles

The city of Los Angeles is noted for many things, but it is perhaps most famous for something that Los Angeles would rather forget—its air pollution. The smog, as it is called, is a mixture of gas and particles manufactured by the sun, chiefly of the combustion products of such organic fuels as gasoline [14:10].

Industry

The contribution to air pollution by industry has been vigorously attacked by the Los Angeles Air Pollution Control District. One weapon of the district, in the battle against air pollution, is the building permit system of the County of Los Angeles. As explained by the Control District:

A permit from the district is required for equipment that may cause air pollution, or which is intended to control air pollution. Before such equipment may be constructed, plans and specifications must be submitted to the Engineering Division for approval. If the proposed construction conforms with requirements, an "Authority to Construct" is issued. Upon completion, application is made for a "Permit to Operate," and the installation is inspected and tested. Permits must be renewed whenever there is a change in ownership, the equipment is rented, changes are made, or the equipment is relocated. The equipment is inspected periodically by inspectors of the Enforcement Division to determine that it is continuing to operate properly. Failure of the equipment to continue to meet requirements results in revocation of the permit [2:1].

The close supervision of Los Angeles over the possible industrial sources of pollution has made industries' contribution to the air pollution dilemma a negligible one.

Power Plants

In Los Angeles, the power plants contribute 16 percent of the nitrogen oxides and 24 percent of the sulfur dioxide in the air. The fuels used for the generation of electricity are oil and gas [1:33].

In a report to the Federal Power Commission the West Regional Advisory Committee showed for California as a whole a projected reduction in natural gas use from 719 billion cubic feet in 1970 to 478 billion cubic feet in 1980. The energy loss represented by the decline in gas consumption will be made up by nuclear plants in the area [31: table 9].

Table 18-1
Fuel Consumption by Power Plants in Los Angeles

	Barrels of Oil (Thousand)	Million Cubic Feet of Gas (Thousand)
1959	3,100	37
1969	4,000	94
1980	4,900	94

Source: Data from STEAM-ELECTRIC PLANT FACTORS for 1959, 1969, projected to 1980 assuming a straight-line increase.

Table 18-2
Pollutants Produced by Los Angeles Power Plants

	Oil (Thousand Tons)				
	Carbon Monoxide	Hydrocarbon	Nitrogen Oxides	Sulfur Oxides	Particulate
1959	neg.	a	7	21	1
1969	neg.	a	9	27	1
1980*	neg.	a	2	8	a

[a]Less than one-half.
*Reduction because of local regulation requiring less than .5 percent sulfur.

	Gas (Thousand Tons)				
	Carbon Monoxide	Hydrocarbon	Nitrogen Oxides	Sulfur Oxides	Particulate
1959	neg.	neg.	6	neg.	a
1969	neg.	neg.	15	neg.	1
1980	neg.	neg.	15	neg.	1

[a]Less than one-half.

Source: Computed from data in Table 18-1, multiplied by emission factors contained in Table 15-3.

Nuclear power can provide an answer to air pollution from power plants, and Los Angeles, which has the greatest fuel consumption for power plants of the five cities, may well opt for this path.

The battle against air pollutants from electrical-generating sources is being waged in Los Angeles, the two major weapons being the nuclear power plants and stricter local regulation of the oil used to produce power.

Automobile

In Los Angeles, the automobile is the major contributor to the famous smog. Early antipollution devices on cars have had the following results:

A study by the California Air Resources Board shows hydrocarbon emission control devices installed on 1966 to 1968 models failed ... the state agency adds that the flimsy control devices wear out in less than a year on the road. In Los Angeles, as well as in San Francisco, the state tries in many ways to limit the amount of air pollution from automobiles. One method is the assembly line testing procedure, pursuant to Section 39052 of the California Health & Safety Code. This procedure allows for testing representative vehicles of at least 80 percent of each manufacturer's total sales in the State [1:15].

Governor Reagan has announced a plan for converting state vehicles from gasoline engines to natural gas or liquified petroleum gas engines to reduce motor vehicles emission. The conversion can be made rather easily, and the photochemical reactivity of these gases is relatively low [1:25]. Although this is not the answer to alleviating any considerable part of the pollution in California, it is at least an attempt to find a solution to an ever-increasing problem.

Unfortunately, while California is making great strides in reducing the pollution per automobile, the population of cars in Los Angeles is climbing at a steady pace; and if it continues as it has in the past ten years, the number of cars in 1980 will negate some of the strides made by technicians and environmentalists.

In the south coast air basin, in which Los Angeles is located, the Air Resources Board estimates that automobiles contribute 97 percent of the carbon monoxide, 71 percent of the hydrocarbons, and 66 percent of the nitrogen oxides [1:33]. The figures developed in this chapter suggest that the automobile is responsible for an even larger proportion of these pollutants. These statistics demonstrate the tremendous contribution of the automobile to the polluted air of Los Angeles. Table 18.4 shows the amount of pollutants from the automobile based on two assumptions. Assumption one states that the 1976 federal standards will be met, and assumption two presupposes that the 1976 standards will not be met and the 1971 federal standards will prevail throughout the 1970s. The 1976 standards are those called for in 1975 by the Clean Air Act of 1970, which permitted one year's delay by the administrator.

The 1971 federal standards will appreciably reduce the pollution in 1980, but the continuing increase in the automobile population makes the situation remain a serious one, which might be met by greater strictness, as in the 1975 federal standards, as shown in table 18.4.

Table 18-3
Passenger Car Population in Los Angeles (Thousands)

1959	2,600
1969	3,400
1980	4,200

Source: SPOT TELEVISION RATES AND DATA, 1959, 1969, projected to 1980 assuming a straight-line increase.

Table 18-4
Pollutants Produced by Los Angeles Automobiles (Thousand Tons)

	Carbon Monoxide	Hydrocarbon	Nitrogen Oxides
1959	1,750	350	200
1969	2,850	750	275
1980[a]	800	150	200
1980[b]	1,103	227	410

[a]Assumption One: 1976 national standards met on schedule.
[b]Assumption Two: 1971 national standards highest achieved in period prior to 1980.
Source: Compiled from data in Table 18-3, multiplied by emission factors in Table 15-2.

Space Heating

The space heating of commercial buildings and private homes is a source of air pollution in Los Angeles, but due to the warm climate and the use of gas as the chief fuel, this source is not as important a source of pollutants as power generation or the automobile. In Los Angeles, the number of homes and businesses that use gas, as opposed to oil, are at a ratio of two hundred to one. In fact, in Los Angeles there are more residences with no heat at all than there are which use oil. Therefore, only space heating by gas will be considered.

The problem of space heating is lessened by two factors, the climate and the use of gas as the chief fuel. However, the steadily increasing Los Angeles population can only mean a continuing increase in the nitrogen oxides which contribute importantly to smog.

The city of Los Angeles appears to have stopped the upward trend in air pollution of the 1960s. Since the automobile is such an important factor in the pollution of the air, the new federal standards are a great help to the city, and should reduce the 1980 level of particulates below the 1969 figure, and likewise reduce the sulfur oxide content of the air.

Table 18-5
Gas Consumed for Space Heating by Commercial and Owner-Occupied Dwellings in Los Angeles (Billion Cubic Feet)

	Owner-Occupied Residential	Commercial
1959	17	15
1969	18	31
1980	19	47

Source: Estimate of gas consumption made in accordance with procedure described in Chapter 15, employing data from Tables 15-5 and 15-6 and the CENSUS OF HOUSING.

Table 18-6
Pollutants Produced by Los Angeles' Residential Space Heating (Thousand Tons)

	Carbon Monoxide	Nitrogen Oxide	Sulfur Oxides	Particulate
1959	neg.	1	neg.	a
1969	neg.	1	neg.	a
1980	neg.	1	neg.	a

Source: Compiled from data in Table 18-15, multiplied by emission factors in Table 15-4.
a = less than one-half.

Table 18-7
Pollutants Produced by Los Angeles' Commercial Space Heating (Thousand Tons)

	Carbon Monoxide	Nitrogen Oxides	Sulfur Oxides	Particulate
1959	neg.	1	neg.	a
1969	neg.	1	neg.	a
1980	neg.	2	neg.	a

Source: Compiled from data in Table 18-5, multiplied by emission factors in Table 15-4.
a = less than one-half.

Table 18-8
Air Pollutants from Los Angeles' Power Plants, Automobiles and Space Heating for the Years 1969 & 1980 (Thousand Tons)

	Carbon Monoxide	Hydrocarbon	Nitrogen Oxides	Sulfur Oxides	Particulate	Total
1969	2,850	750	301	27	2	3,930
1980[a]	800	150	220	8	2	1,172
1980[b]	1,103	227	430	8	2	1,770

[a]Assumption One.
[b]Assumption Two.
Source: Compiled from Tables 18-2, 18-4, 18-6, and 18-7.

San Francisco

In San Francisco the weather and topography play a great part in the occurence of air pollution. The amount of air available for diluting the pollutants depends upon several factors in the area's overall weather pattern. Basically, the Bay area's weather is determined by the interplay of the great high and low pressure areas, the continental and maritime air masses [21:11].

The San Francisco Bay area is ringed by hills that lend themselves to the formation of smog and other forms of air pollution. However, the winds that blow through the basin provide a great relief from air pollution in the city of San Francisco.

A summation of the effect of weather on air pollution in San Francisco was given by the local air pollution control board, when it stated: "Normally, the ventilation of the Basin is adequate to disperse most of the pollution. However, when the inversion layer is low, it in effect becomes a lid, sealing the low-lying pollutant-bearing air in the basin" [21:13].

Power Plants

In San Francisco, according to the readings taken in 1968, power plants accounted for 21 percent of the sulfur dioxide emissions and 14 percent of the nitrogen oxides in the air. The power plants are a greater source of air pollution than is space heating, but once again, the chief source of air pollution in this city is the automobile.

The controls over the power plants in this city are the same as those in Los Angeles, since they are state regulations. The fuels used are oil and gas, and the fact that coal is not used is a benefit to the clean air of the city.

While there appears to be a steady increase in the use of gas, there is a more rapid increase in the use of oil, a fuel that pollutes the air more than gas.

The chief problem in this area is the combustion of oil and the oxides of sulfur that are produced. A policy similar to that of Los Angeles on low sulfur fuel would improve the situation in this area.

Table 18-9
Fuel Consumption by Power Plants in San Fransisco

	Barrels of Oil (Thousand)	Million Cubic Feet of Gas (Thousand)
1959	500	24
1969	3,700	33
1980	4,070	42

Source: Data from STEAM-ELECTRIC PLANT FACTORS 1959, 1969, projected to 1980 assuming straight-line increase.

Table 18-10
Pollutants Produced by Power Plants in San Francisco

	Carbon Monoxide	Hydrocarbon	Oil (Thousand Tons) Nitrogen Oxides	Sulfur Oxides	Particulate
1959	neg.	neg.	5	neg.	a
1969	neg.	neg.	7	neg.	a
1980	neg.	neg.	8	neg.	a

aLess than one-half.

Source: Compiled from data on Table 18-9, multiplied by emission factors in Table 15-3.

Automobiles

The Air Resources Board of California states that motor vehicles account for 49 percent of the hydrocarbons, 66 percent of the nitrogen oxides, and 78 percent of the carbon monoxide in the air of San Francisco. The Bay Area Air Pollution Control District states that even when pollution was removed due to air pollution control regulations, the automobile accounts for 72 percent of the total pollutants in the air.

The number of passenger cars in San Francisco has not shown a great jump. Much of the problem in the city of San Francisco is due to the commutation of workers from the outlying areas to the city.

In order to combat this situation, the city of San Francisco is constructing the Bay Area Rapid Transit system, and it began limited operation in 1972. This, together with the conversion of existing railway systems to rapid transit in place of a comparable road improvement program, should aid in reducing the automobile's contribution to air pollution [9:18].

Application of the 1971 standards tends to increase nitrogen oxide emissions, although they reduce the carbon monoxide and hydrocarbon pollutants. The 1976 standards call for a 90 percent reduction in nitrogen oxides given off, which should effect a dramatic improvement if it can be accomplished.

Table 18-11
Passenger Car Population in San Francisco (Thousand)

1959	260
1969	301
1980	341

Source: STANDARD RATE & DATA, 1959, 1969, projected to 1980 assuming straight-line increase.

Table 18-12
Pollutants Produced by Automobiles in San Francisco (Thousand Tons)

	Carbon Monoxide	Hydrocarbon	Nitrogen Oxides
1959	178	59	21
1969	207	65	24
1980[a]	57	14	17
1980[b]	90	20	30

[a]Assumption One.
[b]Assumption Two.
Source: Compiled from data in Table 18-12, multiplied by emission factors in Table 15-2.

Space Heating

In San Francisco, the warm weather means that space heating does not have the impact that it does in other cities. Gas is the chief source of heat in both commercial and residential heating, and, as in Los Angeles, the number of homes without heating outnumber the owner-occupied dwellings using oil, and almost equal all other sources of heat aside from gas; therefore, only gas will be considered as a heating fuel.

The commercial area appears to be rapidly expanding and accounts for most of the increase in pollution due to space heating between 1959 and 1969, and is projected to increase at the same rate to 1980.

Space heating shows itself to be a small part of San Francisco's problem, chiefly due to its use of gas as a fuel.

Table 18-13
Gas Consumed for Space Heating by Commercial and Owner-Occupied Dwellings in San Francisco (Million Cubic Feet)

	Owner-Occupied	Commercial
1959	6,700	6,300
1969	8,600	12,000
1980	10,500	17,700

Source: Estimate of gas consumption made in accordance with procedure described in Chapter 15, employing data from Tables 15-5 and 15-6 and the CENSUS OF HOUSING.

Table 18-14
Pollutants Produced by Residential Space Heating in San Francisco

	Carbon Monoxide	Gas (Thousands Tons) Nitrogen Oxides	Sulfur Oxides	Particulate
1959	neg.	a	neg.	neg.
1969	neg.	1	neg.	neg.
1980	neg.	1	neg.	a

Source: Compiled from data in Table 18-13, multiplied by emission factors in Table 15-4.
a = less than one-half.

Table 18-15
Pollutants Produced by Commercial Space Heating in San Francisco

	Carbon Monoxide	Gas (Thousand Tons) Nitrogen Oxides	Sulfur Oxides	Particulate
1959	neg.	a	neg.	neg.
1969	neg.	1	neg.	a
1980	neg.	2	neg.	a

Source: Compiled from data in Table 18-13, multiplied by emission factors in Table 15-4.
a = less than one-half.

Table 18-16
Pollutants from Power Plants, Automobiles and Space Heating in San Fransisco (Thousand Tons)

	Carbon Monoxide	Hydrocarbon	Nitrogen Oxides	Sulfur Oxides	Particulate	Total
1969	207	65	33	neg.	1	317
1980[a]	57	15	36	neg.	1	122
1980[b]	90	21	49	neg.	1	174

[a]Assumption One.
[b]Assumption Two.
Source: Compiled from Tables 18-10, 18-12, 18-14, and 18-15.

Summation

The main problem in San Francisco is automobiles, many of which are driven to the city by commuters. These cars would add to the figures above to show an even greater threat to clean air. San Francisco's new transit lines may be of some relief, but adherence to the 1976 federal standards is the more likely path to cleaner air in San Francisco.

Seattle

Seattle, Washington, is located in the Puget Sound area. The topography of the area is described:

The Puget Sound forms a basin oriented in a north-south direction and bordered on the east by the Cascade Mountain Range and on the west by the Olympic Mountains. Both of these mountain ranges rise to elevations well over 5,000 feet [12:36].

The Public Health Service also found that during the warm half of the year, a subsidence inversion commonly occurs in the corridor described above.

The city of Seattle is densely populated and located in a basin with other populated areas, which throw pollution into the air. The topography is not favorable to dispersion and there is sometimes a serious air pollution problem.

Power Generation

Air pollution from power plants in the City of Seattle results only from the combustion of oil, and the amount of fuel is small compared to other cities. However, the rate of increase is alarming, and the figures for the population of King County show that in 1968 the population density was 509 residents per square mile, and projections by the Department of Commerce and Economic Development show that in 1985 the population density will be 700 residents per square mile [12:15].

The increased population will have increased needs for power, and additional plants might be built that use oil; therefore, Seattle may soon have a serious pollution problem due to its power plants.

The West Regional Advisory Committee in a report to the Federal Power Commission stated that it is not entirely clear where expanded power generation for the Seattle area will come from, but it seems unlikely that there will be any substantial increase in air pollution [31:54].

Table 18-17
Fuel Consumption by Power Plants in Seattle

	Barrels of Oil (thousand)
1959	.35
1969	.70
1980	1.05

Source: STEAM-ELECTRIC PLANT FACTORS for 1959, 1969, projected to 1980 assuming a straight-line increase.

Table 18-18
Pollutants Produced by Power Plants in Seattle

	Oil (Thousand Tons)				
	Carbon Monoxide	Hydrocarbon	Nitrogen Oxides	Sulfur Oxides	Particulate
1959	neg.	neg.	neg.	neg.	a
1969	neg.	neg.	neg.	neg.	a
1980	neg.	a	neg.	neg.	a

[a]Less than one-half.
Source: Compiled from data in Table 18-17, multiplied by emission factors in Table 15-3.

Table 18-19
Passenger Car Population in Seattle (Thousand)

1959	340
1969	540
1980	740

Source: Data from STANDARD RATE AND DATA, projected to 1980 assuming a straight-line increase.

Table 18-20
Pollutants from Automobiles in Seattle (Thousand Tons)

	Carbon Monoxide	Hydrocarbons	Nitrogen Oxides
1959	232	76	27
1969	355	90	42
1980[a]	115	35	30
1980[b]	185	68	42

[a]Assumption One.
[b]Assumption Two.
Source: Compiled from data in Table 18-19, multiplied by emission factors in Table 15-2.

Automobile

Seattle is second only to Los Angeles in city automobile populations in the five cities examined. Therefore, air pollution from automobiles in this city is significant.

If assumption one is considered, the 1976 federal standards will greatly decrease Seattle's air pollution problem; and if assumption two is used, the problem will be alleviated by 1980 to a lesser extent, especially the nitrogen oxides.

Space Heating

In Seattle, the main fuels for space heating are gas and oil, with gas being the most popular but with a volume of oil used which is too large to ignore. A breakdown of fuel usage for space heating illuminates the situation.

The use of oil as a fuel in space heating has a major effect upon the sulfur oxide content of the air, while gas yields little in comparison.

Table 18-21
Consumption of Fuel for Space Heating in Seattle

	Residential Gas Million cu. ft.	Commercial Gas Million cu. ft.	Residential Oil Thousands of Barrels	Commercial Oil Thousands of Barrels
1959	11,000	9,000	3,095	2,619
1969	23,000	20,000	4,523	4,286
1980	28,000	31,000	5,476	7,143

Source: Estimate of gas and oil consumption made in accordance with procedure described in Chapter 15, employing data from Tables 15-5 and 15-6 and the CENSUS OF HOUSING.

Table 18-22
Pollutants from Residential Space Heating in Seattle

	Gas (Thousand Tons)			
	Carbon Monoxide	Nitrogen Oxides	Sulfur Oxides	Particulate
1959	neg.	1	neg.	a
1969	neg.	1	neg.	a
1980	neg.	2	neg.	a

Table 18-22 (cont.)

	Oil (Thousand Tons)				
	Carbon Monoxide	Hydrocarbons	Nitrogen Oxides	Sulfur Oxides	Particulate
1959	a	a	1	21	1
1969	a	a	1	31	1
1980	a	a	1	37	1

Source: Compiled from data in Table 18-21, multiplied by emission factors in Table 15-4.
a = less than one-half.

Table 18-23
Pollutants from Commercial Space Heating in Seattle

	Gas (Thousand Tons)			
	Carbon Monoxide	Nitrogen Oxides	Sulfur Oxides	Particulate
1959	neg.	1	neg.	neg.
1969	neg.	2	neg.	a
1980	neg.	2	neg.	a

	Oil (Thousand Tons)				
	Carbon Monoxide	Hydrocarbons	Nitrogen Oxides	Sulfur Oxides	Particulate
1959	a	a	1	17	a
1969	a	a	1	19	1
1980	a	a	2	48	1

[a]Less than one-half.
Source: Compiled from data in Table 18-21, multiplied by emission factors in Table 15-4.

Table 18-24
Pollutants from Power Plants, Automobiles and Space Heating in Seattle (Thousand Tons)

	Carbon Monoxide	Hydrocarbon	Nitrogen Oxides	Sulfur Oxides	Particulate	Total
1969	355	90	47	60	2	554
1980[a]	115	35	37	85	3	274
1980[b]	185	68	49	85	3	390

[a]Assumption One: 1976 national standards met on schedule.
[b]Assumption Two: 1971 national standards highest achieved in period prior to 1980.

Summation

In Seattle, the automobile is a great source of carbon monoxide, hydrocarbons, and nitrogen oxides, while the combustion of oil for space heating is the source of the sulfur oxides in the air. An answer to the automobile problem is clearly an adherence to 1976 federal standards, and the problems wrought by space heating may be combated by using a low sulfur fuel oil, or increasing the use of gas as a fuel in space heating.

Salt Lake City

Salt Lake City, Utah, is located in a valley with an elevation of about 4,200 feet above sea level. It is bounded on the east by the Wasatch Mountains and on the west by the Oquirrh Mountains. Both ranges have elevations of about 7,000 feet above sea level.

A description of the weather situation in the Salt Lake City area shows why a pollution problem exists here:

The Salt Lake City area is subject to temperature inversions a relatively high percentage of the time. Data indicates that perhaps 50 percent of the time during the fall and winter, a temperature inversion exists. This percentage is about as high as any in the Western United States [13:39-40].

Sulfur dioxide is a major air pollutant in Salt Lake City, and one of the chief sources of the sulfur dioxide is the copper smelter in Salt Lake County.

The figures in table 18.25 show that strides are being made to control this industrial source of air pollution. The controls have reduced sulfur dioxide when the converse would have been true without these controls. Even controlled, the smelter by itself accounts for sixteen times as much sulfur oxides emitted as all the power plants in the Salt Lake area.

Table 18-25
Example of Control of Sulfur Dioxide from the Copper Smelter in Salt Lake City County

	Sulfur Dioxide Emitted to Atmosphere		
	Tons Per Year With Controls	Potential Tons Per Year Without Controls	Percentage Control
Pre			
1964	277,400	442,380	37.3%
1968	229,950	487,640	52.8%
1970[a]	177,390	581,080	69.5%

[a]Estimated by Utah State Division of Health.
Source: Division of Health, AIR POLLUTION SOURCE INVENTORY (Salt Lake City: Division of Health, 1970), Table 9.

Power Plants

In Salt Lake City, power generation accounts for 20 percent of the oxides of nitrogen in the atmosphere [3: chart 12]. Several of the major industries have maintained air pollution departments and have supported research programs to find better ways to control their effluents. On a recent survey done by the Utah State Division of Health, it was found that over 90 percent of Utah's industries have taken some steps to combat air pollution.

There is a very favorable trend visible in Salt Lake City due to the fact that its power is being generated increasingly by gas, and less and less by the combustion of oil.

In the "Future of Power in the West Region," the report's projections covered the Rocky Mountain area, and expected increased generation using coal. The application of the very broad projection to Salt Lake City and Denver was not clear, but it is felt that straight-line projection of gas and oil consumption was not contradicted [31:51].

The trend in Salt Lake City is beneficial to clean air. The oil is being used less while the combustion of natural gas is increasing, resulting in less pollution projected for 1980 than there was in 1969.

Table 18-26
Fuel Consumption by Power Plants in Salt Lake City

	Barrels of Oil (Thousands)	Million Cubic Feet of Gas (Thousands)
1959	2,100	1,273
1969	1,600	3,923
1980	1,100	6,603

Source: STEAM-ELECTRIC PLANT FACTORS, 1959, 1969, projected to 1980 assuming straight-line increase.

Table 18-27
Pollutants from Power Plants in Salt Lake City

	Oil (Thousand Tons)				
	Carbon Monoxide	Hydrocarbon	Nitrogen Oxides	Sulfur Oxides	Particulate
1959	neg.	a	5	14	a
1969	neg.	a	4	11	a
1980	neg.	neg.	2	7	a

Table 18-27 (cont.)

	Carbon Monoxide	Hydrocarbon	Gas (Thousand Tons) Nitrogen Oxides	Sulfur Oxides	Particulate
1959	neg.	neg.	a	neg.	neg.
1969	neg.	neg.	1	neg.	neg.
1980	neg.	neg.	1	neg.	neg.

[a]Less than one-half.
Source: Compiled by data in Table 18-26, multiplied by emission factors in Table 15-3.

Automobile

In Salt Lake City, automobiles account for 87 percent of the hydrocarbons and 51 percent of the oxides of nitrogen [3: charts X, XI, XII]. Salt Lake City is chiefly bothered by the automobile, and the following statistics show that the only hope for clean air in Salt Lake City is strict observance of the federal limits on automotive emissions.

If the 1976 standards can be met, the pollutants will be substantially reduced, especially the nitrogen oxides.

Table 18-28
The Number of Passenger Cars in Salt Lake City (Thousand)

1959	130
1964	180
1969	210
1980	300

Source: STANDARD RATE & DATA, 1959, 1969, projected to 1980 assuming a straight-line increase.

Table 18-29
Pollutants from Automobiles in Salt Lake City (Thousand Tons)

	Carbon Monoxide	Hydrocarbon	Nitrogen Oxides
1959	90	30	10
1969	155	39	18
1980[a]	50	12	15
1980[b]	79	18	29

[a]Assumption One: 1976 national standards met on schedule.
[b]Assumption Two: 1971 national standards highest achieved in period prior to 1980.
Source: Compiled from data in Table 18-28, multiplied by emission factors in Table 15-2.

Space Heating

In Salt Lake City, the predominant fuel used for space heating is gas, and once again, this fact has been important in limiting the amount of pollution caused by space heating as compared to other cities of the United States. The growth of new industry shows an increase in gas consumption for commercial heating.

Space heating discharges an estimated 10 percent of the oxides of nitrogen found in the atmosphere [3: chart XII].

The chief offense of space heating is the emission of nitrogen oxides into the air, and this condition could only be improved by nuclear power plants, or new technology in filtering pollutants.

Table 18-30
Gas Consumed for Space Heating for Commercial and Owner-Occupied Dwellings in Salt Lake City (Million cu. ft.)

	Residential	Commercial
1959	4,800	3,800
1964	4,600	6,200
1969	5,400	8,100
1980	4,500	13,000

Source: Estimate of gas consumption made in accordance with procedure described in Chapter 15, employing data from Tables 15-5 and 15-6 and the CENSUS OF HOUSING.

Table 18-31
Pollutants from Space Heating in Salt Lake City

	Gas–Residential (Thousand Tons)			
	Carbon Monoxide	Nitrogen Oxides	Sulfur Oxides	Particulate
1959	neg.	a	neg.	neg
1969	neg.	a	neg.	neg.
1980	neg.	a	neg.	neg.

	Gas–Commercial (Thousand Tons)			
	Carbon Monoxide	Nitrogen Oxides	Sulfur Oxides	Particulate
1959	neg.	a	neg.	neg.
1969	neg.	1	neg.	neg.
1980	neg.	1	neg.	a

Source: Compiled from data in Table 18-30, multiplied by emission factors in Table 15-4.
a = less than one-half.

Table 18-32
Pollutants from Power Plants, Automobiles and Space Heating in Salt Lake City (Thousand Tons)

	Carbon Monoxide	Hydrocarbon	Nitrogen Oxides	Sulfur Oxides	Particulate	Total
1969	155	39	22	11	1	226
1980[a]	50	12	19	7	a	88
1980[b]	79	18	34	7	a	138

[a]Assumption One: 1976 national standards met on schedule.
[b]Assumption Two: 1971 national standards highest achieved in period prior to 1980.
[a]Less than one-half.
Source: Compiled from Tables 18-27, 18-29, and 18-31.

Summation

The pattern remains the same in this city: the new automotive standards will reduce the carbon monoxide and hydrocarbon and to a lesser extent the nitrogen oxides, but the combustion of oil adds to the level of sulfur oxides and particulates.

Denver

The city of Denver, Colorado, has a pollution problem that is very different from the other cities studied, especially those along the coast, due to its altitude. While other cities with severe problems were in valleys, Denver is located in the Rocky Mountain area of the country, and its altitude is one mile above sea level. The front range of the Rocky Mountains is twelve miles to the west, and the Great Plains are found to the east [19:34].

In Denver, 67 percent of the sulfur oxides are due to fuel combustion sources other than industrial and mobile sources. The combustion of coal produces 50 percent of the total particulate emissions [11:52].

The danger to clean air in Denver is the combustion of coal in power plants. Oil is also a fuel that is a strong polluter, but it is not used as much as in other cities.

Table 18-33
Amount of Fuel Consumed by Power Plants in Denver

	Coal (Tons)	Oil (Barrels)	Gas (Million cu. ft.)
1959	570,000	2,000	20,300
1964	1,100,000	18,000	19,000
1969	1,500,000	27,000	27,000
1980	2,400,000	43,000	30,000

Source: DATA FROM STEAM-ELECTRIC PLANT FACTORS for 1959, 1969, projected to 1980 assuming a straight-line increase.

Table 18-34
Pollutants from Power Plants in Denver (Thousand Tons)

	Carbon Monoxide	Hydrocarbon	Nitrogen Oxides	Sulfur Oxides	Particulate
		Oil			
1959	neg.	neg.	neg.	neg.	neg.
1969	neg.	neg.	a	a	neg.
1980	neg.	neg.	1	a	neg.
		Coal			
1959	a	neg.	6	22	a
1969	a	a	15	58	1
1980	1	a	25	94	2
		Gas			
1959	neg.	neg.	4	neg.	a
1969	neg.	neg.	5	neg.	a
1980	neg.	neg.	6	neg.	a

aLess than one-half.

Source: Compiled from data in Table 18-33, multiplied by emission factors in Table 15-3.

Automobile

Automobile pollution has a great effect upon Denver because, in high altitude cities, hydrocarbon emissions are 30 percent greater, carbon monoxide 60 percent greater, and oxides of nitrogen 50 percent less than Los Angeles [8:51]. This means automotive air pollution is a very serious problem in this city, more so than in Pacific coast cities.

There appears to be a steady increase in the number of automobiles in this city. The pollution that these vehicles cause can be seen in table 18.36.

Denver has a serious problem in automotive air pollution, a problem that can be met by a strict adherence to federal standards followed by close supervision of local authorities.

Table 18-35
The Number of Passenger Cars in Denver (Thousand)

1959	193
1969	230
1980	268

Source: STANDARD RATE AND DATA, 1959, 1969, projected to 1980 assuming straight-line increase.

Table 18-36
Pollutants from Automobiles in Denver (Thousand Tons)

	Carbon Monoxide	Hydrocarbon	Nitrogen Oxides
1959	132	43	15
1969	157	50	18
1980[1]	44	11	14
1980[2]	70	16	26

Source: Compiled from data in Table 18-35, multiplied by emission factors in Table 15-2.
[1] Assumption One: 1976 national standards met on schedule.
[2] Assumption Two: 1971 national standards highest achieved in period prior to 1980.

Space Heating

In Denver, as in most of the cities of the West, gas is the chief fuel used for space heating, and this proves to be a large plus in the elimination of air pollution from this source. The ratio of gas fuel to oil fuel used for space heating is about 160 to 1, and the amount of oil used is thus too small to be considered for this study.

Because gas is the chief fuel used, the problem in Denver is not as serious as in other cities. However, the increase in the oxides of nitrogen and the particulates should not be ignored.

Table 18-37
Gas Consumed for Space Heating by Commercial and Owner-Occupied Dwellings in Denver (Million cu. ft.)

	Residential	Commercial
1959	11,600	7,500
1969	15,000	19,500
1980	17,000	22,500

Source: Estimate of gas consumption made in accordance with procedure described in Chapter 15, employing data from Tables 15-5 and 15-6 and the CENSUS OF HOUSING.

Table 18-38
Pollutants from Commercial Space Heating in Denver

	Carbon Monoxide	Gas (Thousand Tons) Nitrogen Oxides	Sulfur Oxides	Particulate
1959	neg.	1	neg.	neg.
1969	neg.	1	neg.	a
1980	neg.	2	neg.	a

aLess than one-half.
Source: Compiled from data in Table 18-37, multipled by emission factors in Table 15-4.

Table 18-39
Pollutants from Residential Space Heating in Denver

	Carbon Monoxide	Gas (Thousand Tons) Nitrogen Oxides	Sulfur Oxides	Particulate
1959	neg.	1	neg.	a
1969	neg.	1	neg.	a
1980	neg.	1	neg.	a

aLess than one-half.
Source: Compiled from data in Table 18-37, multipled by emission factors in Table 15-4.

Table 18-40
Pollutants from Power Plants, Automobiles and Space Heating in Denver (Thousand Tons)

	Carbon Monoxide	Hydrocarbon	Nitrogen Oxides	Sulfur Oxides	Particulate	Total
1969	157	50	40	58	1.6	307
1980a	45	11	48	94	2.3	200
1980b	71	16	60	94	2.3	243

aAssumption One: 1976 national standards met on schedule.
bAssumption Two: 1971 national standards highest achieved in period prior to 1980.
Source: Compiled from Tables 18-34, 18-36, 18-38, and 18-39.

Summation

The chief problem in Denver is the use of coal as a fuel for power production. The conversion should be made to gas, but even oil would be better in terms of air pollution. The projections show that the city should stop using coal or face an increasing pollution problem from sulfur and nitrogen oxides and particulates in the air.

Notes

Documents

1. Air Resources Board, *Air Pollution Control in California* (Sacramento: The Resources Agency, 1970).
2. County of Los Angeles – Air Pollution Control District, *The Air Pollution Control District Permit System,* October 2, 1969.
3. Division of Health, *Air Pollution Source Inventory* (Salt Lake City: Department of Social Services, 1970).
4. Secretary of Health, Education and Welfare, *The Cost of Clean Air,* Document #91-40 (1969).
5. United States Congress, Committee on Interior and Insular Affairs and the Committee on Science and Astronautics. Joint House – Senate Colloquium, *To Discuss a National Policy For The Environment,* Joint Committee Print, Study Paper 8 (Washington, D.C.: Government Printing Office, 1968).
6. United States Congress, House Subcommittee of the Committee on Government Operations, *Federal Air Pollution R. and D. on Sulfur Oxides Pollution Abatement,* 90th Congress, 2nd. Sess., 1968.
7. United States Congress, Joint Committees on Commerce and Public Works, *Automobile Steam Engine and Other External Combustion Engines,* Joint Committee Print, Serial #90-82 (Washington, D.C.: Government Printing Office, 1968).
8. United States Department of Health, Education and Welfare, *Compilation of Air Pollutant Emission Factors,* Public Health Service Publication #999-AP-42 (1968).
9. United States Department of Health, Education and Welfare, *Report for Consultation on the San Francisco Bay Area Air Quality Control Region* (Washington, D.C.: Government Printing Office, 1970).
10. United States Department of Health, Education and Welfare, Public Health Service, *Control Techniques for Carbon Monoxide, Nitrogen Oxides and Hydrocarbon Emissions From Mobile Sources* (Washington, D.C.: Government Printing Office, 1970).
11. United States Department of Health, Education and Welfare, Public Health Service, *Report for Consultation on the Metropolitan Denver Air Quality Control Region* (Washington, D.C.: Government Printing Office, 1968).

12. United States Department of Health, Education and Welfare, Public Health Service, *Report for Consultation on Puget Sound Air Quality Control Region* (Washington, D.C.: Government Printing Office, 1969).

13. United States Department of Health, Education and Welfare, Public Health Service, *Report for Consultation on the Wasatch Front Interstate Air Quality Control Region* (Washington, D.C.: Government Printing Office, 1970).

14. United States Department of Health, Education and Welfare, Public Health Service, *The Effects of Air Pollution*. Publication #1556 (Washington, D.C.: Government Printing Office, 1967).

15. United States Department of Health, Education and Welfare, Public Health Service, *Today and Tomorrow in Air Pollution,* Publication #1555 (Washington, D.C.: Government Printing Office, 1968).

Books

16. John Esposito, *Vanishing Air* (New York: Grossman Publishers, Inc., 1970).

17. Hans Landsberg, Leonard Fischman, and Joseph Fisher, *Resources in America's Future* (Baltimore: Johns Hopkins Press, 1963).

18. National Tuberculosis and Respiratory Disease Association, *Air Pollution Primer* (New York: National Tuberculosis and Respiratory Association, 1969).

19. The Universal Standard Encyclopedia, 1957 ed., "Denver."

Pamphlets

20. *A Joint Report to the Legislature on Air Pollution Health Effects and Emergency Actions* (Sacramento: Air Resources Board, Department of Public Health, 1970).

21. Bay Area Air Pollution Control District, *Air Pollution and the San Francisco Bay Area* (Sausality-Hansen Lithograph Company, 1970).

Periodicals

22. "Air Pollution Control News," *American City,* August 1968.

23. "Los Angeles Officials Insist the Air Is Getting Cleaner," *New York Times,* August 7, 1972, p. 54.

24. "Pollution Control Is Tough Problem," *Automotive Industries,* March 1, 1970.

25. "Pollution Fighters' Newsletter," *Popular Mechanics,* November 5, 1970, p. 94.

26. *Spot Television Rates and Data* (New York: Standard Rate and Data Service, October 1969).

27. *Spot Television Rates and Data* (New York: Standard Rate and Data Service, January 15, 1964).

28. *Spot Television Rates and Data* (New York: Standard Rate and Data Service, Janaury 16, 1959).

29. *Steam-Electric Plant Factors/1959* (Washington, D.C.: National Coal Association, July 1960).

30. *Steam-Electric Plant Factors/1965* (Washington, D.C.: National Coal Association, September 1965).

33. Federal Power Commission, *The 1970 National Power Survey* Part III, The West Regional Advisory Committee Report (Washington: Government Printing Office, 1970).

19 Global Pollution of the Upper Atmosphere

CARL A. KATZ

In the West, our desire to conquer nature often means simply that we diminish the probability of small inconveniences at the cost of increasing the probability of very large disasters.

Kenneth Boulding

Introduction

Technological developments almost always lead to increased economic activity [8:198]. The industrial revolution clearly illustrates this relationship. The invention of the steam and internal combustion engines, mass production of steel, and aluminum and plastics are just some of the technological advances that are the basis for gigantic industrial development. Benefits to the human race have accrued; better communications, faster modes of travel, and more comfortable living conditions. The discomforts of dirt, haze, and smog, and the physical and health problems that arose were considered the price to be paid for this progress. No serious continuing studies of the environmental effects of pollution were made until recently, and only very recently has any serious attention been paid to the complaints and warnings of scientists and others concerned with the situation.

If the rise in pollution could be linked to any one variable it would be economic growth [3:47]. Today this rate of growth is greatly accelerating for the majority of the world's population (see table 19.1). This economic growth will inevitably breed pollution through the rise in demand for electric power, automobiles, and industrial products.

This chapter looks at atmospheric pollution on a global scale. An attempt is made to show the activities that contribute most significantly to the pollution, what the polluted effluents are, and the potential volume of generation in the next decade. The pollutants are looked at assuming no output reduction through the use of controls.

Projections were made on the basis of at least five years of data. Detected trends were used to project the data to 1980. Barring major technological innovations, economic disruptions of a severe nature, or wars, the trends should not shift radically in the next ten years.

World electricity demand was projected on a per capita basis because of its strong relationship to population. World automobile population was also projected in this way. Projections for the industrial areas examined were made

Table 19-1
Rise in Gross National Product Per Capita for 42 Selected Countries: 1958 to 1968

	GNP/Capita Growth in Percentage		
	30	40	50
Number of countries in which growth rate was exceeded	32	27	17
% of countries in which growth rate was exceeded	76	64	40

Source: United Nations Statistical Office, UNITED NATIONS STATISTICAL YEARBOOK, 1969 (New York, N.Y.: United Nations Statistical Office, 1970), Table 183, pp. 563-65.

on the basis of demand over the years 1962 to 1968. Overall world demand for fossil fuel and the projections of future demand are discussed in the section headed "World Fossil Fuel Consumption."

The coefficients of pollution, i.e., the pollutant output per unit of fuel consumed, were determined only for the United States [31]. For purposes of computation it was assumed these coefficients were sufficiently applicable the world over and would remain constant over the projection period used. Some of the weaknesses of these assumptions are:

1. The nature of the fuels used varies the world over, and will change with time.
2. Combustion techniques for the United States aren't universal and will change with time.
3. Maintenance of combustion equipment is not universally uniform, and is variable over time.
4. Rate of fuel consumption isn't universally uniform, and such rates affect pollution levels.

The Work Group on Energy for the 1970 Study of Critical Environmental Problems claimed:

As far as pollutants such as sulfur oxides, nitrogen oxides, hydrocarbons, carbon monoxide, and particulates are concerned, virtually nothing can be said with any assurance about their world-wide emission rates from energy production ... it is not possible to calculate world-wide emissions from simple fuel-by-fuel breakdowns of world energy consumption that are now available, because the precise type and quantity of emissions associated with any energy production process depend upon innumerable variables in the fuel, the processing equipment, the scale and speed of operation, the effects of various control measures already being taken, the materials already present in the atmosphere, and a wide variety of other factors [23:297-98].

Despite the extreme difficulty in assessing world pollution levels and making one-decade projections, the effort has been made. The accuracy of the figures is assumed to be within one order of magnitude. United Nations data as to world consumption of energy and industrial production should be accurate within this range. The inaccuracies associated with the coefficients used (they are over five years old and were determined only in the United States) are very difficult to assess. Nevertheless, they are the only coefficients available. Present pollutant magnitudes, and the potential growth of these magnitudes, had to be examined within the limits of the available information.

World Fossil Fuel Consumption

Carbon-based fuels are categorized herein as coal and lignite (solid), liquid hydrocarbons, and natural gas.

An indication of the rise in the world's demand for fossil fuel energy since the turn of the century is given by comparing the consumption over the 1900-1909 decade, versus the single year 1966, in table 19.2. Table 19.3 shows world fuel consumption over the 1955-1967 period. The approximate average yearly rise in each fuel's rate of use (computed from table 19.3) was: coal, 1.5 percent; lignite, 4.5 percent; liquid Hc., 7.0 percent; natural gas, 7.5 percent.

M. King Hubbert claims that since World War II the yearly increase in combined coal and lignite usage has been about 3.6 percent; for crude oil, about 6.9 percent. The latter was closely paralleled by natural gas [2:163]. The four computed rates above are fairly close to those of Hubbert; however, Hubbert's rates for coal and lignite are combined, his rate for crude oil doesn't cover all hydrocarbon liquids, and he provides no specific rate for natural gas. Therefore, the rates computed from table 19.3 data are used to project fossil fuel consumption to 1980 (see table 19.4).

By 1980 the world's electric power generation technology should be at a threshold; that of the deployment of nuclear breeder reactors on a wide scale. Two significant factors leading to accelerated efforts to develop such reactors are the limitation on the world's fossil fuel reserves, and the pollution generated through the use of fossil fuels. Breeder reactors create more fissionable fuel than they consume. However, they must be deployed before the increasing demand for nuclear power dissipates the present reserves of U_{235}, rendering the potentially convertible uranium and thorium supplies useless.

The next decade should see a shift in the proportion of the usage of each type of fossil fuel in total energy production. This proportional shift is inherent in the projections that have been made, coal and lignite having a steadily decreasing share.

Projections from 1980 to 2000 are of a more tenuous nature, but were made based on the following assumptions:

1. Despite displacement of coal and lignite in some areas, overall increases in energy demands and industrial uses will allow them to grow at prior rates. Significantly, these fuels are in great abundance relative to liquid hydrocarbons and natural gas [2:160-68, 194-205].
2. Liquid hydrocarbon demand should decrease. Factors tending to increase demand should at least be equaled by more efficient consumption, replacement by other fuels, and limitations of supply.

Table 19-2
Fossil Fuel Demand: 1900-1909 Decade vs. 1966 (million metric tons)

Fuel	1900-1909(a)	1966(b)	% of 1900-1909 Demand in 1966
Coal	3050	1845	60
Lignite	385	1050	270
Liquid Hydrocarbons (Including liq. nat. gas)	170	1525	900
Natural Gas(c)	40	762	1900

Sources:
(a) RESTORING THE QUALITY OF OUR ENVIRONMENT, REPORT OF THE ENVIRONMENTAL POLLUTION PANEL, PRESIDENT'S SCIENCE ADVISORY COMMITTEE, John W. Tukey, chairman (Washington, D.C.: U.S. Government Printing Office, 1965), p. 117.
(b) MAN'S IMPACT ON THE GLOBAL ENVIRONMENT, (Cambridge, Mass.: MIT Press, 1970), p. 303.
(c) RESTORING THE QUALITY OF OUR ENVIRONMENT, op. cit., p. 117. (Assumed gas density of 0.8 metric ton/1000 meters.)

Table 19-3
World Fossil Fuel Consumption: 1955-1967 (coal, lignite and liquid hydrocarbons in million metric tons; nat. gas in billion cubic meters)

Year	Coal	Lignite	Liquid Hydrocarbons (Hc.)*	Natural Gas
1955	1,500	630	705	300
1962	1,675	905	1,115	440
1967	1,750	1,040	1,630	816

Source: MAN'S IMPACT ON THE GLOBAL ENVIRONMENT, (Cambridge, Mass.: MIT Press, 1970), Table 7.A.1, p. 303.
*Includes liquified natural gas.

Table 19-4
Projection of World Fossil Fuel Consumption: 1967-1980 (coal, lignite and liq. hydrocarbons in million metric tons; natural gas in billion cubic meters)

Year	Coal	Lignite	Liquid Hc.	Natural Gas
1967	1,750	1,040	1,630	816
1980	1,950	1,830	3,940	2,080

Source: Derived from data of Table 19-3, and the computed growth rates.

3. The reasoning for liquid hydrocarbons applies to natural gas as well. The projections in table 19.5 were made on the basis of the above assumptions.

Total amounts of each of the fossil fuels projected to be consumed between 1968 and 2000 are:

coal:
24,572 × 10^6 m.t. to 1980; 65,691 × 10^6 m.t. to 2000
lignite:
18,577 × 10^6 m.t. to 1980; 68,639 × 10^6 m.t. to 2000
liquid Hc.:
35,184 × 10^6 m.t. to 1980; 208,864 × 10^6 m.t. to 2000 @ 7 percent
                           171,464 × 10^6 m.t. to 2000 @ 5 percent
                           144,044 × 10^6 m.t. to 2000 @ 3 percent
natural gas:
18,219 × 10^9 m.^{3} to 1980; 115,519 × 10^9 m.^{3} to 2000 @ 7.5 percent
                          90,019 × 10^9 m.^{3} to 2000 @ 5 percent
                          76,689 × 10^9 m.^{3} to 2000 @ 3 percent

Table 19.5
Projection of World Fossil Fuel Consumption to the Year 2000 (coal, lignite, liquid Hc., in million metric tons; nat. gas in m.^{3} × 10^9)

	Coal	Lignite	Liquid Hc.			Nat. Gas		
Projection Rate	1.5%	4.5%	7.0%	5.0%	3.0%	7.5%	5.0%	3.0%
	2,160	4,400	15,300	10,400	7,100	8,400	5,150	3,810

Source: Derived from Table 19-4 using the following yearly growth rates:
    coal–1.5 percent/year
    lignite–4.5 percent/year
    liquid hydrocarbons–7.0 percent, 5.0 percent, 3.0 percent/year
    natural gas–7.5 percent, 5.0 percent, 3.0 percent/year.

Carbon Dioxide (CO_2) from Fossil Fuel Combustion

If it is assumed that all fossil fuel carbon oxidizes during combustion, carbon dioxide generation per unit of fuel can easily be estimated. Assuming carbon content levels (by weight) of 75 percent for coal, 45 percent for lignite, 86 percent for liquid hydrocarbons, and 70 percent for natural gas (assumed density of 0.8 m.t. per 1,000 m.^{3}) [24:114], the following coefficients were computed:

        coal   – 6.1 × 10^3 lb. CO_2/m.t. coal burned
        lignite   – 3.7 × 10^3 lb. CO_2/m.t. lignite burned
        liquid Hc. – 7.0 × 10^3 lb. CO_2/m.t. liquid Hc. burned
        nat. gas   – 4.6 lb. CO_2/m.^{3} gas burned

Table 19.6 projects the amount of carbon dioxide that might be injected into the atmosphere. (Totaling the CO_2 data of table 19.6 yields table 19.7.)

The rise in atmospheric concentration of CO_2 has been due entirely to fossil fuel combustion [7:144]. Not all of the CO_2 so released is retained in the atmosphere. Estimates of retention rates range from 33 to 50 percent; the latter being the most likely [7:9, 1-44]. Using these rates, and the data in table 19.7, estimates of the rise in atmospheric CO_2 to 1980 and 2000 were made as shown in table 19.8. Note that in 1950, the world's atmosphere contained about 2.35×10^{18} gm. of CO_2 [24:175].

Atmospheric CO_2 is an affector of the earth's radiation budget. Variations in concentration can possibly affect the temperature and climate of the earth. These effects will be discussed more fully in sections entitled "Upper Atmospheric Pollution" and "Effects of the Atmospheric Pollutants."

Table 19-6
Atmospheric Carbon Dioxide Injection Due to World Fossil Fuel Consumption: to 1980 and 2000

CO_2 Source	to 1980			to 2000	
	lb. CO_2	gm. CO_2		lb. CO_2	gm. CO_2
Coal	149.9×10^{12}	68.3×10^{15}		400×10^{12}	18.2×10^{16}
Lignite	68.7×10^{12}	31.2×10^{15}		255×10^{12}	11.6×10^{16}
Liquid Hc.	246.3×10^{12}	111.8×10^{17}	7.0%	1462×10^{12}	66.4×10^{16}
			5.0%	1200×10^{12}	54.5×10^{16}
			3.0%	1008×10^{12}	45.8×10^{16}
Nat. Gas	83.8×10^{12}	38.0×10^{15}	7.5%	531×10^{12}	24.1×10^{16}
			5.0%	414×10^{12}	18.8×10^{16}
			3.0%	353×10^{12}	16.0×10^{16}

Source: Derived from Table 19-4, Table 19-5, and the following coefficients:

coal—6.1×10^3 lb. CO_2/m.t. coal burned
lignite—3.7×10^3 lb. CO_2/m.t. lignite burned
liquid Hc.—7.0×10^3 lb. CO_2/m.t. liquid Hc. burned
nat. gas—4.6 lb. CO_2/m.^{3} gas burned

Table 19-7
Summary Totals of Carbon Dioxide Injected into the Atmosphere from World Consumption of Fossil Fuels; to 1980 and 2000

To 1980			To 2000	
lb. CO_2	gm. CO_2	Code*	lb. CO_2	gm. CO_2
548.7×10^{12}	249.3×10^{15}	7/7.5	264.8×10^{13}	1.20×10^{18}
		7/5	253.1×10^{13}	1.15×10^{18}
		7/3	247.0×10^{13}	1.12×10^{18}
		5/7.5	238.6×10^{13}	1.08×10^{18}
		5/5	226.9×10^{13}	1.03×10^{18}

Table 19-7 (cont.)

	Code*	To 2000 lb. CO_2	gm. CO_2
	5/3	220.8×10^{13}	1.00×10^{18}
	3/7.5	219.4×10^{13}	0.99×10^{18}
	3/5	207.7×10^{13}	0.94×10^{18}
	3/3	201.6×10^{13}	0.92×10^{18}

Source: Table 19-6.
*Because different growth rates were used for liquid hydrocarbons and natural gas, nine possible sum combinations for CO_2 to 2000 resulted. The code denotes the rate of growth combination used, and can be read as follows:
Percent rate used for liquid Hc./percent rate used for natural gas.

Table 19-8
Estimate of Rise in Atmospheric Carbon Dioxide; 1950 to 1980 and 2000

Possible CO_2 Retention Rates	Rise in Atmospheric CO_2 Over 1950 Level to 1980	to 2000	
		Minimum	Maximum
33%	3.4%	13.0%	16.9%
40%	4.3%	15.7%	20.4%
50%	5.3%	19.6%	25.5%

Source: Table 19-7 and estimates of CO_2 retention rates from S.F. Singer GLOBAL EFFECTS OF ENVIRONMENTAL POLLUTION (New York: Springer Verlag, 1970), p. 9 and 144.

Upper Atmospheric Pollution

From 1880 to 1940, the rise in the observed mean temperature of the earth has largely been explained by the concurrent rise in atmospheric CO_2. However, since 1940, and although CO_2 production from fossil fuels has been accelerating, there has been a net drop in earth temperature [7:142].

Some scientists claim that earth cooling due to increasing upper atmospheric turbidity has countered the earth surface warming effect of the CO_2 [9]. There are disagreements as to whether the turbidity is mainly from human or natural sources, but that such turbidity can affect climate is not disputed. Thus, it is reasonable to examine what is known about its nature and possible sources.

Aerosols absorb and reflect solar radiation. For the size range of particles most common in the atmosphere (.1 to 2.5 microns), most of the reflection by the aerosols is in the forward direction [23:89]; i.e., the direction in which the radiation was originally moving. Absorbed radiation has the effect of heating the scattering medium (aerosol) from within, as well as heating the air below. At middle and lower tropospheric altitudes, depending on the relative reflectivity and absorption occurring in an aerosol layer, the earth's surface can be more warmed or cooled than if the layer wasn't present [23:89-90]. However, for stratospheric aerosol layers, the surface of the earth, hypothetically, is likely to be cooled whatever the level of radiation absorption. This conclusion seems to be verified by observations made after the Agung eruption in Bali in March, 1963. Many millions of tons of dust were spewed into the upper atmosphere, and stratospheric temperatures rose as much as seven and eight degrees C. [15:40-41]. Despite the very large upper altitude temperature rise, earth surface temperatures moved primarily in a downward direction [23:90, 104, 105], tending to confirm the hypothesis.

The long residence time (about two years on the average) of upper atmospheric particles is mainly due to the lack of precipitation at high altitudes that would directly wash them out; the lack of water vapor at high altitudes that could condense on the particles and precipitate them out; the overall low density of particles at high altitudes, minimizing the chance of particle coalescence and removal by sedimentation. Long residence time allows for particle accumulation and global dispersion.

A number of scientific observations and studies give evidence of rising atmospheric turbidity. Of significance are those made by McCormick and Ludwig (turbidity measurements in Washington, D. C., and Davos, Switzerland) [14:1358]; by Ludwig, Morgan, and McMullen in a study of twenty nonurban U. S. cities [22]; by Vincent Schaeffer in studies made in Yellowstone Park (where some of the cleanest air in the United States is found), in Flagstaff, Arizona, and in Schenectady, New York [7:159]; by M. I. Budyko in studies of solar radiation in North America and Europe [11].

These observations indicate that as man's activity, industrial and otherwise, has been increasing, atmospheric particulate loading has risen measurably. The evidence of causality is circumstantial, but requires concerned attention and much more extensive investigation.

Stratospheric airplane and balloon samplings have shown that there is a uniformly distributed aerosol cloud surrounding the globe at an altitude of about 11.5 miles, between 60°S and 70°N latitude. The particles, largely sulfur-derived, range in size from 0.1 to 2.0 microns. Analysis of the stratospheric aerosol's composition has shown that in excess of 85 percent are sulfates $[(NH_4) SO_4$ and $H_2SO_4]$. They appear to have been formed by the oxidation of SO_2 and H_2S that move to the stratosphere by vertical mixing. The particles have a residence time of about six months, shorter than the average for this altitude. They appear to grow on smaller particles that act as condensation nuclei; on reaching a size of approximately 2.0 microns they move down and out by sedimentation [5:196]. (This is in contradiction to the general rule in the stratosphere.)

Although volcanoes contribute great quantities of atmospheric sulfur dioxide (SO_2) on a random basis, the primary contributing source appears to be man's constant industrial and power generating activities. (Volcanoes also throw out great amounts of nitrogen oxides and soot [25:39].)

Approximately 65 percent of the SO_2 emanates from coal burning, 13 percent from the burning of fuel oils, and about 13 percent from the smelting of copper [6:36]. Of all SO_2 generated in the troposphere, less than one-thousandth reaches the stratosphere [5:197]. However, if surface generated SO_2 is to continue to increase as it has been (the rise being directly proportional to the increased use of sulfur-bearing fuels such as coal and oil), the amount filtering up to the stratosphere, and hence the density of the Junge dust cloud, should increase. The climatological effects of the layer might then become much more significant than at present [15:40].

Up to this point, only surface-generated pollutants have been discussed, but man appears to be about to begin introducing stratospheric pollutants directly, through the mechanism of high altitude supersonic transports (SSTs).

SSTs will release CO_2, water vapor, CO, nitrogen oxides, SO_2, soot (carbon particles), and hydrocarbons. Precisely how much of each will be produced is truly determinable only through actual flight testing. Federal Aviation Authority estimates for the 1985-1990 population of SSTs indicate that about 500 of them will be flying; 334 of U. S. origin, using 4 GE-4 engines each, and 166 of other origin, using the equivalent of 2 GE-4 engines each. An average operating time of seven hours per day is projected [23:71]. Table 19.9 provides Federal Aviation Administration estimates of SST by-product output. Congressional rejection of SST funding may call for some downward revision of these estimates.

Table 19-9
Federal Aviation Administration Estimates of Yearly Volume of Supersonic Transport By-Product Output Injected into the Stratosphere by SSTs; 1985-1990 Period

By-Product	Volume (million metric tons)*
CO_2	200.00
Water vapor	80.00
CO	2.70
NO_x	2.70
SO_2	0.10
Soot	0.01
Hc.	0.03

Source: MAN'S IMPACT ON THE GLOBAL ENVIRONMENT (Cambridge, Mass.: M.I.T. Press, 1970), Table 1-4, p. 72.
*The values are based on computations for a Boeing 2707-300, cruising at 65,000 feet at Mach 2.7, under standard weather conditions. Fuel sulfur content assumed to be 0.05 percent. Per the F.A.A. projections for SST population, 1668 GE-4 engines, operating 7 hours per day were assumed.

Of serious concern are the stratospheric particulates and water vapor. The former can occur through direct injection (SSTs), or by photochemical oxidation reactions. Possible climatological effects can be classified under the following three headings:

1. *Greenhouse effect.* This effect is defined as the ratio of earth surface temperature when not all incoming solar radiation is reflected back into space, to the surface temperature that would occur when all incoming radiation is reflected back. Manabe and Wetherald have calculated that, for a projected world increase of stratospheric water vapor of 0.2 ppm (parts per million), the earth's surface temperature would rise less than 0.1 C. [13:242]. Where SST flights would be concentrated, i.e., the northern hemisphere, water vapor content could rise by as much as 2.0 ppm. This, according to the Manabe model, would increase the surface temperature by as much as 0.5 C., not an insignificant amount. Note that high altitude water vapor and clouds have the opposite effect of aerosols.
2. *Ozone depletion.* The addition of hydrogen (via water vapor) would lower the ozone equilibrium level in the stratosphere. This layer acts as an ultraviolet filter, and a significant depletion of it would be damaging to life on the earth. However, projected ozone variations due to SST flights are not significant [23:103].
3. *Cloud formation.* High cloud formation would be enhanced by the addition of water vapor and particulates, with the concomitant effects on the surface temperature of the earth [23:92,103].

Whether SST contributions to stratospheric particulates will be climatologically significant is not known at present. However, the possible results of an increase in particle cloud density is somewhat determinable through the analogy of the Agung eruption; i.e., the effect could be to raise stratospheric and lower earth surface temperatures.

Dr. Reid Bryson claims that heat, moisture, and particulate produced contrails (i.e., condensation trails produced by aircraft) can have a significant effect on the earth's albedo or reflective power. If the conditions are right, contrails can cause the formation of cirrus clouds. These, however, are the least likely to affect the planet's heat balance, unless they are extremely dense, high in altitude, and persistent in nature [10:94]. The latter statement is supported by the work of Manabe and colleagues [7:156]. How significant is this in relation to the planet's radiation budget? Manabe's model indicates that the effect (given a 1 percent global sky coverage) would be marginal, raising the earth's equilibrium temperature about 0.3 C. [7:157].

The mass of water entering the atmosphere from natural sources is about 3½ times as great as that produced by 500 supersonic planes. However, with these planes flying at 70,000 feet, the water is introduced above the cold trap at a rate that is of the same order of magnitude as the natural rate and "is clearly going to lead to a significant increase of water vapor content at high levels" and this "would be expected to produce cloudiness and therefore a change in the albedo." However, Newell cautions that a "full dynamical numerical model is required to study the possible changes [15:37].

The SST potential for creating cloud cover is not a point of great debate; the extent and effect of the cloud cover is. Manabe's model suggests that the effects would be marginal, but he feels results are highly dependent upon "the reflectivity of the solar radiation assumed for computation" and this in turn requires further study [17:157].

The issue therefore is whether to take the risk of proceeding on a large scale now, and learning of the consequences after the fact.

Major Sources of Atmospheric Pollutants

This section will examine major sources of air pollution and estimate the pollutant volume produced. The sources to be dealt with are electric power generation, the internal combustion engine powered automobile, and four highly polluting industrial areas. Each category has its own pollution-generating characteristic, a rough indication of which can be drawn from the data for the United States in table 19.10.

Table 19-10
Approximate Percentage of Total Yearly Atmospheric Pollution (with Controls Used) Contributed by Each Source Area in the U.S.

Source Area	CO	SO$_2$	Hc.	NO$_2$	Particulates
Electric Power	2.5	34.5	22.2	16.6	54.6
Autos	87.0	4.0	61.0	42.0	9.0
Industry	4.0	45.5	5.6	25.0	18.2
Space Heating	4.0	11.5	5.6	8.2	9.1
Other	2.5	3.5	5.6	8.2	9.1
Total	100.0	100.0	100.0	100.0	100.0

Source: S. Fred Singer, "Human Energy Production as a Process in the Biosphere," SCIENTIFIC AMERICAN 223, 3 (Sept. 1970), p. 188.

The World Electric Power Industry

In 1964, approximately 75 percent of the world's electric power was produced through the use of fossil fuels, 24 percent by hydroelectric power, and the remainder by nuclear or geothermal power [4:6]. Of that produced by fossil fuels, about 68 percent was with coal, 18 percent with gas, and 14 percent with fuel oil [4:2]. World consumption of fossil-fueled electricity, on a per capita basis, is shown in table 19.11.

Table 19-11
World Consumption of Fossil Fuel for Electricity Generation; 1964 to 1968 (Electricity figures in billion kwh.)

Year	World Population (millions) (1,2,3)	Total Elect. Demand (4,5,6)	Hydro. and Nuclear Elect. Demand (7,8)	Net Fossil Fuel Elect. Demand (8,9)	Kwh. per Capita (10)	% Change Kwh./Capita
1964	3234	3110	856	2144	660	–
1968	3483	4192	1104	3088	890	35

Sources:
(1) United Nations Statistical Office, UNITED NATIONS STATISTICAL YEARBOOK, 1967 (New York, N.Y.: United Nations Publishing Service, 1968), Table 1, p. 22.
(2) United Nations Statistical Office, UNITED NATIONS STATISTICAL YEARBOOK, 1968 (New York: N.Y.: United Nations Publishing Service, 1969), Table 1, p. 23.
(3) United Nations Statistical Office, UNITED NATIONS STATISTICAL YEARBOOK, 1969 (New York, N.Y.: United Nations Publishing Service, 1970), Special Table, p. xxxi.
(4) UNITED NATIONS STATISTICAL YEARBOOK, 1967, op. cit., Table 1, p. 22.
(5) UNITED NATIONS STATISTICAL YEARBOOK, 1968, op. cit., Table 1, p. 23.
(6) UNITED NATIONS STATISTICAL YEARBOOK, 1969, op. cit., Special Table, p. xxxi.
(7) Ibid., Table 9, p. 40.
(8) Figures given in metric ton coal equivalent. Converted to kwh. x 10^9 using U.N. conversion factor: .125 m.t. coal equivalent 1000 kwh.
(9) Obtained by subtracting hydro. and nuclear demand from total demand. The remainder includes geothermal electricity demand, but this is negligible. (See H.B. Guyol, THE WORLD ELECTRIC POWER INDUSTRY (Berkely, Calif: University of California Press, 1969, Table 1, p. 8.)
(10) For fossil fueled electricity only.

Using population as a base from which to project electricity demand is fairly sound. Even in less developed countries, where uses for electric power are very basic (lighting, light industry, etc.), a growing population will demand greater amounts of electricity.

Projections for world population growth are widely varied. In the past they have been almost exclusively in error on the low side [20:5]. For this chapter, a projection rate between the U. N. medium and high projections will be used. This is a rate of 2 percent per year [2:54-55]. Brown states that by the end of this century, it is expected that the rate will reach 3 percent [20:7]. This portends a very large power demand increase by 2000. If for example, one assumes a rate of population increase of 2.5 percent between 1968 and 2000, and just a linear rise in electricity demand per capita, the demand level by 2000 will be four times the 1968 level.

Although table 19.11 data show an average yearly per capita demand rise of 8 percent per year, a growth rate of 6 percent is used herein, to project 1980 demand. The lower rate was arbitrarily chosen to reflect a displacement of fossil fuel by nuclear sources.

Table 19-12
World Fossil Fueled Electricity Demand; 1969 to 1980

Year	World Population (millions)	Fossil Fueled Kwh./Capita	Total Fossil Fueled Electricity Demand (billion kwh.)
1969	3560	944	3360
1980	4440	1800	8000

Source: Derived from Table 19-11 assuming a world population projection rate of 2 percent/year, and a per capita growth rate/year of 6 percent.

Guyol illustrates the shift in the fossil fuel mix used to generate the world's electricity [4:109].

	1958	1964
Coal shifted from	61%	to 55% of total fossil fuel used
Lignite didn't shift	14%	— 14% of total fossil fuel used
Fuel oil shifted from	10%	to 13% of total fossil fuel used
Nat. gas shifted from	15%	to 18% of total fossil fuel used

It is assumed that these shifts occurred in a linear manner, and will continue to do so to 1980. This assumption underlies the determination of the fuel mix as shown in table 19.13. This table shows projected amounts for electricity produced from each fuel type.

The electricity levels shown in table 19.13 are not produced at anything near 100 percent efficiency. Thus, in order to determine the fuel consumed, estimates of efficiency levels were made using Guyol's 1964 world data [4:119]. It was assumed that they would be representative through 1980.

Fuel consumption projections are presented in table 19.14.

If the assumptions made to this point are valid, the 1980 consumption level of fossil fuel should be more than double the 1969 level. A proportional rise in pollution would result, all things being equal. Changes in combustion efficiencies will occur, new improved power plants will rise, and old ones will deteriorate. Power generating stations, especially in the United States, will be forced to reduce pollutant output. Others, particularly in countries just developing extensive generating capacities, will see few if any restrictions on polluted effluent. The net effect of these and other variables is highly unpredictable. The following data and computations provide a rough idea of the magnitude of potential pollutant generation to 1980.

The emission factors in table 19.15 were gathered in 1965, and only in the U.S. The potentially critical pollutants due to fossil-fueled power plant operation, SO_2 and particulates, can possibly double in the next decade as shown in table 19.16 derived from data in tables 19.14 and 19.15.

Table 19-13
World Electricity Production from Fossil Fuel; 1969 to 2000

Year	Total Fossil Fueled Elect. Demand (billion kwh.)	Percentage from Each Fuel				Electricity from Each Fuel Source (billion kwh.)			
		Coal	Lig.	Oil	Nat. Gas	Coal	Lig.	Oil	Nat. Gas
1969	3360	50	14	15.5	20.5	1680	470	520	690
1980	8000	39	14	21.0	26.0	3120	1120	1680	2080

Source: Table 19-12 and assuming continued linear shift in fuel usage at rate for 1958-1964 given in WORLD ELECTRIC POWER INDUSTRY, p. 109.

Table 19-14
Consumption of Fossil Fuel for World Electricity Production, 1969 to 1980

	Elect. Demand From Each Fuel (billion kwh.) (1)				Total Fuel Consumed (billion kwh. equiv.) (2)				Total Fuel Consumed (million m.t.) (m.^{3} x 10^9) (3)			
Year	Coal	Lig.	Oil	Gas	Coal	Lig.	Oil	Gas	Coal	Lig.	Oil	Gas
1969	1680	470	520	690	5,040	1,410	1,040	1,590	630	209	100	150
1980	3120	1,120	1,680	2,080	9,360	3,360	3,360	4,800	1,170	498	322	453

Sources:
(1) Table 19-13.
(2) Derived using following efficiency factors:
    coal–33%; 3 kwh. energy in for each kwh. of electric output.
    lignite–33%; 3 kwh. energy in for each kwh. of electric output.
    fuel oil–50%; 2 kwh. energy in for each kwh. of electric output.
    nat. gas–44%; 2.3 kwh. energy in for each kwh. of electric output.
    Based on H.B. Guyol, THE WORLD ELECTRIC POWER INDUSTRY (Berkeley, Calif. University of California Press, 1969), p. 119.
(3) UNITED NATIONS STATISTICAL YEARBOOK, 1969, (New York: United Nations Statistical Office, 1970), p. 763. Figures derived using the following conversion factors: coal–8000 kwh./m.t.; lignite-6750 kwh./m.t.; fuel oil-10,400 kwh./m.t.; nat. gas-10,600 kwh./1000 m.^{3}.

Table 19-15
Pollution Coefficients for Fossil Fueled Electricity Generation (No Controls)
(coal, oil, and lignite in lb./m.t.; nat. gas in lb./m.^{3} × 10^6)

Fuel Used	Coefficient for				
	SO$_2$	NO$_2$	Hc.	CO	Particulates
Coal	57(a)	20(b)	0.2(c)	0.5(d)	16*(e)
Lignite	57(a)	17(b)	0.2(c)	0.5(d)	16*(e)
Fuel Oil (Residual)	70(g)	31(g)	0.94(g)	neg.	2.9(g)
Nat. Gas	14.1(h)	13.8(h)(i)	neg.	neg.	530.0(h)

Table 19-15 (cont.)

Sources:

(a) U.S. Department of Health, Education, and Welfare, COMPILATION OF AIR POLLUTANT EMISSION FACTORS, (Washington, D.C.: U.S. Government Printing Office, 1968), Table 3, p. 5.

(b) E. Robinson and R.C. Robbins, SOURCES, ABUNDANCE, AND FATE OF GASEOUS ATMOSPHERIC POLLUTANTS, (Menlo Park, Calif.: Stanford Research Institute, 1968), Table VII, p. 69. NO_2 generation is largely a function of heat of combustion. Since lignite has about 85 percent of the energy capacity of coal, its coefficient was computed by lowering that for coal by 15 percent.

(c) Ibid., Table XIII, p. 99.

(d) Ibid., Table V, p. 50.

(e) COMPILATION OF AIR POLLUTANT EMISSION FACTORS, op. cit., Table I, p. 4. Coefficient is for the general type of furnace using pulverized coal.

(g) Ibid., Table 5, p. 7. Given in lb. per 1,000 gallons of fuel burned. Converted to m.t. by using a fuel density of 7.5 lb. per gallon.

(h) Ibid., Table 4, p. 6. Given in lb./ft.$^3 \times 10^6$. Converted to lb./m.$^3 \times 10^9$ using a factor of 35.31 ft.^{3}/m.^{3}.

(i) NO_2 coefficient in lb. $\times 10^3$/m.$^3 \times 10^6$ for nat. gas.

(*) Fly ash content percentage. This percentage is multiplied by 16 to get the proper coefficient. Nine percent is the average fly ash content for the U.S., and is used as representative for the world here.

Table 19-16
Summary of Total Potential Pollution from the World's Fossil Fueled Electric Power Plants; 1969 to 1980 (million Metric tons)

Year	SO_2	CO	Hc.	NO_2	Particulates
1969	25.0	0.20	0.12	9.9	55.5
1980	53.4	0.38	0.29	22.0	110.0

Sources: Table 19-14 and Table 19-15.

Automobiles

The automobile powered by the internal combustion engine is a major source of atmospheric pollution. It is particularly proficient in its production of CO (it contributes about 75 percent of the world's output), as well an NO_2 and hydrocarbons [7:61-62].

In 1968, the world's automobile population was 217 million, having tripled since 1950 [30:453]. On a per capita basis, the change from 1964 to 1968 was from .052 to .062. The rise was rather linear, and is used as the basis for projecting auto per capita figures to 1980 (see table 19.17). It is estimated that approximately 2 gallons of gasoline per day per car is consumed on the average; 730 gallons per car per year. This figure was verified through computations for the years 1966, 1967, and 1968. Data sources were the *Statistical Abstract of*

Table 19-17
World Gasoline Consumption Projection: 1969 to 1980

Year	Autos/Cap. (1)	World Pop. (millions)(2)	World Autos (millions)	Gas. Demand (billion gallons) (730 gal./auto/yr.) X (number autos)(3)
1969	.066	3,560	242	177
1980	.094	4,440	417	304

Sources:
(1) United States Department of Commerce, STATISTICAL ABSTRACT OF THE UNITED STATES, 1970, Washington, D.C.: Government Printing Office, 1969), Table 838, p. 543.
(2) Table 5.1.2, p. 19.
(3) President's Science and Advisory Committee, RESTORING THE QUALITY OF OUR ENVIRONMENT, (Washington, D.C.: Government Printing Office, 1965), p. 67.

the United States, 1970, Table 838, p. 543, for world auto population, and the *United Nations Statistical Yearbook,* 1969, Table 1250, p. 805, for world gasoline consumption. The per capita and gasoline consumption data allow for a projection of world gasoline consumption to 1980.

The coefficients were compiled in 1965 and don't reflect recent changes in engine designs and efficiencies, driving conditions, etc. Whether the net effect of these and other factors has been the lessening or increasing of automotive emissions per gallon of gasoline used (discounting the use of controls) is not known at present. Despite these uncertainties, some idea of the magnitude of potential pollution, and the effects of various levels of pollution control, can be drawn from the data (see table 19.18).

By 1980, the rise in potential pollutant emissions would require reductions of 43 percent in CO, 42 percent in hydrocarbons, 42 percent in NO_2, 41 percent in sulfur oxides, and 49 percent in particulates, just to reduce the emissions to 1969 levels. The question is, will the pollutant volume due only to the increase in the number of cars nullify, or significantly mitigate the effect of pollution controls. If, for example, controls were put into effect immediately, and all the above-discussed pollutants were reduced 75 percent on a world average basis, and this level of control was maintained through 1980, the 1970-1980 projected increase in auto population of about 170 percent would effectively reduce the control level to around 44 percent.

Table 19-18
Total Potential Pollutant Emissions from the World's Automobiles: 1969 to 1980 (million metric tons)

Year	CO	Hc.	NO_2	SO_2	Particulates
1969	186	16	9	0.7	1.0
1980	320	28	16	1.0	1.7

Sources: Table 19-17, p. 23, and the following coefficients in metric tons per thousand gallons of gas, are: 1.05 m.t. of CO, .09 m.t. of Hc., .05 m.t. of NO_2, .004 of m.t. of SO_2, and .005 m.t. of particulates. COMPILATION OF AIR POLLUTANT EMISSION FACTORS (Washington, D.C.: U.S. Government Printing Office, 1968), Table 8-8, p. 206.

The net effectiveness of pollutant reduction from cars will be a function of control effectiveness and maintenance, worldwide adoption rate of controls, effectiveness of monitoring and enforcement by government agencies, and the adoption rate of alternate modes of transportation.

Industrial Processes

Industrial processes are major contributors of atmospheric sulfur oxides, nitrogen oxides, and particulates. Pollutant volume is a function of the nature of the industry; volume and rate of production; efficiency and maintenance of production and control equipment. To estimate pollution output for various world industry sectors, it is assumed that no controls are used, that processes in a given industry don't vary significantly the world over, and that fuels used in a given world industry don't vary significantly.

Petroleum Refining. World crude oil demand from 1963 to 1968 rose from 377×10^9 gallons, to 556×10^9 gallons; a 47 percent increase. On a per capita basis, it rose 31 percent (120 to 157 gallons/capita) [29:40]. Petroleum refining contributes most significantly to SO_2 and hydrocarbon pollution.

It has been estimated that the rate of generation of sulfur dioxide in petroleum refining is 50 short tons per 100,000 barrels of crude refined [25:16] and 56 short tons of hydrocarbons for each 10,000 barrels of crude refined [25:99]. Table 19.19 presents estimates of the potential levels of these two pollutants resulting from petroleum refining using the above coefficients and assuming a continuation of rate of increase in demand for crude oil that obtained from 1963 to 1968.

Table 19-19
Total Potential Pollution from Petroleum Refining: Sulfur Dioxide and Hydrocarbons; 1968 to 1980 (million metric tons)

Year	SO_2	Hc
1968	5.98	74.0
1980	15.15	187.0

Source: Derived assuming continuation of rate world crude oil demand exhibited 1963 to 1968 (See UNITED NATIONS STATISTICAL YEARBOOK, 1969, op. cit., p. 40) and following coefficients for rate of generation: SO_2–50 short tons/10^5 bbl.; Hc–56 short tons/10^4 bbl.

Source: E. Robinson and R.C. Robbins, SOURCES, ABUNDANCE, AND FATE OF GASEOUS ATMOSPHERIC POLLUTANTS, (Menlo Park, Calif.: Stanford Research Institute, 1968), Table II, p. 215, ibid., Table XIII, p. 99.

Copper, Lead, and Zinc Smelting. Copper, lead, and zinc smelting contribute importantly to SO_2 pollution. In 1956 about 13 percent of the world's SO_2 was generated by copper smelting alone [6:28].

The 1964 to 1968 activity of the world's copper, lead, and zinc smelting industries is shown in table 19.20.

Table 19:20 shows an average copper demand/capita growth rate of 1.2 percent per year. This figure is used to project 1980 demand. Beyond 1980, the rates should decrease because of limitations in supply that will increase the diseconomies in the extraction process. New technological substitutes should lessen, or remove the need for copper in some major areas. Even now, aluminum is replacing copper in some applications [8:206].

Table 19-20
World Copper, Lead, and Zinc Smelting Output: 1964 to 1968
(million metric tons)

Year	Copper (1)	Lead (2)	Zinc (3)	m.t./capita $\times 10^{-3}$ Copper	Lead	Zinc
1964	4.85	2.58	3.69	1.50	0.80	1.14
1968	5.49	2.93	4.54	1.57	0.84	1.30

Sources:
(1) United Nations Statistical Office, UNITED NATIONS STATISTICAL YEARBOOK, 1969, (New York: United Nations Statistical Office, 1970), Table 122, p. 281.
(2) Ibid., Table 124, p. 283.
(3) Ibid., Table 127, p. 286.

Table 19-21
Total Potential Sulfur Dioxide Output from the World's Copper, Lead, and Zinc Smelters: 1968-1980 (million metric tons)

Year	Copper	Lead	Zinc	Total
1968	11.0	1.5	1.5	14.0
1980	16.4	2.1	2.6	21.1

Sources: Yearly per capita growth rates of 1.2 and 2.5 percent for lead and zinc respectively, are indicated in Table 19-20. These rates, and that for copper, yield 1980 consumption rates of 8.20×10^6 m.t. for copper, 4.23×10^6 m.t. for lead, and 7.73×10^6 m.t. for zinc.

The per capita figures projected to 1980 are (in m.t. $\times 10^{-3}$) computed as 1.85, 0.96, and 1.82, for copper, lead, and zinc respectively. These figures, in conjunction with population data (Table 19-12) were used to arrive at volume data.

Two tons of SO_2 are generated per ton of copper produced. For lead and zinc the figures are 0.5 ton/ton and 0.33 ton/ton respectively. E. Robinson and R.C. Robbins, SOURCES, ABUNDANCE, AND FATE OF GASEOUS ATMOSPHERIC POLLUTANTS, (Menlo Park, Calif.: Stanford Research Institute, 1968), Table II.

Steel Production. From 1962 to 1968 world steel production rose 47 percent, and per capita production rose 32 percent. In actual figures this was an increase of from 360.2×10^6 m.t. (.115 m.t./capita) to 528.7×10^6 m.t. (.152 m.t./capita) [29:279].

Among the countries of the world, steel production has risen very unevenly, the largest gains occurring (over the 1957-1967 period) in Japan (139 kg./capita to 513 kg./capita) and the U.S.S.R. (263 kg./capita to 415 kg./capita) [8:198]. As with production, the world's consumption of steel is not very evenly distributed, being concentrated in a minority of the richer nations. Eighteen of these countries (680×10^6 people) consume steel at a rate ranging from 30 to 70 times that of the rest of the world (3.0×10^9 people) on a per capita basis [18:206]. This kind of distribution is similar to that for world per capita energy consumption and per capita income. It can be concluded that the richest countries of the world supply an inordinate amount of atmospheric pollution.

If the United States can be assumed typical of industrially developed countries, then as such nations age, their yearly consumption of steel per capita should taper off. United States per capita steel consumption from 1957 to 1967 rose slowly in relation to that for Japan and the U.S.S.R. This slowing could indicate that, for a given country, with its culture, demand for products, and barring any technological innovations or the rising of other significant affectors of demand, there exists a finite demand saturation level. It isn't likely that many other countries will approach such a saturation state within the next ten years. Japanese steel consumption should taper off, since its recent growth rate has been phenomenal; but as a whole, world steel demand through 1980 should increase at about the same average yearly rate as occurred over the 1962 to 1968 period. This rate was 4.8 percent per year. Therefore, by 1980, world steel demand should be approximately 920×10^6 m.t. (.197 m.t./capita), up 66 percent over the 1969 level of 554×10^6 m.t. (.156 m.t./capita).

The steel industry is responsible for a very great amount of particulate generation. Except for the basic oxygen furnace, other furnaces and processes used in steel production produce particles larger than 5 microns in diameter [31:26]. Such particles serve more readily as condensation nuclei for atmospheric moisture, and can have a direct effect on the amount of precipitation in a local area [5:137]. Particle size, as well as smokestack height, exhaust velocity, and meteorological conditions, are the major variables determining particle settling time, and hence particle dispersion [16:271]. Since larger particles (in excess of 2 microns) tend to settle within a matter of hours, the particulates from steel production have a relatively local effect.

Most U. S. mills have some kind of controls (e.g., baghouse, electrostatic precipitators, scrubbers). Data as to the extent of control usage on a worldwide basis is not available. In any case, controls are not considered in this discussion.

Particulate pollution estimates were made using a coefficient of 200 lbs. of particulates per ton of steel produced [32:99] (see table 19.22).

Sulfuric Acid. As might be expected, the major pollutant resulting from the manufacture of sulfuric acid is sulfur dioxide (SO_2).

World production over the 1962-1968 period rose from 53.1×10^6 m.t. to

Table 19-22
Total Potential Particulate Pollution from the World's Steel Mills: 1969 to 1980 (Million Metric Tons)

Year	Particulate Volume
1969	22
1980	37

Source: Assuming following world output (in million metric tons—see above) 1969-554; 1980-920. Assume 200 pounds of particulate per ton of steel produced (WASTE MANAGEMENT, p. 99).

79.6×10^6 m.t. [29:254]. A sharp drop in the yearly rate of growth of sulfuric acid occurred recently, and appears to indicate the beginning of a trend shift. Therefore, projections to 1980 were made on a linear rather than geometric basis. The 1980 estimate of sulfuric acid production is 141×10^6 m.t.

On the average, about 45 pounds of SO_2 are generated for every metric ton of sulfuric acid produced, using a simple average of the range (20 to 70 lb. m.t.) [31:18]. This coefficient, in conjunction with the volume figures for 1969 and 1980, yields pollution levels of 1.73×10^6 m.t. in 1969, versus 2.85×10^6 m.t. in 1980; a rise of 65 percent.

Effects of the Atmospheric Pollutants

Carbon Dioxide. CO_2 is suspected of having a significant effect on the radiation budget of the earth. The well-known "greenhouse effect" explains that since CO_2 is a strong absorber and back radiator of infrared wavelengths, and since such wavelengths constitute a large part of earth-reflected solar energy, atmospheric CO_2 in effect acts as a greenhouse, trapping earth-radiated energy in the troposphere; warming the earth's surface. This theory is supported by the strong correlation of atmospheric CO_2 concentration and temperature over the 1880-1940 period [7:142]. However, the post-1940 period has shown a drop in the earth's observed mean temperature, and yet the CO_2 concentration continues to increase [19:184]. Why has the temperature dropped? Has some effect countered that of the CO_2?

J. Murray Mitchell states that theoretically the CO_2 warming effect should be nearly linear with the range of temperatures and CO_2 concentrations involved, and that "the rate of warming due to CO_2 was probably greater since 1940 than it was at any earlier time. Since world mean temperatures are estimated to have declined by about 0.3 C. during this period, Mitchell concludes that some climatic cooling mechanism coming into operation during the past quarter century has been more influential than the CO_2 effect in the same period [7:145].

Mitchell goes on to maintain that atmospheric particulate loading was what countered the CO_2 effect, but the loading from human sources was insufficient to the task; that mainly volcanic dust led to the earth's cooling [7:146, 153].

The volcanic loading, however, was transient, and the warming effect of CO_2 might again prevail. Man-created dust, though, is increasing at a greater rate than CO_2, and the cooling effect of this steady atmospheric particulate loading promises to overtake the warming effect of CO_2 on a more permanent basis in the future. Nevertheless, should the warming effects of the CO_2 prevail, the potential effects include:

1. *CO_2 runaway*. The primary sink for atmospheric CO_2 is thought to be the oceans, whose waters contain about 60 times that of the atmosphere. CO_2 solubility is greatly affected by water temperature; the higher the temperature, the lower the amount of dissolved CO_2. Therefore, it is feared that if CO_2 warms the air and the oceans' surface layer, it will be driven out of solution. This in turn will cause more atmospheric and oceanic warming, and drive more CO_2 out of solution, etc. In addition, pH level drops with a drop in free CO_2. Hence, the oceans' acidity level will rise, and this can adversely affect marine life.
2. *Melting of the Antarctic ice cap*. It is estimated that a 2 percent change in the earth's radiation budget could occur by 2000 if atmospheric CO_2 increases by 25 percent. If half of the net energy increases were concentrated in the Antarctica, the ice would melt in 400 years' time. If 1,000 years were the melting period, sea level would rise about 4 feet every ten years; 1,000 times the present rate [24:123].

The melting of the ice would increase the world's ocean surface area. This in turn would increase the oceans' potential for CO_2 absorption, thereby countering the CO_2 runaway effect mentioned previously. Predicting the net effect of these counteracting variables is beyond the scope of this chapter.

3. *Sea life*. Oceanic warming would affect sea life due to changes in, among other things, CO_2 and O_2 concentration.
4. *Plant life*. Where CO_2 is the limiting factor on plant growth, such growth would be stimulated [24:124].

Sulfur Dioxide. On a global scale, the effect of SO_2 in the creation of stratospheric dust has already been discussed (see Upper Atmospheric Pollution, SO_2 oxidation to form sulfate dust particles). The apparent relationship between the large transient quantities of dust that were generated by volcanoes, and the drop in the earth's temperature, seems to make clear that more permanent increases due to human sources can have significant climatological effects. The Agung eruption in Bali in March 1963 resulted in the largest climatic change ever observed by man; a stratospheric temperature rise of up to 8° C. [15:40-41].

It is to be noted that over half of the land-produced sulfur compounds (70 × 10^6 m.t. of SO_2 in 1968) are due to human pollution [25:39]. Effects of SO_2 on a local scale are:

1. *Plant damage*. This will occur if the exposure levels exceed 50 pphm (parts per hundred million) for a few hours or more [5:356].

2. *Acid formation.* In moist air, SO_2 may form sulfuric acid, affecting plants, property, and human and animal life. The acid precipitates into bodies of water, lowering the pH and making it difficult for fish to survive. Below a pH of 7.5, many marine animals cannot live. Below a pH of 7.0, carbonate in water will not come out of solution, making the production of any skeletal structure impossible [21:102; 17:100].

Carbon Monoxide. Carbon monoxide (CO) originates almost exclusively from automobiles. It is found mainly in the atmosphere north of the equator. The northern hemisphere, too, is where most of the world's automobiles are located.

With long exposure, CO can be toxic to humans. Damage is proportional to exposure time and atmospheric concentration. The CO that enters the blood stream reduces its capacity to carry oxygen. This can impair mental functions when one is exposed to concentrations of 200 ppm for around 15 minutes, or 50 ppm for 2 hours or more [15:38-39]. Levels of 50 ppm are common in city streets during heavy traffic, and hundreds of ppm are not unusual in tunnels and underground garages [15:38-39].

Nitrogen Oxides. On a global scale, the NO_2 produced in the natural cycle far exceeds that created by man (10×10^8 tons/yr. versus 50×10^6 tons/yr.) [25:68,91].

On a local scale, NO_2 is a major constituent in the creation of photochemical smog. It photodissociates into nitric oxide and atomic oxygen, which then forms ozone and other oxidation products of hydrocarbons [5:357].

Hydrocarbons. There is a great lack of data on the natural cycle of hydrocarbons, and hence on the worldwide effect on man-created hydrocarbon pollutants.

On a local scale, not all hydrocarbons are serious pollutants. Certain olefins (C_nH_{2n}) react with ozone and atomic oxygen and form a number of compounds including formaldehyde. These compounds irritate eyes, lungs, and damage plants [5:358]. The reactive hydrocarbons are released in different quantities by different sources. For example, the reactive percentage in autos is 44 percent; in oil refineries, 14 percent; in coal-fired power plants, 15 percent; forest fires, 21 percent [25:99].

Particulates. Particulates of different sizes and types have many different effects, the more obvious of which are smoke, haze, and dirt.

Local precipitation frequency and character appear to be affected by particulates in the air. Observations have been made that indicate this. Among them are:

1. Vincent Schaeffer, head of the Atmospheric Science Research Center at the State University of New York, in a flight over Africa, noted that particulate from massive brush and forest burning apparently prevented rain from falling from huge cumulus clouds. It seems that so many cloud droplet nuclei were entering the clouds that the coalescing process required for precipitation to begin was impaired [18:200].

2. Lead iodide particles formed from automobile exhaust lead and atmospheric iodine appear to be effective ice condensation nuclei. Under the proper conditions the lead iodide crystals could lead to the formation of ice clouds [18:201].
3. S. A. Changnon, of the Illinois State Water Survey, observed a strong correlation between the increase in precipitation in LaPorte, Indiana, over the past forty years, and the production curves for iron and steel in Chicago and Gary. The latter two cities are about 30 miles west of LaPorte. In 1965 for example, LaPorte had 31 percent more precipitation than nearby weather stations. In addition, its weather closely varied with Chicago's pollution levels. Apparently, the heat and moisture from the manufacturing plants, combined with the cool night and early morning air, stimulated the formation of cumulus clouds. These clouds drifted over Lake Michigan towards LaPorte, picking up moisture on the way. When they arrived, they were seeded by the particulates that had also drifted east from the Chicago and Gary industrial areas, and precipitation (man-made) resulted [12:62].

Conclusion

The projections presented in this chapter indicate that significant increases in potential pollutant output from some of the world's major air polluting sources will occur in the next decade. The data used to arrive at these projections are of a very preliminary nature, often being no better than educated guesses. Nevertheless, they are all that are available at present, and they did allow for computations that show roughly where we may be heading. The question is, what do the pollutant levels mean in terms of the effects on the global environment? Definitive answers aren't yet available. Most overwhelming is the great lack of knowledge in relation to worldwide effects of locally produced air pollutants.

Among the things to be learned are, how global dispersion of the pollutants occurs, what the sinks for the pollutants are, what the true background levels of their naturally produced counterparts are, what the significance of each pollutant is in relation to its volume and its interactions with other pollutants and nonpollutants, and what the properties (optical, chemical, mechanical, and otherwise) of the pollutants are. These and many other questions can only be answered through intensive and extensive future study.

Predictions can be made with dynamic-numerical computer models of the atmosphere. Such models as presently exist are only in their formative stages. Truly representative models will be incredibly complex, and much development work and research work is needed.

The arguments for continued, unabated growth and progress are economically based; the arguments against are environmentally based. The former can be backed by figures and statistics; the latter by hypothetical argument only. Obviously, industrial operations and technological progress that stimulates economic development cannot be halted while environmental effects are examined, but neither can they continue without any intelligent assessment of the probabilities of their effects on the atmosphere of the planet.

Controls are being implemented in the United States and some other countries, but a large percentage of the world, especially the developing nations, will be hesitant to invest their capital in such nonproductive assets as pollution controls. Hopefully, a centralized organization can be utilized to coordinate worldwide pollution control efforts, and plan to minimize the economic disruptions that might result.

The most logical organization to fulfill this role is the United Nations. Already, many of the nations of the world are gravitating toward this body in the initiation of a mutual effort to understand and eventually solve global environmental problems. In 1972, a conference dealing with critical problems of our global environment was held under the auspices of the U. N.

A centrally coordinated, worldwide pollution monitoring effort is the first step that must be taken. This will help to identify precisely the nature of the problems, their sources, and to more accurately predict the consequences.

When the answers exist, the economic arguments will be more effectively counterable, but by then it may be too late. Therefore, it is absolutely necessary for the exercise of restraint in the light of our present lack of knowledge.

Notes

Books

1. Kenneth Boulding, *Human Values on the Spaceship Earth* (New York, N.Y.: National Council of Churches of Christ in the U.S.A., 1966), p. 14.

2. The Committee on Resources and Man of the National Academy of Sciences, *Resources and Man* (San Francisco, Calif.: W. H. Freeman and Company, 1969), 259 pp.

3. J. L. Fisher and N. Potter, *World Prospects for Natural Resources – Some Projections of Demand and Indicators of Supply to the Year 2000* (Baltimore, Md.: Johns Hopkins Press, 1964), 73 pp.

4. H. B. Guyol, *The World Electric Power Industry* (Berkeley, Calif.: University of California Press, 1969), 365 pp.

5. Christian E. Junge, *Air Chemistry and Radioactivity* (New York, N.Y.: Academic Press, 1969), 382 pp.

6. P. L. Magill, F. R. Holden, and C. Ackley (Eds.), *Air Pollution Handbook* (New York, N.Y.: McGraw-Hill, 1956), 856 pp.

7. S. F. Singer, *Global Effects of Environmental Pollution* (New York, N.Y.: Springer Verlag, New York, Inc., 1970), 218 pp.

Articles

8. H. Brown, "Human Materials Production as a Process in the Biosphere," *Scientific American* 223, 3 (Sept. 1970), pp. 194-208.

9. Reid Bryson, "Is Man Changing the Climate of the Earth?" *Saturday Review* (April 1, 1967), pp. 52-55.

10. Reid Bryson, "The Supercivilized Weather and Sky Show," *Natural History* 89, 7 (Aug.-Sept. 1970), pp. 92-113.

11. M. I. Budyko, "The Effects of Solar Radiation Variations on the Climate of the Earth," *Tellus,* Vol. 21, pp.

12. S. A. Changnon, "The LaPorte Weather Anomaly: Factor Fiction?" *Bulletin of the American Meteorological Society* 49, 4 (1967), pp. 58-64.

13. S. Manabe and T. Wetherald, "Thermal Equilibrium of the Atmosphere With A Given Distribution of Relative Humidity," *Journal of Atmospheric Science* 24 (1967), pp. 238-43.

14. R. A. McCormick and J. H. Ludwig, *Science* (1967), pp. 1358-59.

15. R.E. Newell, "The Global Circulation of Atmospheric Pollutants," *Scientific American* 224, 1 (January 1971), pp. 32-42.

16. H. Panofsky, "Air Pollution Meteorology," *American Scientist* (Summer, 1969), pp. 269-85.

17. H. L. Penman, "The Water Cycle," *Scientific American* 223, 3 (Sept. 1970), pp. 98-108.

18. Vincent Schaeffer, "Inadvertent Modification of the Atmosphere by Air Pollution," *Bulletin of the American Meteorological Society* 50, 4 (1968), pp. 199-206.

19. S. Fred Singer, "Human Energy Production as a Process in the Biosphere," *Scientific American* 223, 3 (Sept. 1970), pp. 174-90.

Reports

20. Harrison Brown, *The Next 90 Years,* Report to the Office for Industrial Associations, Pasadena, Calif., March 7-8, 1968 (Pasadena, Calif.: California Institute of Technology, 1968).

21. Legislative Reference Service of the Library of Congress, Report to the Joint Economic Commission of the Congress of the United States, *The Economy, Energy, and the Environment* (Washington, D.C.: Government Printing Office, 1970).

22. J. H. Ludwig, G. B. Morgan, and T. B. McMullen, *Trends in Urban Air Quality,* Paper for the American Geophysical Union, San Francisco, Calif., 1969.

23. *Man's Impact on the Global Environment,* Report to the 1972 United Nations Conference on the Human Environment, Carroll L. Wilson, chairman (Cambridge, Mass.: M.I.T. Press, 1970).

24. *Restoring the Quality of Our Environment, Report of the Environmental Pollution Panel, President's Science and Advisory Committee,* John M. Tukey, chairman (Washington, D.C.: Government Printing Office, 1965).

25. E. Robinson and R. C. Robbins, *Sources, Abundance, and Fate of Gaseous Atmospheric Pollutants,* Report for the American Petroleum Institute, Menlo Park, Calif., 1968 (Menlo Park, Calif.: Stanford Research Institute, 1968).

Miscellaneous

26. The American Philosophical Society, *Proceedings of a Symposium on Atmospheric Pollution: Its Long Term Implications,* Philadelphia, Pa., 1969.

27. United Nations Statistical Office, *United Nations Statistical Yearbook, 1967* (New York, N.Y.: United Nations Statistical Office, 1968).

28. United Nations Statistical Office, *United Nations Statistical Yearbook, 1968* (New York, N.Y.: United Nations Statistical Office, 1969).

29. United Nations Statistical Office, *United Nations Statistical Yearbook, 1969* (New York, N.Y.: United Nations Statistical Office, 1970).

30. United States Department of Commerce, *Statistical Abstract of the United States, 1970* (Washington, D.C.: Government Printing Office, 1969).

31. U. S. Department of Health, Education and Welfare, *Compilation of Air Pollutant Emission Factors* (Washington, D.C.: Government Printing Office, 1968).

32. *Waste Management,* Regional Plan Association, New York, March 1968.

20 Aircraft Noise Pollution and the U.S. Air Transport Industry Between 1970 and 1980

RONALD BRUCE

The adverse potentials of aircraft noise have been recognized for some time. FAA administrator E. R. Quesada, in 1960, anticipating the present problem, stated his belief that "... the very future of our nation's air commerce may very well rest on the success or failure of our efforts to reduce aircraft noise" [25:13]. Since 1960, the situation has deteriorated. The objective of this chapter is to provide a balanced view and measure of the air transport industry's noise pollution problem as it exists now and as it may exist over the next decade. Toward such end, and in the sequence indicated, this report will present:

1. A general description of the air transport product, resources employed, and governmental influence on the industry.
2. A description and forecast of the demand for air transport.
3. A description and forecast of the supply of air transport.
4. A description of the methods of measuring aircraft noise and predicting human annoyance and response.
5. A general description and forecast of the areas and populations exposed to significant aircraft noise levels.
6. A description of the actions taken by populations exposed to aircraft noise.
7. A discussion of methods for reducing human exposure to aircraft noise.
8. Conclusions.

It is hoped that this particular approach will provide useful insights for those who are interested in maintaining the economic viability of air transportation while dealing with environmental pollution in general and aircraft noise in particular.

1. The Product, Resources, and Government Relations

The primary features of the air transportation product, which have distinguished it from competing transport modes, established its marketing opportunities and limits, and have, directly or indirectly, caused or acerbated the noise pollution problem are: (1) safety, (2) speed, (3) comfort, (4) price, and (5) convenience.

Although these factors are of basic importance to the successful marketing of the air transport product, they are often unrecognized, misunderstood, or lightly dismissed as unimportant, by the legions of well-intentioned but naive proponents and critics of the industry. The mishandling of any of these features, in the process of treating aircraft noise pollution, would undoubtedly affect industry profitability and growth potential. Critics of the industry, including those elements of the population offended by aircraft noise, who would propose economically viable solutions to the noise problem, would have to shape and proportion their solutions with a view to the limited resources of the industry.

The total resources necessary for air transportation are now provided by three classes of owners.

1. The airlines, which own aircraft and furnish air transport;
2. The federal government, which owns resources essential to its furnishing of guidance to enroute aircraft;
3. Local governments, which own airports, terminals, and access and egress systems.

The airlines' commitment of resources is large and scheduled to grow rapidly. In 1968 all the major (i.e., federally certified or franchised) airlines together employed 294,000 people, 2400 aircraft, and a great number of pieces of ground and maintenance equipment, and had total capital investment of $6.75 billion [10:82]. By June of 1970, the total investment (plus imputed debt for equipment under lease) of the single largest and exclusively international airline, and the eleven trunk airlines together, alone totaled $10.7 billion [4:F3]. The airlines' investment is, moreover, scheduled and committed (via contracts and/or advance payments to equipment manufacturers) to grow rapidly to meet projected increases in passenger demand. The scheduled expansion of capacity, sufficient to meet increases in passenger demand projected for the decade 1970-1980, will include the acquisition of a total of 443 aircraft of the new B-747, the DC-10, and the L-1011 types. The scheduled expansion is expected to boost the major airlines' total capital investment by about $7.4 billion [46] for new aircraft alone.

The existing commitments for expansion to meet the increase in passenger demand expected by the end of the next decade, combined with possible financial burdens for noise abatement, presently expose the airlines to serious financial risks. The reality of the risk is quite visible to airline managements. The financial situation appears so precarious and governmental regulation so stifling, that many airline managements have expressed their lack of confidence in the industry's capacity to survive under private ownership [30]. People other than airline managements also perceive the gravity of the current financial problem. Altschul, an industry financial analyst, reports that both the holders of airline debt and equity issues are disturbed by the financial weakness of the industry and that "... an earnings recovery is a must for the airlines if the industry is to obtain the necessary funds to help finance its capital requirements or even meet current obligations when due" [4:F3]. Hence it must be expected that any solution of the aircraft noise pollution problem, which increases the financial strains for the airlines, would, as a minimum, seriously limit airline growth while threatening bankruptcy.

The federal government, with over 49,000 employees [36:29] in its Federal Aviation Administration and others employed in its Weather Bureau, develops, owns, and operates a network of manned and unmanned radio facilities and promotes the safety of aircraft navigation between pairs of cities served by the airlines. The commitment of federal resources is crucial to the marketing of air transport, but has to date played a very limited and indirect role in the creation and solution of the aircraft noise pollution problem. The role of local government, however, is and will be the major factor in both defining the problem and selecting the ultimate solutions.

The ability or willingness of local governments to commit resources to solve the aircraft noise pollution problem is doubtful because of the political unpopularity of the expense. This inability or lack of will is not at all unimportant to the air transport industry: the nation's 548 airline airports are the points of sale and essential to the marketing of air transportation—yet are owned by units of local government—except for two airline airports (Washington and Dulles-International), which are owned by the federal government.

Ordinarily industrial growth in America is financed from industrial profits or on the basis of industrial profitability. The air transport industry is somewhat different. While data on the financial returns on airline airport investment are not available, there is evidence suggesting that such ownership is not profitable, and probably dependent upon the ability of local governments to support the costs of airport ownership through local tax revenue. The evidence is simple: there are no entrepreneurs owning airline airports and, in the absence of laws preventing such ownership, it should be assumed that there are little or no profits in airline airport ownership.

Whatever the financial capacities of the owners of airline airports may be, they will be severely strained without expenditures for elimination of noise pollution, because of the need to provide new or expanded facilities in step with rising demand. The burdens of meeting rising demand are far from trivial. The FAA estimates that, because the capacities of some of the existing airports are inadequate to meet projected demands, five new airline-airports will be needed before 1974; and, pending further analyses of the expandability of existing airports in key traffic hubs, as many as a total of twenty-five new airline-airports may be needed by 1980—at a total development cost of $5 billion [16:16]. With burdens of this magnitude for airport expansion alone, and because the public's capacity or willingness to bear additional taxes is limited, any solution of the noise pollution problem that would place additional financial burdens on airport owners is likely to be rejected. Such rejection would, if shifting the entire burden to the airlines, have severe implications for the air transport industry.

Compounding the difficulties caused by limited resources, and inhibiting market-oriented approaches to problems generated by aircraft noise pollution is the paternal relationship between the air transport industry and the federal government. Indeed while the airlines are privately owned they are subject to such comprehensive controls by the Civil Aeronautics Board, an arm of the federal government, that their private enterprise status has been described as more fictitious than real [39:275]. Thus, with the vital features of the air transport product leading to the creation of a noise pollution problem, low industry profitability, and the heavy hand of government dominating the airlines, solutions, if any, will be difficult to find—as we shall see.

2. Projections of Demand for Air Transport

Human exposure to aircraft noise is generally related to the distribution and magnitude of demand for air transport. Since the United States is so highly urbanized with ". . . less than 1 percent of the nation's area housing three out of every four people and producing well over four-fifths of the nation's output . . ." [51:7], one might reasonably expect to find the demand for air transportation similarly concentrated. In fact: it is. The demand for air transport is concentrated in the nation's larger urbanized or metropolitan areas. The Federal Aviation Administration has classified each urbanized area, within which air transport demand is generated, as a large, medium, small, or non-hub, depending upon the percentage of the nation's domestic scheduled air carrier enplanements occurring at airline airports within the area in question. The criteria for hub classification are:

1. Large hubs are those urban areas within which are enplaned 1 percent or more of the domestic scheduled passengers;
2. Medium hubs are those urban areas within which are enplaned between 0.25 percent and 0.99 percent of the domestic scheduled passengers;
3. Small hubs are those urban areas within which are enplaned between 0.05 percent and 0.24 percent of the domestic scheduled passengers;
4. Non-hubs are those urban areas within which are enplaned more than 0 percent but less than 0.05 percent of the domestic scheduled passengers.

The narrowness of the distribution of demand is indicated in table 20.1.

Although the twenty-one large hubs (with a total of thirty-eight airline airports) collectively accounted for 68 percent of all domestic passengers and 79 percent of all domestic air cargo in 1965, the demand was even more concentrated: the largest nine hubs (with a total of twenty-four airline airports) generated over half of all of the scheduled air passenger trips; and the largest three hubs (New York/Newark, Chicago, and Los Angeles, with a total of twelve airline airports) generated more than one-fourth of all the passenger enplanements during the year 1965 [17].

Total demand, in addition to being geographically concentrated, is strongly segmented. The segments of demand, administratively defined by the Civil Aeronautics Board are: the trunk service market; the local service market; and the international and territorial service market. The most remarkable characteristic of each of the market segments is the average distance between the pairs of included hubs or cities:[a]

1. The trunk market averages 395 miles between city pairs;
2. The local market averages 128 miles between city pairs; and
3. The international market averages 1,020 miles between city pairs.

---

[a]Average distances between city pairs are calculated from: Civil Aeronautics Board, CAB Reports to Congress — FY 1969 (Washington, D.C.: Government Printing Office, 1969), p. 93.

Table 20-1
Demand at Large and Medium Hubs as a Percentage of Total National Demand for Domestic Passenger and Cargo Service by Scheduled Aircarriers—Fiscal Year 1965

Hub Group	Percentage of Total Domestic Passenger Demand	Percentage of Total Domestic Cargo Demand
Large Hubs (21 Metropolitan Areas)	68	79
Medium Hubs (34 Metropolitan Areas)	20	16
Totals	88	95

Sources: Department of Transportation, Federal Aviation Administration Airports Service, AVIATION DEMAND AND AIRPORT FACILITY REQUIREMENT FORECASTS FOR LARGE TRANSPORTATION HUBS THROUGH 1980 (Washington, D.C.: Government Printing Office, 1967), p. 1.

The relative importance of each market segment is indicated in table 20.2.

Because the successful marketing of long distance flights is feasible only with the use of high-speed jet-powered aircraft, the combination of geographically concentrated demand and the predominance of demand for long-distance flights between distant city pairs sets the stage for what has been up until now an unavoidable noise pollution problem. If trends of population growth and relocation continue, the prospects for both wider distribution of demand and demand for shorter hauls will be highly unfavorable. Expectations are far from

Table 20-2
Markets for Trunk, Local, International and Territorial Service in Revenue Passenger Miles and Cargo Ton-Miles Held in 1969 by U.S. Certificated Air Carrier Groups

Market Segment	Revenue Passenger Miles (Thousands)	Percentage of Total Market	Cargo Ton Miles (Thousands)	Percentage of Total Market
Trunk	85,604	72	1,666	60
Local	5,914	5	45	2
International and Territorial	27,852	23	1,059	38
	119,370	100	2,770	100

Source: Calculated from Civil Aeronautics Board, CAB REPORTS TO CONGRESS—FY 1969 (Washington, D.C.: Government Printing Office, 1969), pp. 88-91.

favorable. The nation's population is expected to continue to increase and to continue to migrate from farms and small towns to the already large metropolitan areas, creating "... an increasingly barren hinterland and a dozen or so huge metropolitan complexes" [35:82]. Thus human exposure to aircraft noise, now a problem, is likely to increase especially in view of increasing demand for air travel.

The demand for air transport is expected to grow dramatically. Various and authoritative projections of raw demand, without accounting for the potential limitations that may result from public response to aircraft noise, are based on regression techniques relating demand to population, level and distribution of income, and, implicitly or explicitly, the air transport product features. These authoritative projections are shown in table 20.3. (It is interesting to note that the variations between these tabulated forecasts are "... owing to the differences in assumptions concerning rates of investment in aircraft and airport capacities" [40:41-2]). The tabulated projections in table 20.3 are more recent and considerably different than those given by Landsberg, Fishman, and Fisher, who estimated that raw demand, in 1980, will be between 90.4 and 133.9 billion revenue passenger miles [33:135]. Since, based on actual traffic statistics, in 1970 the demand for trunk service alone totaled about 94 billion revenue passenger miles [34:46], the projections summarized in table 20.3 will undoubtedly be closer to the eventual truth. The expectations for continued and intensified concentrations of population, combined with growth in the demand for air transport, auger aircraft noise pollution problems of continuously increasing intensity.

Table 20-3
Forecasts of Domestic Trunk Airlines Revenue Passenger Miles

Year	Airline Forecasts	Aircraft Manufacturer Forecasts	FAA Forecasts
1970	92-101 billion	94-100 billion	90 billion
1980	220-290 billion	200-295 billion	230 billion

Source: REPORT OF THE TRANSPORTATION WORKSHOP, 1967: AIR TRANSPORTATION 1975 and BEYOND: A SYSTEMS APPROACH, Bernard A. Schreiver and William Seifert, Co-chairman (Cambridge: The MIT Press, 1968), p. 38.

3. Projections of the Supply of Air Transport

The supply of the air transport product, which is the immediate cause of aircraft noise pollution, is generally in proportion to the demand. To a large extent, therefore, the location and frequency of human exposure to noise of objectionable intensity will be concentrated in large hub communities, and the exposure will depend upon the type of aircraft used and the frequency of aircraft operations.

While descriptions of the existing fleet are readily available and reliable, estimates of the composition of future fleets must be based on many expectations, including those related to aircraft that are not yet fully designed. In spite of the uncertainties, an aviation fleet forecast for 1980 has been made by the Federal Aviation Administration, and is based upon knowledge of and/or estimates for seating capacities for the various aircraft types, estimated seat occupancy rates (commonly known as load factors), and estimated frequency of aircraft use required to meet forecasts of raw demand measured in terms of revenue passenger miles [23:16-28]. The existing and projected fleets are described in table 20.4.

While the numbers of aircraft in the air carrier fleet are expected to increase by almost 50 percent over the next decade, the numbers of jet-powered aircraft, which are the noisiest or most annoying and most common aircraft types, are expected to nearly double. Jet-powered aircraft, making up 71 percent of the fleet in 1969, are expected to comprise, by 1980, 93 percent of the air carrier fleet. It is obvious, then, that the communities or hubs that are now exposed to objectionable levels of aircraft noise will continue to be exposed and, as will be shown, with greater frequency.

The large hubs, by offering the richest marketing opportunities for air transport, have been the communities most frequently exposed to aircraft noise of objectionable intensity. Projections of the frequency of large hub exposure to jet aircraft noise, based on traffic and equipment utilization forecasts prepared by FAA, are listed in table 20.5.

As shown in table 20.5, the projected frequencies of operations by jet aircraft are, without exception, expected to increase at every hub—with frequencies at most hubs practically doubling to meet the projected demand. Those hubs at which jet operations are expected to undergo the least percentage increase

Table 20-4

The Reported and Forecast Numbers of Aircraft in the Service of United States Air Carriers

Aircraft Type	Numbers of Aircraft Existing (Year 1969)	Numbers of Aircraft Forecast (Year 1980)
2 Engine and 3 Engine Jet	965	2529
4 Engine Jet	816	888
SST Jet	0	85
Subtotal (Jet)	1781	3502
Subtotal (Non Jet)	789	268
Grand Total (Jet & Non Jet)	2570	3770

Source: Department of Transportation, Federal Aviation Administration, Office of Aviation Economics, AVIATION FORECASTS–FISCAL YEARS 1970-1981 (Washington, D.C.: Government Printing Office, 1970), p. 32.

Table 20-5
Actual and Projected Jet Operations by Scheduled Air Carriers at Selected Large Hubs

Region	Hub	Rank Rank Among 21 Large Hubs Based on Domestic Enplanements	Jet Operations		
			Historic Year 1965 (Thousands)	Projected to Year 1970 (Thousands)	Projected to Year 1980 (Thousands)
Eastern U.S.	N.Y./Newark	1	372	761	1396
	Atlanta	4	96	236	574
	Miami	9	134	263	463
	Pittsburgh	11	53	119	248
	Philadelphia	12	74	143	302
Central U.S.	Chicago	2	294	510	925
	Cleveland	14	59	134	264
	St. Louis	15	40	110	222
	New Orleans	19	50	92	205
Western U.S.	Los Angeles	3	244	398	676
	San Francisco/Oakland	6	139	251	505
	Denver	13	56	128	236
	Seattle/Tacoma	20	64	114	248

Source: Calculated from Department of Transportation, Federal Aviation Administration, AVIATION DEMAND AND AIRPORT FACILITY FORECASTS FOR LARGE AIR TRANSPORTATION HUBS THROUGH 1980 (Washington, D.C.: 1967), pp. iii-124.

(slightly less than 100 percent) are the three top-ranking hubs that constitute the heart of the trunk service market and, therefore, the heart of the air transport industry. It is at the three largest hubs, where airport traffic capacities are fully utilized in meeting current demand, that, in order to service further growth in demand, airlines will have to use an equipment mix containing great numbers of the largest aircraft in the airline fleet. It is only with such equipment mix that airlines are expected to be capable of feasibly and/or profitably marketing the air transport product in the large hub communities.

While there are no published and comprehensive origin-destination studies describing demand for travel between the nation's 548 airline airports, the importance of the major hubs to the value of the entire network of air transport, and the success of the industry, can be intuitively appreciated. The high interdependency of all air transport markets makes the future of the entire air transport industry profoundly and extensively dependent upon the conditions for access to the major hubs. All markets would feel the impact of a restricted access to any one of the major hubs. Total escape from such impact would be impossible since "... each airport affects the air traffic of every other airport

... just as in the game of chess ..., the position of each piece directly or indirectly affects all other pieces on the board" [52:76]. The conditions for access to at least the large or major hubs, if not all hubs, will depend upon human response to aircraft noise. To be able to predict human response is, therefore, to be able to predict the future of the air transport industry.

4. Methods for Measurements of Noise and Predicting Human Response

Sound or noise may be measured or described using any one of a number of technical methods. The most useful of these methods are the perceived noise level method [7:309-12] and the modified but similar effective perceived noise level or EPNL method [44]. Both of these methods are based on correlation analyses that relate the objective and physically measurable properties of sound (i.e., frequency pitch, and waveform) to the corresponding subjective and humanly perceived characteristics of sound (i.e., pitch, loudness, and quality) on the basis of degree of human annoyance. Using the superior EPNL method of measurement and description, the intensity of sound or noise is a computed value based upon physically measurable sound pressure levels, with corrections for subjective responses to noise, pure tone content, and duration of sound signal. With the EPNL method, which is based on psychological evaluations of sound intensity, a prediction of human response is possible and the air transport industry's noise problem can be estimated.

Although jet aircraft generate intense noise continuously while under power, they cause human exposure to noise of annoying intensity only during their approach to or departure from an airport runway when the distance between noise sources and human receivers is at a minimum. The intensity of noise in any particular part of the surrounding community, resulting from an aircraft operation, depends upon a complex combination of factors involving: the distance between any area of the community in question and the particular airport runway being used; the aircraft and engine type; and the path of the aircraft. While it would be possible to consider the noise exposures caused by each of the various aircraft types, it is not necessary and would not change meaningfully either our predictions of human response or the conditions for jet access to major hub airports that may result from such response. A useful appreciation of the magnitude of the aircraft noise pollution problem can be gained by considering the noise exposures caused by the existing and numerous four-engine jets, the B-707 and DC-8. It is these aircraft that are the workhorse aircraft serving a large part of the demand for long-distance flights between the most vital major large hub communities. The areas of land exposed to noises of selected intensities, resulting from landings and takeoffs of these aircraft are listed in table 20.6. (The tabulations are valid for the aircraft path associated with aircraft operations at typical gross weights for the particular range of hauls and the environmental conditions of 84° air temperature, no wind, and sea level airport elevations.) An inexact but satisfactory idea of the degree of annoyance that may be associated with aircraft noise having the intensities listed in table

20.6 can be gained by comparison with the intensities of more commonly heard noises listed in table 20.7.

Comparison between the intensities of aircraft noises and the more commonly experienced sounds listed in table 20.7 makes clear that aircraft noise is louder than all other commonly annoying sounds. Fuller appreciation of the differences between tabulated intensities is gained by noting that "an increase of a mere 10 decibels is, on the average, in the subjective judgment of the human ear and brain, a 100 percent increase in the perceived loudness or intensity of sound" [7:42]. The differences between the noise intensities listed in tables 20.6 and 20.7 are therefore nearly in orders of magnitude.

It is somewhat surprising that although aircraft-generated community noise exposures would appear to be overwhelmingly detrimental to human well-being, medical research does not indicate that aircraft noise is damaging to either physical or mental health [7:99-114]. Authorities do nevertheless concede that because human response to sound is quite subjective and various, exposure to noise is capable of prejudicing human health by interference with peace of mind, privacy, the pursuit of work or pleasure, and undisturbed sleep, even if not inflicting measurable physical damage to the auditory system [7:113]. It may generally be assumed however that human response to aircraft noise depends primarily upon the purely psychological annoyance experienced upon exposure to high intensity sound. The degree of annoyance leading to predictable human action depends upon other factors as well as the intensity of the exposure, as we shall see.

There are presently eight well-known methods in use for predicting human response to aircraft noise [28:31-47]. All eight methods, while somewhat similar, differ in the number, kind, or emphasis given to additional factors that have been found to contribute to human annoyance with aircraft noise. All methods recognize that human response is in proportion to annoyance, which is in turn dependent upon both the intensity and frequency of exposure. The most advanced of these methods for predicting human response to aircraft noise was developed under the auspices of the Federal Aviation Administration and is called the Noise Exposure Forecast or NEF method. The NEF method utilizes the EPNL method of describing sound intensity, makes the applicable but minor corrections for aircraft path, and takes directly into account the frequency of noise exposure with separate adjustments for:

1. The number of repetitions of exposure per day (i.e., between 7 A.M. and 10 P.M.);
2. The number of repetitions of exposure per night (i.e., between 10 P.M. and 7 A.M.).

The research and experience indicate that aircraft noise exposures in populated residential areas stimulate three levels of unfavorable and increasingly aggressive response [28:5]. The three significant levels of response are associated with three different ranges of noise exposure:

Table 20-6
Areas of Land Exposed to Noise Levels of Selected Intensities During Takeoffs or Landings by B-707, or DC-8 Aircraft, on Hauls of 2500 Nautical Miles or Less

Effective Perceived Noise Level EPNL (Decibels)	Area Affected During Takeoffs (Acres)	Area Affected During Landings (Acres)
95 or more	8,680	5,000
100 or more	3,460	2,800
105 or more	1,660	1,600
110 or more	860	760
115 or more	440	240
120 or more	200	140

Source: Tabulated areas are estimated by mechanical integrations of noise level contours found in Dwight E. Bishop and Myles A. Simpson, NOISE EXPOSURE FORECASTS FOR 1967, 1970 AND 1975. OPERATIONS AT SELECTED AIRPORTS, A report to the Department of Transportation Federal Aviation Administration (Van Nuys, California: Bolt, Beranek and Newman, 1970), pp. (C-2)-(C-4).

Table 20-7
Typical Intensities of Various Sounds

Sound	Intensity Level (Decibels)[a]
Painful	120
Riveting	95
Elevated Train	90
Busy Street Traffic	70
Conversation in Room	65
Quiet Radio Program	40
Whisper	20
Rustle of Leaves	10
Hearing Threshold	0

Source: George Shortly and Dudley Williams, ELEMENTS OF PHYSICS (New York: Prentice-Hall, Inc., 1955), p. 407.

[a]The decibel units of intensity measure used here are not given in the source or known to the author but are probably in the "A" or "C" scales, which were in common use at the time of source publication. While measures of noise intensity in either "A" or "C" scales differ from those using perceived noise level and EPNL scales, the differences are not so significant as to make direct comparisons seriously faulty.

1. Within areas subjected to noise exposure levels of up to but not more than NEF 30, it is expected that individual residents will complain mildly.
2. Within areas subjected to noise exposure levels of between NEF 30 and NEF 40, it is expected that some individual residents will complain vigorously and that group actions to abate aircraft noise may be commenced.
3. Within areas subjected to noise exposure levels of NEF 40 or greater, it is expected that individuals will make vigorous and repeated complaint and that concerted group action to abate aircraft noise will occur.

It is important to note the possible numbers of aircraft operations and the consequent noise intensities (in EPNL) associated with various significant noise exposure levels, NEF 30, NEF 40, and NEF 50, causing predictable human response. The relationship between NEF and EPNL for various numbers of operations (by B-707 or DC-8 aircraft) are calculated, using the equation [28:A-2] NEF = EPNL + 10 [Number of operations per day + 16.67 × Number of operations per "night"] − 88, and shown in table 20.8.

The data in table 20.8 make it clear that the frequency of daily exposure required to cause significant human response is not high—especially if there are any aircraft operations after 10 P.M. and before 7 A.M. The data also make apparent the importance of the frequency of noise exposure in determining human tolerance of noises of various intensities. In spite of the many uncertainties and variables, it is reasonable to expect, based on the number of jet operations at large hub airports (see table 20.5) and an assumption of the existence of four runways suitable for jet operations per airline airport in each large hub,[b] that the frequency of noise exposures in year 1970 was sufficient to establish the following relationship between noise intensity and human response:

1. Aircraft noise of intensity 95 EPNL occurred with enough frequency to create an exposure level of NEF 30 and stimulate mild complaint.
2. Aircraft noise of intensity 105 EPNL occurred with enough frequency to create an exposure level of between NEF 30 and NEF 40 and stimulate group action.
3. Aircraft noise of an intensity of 115 EPNL occurred with enough frequency to create an exposure level of NEF 40 or greater and stimulate concerted group action.

The briefest reconsideration of information on the areas exposed to noise of intensities of 95 EPNL, 105 EPNL, and 115 EPNL (listed in table 20.6) indicates in some measure the present extent of significant noise pollution existing at each end of each jet runway at each airline airport located in a large hub community. If noise pollution is creating problems for the air transport industry in 1970, by 1980 the problem can only become significantly worse. Even if it is optimisti-

---

[b] For identification of airline airports within large hub communities see: Department of Transportation, Federal Aviation Administration, *Aviation Demand and Facility Forecast for Large Air Transportation Hubs Through 1980* (Washington, D.C.: Government Printing Office, 1967), pp. 1-130.

Table 20-8
NEF Levels Created by Selected Noise Intensities (EPNL) and Various Numbers of Aircraft Operations per 24-hour Period

NEF Level	EPNL (Decibels)	Frequency of Number of Aircraft Operations		
		If Operations are All During Day (7 AM-10 PM)	If Operations are Split Between Day and Night	
		Day	Day +	Night
30	90	630	430	12
30	95	200	100	6
30	100	63	30	2
30	105	20	3	2
40	100	630	430	12
40	105	200	100	6
40	110	63	30	2
40	115	20	3	2
50	105	2000	600	24
50	110	630	430	12
50	115	200	100	6
50	120	63	30	2

Source: Tabulated data are calculated by the author.

cally assumed that human response to noise exposure will be proportioned until 1980 as it was until 1970, the doubling of jet operations (see table 20.5) by 1980 could cause the lowering of important noise intensity response thresholds. The higher frequencies of noise exposures by year 1980 may be sufficient to reestablish the relationships between noise and intensity and human response as follows:

1. Aircraft noise of intensity 95 EPNL will occur with enough frequency to create an exposure level of NEF 30 and stimulate mild complaint.
2. Aircraft noise of intensity 100 EPNL will occur with sufficient frequency to create an exposure level of between NEF 30 and NEF 40 and stimulate group action.
3. Aircraft noise of intensity 110 EPNL will occur with enough frequency to create an exposure level of NEF 40 or greater and stimulate concerted group action.

Although the probability of the thresholds for significant human response being lowered would ordinarily have to be considered small, the current popular concern with environmental quality could easily and severely change the proportions between human response and noise exposure. If noise intensity levels triggering the predictable and significant levels of human response are in time lowered to 95 EPNL, 100 EPNL, and 110 EPNL, the extent of aircraft noise pollution will be increased by between 200 and 300 percent. The conditions for airline access to major hub airports, depending upon human response to aircraft noise, will certainly be affected—and probably in proportion to the numbers of people annoyed by aircraft noise.

5. Areas and Populations Exposed to Significant Levels of Aircraft Noise

The Airport Operators Council International, a group comprised of airport owners and/or operators, estimates (without defining its terms) that by 1975, near the nation's "major" airports alone, 15 million people will live within aircraft noise-sensitive areas comprehending hundreds of square miles [2]. This estimate, if at all accurate (even as to order of magnitude), while giving meaningful definition of the extent of the noise pollution problem also suggests the extent to which political solutions would be popular. Before accepting the Airport Operators Council-International's estimate, it would be wise to make some independent estimates—however crude. The independent estimate is as follows.

The major markets for air transport within the United States are the twenty-one large hubs that contain a total of only thirty-eight airline airports. If it is reasonable to assume for the purposes of this estimate that each airline airport in a large hub community must have at least one runway, and will both require and have (on the average) no more than three runways by 1980, then the total number of runways existing at major airline airports between now and 1980 will be at least thirty-eight, but not more than one hundred and fourteen. Based on the existing traffic capacities of and operations at congested major airports [15], such numbers of runways, if optimally distributed among the twenty-one large hubs, should be capable of sustaining current and forecast operations at the thirty-eight airline airports. Hence, making use of the probable current (95 EPNL, 105 EPNL, 115 EPNL) and future (95 EPNL, 100 EPNL, 110 EPNL) noise intensity thresholds associated with the three levels of significant human response, and assuming the current existence of a mere thirty-eight airline runways, operations by B-707 and DC-8 on hauls of up to 2,500 nautical miles, landings from one direction and takeoffs from the other directions (for each of the thirty-eight runways), the total area currently exposed to significant noise levels is estimated (using data in table 20.6) at:

1. 518,320 acres exposed to noise of an intensity of at least 95 EPNL with such frequency as to cause at least mild complaints from residents.
2. 123,160 acres exposed to noise of an intensity of at least 105 EPNL with such frequency as to cause at least the initiation of resident group action against noise exposure.
3. 25,840 acres exposed to noise of an intensity of at least 115 EPNL with such frequency as to cause vigorous and concerted resident group actions against noise exposure.

Assuming the existence of 114 airline runways, areas of land within which similar response may be anticipated by year 1980 are estimated at:

1. 1,559,520 acres exposed to noise causing at least mild complaint from residents.
2. 713,640 acres exposed to noise causing at least initiation of group action against noise exposure.
3. 371,640 acres exposed to noise causing at least vigorous and concerted resident group action against noise exposure.

Since a square mile contains approximately 650 acres, an area of nearly 800 square miles is currently exposed to aircraft noise levels stimulating at least mild complaint; and an area of slightly less than 2,400 square miles will be similarly exposed by year 1980.

To estimate the population affected by aircraft noise pollution, additional simplifying assumptions must be made in the absence of detailed data on land use surrounding each of the thirty-eight airline airports. The assumption is that the entire area affected by the least but significant level of aircraft noise is or will be used for single-family housing by year 1980. Since the average building lot size in the United States was, in 1964 and 1965, three lots per acre [42:12] and since the average population per housing unit was, in 1960, 3.3 [50], the average population density in residentially developed areas may be taken as approximately ten people per acre. Hence, using the developed estimates of the areas exposed to aircraft noise pollution, the population currently exposed to objectionable aircraft noise pollution, in the vicinity of the airline airports serving the twenty-one large hub communities, is estimated at approximately 5 million. By 1980, the exposed and annoyed population in the vicinity of the same thirty-eight airline airports could total slightly more than 15 million.

The areas exposed to noise, and annoyed resident populations, now large, could become much larger and practically triple by the year 1980. The projected extent of the aircraft noise pollution problem promises a difficult future for the air transport industry. With the various remedies for reducing the extent of aircraft noise pollution being exhausted, inadequate, enormously costly or unpopular, the projected growth in the numbers of people affected by noise is bound to render the air transport industry increasingly vulnerable to politically chosen solutions. Such solutions may be uneconomic and totally ineffective, as we shall see.

6. Actions Taken by Population Exposed to Aircraft Noise

Populations exposed to aircraft noise have sought relief by recourse to the law and/or political action. To date, those aggrieved residents of communities exposed to aircraft noise who have resorted to the courts of law, seeking injunctive relief in actions involving publicly-owned airports, have without exception [38:122] failed to gain the desired remedy. However, by taking alternative adversary courses of action against defendant airport owners, and pleading for the remedy of money damages in lieu of injunctive relief, aggrieved residents have achieved a series of limited but portentous victories [49:928-92]. The victories, although hollow because they have not resulted in the abatement of aircraft noise, have suggested to airline airport owners that future court decisions will ultimately develop their liabilities to unbearable magnitudes.

The victories, flawed in essence, have been, moreover, ones more in principle than of principal. The courts, while wading through legal complexities, esoteric arguments, and numerous litigations, are confounded in the administration of justice: decisions of courts and juries have been stalled or impaired by the technical complexities associated with giving definitions to thresholds of legally significant sound intensities, and the difficulties of relating such sound intensities to actionable damages. Thus, while owners of airline airports have defended against many actions brought by offended residents, involving enormous claims, damages recovered have been astonishingly small. During the ten-year period from 1956 until 1966, for example, only six of the many cases or actions brought to court resulted in awards to the aggrieved, and the awards for damages totalled less than $100,000 [38:123].

If, however, the courts, which have already found merits in the arguments of the aggrieved residents, begin to return judgments closer to the amounts claimed, the fears of airport owners will not have been exaggerated. Although there are no recent comprehensive estimates available, the situation of the owners of the Los Angeles International Airport, who alone currently face $2 billion in claims [24], strongly suggests that the nationwide aggregate of claims pending litigation is quite likely in excess of $10 billion and/or nearly the total capital investment of the air transport industry. Accelerating, intensifying, and compounding the prospective difficulties resulting from legal actions are those resulting from political actions. Indeed, to the extent that the aggrieved public has been frustrated and unable to duel for and gain swift and effective relief via the courts, antagonistic political actions have appeared to be the only alternative recourse.

Political actions, caused by and causing diminished public tolerance of any kind of environmental pollution, although offering opportunities for the reduction of noise exposure by agitation and fiat, threaten both the growth and survival of the air transport industry. This genre of political actions, at first resulting in and now powered by the legislation of Public Law 90-411, of 1968, and the National Environmental Policy Act of 1969, presently menace the air transport industry in two ways:

1. Airline airport owners or prospective owners, practically all of whom are heavily dependent upon federal grants of funds to develop new airports or expand existing airports, must, in order to qualify for such financial aid, consult with dozens of governmental agencies and hold public hearings in order to allow public participation in decisions on proposed developments of airports; and
2. The Federal Aviation Administration (which also administers the airport grant-in-aid program) is preparing, consistent with the intent and allowances of the law (PL90-411) and in response to the desires of the aggrieved public, a federal regulation that will, by prescribing limits for aircraft noise, probably require the fitting of the entire jet-powered fleet with engine silencers of enormous cost.

In the words of James Carr, director of the San Francisco Airport Commission, speaking before the annual meeting of the Airport Operators Council - International (AOCI) "... ecological judgment day has arrived" [24]. Since the development of new airports in, near, or for the cities of London, New York, Tokyo, Miami, Boston, and Stockholm had been blocked by political actions prior to the passage of the Environmental Policy Act, Homer Anderson, president of the AOCI, is not unwise for his concern that airport owners will fail to qualify for federal funds because of their inability to successfully deal with the political actions promoted by the Environmental Policy Act. If recent experience is any guide, the political impediments to airport developments may indeed be formidable: after five months of reviewing 352 applications for federal funds (most of which were considered non-controversial) for the purpose of budgeting grants in fiscal year 1971, the Federal Aviation Administration approved only fifteen [24].

While the failure to develop airports in a timely fashion could limit the growth of the air transport industry, the political solution of imposing noise standards would have a direct and severe industry-wide financial impact. Rather than merely establishing conditions for jet access to the heavily exposed major hubs, noise standards would establish conditions for the operation of every airlines' jet aircraft to any hub regardless of how small. Political solutions, never noted for their economic viability, may nevertheless be the inevitable choice in the politically charged atmosphere surrounding the questions of how to relieve aircraft noise pollution. As we shall see, the number and nature of possible alternative solutions is limited.

7. Methods for Reducing Human Exposure to Aircraft Noise

Methods for reducing or minimizing human exposures to aircraft noise, practiced or advocated, involve the alterations of one or more of the following:

1. Aircraft operating procedures;
2. Land ownership and use;
3. Aircraft design or engine design.

No method practiced or advocated, however, is without consequences for the air transport industry.

Toward minimizing human exposures to aircraft noise, scores of legally and politically besieged airport owners or operators have imposed restrictions on the operation of aircraft with various degrees of success. The restrictions, not binding on the pilot because of their influence on flight safety, have included the following measures:

1. The diversion of aircraft operations to so-called preferential runways, the use of which minimize approach and departure operation over populated areas,
2. The use of so-called noise abatement procedures for the operation of aircraft, during visibility conditions permitting use of visual flight rules (VFR), in order to produce aircraft flight paths and engine power settings minimizing the noise exposure of underlying communities; and,
3. Curfews for the operation of jet aircraft.

While preferential runway and noise abatement procedures are capable of minimizing human exposure to aircraft noise, the reduction of noise intensities in areas over-flown probably average only 5 decibels [38:97]. Such procedures, while yielding less than satisfactory reductions in the intensity of noise exposures, involve increasing risk [38:102] to flight safety by requiring low-speed turns and/or engine power reductions of about 30 percent [26:116], and are accused of delaying traffic and reducing the air traffic capacity of one major large hub airline airport by as much as 30 percent [18:145]. The traffic delays at the same airport are estimated to be costing the airlines $30 million per year in extra operating costs [18:149].

Such procedures, by reducing flight safety, increasing flight time because of traffic delay, and pressuring price increases to cover costs of traffic delay, adversely affect the marketability of air transport. With the increasing flight frequencies required to meet projected growth in demand, moreover, the advantages of such procedures will become increasingly marginal and should entirely disappear. Thus, noise intensity reductions, which now offer only marginal benefits gained by the altering of operational procedures, are unlikely to be capable of providing additional and significant noise exposure reductions required to avert the mounting legal and political conflict between airport owners and resident airport neighbors.

In order to minimize human exposures to aircraft noise and thereby create conditions favorable to the satisfaction of growing demand for air transport, airport owners have attempted by various means to limit the uses of nonairport land to those that are compatible with airport operations. Their efforts to achieve compatible land use have involved:

1. Promotion of and/or passage of zoning ordinances;
2. The acquisitions of easements in land (i.e., the acquisition of the right for aircraft to fly over and make noise over land);
3. The acquisition of fee simple interests in land (i.e., the acquisitions of the right to use the land for whatever purpose the laws permit).

Because the land areas affected by significant noise exposures are quite extensive, such areas are typically under a multitude [13:13] of political subdivisions other than those which own and control the airport and which, therefore, are under no obligation to cooperate in the imposition of zoning ordinances for the benefit of airport owners or users.

Although the passage of airport zoning ordinances is strongly urged by the federal government, the U. S. Supreme Court, by the rendering of adverse decisions, had indicated that airport zoning ordinances are of questionable constitutionality [13:14]. A competent legal authority has understood the courts decisions to mean that:

...apart from the practical difficulties which arise from multiple jurisdictions ... at least as far as existing airports are concerned, zoning cannot be used as a substitute for needed exercise of the power of condemnation and the ensuing obligations to pay for whatever is appropriated... [38:141].

Hence, airport zoning ordinances, which by themselves are incapable of altering existing residential uses of land, and are in most cases impossible of passage because of uncooperative political subdivisions, being of doubtful constitutionality, are hardly to be proffered as panaceas. Legalities and political difficulties aside, the basic flaw of zoning ordinances is that in order to be useful in removing the constraints to airport development and air transportation, they must allow a continuation of intense noise exposures in surrounding communities. Thus, because of their leaving residents exposed to aircraft noise, zoning ordinances are vulnerable not only to the remedies of the court, but to outflanking by antagonistic political action. The flaws of zoning ordinances, in removing constraints to the development of existing airports and the air transport industry, are, hence, fatal.

The legally and politically besieged airport owners, being either unable to obtain or unable to enjoy the imagined benefit of zoning ordinances, have in some instances either fallen upon or contemplated an expensive and inadequate remedy: the acquisition of limited property rights (i.e., easements) permitting the right to fly over and create noise exposures in areas where adverse human response could be anticipated. This resort has at least two flaws. The first flaw is that if the character of aircraft using the airport changes for the noisier, after the acquisition of the easement, the resident may, United States courts have held, bring suit and recover additional money damages from the airport owner [31:15]. This resort, which fails to provide a suitable legal shield and is not without cost, has the additional and fatal liability of being totally ineffective against outflanking by residents taking antagonistic political action. The flaws of easements in removing constraints to the development of existing airports and the air transport industry are thus also fatal.

With zoning ordinances and easements being legally and politically unsatisfactory for shielding airport owners from the necessary liabilities caused by aircraft noise, groaning airport owners are left with the alternatives of either relocating airports well beyond populated areas or acquiring full title to all lands exposed to significant levels of aircraft noise.

Relocations of airports to remote areas, while reducing the convenience of access to air transport and therefore to a greater or lesser extent its marketability, would not, however, eliminate the necessity for the acquisitions of the vast areas that would be exposed to intense aircraft noise since as Deem has found: airports attract certain types of industries and motels, as well as housing and services for people working at the airports [13:12].

Sites possessing characteristics making airport developments feasible are not, moreover, commonly available. Indeed, open sites large enough and otherwise suitable for airports serving burgeoning suburbanizing large hub communities are rare and also in great demand for use in the development of more profitable "shopping centers, industrial parks, and large residential developments" [13:6].

In addition to shopping centers, industrial parks, and residential developments being in greater demand than air transport they are, it may be assumed, both vastly more popular with indigenous populations and less likely to generate either political resistance or the erection of legal barriers having the form of obstructive zoning ordinances. Indeed, distant political subdivisions, however financially able and anxious to relocate their airport functions to less populous sites, since they lack powers of condemnation within areas not under their immediate political jurisdiction, are unable to overcome local resistance and barriers to airport development. It is, therefore, hardly possible that the many difficulties associated with the relocation of airport functions will be surmountable and offer viable alternatives over the foreseeable future.

Airport owners considering the alternative of acquiring full title to lands exposed to significant levels of aircraft noise, for the purpose of altering land use, will face monumental difficulties of every kind: political, legal, and financial.

The acquisition of full title to land, which would eliminate residential and other incompatible land use from within areas exposed to significant noise levels, could in the next five years, for only the ten largest airports in the nation, involve, according to the Airport Operators Council-International (AOCI), total acquisition costs of "as much as $50 billion, and the relocation of hundreds of thousands of people" [2]. Based on the author's independent estimates of the areas and populations affected by noise exposure, and since the median value per one-third acre of residentially developed land was over $15,000 [42:4] in the year 1963, neither the AOCI's estimate of acquisition costs, nor the AOCI's estimate of populations to be relocated, may be rejected.

Some, however, would suggest that the AOCI's estimate of land acquisition costs are incomplete since such estimate fails to account for the revenues that might be derived from "the eventual resale of formerly residential land for ... more compatible uses" [38:7]. Indeed it is also suggested that such resales might result in "final profits to the condemning authority" [38:7]. The suggestions, although provocative, are based on implicit premises that are at best dubious:

1. The land may not be within the jurisdiction of the police power of the airport owner, and if not, may be held by owners unwilling to sell;
2. The alteration of land use would require the rezoning of residential areas, which probably cannot be achieved by fiat—whether or not the land acquired is within the jurisdiction of the condemning authority.

The political, legal, and financial problems associated with the alteration of existing land uses to those uses which are compatible with aircraft noise, via land acquisition, do not engender any hope for air transport.

Given the fragmentation and zealous guarding of political power, the present structure of law, and the limited financial capacities of airport owners, the prospect for achieving the reduction of human exposure to aircraft noise, by the establishment of compatible land use, in time to avoid constraint of or threat to air transportation, is—one may fairly conclude—extremely poor.

Airport owners, being unable to satisfactorily reduce the menace of their liabilities for aircraft noise pollution, via either the alteration of operational procedures or the alteration of land ownership and use, have joined voices and, with others, advocated the alteration of aircraft design and/or engine design. The advocates pin their hopes to:

1. New and technologically advanced aircraft such as the B-747, DC-10, and L-1011;
2. New and technologically advanced silencers to be fitted to existing engines powering jet aircraft;
3. New and technologically advanced engines.

Any new aircraft, if delivered by manufacturers after December 31, 1971, will have to meet noise standards prescribed by the Federal Aviation Administration and devised in accord with the intent of Public Law 90-411 [22]. The new DC-10 aircraft, which were designed to meet the mandatory noise standards prescribed by the Federal Aviation Administration, are reported to be about 10 EPNL quieter [21] than the first generation of jet aircraft (i.e., B-707, DC-8) and therefore will expose much smaller areas to significant aircraft noise than caused by the B-707 and DC-8 aircraft. Indeed, the new and quieter aircraft may be estimated to affect between one-fifth and one-third of the area now affected by the B-707 and DC-8.[c] Projections of the numbers and types of aircraft expected to be in the air carrier fleet over the next decade [23:29] indicate, however, that rejoicing now would be premature. Indeed, the happy prospect of achieving noise exposure reduction through use of new aircraft is moderated by the fact that such new aircraft are not at all intended to replace noisy predecessor aircraft, but to increase the supply of air transport to help meet the perhaps overly optimistic growth of demand.

Given that application of antitrust laws will continue to prevent or delay airline mergers that would permit reduction of excess capacity, until airlines verge on backruptcy, competitive and marketing considerations alone should prevent the rapid and wholesale replacement of the offending jet fleet. The facts are plain: the marketing of air transport, the success of which depends in part upon high flight frequencies, being unprofitable during recent years because half the available seats have been unsold [34:46-8], can hardly be made profitable or even solvent by the use of new and doubly, triply, or quadruply expensive aircraft, operated with reduced frequency to take "advantage" of their seating capacities that are two, three, and four times the seating capacities of the half-empty aircraft now in the airline fleet. Profits require more passengers, not more seats. Such marketing considerations notwithstanding, the preponderance of debt in the capital structures of the airlines makes it unlikely that a wholesale replacement of the existing fleet could be financed on credit of any kind—including lease arrangements. The financing could only be achieved by the massive sale of new equity issues, sold at deeply depressed prices or, possibly, the massive sale of the existing aircraft at favorable prices. Such massive equity financing arranged at distress prices may not be possible; nor is it easy to imagine a massive sale of the tainted aircraft at favorable prices. Where would the buyers be found?

The probability of airport owners gaining relief by airline acquisitions of new and quieter aircraft, during the next decade, is rather low.

---

[c]Estimates are based on data included in table 20.6.

Perhaps recognizing the limited feasibility of airlines replacing their existing fleet, airport operators, joining the political cacaphony of aggrieved residents, now lobby for action by the federal government that would force airlines to fit their aircraft engines with silencers. Their prayers may be answered: the Federal Aviation Administration, under the broad authority of Public Law 90-411, after having prescribed noise standards that must be met by new aircraft delivered after December 31, 1971, is preparing [20:1-3] to either extend the same noise standards to the existing aircraft or to require that such aircraft be fitted with specific sets of acoustical devices that have recently emerged from research and development phases of a government-sponsored noise abatement program. Either means of mandating the reduction of noise levels generated by the existing aircraft fleet ensures that airlines will either comply by fitting engines with silencers or probably be forced to ground their aircraft, with bankruptcy ensuing.

The benefits of adding silencers, although uncertain because unproved by production, installation, and commercial operations, promise to be not at all insignificant.

The particular silencing systems advocated by the Airport Operators Council International [2] have been estimated by the Rohr Corporation, on the basis of its parameter studies, as offering noise reductions of approximately [41:4]:

1. For departing four-engine aircraft (i.e., B-707, DC-8), 3.1 EPNL, measured at ground level and directly under the departure flight path at a distance of 3.5 nautical miles from the start of takeoff roll; and
2. For approaching four-engine aircraft (i.e., B-707, DC-8), 13 EPNL measured at ground level and directly under the landing flight path at a distance of 1 nautical mile from the near end of the runway.
3. For two- or three-engine aircraft (i.e., BAC-111, DC-9, B-727, B-737), 1.1 EPNL and 10.0 EPNL for departing and approaching aircraft, respectively, measured at the same points used to measure sound intensity reductions for operations by silencer-equipped four-engine aircraft.
4. Silencers combined with power cutbacks of only 20 percent, used during takeoffs, would render the two- or three-engine aircraft 5 EPNL quieter, and the four-engine aircraft 8 EPNL quieter, than the currently operating aircraft.

Reilly reports that if noise reductions of only 10.5 EPNL were achieved by the installation of silencers, the areas of land exposed to at least 100 EPNL during landings by four-engine aircraft would be reduced by 85 percent [18:145]. Based on information in table 20.6, during takeoff and with 20 percent power cutbacks, the area exposed to 100 EPNL by four-engine aircraft would be reduced by well over 50 percent.

Thus it is no surprise that airport owners and operators lobby for the fitting of silencers on aircraft: the reduced noise exposures could indeed be useful in stemming the rising waves of public opposition to jet aircraft and airport operations. The proposed silencing of aircraft would be enormously expensive, but since the costs would appear to be borne by the airlines, the remedy is politically popular.

Under the auspices of the Federal Aviation Administration, the Rohr Corporation undertook and recently completed a study of the economic impact of fitting silencing systems to existing jet aircraft. The Rohr Corporation studies have indicated that (for the sets of silencers that the AOCI advocates be used):

1. It would cost $856 million to furnish and install silencing systems in all the existing two-engine (i.e., BAC-111, DC-9), three-engine (i.e., B-727, B-737), and four-engine aircraft (i.e., B-707, DC-8) expected to be flying within, into, or out of the United States by the year 1975.
2. For the aircraft belonging to the eleven trunk airlines, Pan American Airline and three of the largest local airlines alone, the cost of furnishing and installing the silencing systems would aggregate $575 million; the average direct operating costs would be increased by the diminishing amounts of 5.3 percent in 1975, 3.9 percent in 1978, and 2.6 percent in 1982; the pretax return on investment would be decreased by the diminishing amounts of 0.86 percent in 1975, 0.32 percent in 1978, and 0.08 percent in 1982; the fare increases (noncumulative), required to amortize the acquisition cost and cover increased operating costs would be 0.8 percent in 1975, 0.3 percent in 1978, and 0.1 percent in 1982 [41:98,105].

Arguing against the silencer installation advocated by AOCI, on the grounds that "... it would take five years or more to install ... cost up to $1 billion, increase operating costs by 5 percent ... and, in the end might yield such small noise reduction that the change would hardly be noticed on the ground ...", the air transport industry with "... the private support of some key officials within the FAA and the public support of the Chairman of the Civil Aeronautics Board, Secor D. Brown, is vigorously fighting the proposal" [3].

The opposition to a political solution may be too little and too late. Although the airlines have the support of the Civil Aeronautics Board and are skilled in the arts of lobbying, it is unlikely that their technical and economic arguments, however meritorious, will prevail against the political momentum of airport owners who are organized under the banner of the AOCI. Although this means of reducing noise pollution may be effective, it may mean economic and financial havoc for the entire air transport industry whose individual firms are presently groaning under enormous burdens of debt. The only remaining alternative would involve the replacement of existing engines by newer, more economical, and technologically advanced quiet engines. All research and development to date, however, has shown that useful noise reductions are achievable only at prohibitive cost [19]. It would seem likely, then, that the range of alternatives will be, for at least the next decade, narrow, expensive, and politically popular if not economically viable.

8. Conclusions

Although the projections of demand for air transportation are staggering—practically tripling in terms of revenue passenger miles over the next decade—

they are hardly likely to be realized. Since the industry's sales are so highly dependent upon high-powered, high-speed jet aircraft that pollute airport environs with provocative levels of noise, legal and political action will undoubtedly harry the servicing of demand with increasing frequency and effect. The problems caused by noise pollution, which are now formidable threats to industry growth, must be expected to intensify: the number of jet aircraft would have to nearly double to meet the projected demand.

Offended populations, having been denied relief by the courts, are hardly likely not to seize upon any opportunity to reduce the noise exposures of their communities. With the recent politicalizing of questions of environmental quality, and legislation empowering regulatory reductions of aircraft noise, the stage has been set for the forced, indiscriminate, and dangerously expensive silencing of jet aircraft. Such political solution, however effective in reducing jet noise, would unavoidably reduce the immediate financial strength of the industry and reduce growth by diverting scarce capital into economically unproductive silencing systems. In the longer run, the resulting reductions of return on investment would, through free market investment mechanisms, cause at least a reduction of the rate of capital investment in the industry. Such reduction in the rate of investment would have to result in both the retardation of the industry's growth rate and the denial of the benefits of air service to a potentially large segment of the market.

The frustrated American public may be quite willing to forego free market solutions to the problem, by resort to the clumsy instrument of politics, but there will be a price—known or unknown—that all will pay.

Notes

1. Joseph Ackerman, Marion Clawson, and Marshall Harris (Eds.), *Land Economics Research* (Baltimore: The Johns Hopkins Press, 1962).

2. Airport Operators Council International. *Aircraft Noise Can Be Controlled*, (a pamphlet) (Washington, D.C.: Airport Operators Council International, 1970).

3. "A Plan to Muffle Jets Draws Opposition," *New York Times*, February 14, 1971, p. 34.

4. Selig Altschul, "Aircraft Leasing Flying Into Storm," *New York Times*, December 27, 1970.

5. Dwight E. Bishop and Myles A. Simpson, *Noise Exposure Forecast Contours for 1967, 1970 and 1975 Operations at Selected Airports*, a report to Department of Transportation, Federal Aviation Administration, Office of Noise Abatement, September 1970 (Van Nuys, California: Bolt Beranek and Newman Inc., 1970).

6. Board of Trade, *Aircraft Noise – Report of an international conference on the reduction of noise and disturbance caused by civil aircraft, November 22-30, 1966* (London: Her Majesty's Stationery Office, 1967).

7. William Burns, *Noise and Man* (Philadelphia: J. B. Lippincott Company, 1969).

8. Richard E. Caves, *Air Transport and Its Regulators — An Industry Study* (Cambridge, Massachusetts: Harvard University Press, 1962).

9. Angelo J. Cerchione, et al. (Eds.), *Master Planning the Aviation Environment* (Tucson: The University of Arizona Press, 1970).

10. Civil Aeronautics Board, *CAB Reports to Congress — FY-69* (Washington, D.C.: Government Printing Office, 1969).

11. Marion Clawson and Charles L. Stewart, *Land Use Information, A Critical Survey of U. S. Statistics Including Possibilities for Greater Uniformity* (Baltimore: The Johns Hopkins Press, 1965).

12. J. H. Dales, *Pollution, Property and Prices — An essay in policy making and economics* (Toronto: University of Toronto Press, 1968).

13. Warren H. Deem and John L. Reed, *Airport Land Needs* (Washington, D.C.: Communications Service Corporation, 1966).

14. Department of Transportation, Federal Aviation Administration, Airports Service, *Aviation Demand and Airport Facility Requirement Forecases for Medium Transportation Hubs Through 1980* (Washington, D.C.: Government Printing Office, 1969).

15. Department of Transportation, Federal Aviation Administration, Airport Service and Air Traffic Service, *A Suggested Action Program for the Relief of Airfield Congestion at Selected Airports* (Washington, D.C.: Government Printing Office, 1969).

16. Department of Transportation, Federal Aviation Administration, *National Airport Plan — FY-1969-1973* (Washington, D.C.: Government Printing Office, 1968).

17. Department of Transportation, Federal Aviation Administration, Airport Service, *Aviation Demand and Airport Facility Requirement Forecasts for Large Transportation Hubs Through 1980* (Washington, D.C.: Government Printing Office, 1967).

18. Department of Transportation, Federal Aviation Administration, *National Aviation System Planning Review Conference April 14-17, 1970 — Summary Report* (Washington, D.C.: Government Printing Office, 1970).

19. Department of Transportation, Federal Aviation Administration, *First Federal Noise Abatement Plan — FY 1969-70* (Washington, D.C.: Government Printing Office, 1969).

20. Department of Transportation, Federal Aviation Administration, *News — November 1970* (Washington, D.C.: Government Printing Office, 1970).

21. Department of Transportation, Federal Aviation Administration, *News — January 1971* (Washington, D.C.: Government Printing Office, 1971).

22. Department of Transportation, Federal Aviation Administration, *Federal Aviation Regulation — Part 36* (Washington, D.C.: Government Printing Office, 1970).

23. Department of Transportation, Federal Aviation Administration, *Aviation Forecasts — Fiscal Years 1970-1981* (Washington, D.C.: Government Printing Office, 1970).

24. "Environmental Requirements Stir Airport Operations," *Engineering News Record,* November 1970, p. 20.

25. Federal Aviation Agency, *Noise Abatement Procedures* (Washington, D.C.: Government Printing Office, 1960).

26. Federal Aviation Agency, *A Citizens Guide to Aircraft Noise* (Washington, D.C.: Government Printing Office, 1963).

27. Patrick Frangella and Victor John Yannacone, Jr., "Environmental Concern – The Law and Aviation," in *Master Planning the Aviation Environment,* ed. by Angelo J. Circhione, et al. (Tucson: The University of Arizona Press, 1970).

28. William J. Galloway and Dwight E. Bishop, *Noise Exposure Forecasts: Evolution, Evaluation, Extensions, and Land Use Interpretations,* A report to Department of Transportation, Federal Aviation Administration, Office of Noise Abatement, August 1970 (Van Nuys, California: Bolt Beranek and Newman Inc., 1970).

29. Sidney Goldstein, "Jet Noise Near Airports – A problem in Federalism, Property Rights in Air Space and Technology," Unpublished report by the General Counsel of the Port of New York Authority, Port of New York Authority, 1965.

30. William V. Henzey, "Is Nationalization Inevitable," *Airline Management*, January 1971, pp. 20-23.

31. Issac H. Hoover and Donald G. Cochran, "Airport Design and Operations for Minimum Noise Exposure," in *North Atlantic Treaty Organization Advisory Group for Aerospace Research and Development Conference Proceedings No. 42* (London, England: Technical Editing and Reproduction Ltd., 1969).

32. Richard D. Horonjeff and Allan Paul, *A Digital Computer Program for Computation of Noise Exposure Forecast Contours,* A report to Department of Transportation, Federal Aviation Administration, Office of Noise Abatement, June 1970 (Van Nuys, California: Bolt Beranek and Newman Inc., 1970).

33. Hans H. Landsberg, Leonard L. Fishman, and Joseph L. Fisher, *Resources in America's Future – Patterns of Requirements and Availabilities 1960-2000* (Baltimore: The Johns Hopkins Press, 1963).

34. "Management Report: Traffic," *Airline Management,* January 1971, pp. 45-49.

35. Lawrence A. Mayer, "New Questions About the U. S. Population," *Fortune,* February 1971, pp. 80-85, 121-25.

36. National Bureau of Standards, *Report No. FAA-AV-70-1, Estimated Trends of Unit Cost of Government Programs in Support of Air and Highway Travel,* a report prepared for the Federal Aviation Administration, Office of Aviation Policy and Plans (Washington, D.C.: Government Printing Office, 1970).

37. Albert H. Odell, "Jet Noise at John F. Kennedy Airport," unpublished report by Assistant Supervisor of Aeronautical Planning for the Port of New York Authority, Port of New York Authority, 1965.

38. Office of Science and Technology, Executive Office of the President, *Alleviation of Jet Aircraft Noise Near Airports – A report of the Jet Aircraft Noise Panel* (Washington, D.C.: Government Printing Office, 1966).

39. Robert Poole, Jr., "Fly the Frenzied Skies," *The Freeman – Ideas on Liberty,* May 1970, pp. 273-86.

40. *Report of the Transportation Workshop, 1967: Air Transportation 1975 and Beyond: A Systems Approach,* Bernard A. Schreiver and William W. Seifert, Co-chairmen (Cambridge: The M.I.T. Press, 1968).

41. Rohr Corporation, Final Report, *Economic Impact of Implementing Acoustically Treated Nucelle and Duct Configurations Applicable to Low Bypass Turbofan Engines* (Chula Vista, California: Rohr Corporation, 1970).

42. Allan A. Schmid, *Converting Land From Rural to Urban Uses* (Baltimore: The Johns Hopkins Press, 1968).

43. George Shortly and Dudley Williams, *Elements of Physics* (New York: Prentice-Hall Inc., 1955).

44. Society of Automotive Engineers, Inc., *Technique For Developing Noise Exposure Forecasts – Technical Report,* a report to the Department of Transportation, Federal Aviation Administration, Aircraft Development Service (Washington, D.C.: Government Printing Office, August 1967).

45. Alan H. Stratford, *Air Transport Economics in the Supersonic Era* (New York: St. Martins Press, 1967).

46. "The Amazing Future of Air Travel," *U. S. News and World Report,* September 1968, p. 92.

47. Robert L. Thornton, *Airline Management and the Innovative Process* (Ann Arbor: Bureau of Business Research, Graduate School of Business Administration, The University of Michigan, 1967).

48. Poyntz Tyler (Ed.), *Airways of America* (New York: The H. W. Wilson Company, 1950).

49. U. S. Congress, Senate, Committee on Public Works and Committee on Commerce, *Air Pollution – 1970 – Part 3 Joint Hearings* before the Subcommittee on Air and Water Pollution, United States Senate, Ninety-First Congress, 2nd Session on S.3229, S.3466, and S.3546. March 24 and 25, 1970 (Washington, D.C.: Government Printing Office, 1970).

50. U. S. Department of Commerce, Bureau of the Census, *County and City Data Book – 1962.* (A Statistical Abstract Supplement) (Washington, D.C.: Government Printing Office, 1962).

51. Lowdon Wingo, Jr. (Ed.), *Cities and Space – The Future Use of Urban Land – Essays From the Fourth RFF Forum* (Baltimore: The Johns Hopkins Press, 1963).

52. John Walter Wood, *Airports and Air Traffic – The Airport Needs of Your Community* (New York: Coward-McCann Inc., 1949).

21 Summary of Findings on Air Pollution

ALFRED J. VAN TASSEL

The unfolding story of America's continuing love affair with the automobile will tell a lot about the outlook for air pollution in 1980. There is every current indication that the number of cars in use will increase rapidly in all parts of the country. The projections in this study were made assuming a continuation to 1980 of the rate of increase in the number of automobiles that prevailed from 1959 to 1969. This increase in the number of cars on the road in the fifteen cities studied means a rise in the amount of pollution emitted if the rate per car remains the same. But federal regulations for 1971 cars provided for sharp reductions in the amounts of pollutant emitted; and the Clean Air Act of 1970 calls for further reductions to take place in 1975 and, in any event, not later than 1976. If the provisions of the Clean Air Act of 1970, known as the Muskie bill, can and will be strictly enforced, the prospect for generally cleaner skies seems bright, indeed; the quality of the air still may not be satisfactory in many cases, but it will be better in 1980 than it is now. After all, the Clean Air Act decrees a flat 80 to 90 percent reduction in emissions of carbon monoxide and hydrocarbons from cars below the amount permitted under standards of 1971, and 1971 was a large advance over earlier standards. Since the automobile is the main source of the two pollutants mentioned, progress is certain in respect to them if its emissions are tamed as much as is called for.

But so far the automobile industry has been hauled kicking and screaming, like a balky youngster, into the new age. Detroit protests that the 1976 standards for automobile emissions are arbitrary and contradictory since if you cut down on some emissions to the extent legislated for, you necessarily increase others. Moreover, the industry has won support for its point of view from a committee appointed by the White House Office of Science and Technology, which found that the Muskie bill standards might be met but at a cost to consumers out of line with the benefits received. Since there seemed to be some doubt whether it would prove possible to attain the standards of the Clean Air Act (designated in this text as the "1976 standards"), two sets of projections were made in Chapters 16-18, one assuming 1976 standards reached and a second assuming that the 1971 standards are the highest achieved.

Whatever the outcome of the fight over the enforcement of the 1976 standards under the Clean Air Act, it seems evident that the air in our fifteen cities will be freer of carbon monoxide and hydrocarbons in 1980 than it was in 1969. Without exception, in all the cities studied, the projected emissions in 1980 of carbon monoxide and hydrocarbons were markedly lower under either assumption employed than in 1969, despite the expansion in automobile populations. If the Muskie bill standards are met, the total volume of carbon monoxide from all sources in the fifteen cities studies will be reduced by 1980 to 25 percent of the level in 1969, and even if only the 1971 standards prove feasible the 1969 emissions will still be cut by more than half to 42 percent of 1980 emissions. Similarly for hydrocarbons, if the 1976 standards prevail in 1980, total tonnage of hydrocarbon pollutants in the fifteen cities will be 36 percent of 1969 levels, and if only 1971 standards are achieved hydrocarbons will nevertheless be reduced by half by 1980. The automobile is far and away the most important source of carbon monoxide and hydrocarbons, and so their projected reduction by 1980 in all fifteen cities studied suggests the likelihood of a major victory for technology over pollution.

More Nitrogen Oxides

But the outlook is much dimmer if we look at nitrogen oxides. Even if the Clean Air Act standards for 1980 emissions by automobiles are rigorously maintained, nitrogen oxide emissions will be lower than in 1969 in only a fifth of the fifteen cities studied. In this case, the increase in nitrogen oxide emissions from 1969 to 1980 will be greater than 20 percent in all cities studied and more than 100 percent in two. If only the 1971 standards for automobiles are achieved by 1980, the smog in that year will be denser than ever in every one of the fifteen cities, assuming that the projections are correct and that nitrogen oxides are the main pollutants producing smog. The projected increase in nitrogen oxide emissions by 1980 assuming 1971 standards for automobiles is dramatic in all of the cities studies—including several of those most plagued by air pollution problems such as Los Angeles, New York, and Chicago. The 1969 to 1980 increase in nitrogen oxides assuming 1971 standards is greater than 20 percent in all cities studied, over 50 percent in more than half, and exceeds 150 percent in two cities. How can this be? Will we win the battle against carbon monoxide and hydrocarbons in 1980 and lose the war against smog? If true, this would be a startling result.

There is no mystery involved, however. Consider first the finding that even if the nitrogen oxide emissions from automobiles are reduced as called for, in all but three of the fifteen cities studied the nitrogen oxide emissions are projected to be higher in 1980 than in 1969. The reason is simple. The automobile is by no means the only source of nitrogen oxides; gross nationwide emissions from stationary sources are almost equal to those from internal combustion engines [1:4]. Moreover, the combustion of one source of nitrogen oxides has been rising. The use of natural gas has been increasing rapidly in most of the cities studied, is the main fuel for space heating in about half, and its usage is projected to increase further by 1980. This is mostly in response to market forces, but to some degree the use of gas has been encouraged by environmental officials to reduce the even more harmful sulfur oxide concentrations. But, as in so many aspects of pollution control, there is a tradeoff involved; while natural gas is ideal from the standpoint of sulfur oxides and particulates, its combustion releases large quantities of nitrogen oxides. Thus, much of the projected increase in nitrogen oxides between 1969 and 1980 involves major sources other than the automobile.

Somewhat similar considerations led the Ad Hoc Committee of the Office of Science and Technology to this gloomy conclusion:

Barring unpredictable changes in atmospheric chemistry, the NO_x (nitrogen oxides) air quality standard will not be met in most major cities despite the anticipated reduction of NO_x emissions from existing stationary sources by 20 percent and of new sources by 50 percent [1:8].

There can be little doubt that in 1980 nitrogen oxide concentrations will be higher than they were in 1969 in most of the cities studied. This is a very conservative statement and it is probably likely that the increase in nitrogen oxide concentration will be greater than 20 percent in all of the cities studied and over 50 percent in more than half. Does this mean that we are justified in advancing the provocative paradox: In 1980, there will be cleaner air and more frequent smog? Probably not, without extensive qualification. Why not? Certainly, it is known that "nitrogen dioxide . . . interferes with the visibility of distant objects" [1:I-F6], a sort of working definition of the haze associated with smog. Also, nitrogen dioxide has "been associated with increased incidence of acute bronchitis in infants and school children and acute respiratory disease in entire family groups" [1:I-F6]. Accordingly, we may assume that nitrogen oxides qualify as full-scale air pollutants in their own right, operating independently.

On the other hand, it is also known that: "hydrocarbons represent the major class of reactive organic matter in the atmosphere that is responsible for photochemical smog" [1:I-F6]. Thus, one important way to reduce smog is to reduce the concentrations of hydrocarbons in the atmosphere. How much reduction in hydrocarbon concentration is necessary to effect a significant reduction in smog in the presence of a surplus of nitrogen oxides? Judging by the Office of Science and Technology report [1:I-F6], there does not appear to be a precise answer. On the one hand, the Office of Science and Technology report, which seeks throughout to prove that control of nitrogen oxides is an expensive and ineffective means of air pollution control, attempts to establish that control of hydrocarbons might offer a better route [1:I-A15]. On the other hand, the report cites data demonstrating that just two hydrocarbons emitted from natural sources—methane and the volatile terpenes—enter the atmosphere at a rate of 7.7×10^8 tons per year [1:I-A6]. This rate of natural emission of hydrocarbons was over 100 times the rate of pollution cited in the same source from nitrogen oxides emitted by automobiles at the rate of 7×10^6 tons per year [1:I-A6]. Such natural emissions might not occur at locations where they could enter into reactions with other constituents of smog. However, in view of the large quantities of both nitrogen oxides and hydrocarbons emitted to urban atmospheres, it seems hazardous to assume that reduction in the concentration of either pollutant in the presence of a supply of the other can reduce the formation of smog.

With any such complicated question, no shorthand formulation is likely to be fully satisfactory, and yet it may help to dramatize the situation to state a short conclusion flatly and with full knowledge of its theoretical inadequacies. Accordingly, the striking paradox: 1980 will see cleaner skies and more frequent smog!

Progress Slow on Sulfur Oxides

The fight to reduce sulfur oxides, like that to reduce nitrogen oxide pollutants, has proved a difficult one, but for different reasons. Natural gas has gained ground relative to fuel oil with a beneficial result as far as sulfur oxides are concerned. Sulfur oxide emissions have also been reduced by municipal ordinances requiring use of low sulfur oils, an especially important restriction in commercial-industrial space heating, a category that includes apartment houses. Such specification of sulfur content has also been employed to control sulfur oxide emissions from electric utilities. Electric utilities remain a major source of sulfur oxide pollutants, especially since most of them still prefer coal as fuel for reasons of economy, and no economically feasible method of sharply curtailing sulfur oxide emissions from utilities burning sulfurous coal has yet been discovered. Power production is one of the most rapidly expanding sectors of the economy and the major source of sulfur oxides. It is therefore not surprising that sulfur oxide emissions were projected to increase from 1969 to 1980 in ten of the eleven cities that had appreciable sulfur oxide emissions in 1969. (In four of the fifteen cities studied, sulfur oxide emissions were negligible in 1969 and projected to remain so in 1980.) Only in New York, the worst offender in sulfur oxide pollution in 1969, was a really sizable reduction in the dangerous sulfur oxide emissions expected by 1980, a direct result of vigorous effort by local pollution control officials.

In the study of air pollution in Salt Lake City, a single copper smelter was estimated in 1970 to have the potential for emitting 581,000 tons of sulfur dioxide, an amount greater than that projected for the entire New York SMSA in 1980. With proper controls, it was felt, this could be held to 177,000 tons annually. This case is an indication of how important the industrial sector is with respect to sulfur oxides. Our estimates of sulfur oxide emissions are gravely weakened by omission of the industrial sector.

Sulfur dioxide constitutes the most abundant form of sulfur oxide entering the atmosphere, and this forms the relatively weak sulfurous acid upon solution in water. It was noted earlier that the volume of nitrogen oxides emitted to the atmosphere was increasing. Nitric oxide, one of the commonly occurring forms, is the chemical used in oxidizing sulfur dioxide to sulfur trioxide in the manufacture of sulfuric acid—a powerful and corrosive chemical. Thus, nitrogen oxides combine with sulfur oxides in the atmosphere to make the latter more dangerous.

Two broad methods have been used to control sulfur oxide emissions. Most successful with space heating installations has been the use of input controls specifying the maximum permissible sulfur content of the fuel—usually fuel oil—which is employed. Such controls are unnecessary for most individual home heating because the oil employed is low in sulfur even in the absence of controls. The sulfur content of natural gas is negligible and requires no control.

For control of sulfur oxide emissions from large-scale sources such as power plants and major industrial users, output controls will probably prove most effective in the long run. Processes for recovery of either sulfuric acid or elemental sulfur from flue gases have been experimented with, but appear to have found limited practical application thus far [2:179-81].

It would appear reasonable to suppose that if enforcement of rigorous emission standards for large-scale fuel users were sufficiently vigorous, by-product recovery from stack gases might become increasingly attractive from an economic standpoint.

Thus far, however, processes for absorption of sulfur oxides from stack gases have not proven themselves commercially feasible according to the Federal Power Commission [3:II-4-25, III-1-14].

Perhaps the most important measure employed to hold down sulfur oxide emissions has been the increased use of natural gas, which is virtually sulfur-free, for space heating and electric generation. There is no immediate reason to anticipate a decline in the rate of adoption of gas for space heating, but the Federal Power Commission foresees the possibility of projected use of natural gas overtaking additions to reserves by about 1985 in the western states [3:iii-3-51].

Particulates — Soot

Perhaps the most visible cleanup of the air between now and 1980 will be in the reduction of soot, or particulates to use the more general term usually employed in describing the phenomenon. For the five far western cities and the three southern cities studied, soot has never been a problem. Four cities are projected to cut their soot emissions by 1980 to about a fifth of 1969 levels. This group includes Chicago, which expects to reduce from more than 155 thousand tons per year in 1969 to about 30 thousand tons per year in 1980. Three others—Pittsburgh, Cleveland, and St. Louis—are projected to cut 1969 soot emissions of about 50 thousand tons per year to about one-fifth that amount in 1980. Two others—Philadelphia and Atlanta—expect to hold the line, neither gaining appreciably nor losing, with annual emissions in thousand tons at about 50 and 25 respectively. Only in New York, of the fifteen cities studied, is particulate matter expected to increase by about 6 percent between 1969 and 1980 to 88 thousand tons annually. Thus cleaner—read less sooty—skies seem to be on the agenda for 1980 in most American cities. For well over half the cities, this is a matter of use of natural gas, which forms little soot but much nitrogen oxide.

In terms of total tonnage of all forms of pollution emitted computed on a per square mile basis, New York was by far the dirtiest among the eastern cities. In the midwest, St. Louis had the highest ratio of pollutants per square mile, with

Chicago and Cleveland close behind. It should be noted that Mr. Caulson made his calculations throughout for the central city county only, as did Mr. Fennimore, whereas Mr. Goffin made his calculations in terms of the entire SMSA which has, of course, a much greater area. No firm rule was laid down for this because none seemed called for, but it clearly limits the comparability of results on minor points such as amount of pollution per square mile.

Global Pollution of the Upper Atmosphere

In an extraordinarily careful and detailed study, Carl Katz has in masterful fashion reviewed current research in regard to the pollution of the upper atmosphere and has summarized what appears most likely to happen in view of current rates of output and consumption. Without the detailed qualifications and statements of assumptions, there is a considerable risk of oversimplification. Bearing that risk in mind, we will attempt here to summarize his principal findings.

1. There can be no question concerning the enormous increase in the formation of carbon dioxide that will occur given the rising curves of consumption of fossil fuels. Mankind has become a geological force reaching back millions of years to unlock and loose on the atmosphere the combustion products of fossil fuels, of which carbon dioxide is the most important.
2. The question is not the amount of carbon dioxide formed, which will be enormous, but the rate of its retention in the atmosphere, which depends importantly on the area of ocean available to absorb carbon dioxide from the atmosphere.
3. If carbon dioxide were retained in the atmosphere, the theoretical basis for assuming a warming trend in the earth's atmosphere seems firm and, in fact, such a trend was observed from 1880 to 1940. But since 1940, earth temperature has fallen despite an accelerated increase in carbon dioxide formation.
4. If the warming trend had continued unabated, some scientists have hypothesized a melting of polar ice caps with a fairly rapid rise in sea levels and perhaps a "runaway" carbon dioxide effect due to warming of the ocean surface layer, thereby driving additional carbon dioxide out of solution.
5. Actually, as noted, the trend in the earth's temperature has been downward since 1940 and this is often attributed to another effect of upper atmospheric pollution, viz the formation of aerosols and particulates that form clouds, shielding the earth's surface from the rays of the sun and permitting less energy to enter the earth system. The magnitude of the recent drop is believed to be too great to be explained by man-made soot and aerosols, and is attributed rather to a high rate of volcanic activity.
6. The one thing that all can agree on is that the changes are of such magnitude and such potential worldwide importance that it is essential that we know about changes in time to take needed corrective action. This argues for

international scientific cooperation through such an agency as the United Nations, and such cooperation was initiated by the convening of the U. N. Stockholm meeting on the environment in 1972.

Noise Pollution – Another Dimension

The air may be polluted by more than noxious chemicals. Ronald Bruce has chosen to investigate the problems that have arisen for the air transport industry by the noise pollution created by ever larger planes.

As planes have grown in size, the noise associated with takeoff and landing has increased as well. As the volume of air traffic has grown, the number of takeoffs and landings has risen. In a carefully reasoned piece, Mr. Bruce makes clear how this may lead to crushing legal judgments against airports that will have to be passed on in one way or another to the carriers. For an industry with strained financial resources, still adjusting to the rapid technological changes of recent years, the litigation associated with noise abatement may be a disastrous burden. No easy solutions are in sight.

Notes

1. Office of Science and Technology, *Cumulative Regulatory Effects on the Cost of Automotive Transportation* (Washington, D.C.: Final Report of the Ad Hoc Committee, February 28, 1972).

2. Alfred J. Van Tassel, *Environmental Side Effects of Rising Industrial Output* (Lexington, Mass.: D.C. Heath and Company, 1970), 548 pp.

3. Federal Power Commission, *The 1970 National Power Survey* (Washington, D.C.: Government Printing Office, 1970), 4 volumes.

Part 3
Our Environment: Solid Waste Disposal

22 Some Aspects of Solid Waste Disposal

JOHN A. BURNS AND MICHAEL J. SEAMAN

There was no common methodology followed by the authors of the chapters on solid waste disposal. Instead, each author took the approach that seemed to fit best the data available to him. However, all three chapters benefited enormously from computer print-outs of preliminary data from the *1968 National Survey of Solid Waste Disposal Practices,* prepared by the U. S. Department of Health, Education and Welfare, Public Health Service. Arrangements for the use of these print-outs were made by John A. Burns. —Ed.

The decade of the sixties may ultimately be looked upon as the era in which the general populace became aware to some degree of the consequences of our consumption society. This discovery first manifested itself with the recognition of air pollution in large metropolitan areas. Later, conservationists brought our attention to the reality of water pollution. Since this time, much attention and publicity have been given to these two very important areas.

However, there is a "third pollution." One that has gone, for the most part, unrecognized through much of the sixties—that of land pollution. Translated in terms of origin, we may refer to this "third pollution" as bearing a direct relationship to our ability to dispose of our solid wastes. But our ability (or inability) to dispose of our solid wastes is only one side of this very complex challenge facing us today. It may be well for us to examine our more recent industrial history to gain some insight in how the problem of solid waste disposal grew to its present state.

That we are a consumption-oriented people is reflected in our abiding concern with our gross national product and so-called rate of economic growth. During the sixties, the annual rate of growth for this latter category was slightly under 5 percent. It does seem logical, then, to hypothesize that growth of production, spurred by our need to consume, is the underlying cause of our present-day pollution. Nowhere is this relationship more apparent than in the area of solid waste (land) pollution.

Types of Solid Wastes

There is a common tendency to associate solid waste with garbage, although garbage only constitutes a small part of the problem. A classification of refuse materials, reprinted from *Refuse Collection Practice* (table 22.1), may be used to standardize definitions of various components of solid wastes.

Household solid wastes are always collected locally and the amount collected is dependent upon the size and consumption habits of the population. It is disposed of on either public property or private contractors' sites.

Commercial wastes, those generated by commercial establishments, can be collected by any number of agencies, but are usually disposed of by public facilities. In the recording of total collections, commercial waste may be measured as a separate category or combined with household waste.

Industrial wastes are those generated by various manufacturing firms. Special factors involved are:

1. Individual industries generate wastes peculiar to that industry, and all industries generate plant trash.
2. Most industrial wastes are homogeneous and sorted, so recycling is utilized to a large extent. There is usually an established salvage industry.
3. Information concerning industrial wastes is hard to procure.
4. These wastes are not usually collected by public agencies.
5. The responsibility of disposal usually is on the firm, although public facilities may be used for wastes.

The volume of construction and demolition waste will vary with the level of building activity, urban renewal, and highway construction. As such, these wastes are difficult to project, and most estimates of waste generation or collection are exclusive of this type of waste. The nature of these wastes is such that they are difficult to incinerate without preprocessing and can create voids in sanitary landfill.

The amount of agricultural and mineral wastes generated is truly staggering. However, since they do not usually occur in proximity to the urban areas under analysis in this study, they need not concern us here.

Other wastes, pathological, radioactive, sewage sludge, etc., are usually given

Table 22-1
Classification of Refuse Materials

Name		Content	Source
Garbage		Wastes from the preparation, cooking, and serving of food Market refuse, waste from the handling, storage, and sale of produce and meats	
Rubbish	Combustible (primarily organic)	Paper, cardboard, cartons Wood, boxes, excelsior Plastics Rags, cloth, bedding Leather, rubber Grass, leaves, yard trimmings	From: households, institutions, and commercial concerns such as: hotels, stores, restaurants, markets, etc.
	Noncombustible (primarily inorganic)	Metals, tin cans, metal foils Dirt Stones, bricks, ceramics, crockery Glass, bottles Other mineral refuse	

Table 22-1 (cont.)

Name	Content	Source
Ashes	Residue from fires used for cooking and for heating buildings, cinders	
Bulky Wastes	Large auto parts, tires Stoves, refrigerators, other large appliances Furniture, large crates Trees, branches, palm fronds, stumps, flotage	
Street refuse	Street sweepings, dirt Leaves Catch basin dirt Contents of litter receptacles	From: streets, sidewalks,
Dead animals	Small animals: cats, dogs, polutry, etc. Large animals: horses, cows, etc.	alleys, vacant lots, etc.
Abandoned vehicles	Automobiles, trucks	
Construction & Demolition wastes	Lumber, roofing, and sheathing scraps Rubble, broken concrete, plaster, etc. Conduit, pipe, wire, insulation, etc.	
Industrial refuse	Solid wastes resulting from industrial processes and manufacturing operations, such as: food-processing wastes, boiler house cinders, wood, plastic, and metal scraps and shavings, etc.	From: factories, power plants, etc.
Special wastes	Hazardous wastes: pathological wastes, explosives, radioactive materials Security wastes: confidential documents, negotiable papers, etc.	Households, hospitals, institutions, stores, industry, etc.
Animal and Agricultural wastes	Manures, crop residues	Farms, feed lots
Sewage treatment residues	Coarse screenings, grit, septic tank sludge, dewatered sludge	Sewage treatment plants, septic tanks

Source: American Public Works Association, REFUSE COLLECTION PRACTICE, Public Administration Service (Chicago, Illinois: 1963), p. 15.

special treatment and they, like agricultural and mineral wastes, generally do not fall within municipal collection or disposal facilities. Due to the specialized nature of these wastes, technological advances and innovations will probably first manifest themselves in these areas.

Major Categories of Matter Involved in Household Waste Disposal

Packaging wastes account for over 13 percent of all solid wastes discarded in this country. The impact of these wastes in the household sector is even greater.

It has been established that packaging wastes constitute 19.9 percent of household wastes [9:110]. Further, it can be assumed that these wastes will not diminish as a major contributor to solid wastes during the 1970s.

There is every indication that the contribution of packaging to solid waste will continue to increase through the 1970s. Packaging has become a "point of sales" tool as well as performing a utility function.

At this time, and continuing through the 1970s, paper and paperboard constitute the bulk of all packaging consumed in this country. (Paper products are employed as both primary and secondary containers.) Use of paper and paperboard most commonly used for packaging is growing at an annual rate of "close to 4.5 percent" [4:8]. Although subject to the same overall economic restraints as other industries, the companies engaged in producing paper and related products have estimated capital expenditures of $1.62 billion for 1970 [4:3], which indicates that they are expecting continued growth of their product line through the 1970s.

The use of metal cans as containers for food (including pet food) began to grow rapidly in 1963. Consumption of cans for these purposes is expected to increase by over 40 percent through 1976 [9:51], or at an annual rate of 3.4 percent.

The greatest increase has been and will continue to be in the area of metal containers for beer and soft drinks. Continued increase of metal containers bears a direct relationship to mass marketing and distribution techniques employed by individual companies as well as industry trade associations.

The glass industry has fought the competition of metal cans by introducing the nonreturnable bottles and other innovations, such as the twist-off cap. Use of nonreturnable bottles is expected to grow at an annual rate of 12 percent through 1970 [9:41]. It is estimated that each returnable bottle can be used approximately nineteen times before it is rendered useless [9:41], as against nineteen nonreturnable containers requiring collection and disposal. Table 22.2 details the growth in output of all nonreturnable containers since 1958.

The use of plastics in packaging did not really develop until the 1960s. This was due to price considerations as well as competitive resistance from other forms of packaging. However, it appears that the growth of plastics in all descriptions will continue at an *accelerated rate through the 1970s.*

Each of the four areas mentioned above will enjoy continued growth during the 1970s. Some of this increase will be at the expense of other categories. For example, some increase in output of nonreturnable bottles occurs at the expense of canned beer. However, a considerable amount will be as a result of population growth and increased per capita income.

However, a considerable amount will be as a result of population growth and increased per capita income.

The abandoned automobile is another source of urban solid waste. In any given year, we can expect approximately 15 percent of all those cars going out of service to be simply abandoned. During the late 1960s, an annual average of over seven million cars and trucks per year were scrapped [34]. There are various reasons for abandonment, but the most common centers on economics. It is just not profitable for the owner to undertake a repair bill in excess of the

Table 22-2
Nonreturnable Beer and Soft Drink Container Production by Type of Container and Use (In Millions of Units)

	1958	1966	1976 (Estimated)
Bottles			
Soft Drink	192	1,980	13,500
Beer	1,239	5,031	8,600
TOTAL	1,431	7,011	22,100
Cans			
Soft Drink	409	5,612	17,000
Beer	8,337	12,947	19,000
TOTAL	8,746	18,559	36,000
Total All Nonreturnables	10,177	25,570	58,100

Source: THE ROLE OF PACKAGING IN SOLID WASTE MANAGEMENT 1966 TO 1976, U.S. Department of Health, Education, and Welfare. Bureau of Solid Waste Management, Rockville, Maryland, 1969.

worth of the car. Ultimately, these cars will move into wrecking yards where, depending on the condition of the scrap metal market, they will remain for some time. In the final analysis, the price that steel mills are willing to pay the scrap processor will determine the rapidity with which all discarded cars will be recycled. The increased use of electric furnace steel and the development of the localized "mini-mill" may be one way for our economy to digest the millions of cars that become scrap each year.

Oversize Burnable Waste (O.B.W.) consists mainly of lumber, 77.5 percent of the estimated United States total; furniture and fixtures, 12.9 percent; plus Christmas trees, driftwood, stumps, etc. Their estimated growth rate is between 2-3 percent [23:107]. Building activity, which was at a low level during the 1960s, accounts for much of the O.B.W. These wastes cannot be incinerated without prior processing such as shredding, and are generally unsuitable for sanitary landfills because they tend to produce voids that can harbor disease vectors, promote uneven settlement, and increase the possibilities of fire. They also require specialized collection vehicles and schedules. Hence, they are usually disposed of by open burning, at heavy expense to air quality.

Historical Trends of Solid Wastes

In discussing the historical behavior of solid wastes, a distinction must be made between the amount *generated and collected*.

It is estimated that urban and industrial wastes generated in the United States average 10 lbs. per capita per day, whereas collected wastes average only 5.12 lbs. per capita per day, only 51% of the total generation. The uncollected balance is generally an unknown factor in urban life" [10:15].

If we accept a .51 collection/generation factor, then we must admit wide variances are possible by using collection data, and the possibility that changes over a period of time are more likely to be influenced by collection practices than the underlying fundamental generation factors. For example, deteriorating services are likely to reflect a lower per capita collection rate than would be justified by waste generation.

In 1963, the American Public Works Association estimated that growth in per capita refuse production had risen from 2.8 pounds to 4.5 pounds in the 1920-1965 period. This is an annually compounded growth rate of 1.2 percent, and they projected a 1.4 percent growth rate from 1965-1980, to 5.5 pounds/capita/day [30:25]. In retrospect, this estimate appears to have considerably underestimated our capacity to generate solid waste.

The 1968 National Survey of Community Solid Waste Practices, covering 46 percent of the total population of the United States, arrived at a national figure of 5.32 pounds/capita/day (5.72 for urban areas) *collected* [37].

Another significant relationship over the century has been the increase in the volume of solid wastes. In fact, largely because of the increase in paper consumption, the uncompacted *volume* of solid wastes has been increasing at a faster rate than the growth in weight [30:27]. Since the vast bulk of total solid waste expenditures concern collection, the increased volume of solid wastes expected over the next decade will make collection more difficult and costly.

The composition of the "solid waste mix" has undergone a profound change and this accounts for the accelerating per capita waste generation growth. An interesting study can be made from samples taken from the solid wastes of three cities—New York, London, and Berlin—made around the turn of the twentieth century.

It appears, then, that the average per capita weight and volume of solid wastes generated has shown a moderate growth rate from the 1920s to 1960s primarily because of the use of oil, gas, and electricity for heating and cooking purposes, with the decline in the percentage by weight of ash being overwhelmed by the tremendous increase in paper consumption. Inasmuch as ash has a higher density than paper, we can no longer look to ash to act as a moderating influence on the growth of per capita production of solid wastes.

The manner in which an area processes its wastes will affect the overall level of solid wastes collected. For instance, if a city has a large proportion of installed garbage-grinders, that city's solid wastes are transformed into a liquid waste problem, and collection and disposal of the remaining wastes will follow a different pattern than that of a city not using grinders. Another obvious situation is when incineration is used. Here we reduce the solid wastes prior to disposal albeit at the expense of increased air pollution. The process also works in reverse. Over the next decade, water pollution controls and regulations are likely to create solid wastes from hitherto liquid wastes.

There are two basic methods of estimation and measurement: weight and volume. Cities that do keep records of the amount of wastes collected can do this by the use of scales, to arrive at the tonnage, or by measuring the load by the volume of the collection vehicle. When the latter method is used, wide variations can occur. An open truck can carry between 150-300 lb./cu yd., while the same load, properly loaded and compacted in a modern compaction vehicle, may have a density of more than 500 lb./cu yd. [30:23].

Table 22-3
Selected Components of Solid Waste, by Weight 1900-1965

	1900			1965		
Component	New York	London	Berlin	United States	U.K.	W. Germany
Ash	80.5	84.0	53	10.0	30.-40.0	30.0
Paper	5.5	4.3	4.3	42.0	25-030.0	18.7

Source: 1900–W.F. Goodrich, THE ECONOMIC DISPOSAL OF TOWN'S REFUSE (London: P.S. King and Son, 1901), as cited by U.S., Department of Health, Education and Welfare, SOLID WASTE MANAGEMENT: ABSTRACTS AND EXCERPTS FROM THE LITERATURE, Public Health Service Publication, no. 2038 (1970), Vol. 1, pp. 112-114.
1965–"World Survey Finds Less Organic Matter," REFUSE REMOVAL JOURNAL, September 1967, p. 26.

Major Coefficients of Solid Waste

Population. Regardless of any *percentage* decline in the rate of population growth, the *absolute* increase will continue and will no doubt be substantial. Those persons born during the high birth rate years of the late 1940s and the 1950s will be reaching the age of maximum demand for consumer products, both durable and nondurable, during the seventies and eighties.

Of primary importance to per capita coefficients of solid waste are the *economic trends* applicable to the area under study and their relationship to solid wastes generated.

Economic growth rates are merely the point of departure. They must be adjusted for the effects of inflation. We generally measure inflation against the amount of real goods and services that can be purchased with a given amount of money. For example, inflation could be said to exist if we can buy less meat for the dollar than we could in the previous period, but does this imply we are getting less wrapping on the meat? If we get more packaging and less primary goods for more dollars, "inflation" is not applicable to the solid waste problem, and the basic growth rate should be adjusted upward, not downward.

Density will obviously influence the collection of solid wastes, and the productivity achieved (traveling time is unproductive), and since the density will reflect the degree of an area's development, it will also influence the type and costs of the disposal methods used, whether this involves sanitary landfill, incineration, transfer stations, or any other type of waste processing. In addition, there is some evidence, at least for rural areas, that the density will affect, in a direct relationship, the unit generation of solid wastes [14:86-87].

Some hypothetical reasons for this relationship are:

1. Density will reflect single vs. multi-family housing patterns, and influence the type and extent of on-site storage and disposal of solid wastes.
2. Density may affect the "turnover" of goods beyond their primary utility. A sofa no longer suitable for the living room is likely to be immediately discarded by the apartment dweller, while it may languish indefinitely in a private home.

Of more interest than total population growth is the movement from the central city. This has had economic as well as political and jurisdictional effects on the city itself and has magnified its solid waste disposal problems. The development of the semi-rural buffer zone has raised land values and decreased the amount of land suitable for the disposal of wastes.

The 1968 National Survey of Community Solid Waste Practices was an immense contribution to the stock of information regarding solid wastes, but there is a wide range of surveys and samples conducted by the federal, state, and local governments, the American Public Works Association, universities, and various other competent sources, at various times, which leads to some contradiction. Most of these collected data are limited to a varying degree by several factors. These include: the competence of the data collectors, the seasonal variation of solid wastes in general, moisture content and weather factors, the presence or absence of measuring devices, variances in reported results, consistency in reporting categories, and, most importantly, the lack of uniformity in the services and facilities provided by the agencies responsible.

Thus, in examining the solid waste situation over the next decade, we project:

1. A greater increase in per capita weight generation that was anticipated in earlier studies.
2. A still greater increase in per capita volume generation relative to weight, which will serve to increase the cost and difficulty of collection, transportation, and disposal. The volume factor appears to be more critical to cost and visibility than weight, per se.
3. No countervailing decrease in any major component of the solid waste mix (e.g., ash) is likely to moderate the growth rate as happened in prior decades.

Sanitary Landfill: Current State of the Art

Sanitary landfill offers our urban areas a method of solid waste disposal capable of accommodating increasing amounts of solid waste forecast through 1980.

At the outset, however, we must enumerate the five basic requirements of the sanitary landfill (as opposed to an "open dump"). These criteria are: (1) solid wastes must be deposited in a controlled manner; (2) solid wastes are spread in thin layers; (3) daily ground cover of not less than six inches is applied daily; (4) no open burning; (5) no factors present that would allow deposited wastes to contribute to water pollution.

The 1968 national survey [38] found that true sanitary landfill existed in only 6 percent of the 6,000 sites surveyed. This led H. Lanier Hickman, Jr., Chief, Technical Services, Solid Waste Program, to say that "94 percent are unacceptable and represent disease potential, threat of pollution, and land blight" [38:44].

A high degree of opposition and resentment often accompanies announced plans for a sanitary landfill. This is especially true when some or all of the waste destined for the site is generated in an adjoining, usually large, city. Only in those cases where there has been a definitive plan of action involving planned solid waste management has there been any real success in the establishment of sanitary landfills.

Planned management with respect to sanitary landfills is begun with analysis of soil conditions surrounding a proposed site. A ravine or man-made gravel pit generally is sought as a sanitary landfill site. Soil conditions surrounding the site are significant to the ultimate success of the landfill. Fine-grained soils are more useful in containing unwanted gas and water movements outside the landfill area. The type of soil will also ultimately determine how the landfill will be utilized once completed.

Within the landfill itself, unwarranted gas and water movements can adversely affect its utility. The production of gases (usually either methane or carbon dioxide) within a landfill is a function of matter placed there for disposal. Control of these gases can usually be maintained by insuring that soil immediately surrounding the landfill is not porous. Except for the ease with which it disperses itself away from the landfill, containment of water within the landfill presents much the same problems as gas. In order to avert water pollution, the lowest level of the landfill should be no less than ten feet above expected high water table level. In addition, proper surface drainage should be established and maintained in order to prevent waters from washing through the rubbish and reentering existing water routes.

This cursory probe into the considerations of sanitary landfill preparation and management should at least alert the reader to the magnitude of the work to be done when measured against disclosure of the fact that 94 percent of land disposal sites in use today fail to meet one or more of the standards for sanitary landfill.

The decade of the seventies will no doubt witness the continued upgrading of our seemingly ubiquitous open dumps and landfills to conform to sanitary landfill specifications. Stronger local laws regulating landfill operations, as well as stringent air and water pollution laws, will serve to make these operations more efficient. These actions, while beneficial to overall land management, are critical in urban areas. As the urban expansion continues to move outward from the central city, land suitable for sanitary landfill becomes increasingly scarce and sometimes inaccessible. Technological benefits such as those accruing from more efficient handling of solid wastes can offset this disadvantage to some degree. Establishment of central collection or transfer stations will enable cities to ultimately transport solid wastes to more distant disposal sites more efficiently. Depending on the degree of density surrounding the urban area, it is not improbable that the 1970s will see the use of rail transportation as a means of moving solid wastes to distant sanitary landfill sites. Several feasibility studies have already been undertaken to determine the possibility of using rail transportation by Chicago, Denver, and Los Angeles. To date, no major city has embarked on a plan of action involving rail transportation.

Although the day-to-day costs of operating a sanitary landfill vary from one location to another, they are generally acknowledged to be greater than incineration. However, in terms of final costs to the community, they are probably no more, and even less. For the planned sanitary landfill, conceived and administered with modern solid waste management, can provide the community with an annuity in the form of future land sites.

Ultimately, our land will have to accept virtually all solid wastes generated by man. Whether intermediate steps of incineration or compaction are employed, the final disposition of our wastes remains unchanged. What is altered, however, is the efficiency with which the land can be managed through the utilization of a solid waste management plan.

The current state of the art, then, calls for appraising the current condition of landfill operations, recognizing what needs to be done to conform to sanitary landfill standards and moving towards the improvement that must come through implementation of solid waste management programs at the community level.

Incineration: Current State of the Art

Incineration of solid wastes serves the function of reducing the volume of these wastes from their raw or collected states to more manageable levels, thus obtaining lower transportation costs to the ultimate disposal site, and accommodating the wastes of a greater number of people for a given acreage available for landfill. The two main disadvantages of the technique are that both capital and operating costs are considerably higher than sanitary landfills, and without adequate and expensive equipment there is a considerable cost to the quality of the air.

Incineration has had a cyclical existence. First used in 1885, incinerators rose and fell in popularity, so that by 1952 over one-third of the cities operating incinerators had phased them out. One of the main problems was that incineration furnaces were modified from existing manufacturer's design, rather than being specifically designed for solid wastes [20:83]. However, since 1950 there has been a resurgence of incineration. Table 22.4 shows the growth in number and capacity of operating incinerators in the United States.

Two types of incinerators are generally used for unsorted wastes. The *batch-type* plant is manually stoked and has a relatively small rated capacity. Operation is intermittent (hence the name "batch") so there is a lack of uniform burning temperatures. This promotes inadequate combustion and an attendant heavy output of particulate matter, less than optimal volume reduction, and an unstable residue often containing putrescibles. These units are generally unsuitable for the large urban centers, although they are used in the smaller, surrounding communities. The *continuous feed* plant has larger storage bins, automatic feed hoppers, and a variety of types of moving grates and ash removal systems. The unit maintains a uniform combustion temperature range, can be fitted with pollution-control devices, albeit at a higher cost, and generally yields a stable residue [36:19].

The reason incineration is experiencing rapid growth is the volume reduction that can be achieved in a well designed and operated incinerator with controlled furnace temperature in the 1,400-1,800° F. range. A ton of refuse, at the generating source, occupies 13.3 cubic yards—a density of 150 lb./cubic yard. After substantial compaction in the collection vehicle, and further compaction in a sanitary landfill, it will occupy 2.22 cubic yards and its density will be 900 lb./cubic yard. However, if the refuse is incinerated and then landfilled, after one

Table 22-4
Operating Incinerators and Rated Capacity in the United States 1950-1966

Year	Number of Installations	Tons Rated Capacity	Capacity/Installation
1950	90	22,932	255
1960	205	57,044	278
1966	254	74,578	294

Source: Combustion Engineering Company, TECHNICAL-ECONOMIC STUDY OF SOLID WASTE DISPOSAL NEEDS AND PRACTICES, vol. I, part 3, pp. 10-17.

year the original ton will only take up .194 cubic yards of space and have a bulk density of 2,700 lb./cubic yard. With land growing increasingly more scarce, more costly, and available only at longer distances from the collection points, the value of incineration is that it cuts the land requirement to one-third of that required if the refuse was only landfilled. Moreover, a stable residue cuts drastically, or eliminates entirely, the need to import cover material to the landfill site [6:97-100].

Public acceptance of incineration is retarded by the existence of a substantial number of old and inefficient (mainly batch-type) units. They frequently operate substantially below their rated capacities, are visibly dirty in particulate emissions, and are uneconomical to modify to meet emission standards. A well-run, modern, large-capacity incinerator is not a major source of air pollutants. In an ordinary ton of mixed refuse, sulfur constitutes as little as one-tenth of one percent, and Elmer Kaiser has said, "... we have in refuse the sweetest fuel this side of natural gas" [6:126].

In a study of major atmospheric contaminants in several large U. S. cities, it was found that particulates emitted from incinerators accounted for 18.2 percent of total particulates from all sources. However, two devices, the wet-scrubber and electrostatic precipitator, can effect removal of between 90-99 percent of incinerator pollutants, and even at 90 percent efficiency the result of fitting these devices would be to cut total emissions of particulates by 13.9 percent. Marginal costs of efficiency are high. Increasing the efficiency of a precipitator from 90 to 99 percent can increase the cost of the unit by 50 percent. Scrubbers, although lower in cost, have a high water consumption, emit unsightly vapor plumes, and create unwanted pressure drops in the furnace. Most European installations (which, as a rule, are quite clean in emissions), and most planned modifications to existing domestic units, are expected to use precipitators. Other types of pollution control equipment, such as subsidence chambers, baffling systems, centrifugal collectors, and cloth filter have either low efficiencies or high space requirements that render them impractical for use on incinerators [21:100-103].

Development of incinerators in the immediate future will probably tend towards large units of water-cooled steel walls, rather than refractory-lined design. The reason for this is that precipitators and other pollution control devices require cooler raw gas temperatures while permitting higher burning temperatures. When higher furnace temperatures are achieved in refractory-lined units, melting glass adheres to the walls and forms a bond stronger than that holding the fire-brick together, and serious damage can occur. Another advantage of the water-cooled units is that the waste heat of combustion can be utilized. Recent installations have used the recovered waste heat for steam generation that provides space or district heating, electricity, and desalinization of sea water. While these installations are still a rarity in the United States, they are almost standard in large European units, and there exists a wealth of technology to make such installations an expected near-term reality in this country. Mixed refuse has a thermal value of between 3,500 to 5,500 BTU/lb., or about 37 percent that of high-grade coal; and the trend is towards higher heat values for refuse, especially if O.B.W.s are shredded prior to incineration. As for European practice, Rogus has said, "There are some indications that this trend [towards waste-heat recovery] may not continue should the natural gas resources recently opened in the Netherlands be made economically available in Central Europe" [22:117]. Thus, waste heat recovery modifications should be weighed against alternate energy sources and their effect on the environment.

A major problem for water-cooled installations is the increasing proportion of polyvinyl chloride (PVC) in the refuse "mix." Today, PVC accounts for between 25-35 percent of total plastics, and about 2 percent of all unsorted refuse. The heat value of PVC is about one-half that of other plastics (which have a heat value approximating fuel oil) and, when incinerated, PVC forms hydrogen chloride gas, a toxic emittant, and one that can rapidly corrode the metal walls. The rapid growth of plastics could lead to a dangerous concentration of 5-6 percent PVC of total mixed refuse, at which time the PVC would have to be removed prior to incineration, thus increasing the already high costs of incineration" [16:148, 150].

There are two technological innovations in the incineration art, both of which have yet to be evaluated beyond their initial promise, and both not yet in production. As with some projects in the past, whose expectations failed with production, a great deal of caution should be used in their evaluation as alleviants to the solid waste situation.

The first unit, the CPU-400, is an outgrowth of aerospace technology. It is designed as a turboelectric generator plant using solid wastes as fuel. Unsorted wastes are passed through a shredder, dried by the hot gases discharged from the combustor, and injected into the combustor at high pressure. Then the wastes are "fluidized" as they pass through the chamber. The plan is that the hot gases produced will drive a gas turbine which will drive an electrostatic precipitator, and the hot gases will be expelled at atmospheric pressure. The unit is currently in the development stage, and a full-scale, 15,000 kw. prototype unit is expected in 1972 [39].

The second unit, the "Melt Zit," slag-tap process, derives from blast furnace technology. The idea here is to melt everything, as the glass component of solid wastes is liquid at 1,800 F. and most of the ash molten at 2,350 F. The resulting magma can either be molded into large blocks, or run into water to produce a coarse type of sand suitable for road or concrete aggregate. In Germany, the Volkswagen Works operates a slag-tap plant, but there are no estimates on when this process will become commercially feasible here [6:100, 124-25].

Social and Political Aspects of Solid Waste

"Of the more than 12,000 governmentally controlled land disposal sites in the United States, 94 percent are unacceptable and do not meet even the minimum standards for sanitary landfills" [17:24]. This refers to open dumps. In urban areas, other factors, such as housing and crime, can magnify the role of solid wastes in contributing to urban decay. The open dump in an abandoned quarry is quite different from its counterpart in an abandoned urban building. A short drive through a central city ghetto area will confirm this contrast.

Robert H. Finch, the Secretary of Health, Education and Welfare, described the situation this way:

It is estimated that at least 14,000 people suffered rat bites in 1968. Most of these people live in areas where mismanagement of waste offers food and harborage to these vermin. Rats, flies, and fleas are common in neighborhoods that do not have adequate waste disposal services; these pests serve as disease vectors, spreading illnesses such as diarrhea and dysentery. The results of one of our recent studies indicate a relationship between solid waste and no less than 22 human diseases prevalent in this country, including encephalitis and hepatitis [18:28].

The bulk of expenditures for solid waste is accounted for by the costs of collection. Collection is a labor intensive industry or service. The job of the sanitation man is difficult and unattractive and has a high incidence of injury [19:10-11].

The problem here is further exaggerated by ethnic, racial, political, and union relationships that tend to create a matrix of conflicts and roadblocks to the resolution of disputes.

The late 1960s were witness to instances of disruption of sanitation services and urban disorder, as well as huge pressures on wages due to extensive wage increases in the private sector. It is believed that both these trends will continue for the next decade. Unless job improvement and productivity increases are instituted in the near future, rapid cost increases and the deterioration of existing services are likely to contribute to crisis conditions.

The advantages of regional planning are numerous: Land requirements and facilities can be planned on a time scale of forty to fifty years, sufficient to provide for foreseeable developments. Efficiencies of scale in the planning, engineering, and facility fields can be utilized whereas individual communities have neither the resources or experience necessary. Developments can be planned and coordinated to take maximum advantage of federal and state incentives. However, to be effective, a regional agency needs the cooperation of its constituent units. It needs to develop and enforce standards, and maintain the authority, right of eminent domain, and revenue sources needed to accomplish its ends. This may be difficult to accomplish in the absence of a true, areawide crisis of major proportions. Accordingly, true regional agencies will probably develop only out of expediency.

Political difficulties begin at the local unit of government that usually is responsible for collection. As one elected official has said,

I would say that if our community—the one I live in—is any example, if you were to ask the citizens to segregate and separate out their refuse, we would have a rough time on our hands. I wouldn't be standing here, I wouldn't be elected, I can assure you [6:67].

Local citizenry are less responsive to accepting waste generated outside their communities. Thus rail haul proposals have hardly advanced past the planning stage. Other ideas that are technically feasible, such as pipeline transport of solid wastes, are hampered as much by property rights considerations as the high initial costs. The idea of dumping treated wastes at sea is restrained by legal and international constraints, as well as cost and environmental constraints. One must also recognize that the disposal field is rife with partisanship. Advocates of one form of disposal tend to promote their technology to the exclusion of other methods.

Solid wastes traverse political boundaries; products are very rarely disposed of in their area of manufacture. The government, through the I.C.C., regulates freight rates crucial to both the disposal of treated solid wastes and the economies of recycling and salvage. Should government subsidies be employed to advance recycling? Or should the producer be taxed according to the disposability of his product? If so, who is to determine the taxation? What is the effect of increased controls in employment, on our system of letting the consumer make his choice in the marketplace? Darney and Franklin have found that the most formidable barrier to waste reduction objectives are those caused by our cultural free-enterprise philosophy [9:167].

Or, do we as free individuals have the right to destroy our waters; to carpet our land in refuse? The classic, conservative political-economic argument relates to whether we have the right to saddle future generations with our debt; the conservation argument concerns our right to despoil the environment of future generations with our current consumption. Of one thing there can be no doubt: the issue of the environment, and the relationship of the environment to solid as well as other states of waste, may well be the *political* domestic issue of the seventies. The problem, of course, is inertia. Inertia on all levels, the consumer, the producer, the endless tiers of government, the endless studies, surveys, and proposals.

Despite many unanswered questions, we have witnessed some positive measure of progress. At the time the Solid Waste Disposal Act (PL-89-272) of 1965 was enacted, there were no solid waste agencies at the state level and only five states had any personnel whatsoever assigned to solid waste management. Consider now, four and a half years after the first grant for technical assistance was awarded, that forty-four states have active solid waste programs.

The Resource Recovery Act of 1970 (PL 91-512) enables the Bureau of Solid Waste Management to provide regions and municipalities with technical and financial assistance in planning, developing, and conducting a solid waste program. This will enable closer cooperation with the operating sector of these governments and allow utilization of concepts developed at the state level.

Notes

Government Publications

1. U. S. Department of Health, Education and Welfare, *Preliminary Data Analysis: 1968 National Survey of Community Solid Waste Practices – An Interim Report,* 1970, Cincinatti, Ohio.
2. U. S. Department of Health, Education and Welfare, *Solid Waste Processing, A State-Of-The-Art report on Unit Operations and Processes* (Washington, D.C.: Government Printing Office, 1969).
3. U. S. Department of Health, Education and Welfare, *Summaries Solid Wastes Demonstration Grant Projects 1969,* (Washington, D.C.: Government Printing Office, 1969).
4. U. S. Department of Commerce, *Pulp, Paper and Board Quarterly Industry Report,* (Washington, D.C.: Government Printing Office, October 1970).
5. U. S. Department of Health, Education and Welfare, *Comprehensive Studies of Solid Waste Management,* Public Health Service Publication No. 2039, by C. G. Coluuke and P. H. McGauhey (Washington, D.C.: Government Printing Office, 1970).
6. U. S. Department of Health, Education and Welfare, *Proceedings The Surgeon General's Conference on Solid Waste Management for Metropolitan Washington,* Public Health Service Publication No. 1729, ed. Leo Weaver (Washington, D.C.: Government Printing Office, 1967).
7. U. S. Department of Health, Education and Welfare, *Preliminary Data Analysis: 1968 National Survey of Community Solid Waste Practices,* Public Health Service Publication No. 1867, by Anton J. Muhich, Albert J. Klee, and Paul W. Britton (Cincinatti: Government Printing Office, 1968).
8. U. S. Department of Health, Education and Welfare, *Solid Waste Management: Abstracts and Excerpts from the Literature,* Public Health Publication No. 2038, Vols. I and II, by C. G. Golueke and the Staff of the College of Engineering, University of California (Washington, D.C.: Government Printing Office, 1970).
9. U. S. Department of Health, Education and Welfare, *The Role of Packaging in Solid Waste Management 1966-1976,* Public Health Service Publication No. 1855, by Arsen Darney and Willian E. Franklin (Rockville, Maryland: Government Printing Office, 1969).
10. U. S. Office of Science and Technology, *Solid Waste Management: Comprehensive Assessment of Solid Waste Problems, Practices and Needs,* Executive Office of the President, by the Ad Hoc Group (Washington, D.C.: Government Printing Office, May 1969).
11. Combustion Engineering Company, *Technical-Economic Study of Solid Waste Disposal Needs and Practices,* 4 vols. prepared for the Bureau of Solid Waste Management, U. S. Department of Health, Education and Welfare (Washington, D.C.: Government Printing Office, No. 1886, 1969).

Periodicals

12. "Welcome to the Consumption Community," *Fortune,* September 1, 1967, pp. 118, 120, 131, 136, 138.

13. "The Enemy Is Us," *The Saturday Review,* March 7, 1970, pp. 58, 59.

14. Garret P. Westerhoff and Robert M. Gruninger, "Population Density vs. Per Capita Solid Waste Production," *Public Works,* February 1970.

15. "World Survey Finds Less Organic Matter," *Refuse Removal Journal,* September 1967.

16. Ake V. Bjorkman, "Plastics Needn't be a Problem," *The American City,* September 1970.

17. Hugh H. Connolly, "Solid Waste Management: A Challenge to the Environmentalist," *Environmental Control Management,* December 1969.

18. Robert H. Finch, "The Primary Federal Role in Solid Wastes Control," *Environmental Control Management,* December 1969.

19. "Injury Record Tops All Others in the Country," *Solid Wastes Management/Refuse Removal Journal,* January 1969.

20. Abraham Michaels, "What Good Incineration Means," *The American City,* May 1968.

21. Casimir A. Rogus, "Control of Air Pollution and Waste Heat Recovery from Incineration," *Public Works,* June 1966.

22. Casimir A. Rogus, "European Developments in Refuse Incineration," *Public Works,* May 1966.

23. Casimir A. Rogus, "Collection and Disposal of Oversized Burnable Wastes," *Public Works,* April 1966, p. 107.

24. "Composting Operation Handles Refuse and Sludge," *Public Works* 99, 3 (1968): p. 84.

25. J. S. Wiley, "A Look at European Composting," *Public Works* 92, 2 (1961): p. 107.

26. J. A. Fife, "European Refuse Disposal," *The American City,* September 1966, p. 125.

27. C. G. Golueke, "Composting Refuse at Sacramento, California," *Compost Science* 1, 3 (1960): p. 12.

28. W. F. Reynolds, "The Bureau of Mines Looks at Refuse Disposal and Recovery Possibilities," *Public Works,* v. 99, December 1968, p. 85.

29. *Chemical Engineering Progress,* February 1970, p. 55.

Books

30. American Public Works Association, *Refuse Collection Practice* (Chicago, Illinois: Public Administration Service, 1963).

31. S. M. Blumenreich, et al., *Aluminum Statistical Review – 1968* (New York: The Aluminum Association, 1969).

Reports – Published

32. The First National Conference on Packaging Wastes, *Solid Waste Management and the Packaging Industry,* San Francisco, California, September 1970.

33. The National Conference Board, Inc., *A Guide To Consumer Markets 1970,* New York, New York, 1970.

34. Automobile Manufacturers Association, Inc., *1970 Automobile Facts and Figures,* Detroit, Michigan, 1970.

35. Regional Plan Association, *Waste Management,* New York, March 1968.

Reports – Unpublished

36. Tri-State Transportation Commission, *Solid Waste Management in the Tri-State Region: Alternatives,* Interim Technical Report 4125-1620, New York, August 1969.

37. R. J. Black, A. J. Muhich, A. J. Klee, H. L. Hickman, Jr., and R. D. Vaughan, *The National Solid Wastes, an Interim Report,* Presented at the 1968 Annual Meeting of the Institute for Solid Wastes of the American Public Works Association, Miami Beach, Florida, October 24, 1968.

Other

38. U. S. Department of Health, Education and Welfare, *The Interim Report of Community Solid Waste Practices,* Washington, D.C., 1970.

39. Staff of Combustion Power Company, Inc., *Combustion Power Unit-400,* for the Bureau of Solid Waste Management, U. S. Department of Health, Education and Welfare, Public Health Service (Rockville, Maryland, 1969), as cited in U. S. Department of Health, Education and Welfare, *Solid Waste Management: Abstracts and Excerpts From the Literature,* Public Health Service Publication No. 2038 (1970), vol. II, p. 81.

23 Solid Waste Disposal in Four Eastern Cities

MICHAEL J. SEAMAN

Increasingly, the ability of a city to manage its solid wastes diminishes. Litter, although only a small portion of total generated wastes, has lead to a generally accepted belief that our affluence, which has done so much damage to our air and water, actually threatens the land itself. But even more than offending our aesthetic senses, solid wastes require land for disposal, and land is generally limited in quantity.

Waste, by definition, is devoid of value, yet the costs of collecting, transporting, and disposing of this valueless material are very real. In our continual quest for economic growth and "convenience" we sometimes forget the side effects that occur. Solid wastes are one of these side effects. Ultimately, through either taxes, higher prices, or a decreasing level of services, the invisible costs of consumption are passed on to the consumers themselves.

This chapter will attempt to examine the effects of solid waste generation for four eastern cities. First the solid waste output for each area is related to economic activities of the area under different growth assumptions. Rapid and unexpected growth in waste output can overload even a sophisticated waste system.

Next, similarities and differences among solid waste systems will be examined. The costs of waste management will be explained in terms of variances from an ideal system, and judgments will be made about the adequacy of these systems for the present and the next ten years.

Finally, since all solid wastes are eventually disposed of on the land, we will examine the broad patterns of land development and disposal requirements. Can we reconcile our needs for land with the need to devote land to the purpose of waste disposal? Do current disposal practices increase or decrease the ultimate value of the land used for this purpose? Is there enough land available to meet each area's needs, and what techniques are there for conserving this increasingly scarce land? These are some of the questions that will face each area, and we will attempt to answer these questions for a limited (ten year) time period.

Solid Waste Output Model

All solid wastes are generated either by *productive* activity or *consumption*. All residents of an area will generate consumption wastes; only those engaged in productive activity will generate productive wastes.

The purpose of the projection is to obtain a range of total waste quantities that enter the disposal stream. The basic assumptions are:

1. The manufacturing employee will generate the most waste.
2. The commercial employee will generate less waste than will the manufacturing employee, but more waste than the average resident.
3. The resident population will generate a certain level of consumption waste.

Productive waste output will be measured by the trend of employment in manufacturing and commercial activities. Consumption waste output will be measured by the trend of food sales, because of their impact on the waste management system and their rapid and continual turnover, and effective buying income (EBI). EBI will reflect the "affluence" factor that influences discretionary spending patterns primarily for durable goods. Because EBI represents potential rather than actual spending, it will be given a lesser weight than food sales, so consumption waste output will equal:

$$\frac{2 \text{ (trend of food sales)} + \text{trend of EBI}}{3}$$

All trends—manufacturing and commercial employment, food sales, and EBI—will be represented as annually compounded rates of change for the three time periods for which projection parameters are calculated. Data from the five-year period from *1959-1964* are used for calculating the *low* projection possibility; from the five-year period from *1964-1969* for the *high* projection possibility; and from the ten-year period from *1959-1969* for the *medium* projection possibility.

Table 23.1 depicts the percentage of solid waste generated by source,

Table 23-1

Two Recent Estimates of Urban Waste Generation, by Contributing Source (Expressed as percentages of total per capita output.)

Source of Solid Waste	Office of Science and Technology	Combustion Engineering Company, Inc.
Consumption	35	33
Commercial	23	19
Industrial	30	44
Other	12[1]	04[2]
Total	100	100

Source: Ad Hoc Group, Office of Science and Technology, SOLID WASTE MANAGEMENT, Executive Office of the President (Washington, D.C.: U.S. Government Printing Office, 1969), p. 7.

Combustion Engineering Company, Inc., TECHNICAL-ECONOMIC STUDY OF SOLID WASTE DISPOSAL NEEDS AND PRACTICES, U.S. Department of Health, Education and Welfare, Public Health Service Publication no. 1886 (Rockville, Md.: U.S. Government Printing Office, 1969), Vol. 1, p. 12.

[1] Classified as municipal waste; probably would fall into consumption and commercial categories.

[2] Classified as bulky waste; probably would fall into industrial category.

according to two recent estimates. Note that especially for productive wastes, the amount of waste generated exceeds the amount that will actually enter the disposal stream. This is because a good deal of productive wastes generated are diverted by salvage, recycling, and in-plant disposal and, thus, do not enter the disposal stream. The Combustion Engineering Company estimates that only 56 percent of generated industrial wastes end up in municipal facilities [4:II,13-14].

Basic Waste Output Model

Because of recycling, and on-site disposal of productive solid wastes, we must adjust downward their impact on the entire disposal stream. We will assume that the area covered by the basic model (and each of the areas covered by the modified models) has a current per capita waste output of 5 pounds daily. The basic model is shown in table 23.2.

Modifications to the Basic Model

The basic model cannot be considered to be universal in scope, since it would imply that each of the eastern cities under study has the same employment distribution. Table 23.3 shows the relation of manufacturing and commercial employment to the resident population. Since the total amount of wastes entering the disposal stream must equal 100 percent from all sources, the modifications should conform to each area's productive activity pattern.

Philadelphia and Pittsburgh (Allegheny County)

Both the Philadelphia and Pittsburgh areas show a higher proportion of manufacturing and a lower percentage of commercial employment than in the basic, New York City, model.

Atlanta (Fulton County)

Manufacturing employment in Fulton County is the same percentage of the total as in the basic model, but there is a greater percentage of commercial employment (relative to the resident population) in Atlanta than in New York.

Land and Density Patterns

The final repository for solid wastes is the land. As the population of an area and the per capita waste generation increase, the land requirements will also increase. The very same process of growth that leads to increased waste production

Table 23-2
Assumed Distribution by Source of New York City Solid Waste

Waste Source	Percentage of Total Wastes Entering the Disposal Stream	Daily Pounds/Capita
Consumption	50	2.50
Commercial	20	1.00
Industrial	30	1.50
Total	100	5.00

Source: Based upon activity pattern data of Table 23-3.

Table 23-3
Productive Activity Patterns, 1969

Area	(1) Resident Population (000.)	(2) Manufacturing Employment (000.)	(3) Commercial Employment (000.)	Mfg. Emp. Pop. Percentage	Com. Emp. Pop.
New York City	8,080.	899.9	1,998.6	10.2	22.7
Philadelphia	2,041.	265.8	415.8	13.0	20.4
Allegheny (Pittsburgh) County	1,600.	200.3	283.1	12.5	17.7
Fulton (Atlanta) County	629.	71.9	216.3	11.4	34.4

Source:
(1) "Annual Survey of Buying Power," SALES MANAGEMENT, June 10, 1970. Various pages.
(2) U.S. Department of Commerce COUNTY BUSINESS PATTERNS, 1969. Various Volumes and pages.
(3) Ibid.
*Commercial includes wholesale, retail, finance and service employment.

Table 23-4
Assumed Distribution by Source of Philadelphia and Pittsburgh Solid Waste

Waste Source	Percentage of Total Entering the Disposal Stream	Deviation from Basic Model	Daily Pounds/Capita
Consumption	45	−5	2.25
Commercial	15	−5	.75
Industrial	40	+10	2.00
Total	100	0	5.00

Source: Based upon Table 23-2 as modified by activity pattern data of Table 23-4.

Table 23-5
Assumed Distribution by Source of Atlanta Solid Waste

Waste Source	Percentage of Total Entering the Disposal Stream	Deviation from Basic Model	Daily Pounds/Capita
Consumption	45	−5	2.25
Commercial	25	+5	1.25
Industrial	30	−	1.50
Total	100	0	5.00

Source: Based upon Table 23-2 as modified by activity pattern data of Table 23-4.

consumes land. So, as development progresses the need for land for waste disposal increases as the amount of available land decreases.

An area's waste management system is developed around the area's land availability. Incineration, an extremely costly waste processing technique, is used only because of the drastic reductions in land requirements that can be achieved. The distance between collection points and disposal areas will have a major effect on the system's cost. And, since most sanitary landfill of consumption wastes will not support much building, that land which is used for waste disposal will rarely provide a stream of revenues back to the community. Stable fill, which would include certain types of construction, demolition, and industrial wastes, as well as the residue of good incineration, can support buildings and is quite valuable in the development of marginal land in the central city.

One area of American life that is virtually statistic-free is how we use our land, and the prospects of obtaining such data are virtually nil [27:76]. Therefore, some broad measure of land availability must be devised. Density, the population per square mile, will be used as a criterion of land availability. The higher the density of an area, the higher will be the cost (either a real cost, or an opportunity cost for alternate land uses) of land. In certain instances, the raw density patterns will be adjusted for housing patterns.

Another criterion of *future* land availability will be the "green-belt." A "green-belt" will be said to exist if a significant portion of the *area's* (not central city) total land area is devoted to agriculture. The *County and City Data Book*, which will be the main source of land and density information, lists for each county the percentage of land classified as farmland.

Whether or not a central city will be able to use a portion of land classified as agricultural will depend on the political consent of the area, and its distance and access from the city. It is not something that can be projected with any degree of reliability whatsoever, but given an eventual regional waste management system and a significant "green-belt," this type of relationship between the city and its surroundings should not be overlooked.

Solid Waste Management

The *1968 National Survey of Community Solid Waste Practices*, which will be the main source for this section, has revealed a great deal about each city's waste management system. This information can lead to conclusions as to the current costs of waste management, and the effects of output growth on future costs. For each area the following will be examined:

1. The degree of regulation regarding the on-site storage and open burning of wastes.
2. The role of the public waste agency. What types of waste are collected by the public agency? What provisions are made for the disposal of various types of waste?
3. Cost breakdowns. What are the total and per capita costs of public waste management? How are these costs spread between collection and disposal; between capital and operating expenses?
4. Landfill acreage analysis. How much current acreage available for landfill does each city have; what is the life expectancy of these sites? How much of such land is located within the city? What is the distance between collection points and disposal areas? What is the pattern of the city's landfill operations; what is excluded from the sites? Finally, how much of the landfill is considered sanitary; are sites planned for after-use?
5. Planning. Which agencies, if any, include solid waste management in their comprehensive planning?

In addition, information obtained from the Combustion Engineering Company's incineration survey will be used to obtain the number of operating incinerators, and their rated capacities in each area. By estimating the amount of solid waste currently generated, and the total capacity of each city's incinerators, it will be possible to determine what percentage of a city's wastes is incinerated. The actual amount of wastes incinerated will depend upon how well each incinerator operates; the variances between actual performance and rated capacity are great.

Certain relationships of a waste management system can be assumed:

1. The larger the role of the public waste agency, the more inclusive will be the costs.
2. The greater the incidence of incineration, the higher will be the system's costs.
3. The further the distance between collection points and disposal sites, the higher will be the system's costs.
4. The higher the percentage of sanitary landfills, and the higher the percentage of sites with a planned after-use, the higher will be the system's qualitative rating. Cost savings by operating a number of offensive sites are false savings.

Table 23-6
New York City Analysis: Economic Variables in New York City, 1959-1969, Expressed As Annually Compounded Rates of Increase[a]

Variable	1959-1964 Low	1964-1969 High	1959-1969 Medium
Food Sales	1.5	5.5	3.5
EBI	2.7	5.8	4.7
Consumption = $\frac{2 \text{(Food Sales)} + \text{EBI}}{3}$	1.9	5.6	3.9
Commercial Employment	0.2	3.0	1.6
Manufacturing Employment	−0.8	0.2	−0.5

Source: Food sales and EBI—"Annual Survey of Buying Power," SALES MANAGEMENT, various years. Employment—U.S. Department of Commerce, COUNTY BUSINESS PATTERNS, various years.
[a]Minus sign = decrease.

Table 23-7
New York City: Projected Range of Solid Waste Output, 1969-1979

Source of Waste	Output, in Pounds per Capita per Day			
	1969	1979-Low	1979-High	1979-Medium
Consumption	2.50	2.74	4.31	3.67
Commercial	1.00	1.02	1.34	1.17
Manufacturing	1.50	1.38	1.53	1.43
Total	5.00	5.14	7.18	6.27

Source:
1969—Table 23-2.
1979—Table 23-2 data as modified by growth rates shown in Table 23-6.

New York City

Analysis

New York City's population has not exhibited much growth in the past twenty years. In the 1950-1960 period, population actually decreased at an 0.1 percent annual rate [14:524]. However, during the 1959-1969 period, population increased at a 1.0 percent annual rate; in the five-year 1964-1969 period, the rate dropped to 0.9 percent.

Almost all per capita waste growth over the past ten years can be attributed to consumption wastes. From a moderate 1.9 percent annual increase in the 1959-1964 period, consumption wastes grew at a 5.6 percent rate during the latter five years. Little wonder the public collection agency was under extreme pressure in the late 1960s.

A declining amount of manufacturing activity in the city will serve to moderate the total amount of wastes in the city. Both the low and medium projections assume less industrial wastes will be generated in 1979 than were in 1969. During the high growth, 1964-1969 period, New York City gained approximately 13.3 thousand manufacturing jobs; the suburban areas gained 37.4 thousand jobs during the same time period, and this despite the fact that in 1969, over 80 percent of all manufacturing employment in the New York SMSA was located within the central city [29]. Thus, suburbanization in the New York area is leading to a redistribution of productive, as well as residential, patterns. If these trends continue, there will be further moderation in the amount of total wastes generated in New York City.

Land and Density Patterns

Density, in New York City, is extremely high; and land, relative to the population's needs, is a commodity in short supply. What is surprising is the amount of land in New York City that is currently utilized for waste disposal.

Table 23.8 depicts the density (population per square mile) and landfill intensity (acres of current sanitary landfill located within the central city per square mile), of each of the four eastern cities. New York City has approximately 7½ times as many people per square mile as Atlanta has, but it also devotes over 38 times as much land to waste disposal as does Atlanta.

Certain qualifications regarding landfill intensity are in order:

1. The vast majority of landfill acreage is located in Richmond County (Staten Island), the least populous and dense county of the five counties that comprise New York City. Yet density in Richmond County exceeds the density of the city of Atlanta. While no landfilling of *consumption* wastes is being practiced in Manhattan, the landfill intensity in each of the other four counties exceeds that for any other eastern city.
2. Almost all of the city's landfill is located in the wetlands areas, thus extending the city's land area. The record of creating and upgrading the total land area of New York City by landfill is excellent. Such facilities as Kennedy International Airport, Flushing Meadow Park (site of the two World's Fairs), and the World Trade Center have been developed on filled land. One-half of Governor's Island was constructed on fill from subway construction [30:48]. Sanitary landfill of consumption wastes will not support intensive building, so most landfill is planned for recreational development. Given the high density patterns in the city, and the ever-increasing amounts of leisure time, the recreational development of completed landfill serves a needed public

purpose. Yet, increasing public recognition of the ecological purpose of the wetlands can be expected to increase the pressure to leave these areas in their natural state as wildlife refuges. For the conservationist point of view, see an article by Michael Harwood [35].
3. With over 8 million people and 300 square miles being served by one public agency, waste management in New York City may be considered to be regional, albeit on a small scale. Yet even at the present time, all of the city's wastes are not disposed of in the city. A large amount of productive wastes, which the city doesn't collect, end up in the New Jersey Meadowlands, a massive 18,000-acre tract closer to the central business district than the city's own sites. The city does provide for the disposal of productive wastes. In 1963, when the city raised the dumping charge at public sites, a drop of 17 percent in the amount of productive wastes brought to the public sites was experienced. This was the first time any such decline had occurred [13:40], and it illustrates the fact that private collection firms are much less limited by political jurisdictions, and they are also much more responsive to costs.

Efforts to export publicly collected wastes to outside communities, primarily by rail-hauls, have been frustrated in New York (and every other area where these proposals have been considered) by the difficulty of finding a community willing to accept these wastes. A strong tradition of "Home Rule" is the primary obstacle to the development of regional approaches to waste management. "Home Rule" refers to the grant of powers to local governmental units, and restrictions upon state legislature intervention in local affairs. Even when outside communities are given positive economic incentives to accept urban wastes, political opposition has been intense [28:D-3,A-11].

Table 23-8
Population Density and Landfill Intensity, Selected Eastern Cities

Area	(1) Population Per Square Mile	(2) Acres of Landfill Per Square Mile
New York City	29,300	10.30
Philadelphia	15,850	0.67
Pittsburgh	9,800	0.27
Atlanta	3,770	0.27

Source:
(1) 1969 population estimates from Sales Management, OP. CIT., various pages.
(2) Landfill area obtained from 1968 NATIONAL SURVEY OF COMMUNITY SOLID WASTE PRACTICES, LAND DISPOSAL SITES for sites located and surveyed within the cities' limits.

Table 23-8A
New York City Waste Outlook Analysis

Population (1969): 8,080,000	
Daily Waste Output (at 5 lbs./capita/day): 20,200 tons	
Population (1979): 8,750,000 Implicit Annual Growth Rate: 0.8%	
1979 Projected Daily Waste Output:	High — 31,400 tons
	Low — 22,350 tons
	Medium — 24,250 tons

Source: Population-Table 23-3 projected as shown. Waste projections population times daily per capita waste output estimates of Table 23-2.

Table 23-9
Waste Management in New York City

On-site waste storage:	Regulated and enforced for all types of wastes.
On-site waste burning:	Prohibited, except for apartment house incineration.
Waste Collection:	Virtually all *consumptive* wastes are collected by the public agency. Private, unregulated firms, or self disposal, is used for the collection of most *productive* wastes.
Waste Disposal:	Public funds have been budgeted to provide for the disposal of all wastes, consumptive, and productive.
Solid Waste Costs:	In 1967 a total of over 165 million dollars, or $20.40 per capita, was budgeted for waste management. 82 percent of these funds were for the collection of waste, and 11 percent represented capital expenditures.

Landfill:	New York City utilizes 9 different sites with a total disposal area of 3,088 acres. All these sites are located within the city. Two sites, in Richmond County, account for 70 percent of total acreage, and 98.2 percent of all acreage is located in tidal areas.
	    One major influence on the high cost of waste management is that the Richmond County sites are located 20 miles from collection points.
	    The city's sites are atypical of the national experience since 7/9 sites, and 98 percent of total acreage, is considered sanitary. Only one site has no planned afteruse, and most of the acreage is slated for recreational use.
	    Considering the large current acreage in use, it is surprising that the average expected life, per site, is only 8.45 years; weighted by acreage, the average life of the city's landfill is 9.45 years. Thus, current landfill will be exhausted by 1977, at the latest.

Source: Tabulation at Hofstra University from computer printout of U.S. Department of Health, Education and Welfare, Public Health Service, PRELIMINARY DATA ANALYSIS, 1968 NATIONAL SURVEY (Washington, D.C: 1968).

Two interesting situations are revealed in the New York City survey. The first involves the cost of clean streets. The city keeps measured records of annual solid waste collections. 3.04 million tons of *household* wastes, 99 percent of which is collected by the public agency, were measured annually. *Street and alley* wastes, all collected by the public agency, amounted to 90 thousand tons. Almost 34 times as much household wastes was collected for each ton of streetcleanings. But, in terms of *manpower,* a different situation emerges. There were 7,012 collectors and drivers employed by the public agency. Twenty percent of this manpower was devoted to streetcleaning. Since labor costs represent the bulk of collection costs, the collection costs are 82 percent of total costs; the costs of clean streets (or the cost of litter, on a unit-cost basis) are astronomical. Considering *household* and *streetcleaning* wastes together, 20 percent of available manpower is required for less than 3 percent of collected wastes [37].

And what is the primary cause of litter? A recent article blamed most litter on food containers and accessories, and food takeout establishments. It also went on to say, much to the chagrin of those who feel litter (and solid wastes in general) can be legislated out of existence, that in studies between cities with a high and low percentage of no-return packaging, "Redeemability does not appear to be a major factor in littering" [36].

The second situation involves the role of private collection firms. Ninety percent of all commercial and industrial wastes are collected privately (the rest is done by the individual). These private firms employ 2,747 collectors and drivers, as opposed to 5,600 employed by the city (not including street cleaners). However, 1,300 compactor trucks (a specialized vehicle for waste collection) are owned privately. The city only owns 1,402 compactor trucks. Thus, there is more capital collection equipment per collector and driver in the private collection sector, much more. Again, the private sector appears to be more capital intensive and hence presumably more efficient in waste management than is the public sector.

Incineration

As of 1966, New York City operated eight incinerators with a total rated operating capacity of 9,500 tons per day. The largest single plant is also the oldest [4:I,12-13]. If we assume that per capita wastes entering the disposal stream are 5 lbs./capita/day, and that the incinerators operate at rated capacity, then roughly 47.5 percent of the total of 20,000 tons per day could be incinerated. Even this is an overly optimistic hypothesis since a number of incinerators have been shut down because of age, poor design, and the high cost of modifying these units to meet air pollution standards. For instance, the cost of modifying one 1,000 ton per day plant with an electrostatic precipitator is between $400,000 and $500,000 [8:42].

Since land requirements and transportation costs are much more for unincinerated wastes, the city has embarked on a program to enlarge incinerator capacity. Five "super-incinerators" are planned with a total rated capacity of

19,500 tons per day; the largest individual plant to be located in the Brooklyn Navy Yard and utilize and sell waste heat. However, the program has not progressed as rapidly as the need for incineration would warrant. The site of one plant was acquired over ten years ago; final plans are still not ready and there is little hope that it will be on stream until 1974. Meanwhile, its cost has risen from $20 million to $80 million [34]. The non-partisan Citizen's Budget Commission has estimated the cost of "super-incineration" to be close to $1 billion, with annual operating costs of $50 million, if and when (the late 1970s) they are on stream. Instead, they recommend scrapping the program in favor of mounding (the creation of landfill plateaus) and the acquisition of an additional 6,000 acres from Richmond County [33].

Of course, the mounding of incinerator residue, which would require less imported cover material, could conserve a great deal of space. The probability that the city will not be able to acquire more landfill area until the mid-1980s is quite remote.

Conclusion

Two main influences on the high costs of waste management in New York City are:

1. The reliance on incineration;
2. The long (20-plus miles to the Richmond sites) distances between collection points and disposal sites.

Future developments in the city will almost certainly intensify the already high costs, since future land sites will be acquired in Richmond County and a massive program of incineration (or other techniques of waste volume reduction, such as pyrolysis) will have to be instituted.

Two developments threaten to increase the load on the city's already burdened disposal sites. One would be the elimination of apartment house incineration because of air pollution considerations, and the other would be any development of the New Jersey Meadowlands, which would have the effect of raising the dumping fee relative to the cost of disposing these *productive* wastes in public sites.

Clearly, from a qualitative standpoint, New York City's disposal sites are far superior to those in other areas. The large percentage of sanitary filled sites and good planning as to site after-use show the advantages of a large, centralized waste management system. But the enormous quantities of waste generated, the interminable delays and resulting cost escalations of new processing facilities, and the lack of any close-in future sites all guarantee that the "crisis" of solid waste management in New York City will be as continual and massive as are the wastes that are the cause of this "crisis."

Undoubtedly the output of wastes in New York City will reflect the degree of relocation of manufacturing activity from the city. Not only will this minimize productive waste output, but also influence population growth within the city.

An unfortunate side effect of this relocation will be the loss of revenues that will accompany industry's departure, and the cost of required facilities to handle consumption wastes may be beyond the city's ability without assistance.

One alternative that should be explored by the city, although it is a politically explosive one, is to increase the role of private waste management. Private firms have demonstrated that they are more flexible in their ability to use regional, rather than local, disposal sites. Also, their ability to substitute capital for labor, their rapid responsiveness to changing conditions, and the effects of competition all might lead to a decrease in the amount of collection costs, thus freeing funds for the more pressing problem of acquiring land and waste processing facilities for the disposal of wastes. Such a conversion of the collection function to private operation would have to include provisions regarding minimum standards of service. At the present time, private collectors are totally unregulated.

Table 23-10
Philadelphia Analysis: Economic Variables in Philadelphia, 1959-1969, Expressed as Annually Compounded Rates of Increase[a]

Variable	1959-1964 Low	1964-1969 High	1959-1969 Medium
Food Sales	1.0	4.1	2.5
EBI	1.9	6.0	3.9
Consumption = $\frac{2 \text{(Food Sales)} + \text{EBI}}{3}$	1.3	4.7	3.0
Commercial Employment	0.2	3.0	1.6
Manufacturing Employment	−2.1	0.4	−1.7

Source: Food sales and EBI—"Annual Survey of Buying Power," SALES MANAGEMENT, various years. Employment—U.S. Department of Commerce, COUNTY BUSINESS PATTERNS, various years.

[a]Minus sign = decrease.

Table 23-11
Philadelphia Projected Range of Solid Waste Output, 1969-1979

Source of Waste	Output, in Pounds per Capita per Day			
	1969	1979-Low	1979-High	1979-Medium
Consumption	2.25	2.56	3.56	3.02
Commercial	.75	.77	1.01	.88
Manufacturing	2.00	1.62	2.08	1.69
Total	5.00	4.95	6.65	5.59

Source:
1969—Table 23-4.
1979—Table 23-4 data as modified by growth rates shown in Table 23-10.

Table 23-12
Philadelphia Waste Outlook Analysis

Population (1969): 2,040,000
Daily Waste Output (at a total of 5 lbs./capita/day): 5,100 tons
Population (1979): 2,040,000   Implicit Annual Growth Rate: Nil
1979 Daily Waste Output: High   -·6,780 tons   Low   - 5,050 tons   Medium - 5,700 tons

Source: Population-Table 23-3 projected as shown. Waste projections population times daily per capita waste output estimates of Table 23-4.

Philadelphia

Analysis

The past twenty years has witnessed a continual decline in Philadelphia's population. From 1950-1960, population declined at a .3 percent annual rate [14:544], and during the 1959-1969 period the decline continued at a .4 percent rate; however, in the 1964-1969 period the population remained the same [2]. Thus, any increases in the total amount of wastes entering the disposal stream will result entirely from an increased level of per capita generation. Like New York, the amount of generated productive wastes should decrease, reflecting the exodus of manufacturing from the city. So, increased output will primarily be from consumption wastes.

Like New York City, Philadelphia's public waste management costs are unduly influenced by litter, or street sweepings. Of the 2,100 public collectors and drivers, 35 percent were engaged in street cleaning. Street and alley cleanings accounted for only 5.9 percent of measured (in tons) household, combined, and street and alley wastes, collected annually [37].

Incineration

In 1966, eight incinerators with 3,150 tons of daily rated capacity were operated in the city [4:I-13]. This indicates that 62 percent of current waste output is incinerated. Actually, a higher percentage of publicly collected wastes is incinerated, since the city does not collect much productive wastes, and the public disposal sites exclude most raw wastes (i.e., garbage and putrescibles). In light of an absence of close-in disposal land and a good prospect for increased publicly collected wastes, more incineration capacity will have to be added over the next decade.

Table 23-13
Waste Management in Philadelphia

On-site waste storage:	Regulated and enforced for all types of waste.
On-site burning:	Regulated and not practiced.
Waste Collection:	The public agency collects all *Consumption* waste and very little *productive* waste. Private firms are regulated and are important. Records are kept of waste measurement.
Waste Disposal:	Funds budgeted in 1967 were almost exclusively for consumption wastes. There are no provisions for the disposal of industrial wastes in public sites.
Solid Waste Costs:	In 1967 almost 18.8 million dollars was budgeted for waste management; a per capita cost of $9.12. 85 percent of these funds were for waste collection; 4.6 percent represented capital expenditures.
Landfill:	Six public disposal sites are located within the city. All are under 35 acres in size; the oldest placed into operation in 1959. While only 2/6 sites are considered sanitary, 5/6 have a planned after-use. This is good because on an acreage basis, all current space will be exhausted before 1972. Many materials, including garbage and putrescibles, as well as industrial wastes, are excluded from public sites. A good deal of the city's waste is exported to surrounding counties. Four sites with a total of 255 acres (as opposed to 80 acres within Philadelphia) have been identified as serving the city. The largest site, 200 acres and a 10 year life, serves four counties. None of the outside sites is considered sanitary.
Planning:	Local and regional agencies include the city's waste within the framework of their comprehensive planning.

Source: Tabulation at Hofstra University from computer printout of U.S. Department of Health, Education and Welfare, Public Health Service, PRELIMINARY DATA ANALYSIS, 1968 NATIONAL SURVEY (Washington, D.C.: 1968).

Land and Density Patterns

Excluding New York City, Philadelphia has the largest population density, spread over 130 square miles, of the eastern cities under study. Unlike New York, Philadelphia is an inland city and is unable to extend its land area by landfill. Also, housing patterns in Philadelphia would mitigate against any likelihood of the city finding its land requirements within its boundaries. The percentage of the city's residents living in one-family homes is 73.6 percent; by contrast, Queens County (in N.Y.C.), which has a slightly higher density, has only 31.4 percent of its population in one-family housing, and even Richmond

County (where most N.Y.C. landfill is located) has only 58.1 percent [14:1967,254,304].

What Philadelphia has, and New York City lacks, is a good deal of surrounding "green-belt" and an amicable working arrangement with its surrounding counties. In 1964, 39.5 percent of the land area of the Philadelphia SMSA (only those counties located in Pennsylvania, since three SMSA counties are in New Jersey) was classified as farmland. In the New York City SMSA, which is comparable in size, only 7.6 percent of total land was so classified [14:260,310,320]. This "green-belt" is being developed at a fairly rapid rate; in 1954, 53.4 percent of this land area was classified as farmland [15]. Thus, the development of hitherto rural land into suburban land intensified both the demand for land (to meet the needs of the city and suburbs) while potential disposal sites disappear.

Recognition of the land requirement problem in this region predates the 1964-1969 waste output crisis. Delaware County (the second most dense in the region) initiated a countywide waste management approach in 1954 with the creation of the Delaware County Incinerator Authority. In 1958, the Authority was absorbed by the Delaware County Commissioners, and three incinerators, with over 1,000 tons daily rated capacity were constructed [31:100]. Thus, the combined efforts of a central authority accomplished what was beyond the resources of any one community.

The waste management analysis indicates the advanced state of regional planning in Philadelphia. It is the only eastern city under study in which a regional, as well as county or local, agency is involved with waste management as a part of its comprehensive planning, and one might note the high incidence of sites serving the city, which are located outside the city's boundaries.

Conclusion

A fairly low rate of projected solid waste growth over the next decade will give Philadelphia a slight respite in planning to avoid a major crisis. The probability of Philadelphia finding its disposal sites within the city is extremely low, even for ten years, but an advanced system of intercounty and intracounty waste cooperation and a still sizable amount of rural land indicate that the land requirements will be met for the region.

Costs for waste management in the city are likely to rise substantially due to the need for increased incineration, the high rate of consumption wastes projected (relative to productive wastes, for which the city is not responsible for the most part), and the exhaustion of existing close-in sites, which will increase waste transport costs.

An operational rail-haul, the current status of which is not clear, is a distinct possibility in Philadelphia, especially in light of the advanced regional recognition of the waste problem. Authorization to enter into a contract with the Eastern Land Reclamation Company was obtained from the Philadelphia City Council in 1968. Two transfer stations, located next to existing incinerators and the Reading Railroad tracks, are to be built. There the refuse will be

Table 23-14
Fulton County (Atlanta) Analysis: Economic Variables in Fulton County, 1959-1969 Expressed as Annually Compounded Rates of Increase

Variable	1959-1964 Low	1964-1969 High	1959-1969 Medium
Food Sales	1.6	4.7	3.1
EBI	4.6	6.1	5.4
Consumption = $\frac{2 \text{ (Food Sales)} + \text{EBI}}{3}$	2.6	5.2	3.9
Commercial Employment	4.5	5.2	4.8
Manufacturing Employment	1.1	3.4	2.2

Source: Food sales and EBI—"Annual Survey of Buying Power," SALES MANAGEMENT, various years. Employment—U.S. Department of Commerce, COUNTY BUSINESS PATTERNS, various years.

Table 23-15
Atlanta Projected Range of Solid Waste Output, 1969-1979

Source of Waste	Output, in Pounds per Capita per Day			
	1969	1979-Low	1979-High	1979-Medium
Consumption	2.25	2.90	3.72	3.30
Commercial	1.25	1.94	2.08	2.00
Manufacturing	1.50	1.68	2.10	1.87
Total	5.00	6.52	7.90	7.17

Source:
1969—Table 23-5.
1979—Table 23-5 data as modified by growth rates shown in Table 23-14.

Table 23-16
Atlanta Waste Outlook Analysis

Population (1969): 629,000

Daily Waste Output (at 5 lbs./capita/day): 1,570 tons

Population (1979): 710,000   Implicit Annual Growth Rate: 1.2%

1979 Daily Waste Output: High - 2,800 tons
                                  Low - 2,300 tons
                                  Medium - 2,530 tons

Source: Population-Table 23-3 projected as shown. Waste projections population times daily per capita waste output estimates of Table 23-5.

shredded and baled and disposed of in strip mines in upstate Pennsylvania. The cost to the contractor for these services is expected to be $5.39 per ton; a distinct savings from the city's cost of $7.50 per ton, as well as the possibility of reclaiming some of the damage done by strip mining [32].

Atlanta

Analysis

Population growth in Fulton County, which grew at a 1.1 percent annual rate during the 1950-1960 period [14:62], increased at a 1.4 percent annual rate during the 1959-1969 period; the rate of increase slackened to 1.0 percent during the latter five-year period [2]. The potential for a significantly higher level of waste output is greatest for the central city of Atlanta than for any of the eastern cities under study.

Rather than use comparisons of the economic variables during the *High* period (1964-1969), Fulton County grew a good deal faster during the Low growth period, and for this reason it is expected that waste in Fulton County will grow at a faster rate than the other cities studied. Also, the growth in Fulton County will result in a higher level of productive wastes as well as consumption wastes.

Land and Density Patterns

Relative to the northern cities, density in Atlanta (3,770 people per square mile) and Fulton County (about one-third that of Atlanta) is extremely low. Combined with intensive incineration, the low density of both the city and county should rule out any "crisis" of waste disposal.

The cost of these disposal sites is another story. There is a literal explosion of development in the region. In the Atlanta SMSA, reported "farmland" decreased from 900.5 to 441.7 square miles in the ten-year 1954-1964 period. An equivalent "green-belt" shrinkage occurred in Fulton County. In Fulton County the reported value per acre of this farmland rose from $157 to $578 during this period [14], [15].

The ill effects of urban sprawl must be taken into account or the long-term fate of Atlanta will be similar to that of the northern cities. Much urban growth, as William Whyte points out, proceeds in leapfrog fashion, leaving within big cities a surprising amount of open land. "But it is scattered; a vacant lot here, a dump there—no one parcel big enough to be of much use" [27:76].

Incineration

Relative to current waste output, Atlanta has ample incineration capability. The 513,000 people in Atlanta would generate 1,280 tons per day (at 5 lbs. per capita); in 1966 the three incinerators had a rated capacity of 1,180 tons [4:I-10], so on this basis 92 percent of all current wastes could be incinerated.

Table 23-17
Waste Management in Fulton County

On-site Waste Storage:	Regulated and enforced for all types of wastes.
On-site Burning:	Regulated and not practiced.
Waste Collection:	The City of Atlanta provides a relatively comprehensive system of public collection. All *consumption* wastes, 80 percent of commercial wastes, and 10 percent of industrial waste is publicly collected. However, waste quantities are not measured, and no information regarding the role of private collection firms is provided.
Waste Disposal:	Public funds were budgeted in 1967 for the disposal of all types of waste.
Solid Waste Costs:	In 1967, a total of over 4.8 million dollars was budgeted for waste management. However, capital collection expenditures were not reported, which will have the effect of understating total costs. 80 percent of total funds were for waste collection; capital expenditures accounted for only 2.2 percent of total costs. Per capita costs of waste management were $9.37.
Landfill:	The pattern of Atlanta's landfill is similar to Philadelphia's. There are 8 sites located in the city. All are small (10 acres or less) with a limited life (6 years or less). All have been placed in operation after 1964, and all are close to collection points (the mean distance is 5.7 miles). All are privately owned and publicly operated. Not one site is considered sanitary, and only one site has a planned afteruse. The city's landfill is augmented by two massive county owned and operated sites. These sites total 259 acres (opposed to 37 acres within the city) and have a mean life of 42.5 years. The larger of the two sites is considered sanitary, and both are planned for after-use.
Planning:	No agency is responsible for including solid wastes in their comprehensive planning.

Source: Tabulation at Hofstra University from computer printout of U.S. Department of Health, Education and Welfare, Public Health Service, PRELIMINARY DATA ANALYSIS, 1968 NATIONAL SURVEY (Washington, D.C.: 1968).

Conclusion

The high per capita costs of waste management in Fulton County (and the City of Atlanta) can be attributed to the large role of the public collection agency (particularly in the amount of commercial wastes that are publicly collected) and the intensive use of incineration. The cost of upgrading unsanitary sites and instituting a long-term planning agency would add little to current costs and could greatly reduce future costs, mainly in land acquisition.

Yet, despite a high projected rate of waste growth, the low population density, large amount of "green-belt," and ample incinerator capacity should rule out any major waste problem in either the city of Atlanta, or Fulton County in the foreseeable future.

Pittsburgh

Analysis

Population in Allegheny County, which increased at an annual rate of 0.7 percent in the 1950-1960 period [14:302], decreased at a 0.2 annual rate during the 1959-1969 period, while the decline intensified to a 0.8 percent rate during the 1964-1969 period [2].

Note, however, that employment in the productive sector increased at a rapid rate during the latter five years; a turn-around from a decreasing rate of the first five years. Of all the cities under study, Allegheny County has the largest range of waste outputs, based upon the experience of the 1960s.

Another point worth noting is that Allegheny County's growth is far in excess of the growth in the city of Pittsburgh. In 1950 Pittsburgh had over 675,000 residents [14:544]; by 1969 population in the city had declined to 544,000 [2].

Table 23-18
Allegheny County (Pittsburgh) Analysis: Economic Variables in Allegheny County, 1959-1969, Expressed as Annually Compounded Rates of Increase

Variable	1959-1964 Low	1964-1969 High	1959-1969 Medium
Food Sales*	−2.0	4.7	1.7
EBI*	2.5	5.1	3.8
Consumption = $\frac{2 \text{(Food Sales)} + \text{EBI}}{3}$	−0.5	4.8	2.4
Commercial Employment	−0.5	4.7	2.1
Manufacturing Employment	−1.6	2.1	0.2

Source: Food sales and EBI–"Annual Survey of Buying Power," SALES MANAGEMENT, various years. Employment–U.S. Department of Commerce, COUNTY BUSINESS PATTERNS, various years.
*For City of Pittsburgh.

Table 23-19
Pittsburgh Projected Range of Solid Waste Output, 1969-1979

Source of Waste	Output, in Pounds per Capita per Day			
	1969	1979-Low	1979-High	1979-Medium
Consumption	2.25	2.29	3.63	2.88
Commercial	.75	.73	1.17	.95
Manufacturing	2.00	1.70	2.46	2.04
Total	5.00	4.72	7.26	5.87

Source:
1969-Table 23-4.
1979—Table 23-4 data as modified by growth rates shown in Table 23-19.

Table 23-20
Pittsburgh Waste Outlook Analysis

Population (1969): 1,600,000

Daily Waste Output (at a total of 5 lbs./capita/day): 4,000 tons

Population (1979): 1,600,000   Implicit Annual Growth Rate: Nil

1979 Daily Waste Output: High   - 5,810 tons
                          Low   - 3,880 tons
                          Medium - 4,700 tons

Source: Population-Table 23-3 projected as shown. Waste projections population times daily per capital waste output estimates of Table 23-4.

Land and Density Patterns

Pittsburgh is a small city, with only 55 square miles of area in a large county (Allegheny), which has 728 square miles of land. So density, which is moderate to high in Pittsburgh (9,880 people per square mile) is quite low (2,200 people per square mile) in the county. Since a good deal of the city's wastes are disposed of outside the city, and there appears to be ample land available, the disposal "crisis" is ruled out.

There is also a good deal of agricultural land in the SMSA. In 1964, over 35 percent of the SMSA's land, and 11.8 percent of Allegheny County land was classified as farmland [14:310].

Table 23-21
Waste Management in Pittsburgh

On-site Waste Storage:	Regulated and enforced only for household garbage. No regulation for productive wastes.
On-site Burning:	Prohibits only the burning of household waste. Burning of productive wastes is practiced.
Waste Collection:	The public agency is responsible for the collection of 80 percent of consumption wastes. The private sector is responsible for the collection of all productive wastes and 20 percent of household wastes.
Solid Waste Costs:	Over 5.3 million dollars was budgeted in 1967, which is equal to a per capita cost of $9.38. 77.4 percent of total costs were for waste collection; 8.2 percent of total costs represented capital expenditures.
Waste Disposal:	Funds budgeted in 1967 were exclusively for the disposal of consumption wastes. There are no provisions for the public disposal of productive wastes.
Landfill:	The city owns and operates only one small, unsanitary site, which has a 15-year life expectancy. Two other sites, located in Allegheny County, serve Pittsburgh. These two unsanitary sites, with no planned after-use, total over 350 acres (the city site is 15 acres) and their mean life is over 32 years. All the sites are located a good distance from the collection points; the smallest site is 8 miles away, the largest is 17 miles.
Planning:	Local and county agencies consider solid waste in their comprehensive planning.

SOURCE: Tabulation at Hofstra University from computer printout of U.S. Department of Health, Education and Welfare, Public Health Service, *Preliminary Data Analysis, 1968 National Survey* (Washington, D.C.: 1968).

Incineration

One incinerator with 400 tons per day rated capacity was operated in Pittsburgh in 1966 [4:I,13]. Of the 1,360 tons per day generated in Pittsburgh (at the assumed rate of 5 lbs./capita), only 29.5 percent could be incinerated. Clearly more capacity is needed, but with ample landfill available the main advantage of increased incineration would be in lower transportation costs.

Conclusion

Pittsburgh is a prime example of an area where the gap between wastes collected and wastes generated is responsible for a nominally secure future, in that there appears to be no shortage of disposal sites. Yet this is the only eastern city under study that doesn't regulate the ways by which industry stores and disposes of its waste. For example, it has been recently estimated that 100 million tons of blast furnace and open hearth wastes are accumulated in the Pittsburgh area [23:36]. Note that if there was a current 5 pounds per day generation of only consumption wastes (highly unlikely for the next decade), then Allegheny County would generate 400,000 tons of landfill material per year. At this rate it would take 225 years of constant consumption generation to equal, by weight, the accumulated slags in the area!

Thus we have come the full cycle. From New York City, which has the highest rated waste system qualitatively, and must scramble for ample land, and reduction facilities just to make it to 1980; to Pittsburgh, where services are minimized, regulation (especially for productive wastes) is nonexistent, facilities violate even the most basic health and water pollution standards, and landfill needs can be easily met.

Certainly there is no reason why any American city should have a legitimate crisis. Money is needed; the ability of the cities to raise funds for their current needs is in serious doubt; for future needs it is almost impossible. Whatever the source of future waste management funding, increased collection efficiency, increased use of alternate capital equipment and nonpolluting waste reduction facilities, and increased land acquisition will require money.

Especially in those urban areas whose surrounding countryside is either developed, or in the process of development, the land needs of both the city and its suburbs will increasingly have to turn to outside areas and rail transport to meet long-term disposal needs. Thus, a second precondition of crisis avoidance is regionalism, whether this be by creating a regional governing unit or cooperating with the private sector, whose role in waste management is greater than that of the municipalities.

Finally, we must recognize the effects of our present economic growth and its effect on waste management systems. Clearly the haphazard growth during the late 1960s, which was largely unexpected and welcomed at the time, overloaded current facilities. Even more fundamental to our economic system is our attitude towards waste. We consider our discards "waste" because the cost of transforming these wastes is more than the value of the raw material itself. At our current rate of consumption, as we deplete our natural resources the value of our waste will rise, and recycling may become economically viable to a greater extent than it is now.

Notes

1. American Public Works Association, *Refuse Collection Practice* (Chicago, Illinois: Public Administration Service, 1963).

2. "Annual Survey of Buying Power," *Sales Management,* June 1970, 1965, and 1960.

3. Ake V. Bjorkman, "Plastics Needn't Be a Problem," *The American City,* September 1970.

4. Combustion Engineering Company, *Technical-Economic Study of Solid Waste Disposal Needs and Practices,* 4 vols., prepared for the Bureau of Solid Waste Management, Department of Health, Education and Welfare (Washington, D.C.: Government Printing Office, No. 1886, 1969).

5. Hugh H. Connolly, "Solid Waste Management: A Challenge to the Environmentalist," *Environmental Control Management,* December 1969.

6. Robert H. Finch, "The Primary Federal Role in Solid Wastes Control," *Environmental Control Management,* December 1969.

7. "Injury Record Tops All Others in the Country," *Solid Wastes Management/Refuse Removal Journal,* January 1969.

8. "Install Precipitation Units in New York City Incinerators," *Solid Wastes Management/Refuse Removal Journal,* July 1968.

9. Abraham Michaels, "What Good Incineration Means," *The American City,* May 1968.

10. Regional Plan Association, *Waste Management* (New York, March 1968).

11. Casimir A. Rogus, "Control of Air Pollution and Waste Heat Recovery from Incineration," *Public Works,* June 1966.

12. Casimir A. Rogus, "European Developments in Refuse Incineration," *Public Works,* May 1966.

13. Tri-State Transportation Commission, *Solid Waste Management in the Tri-State Region: Alternatives,* Interim Technical Report 4125-1620, New York: August 1969.

14. U. S. Department of Commerce, Bureau of the Census, *County and City Data Book,* 1967.

15. U. S. Department of Commerce, Bureau of the Census, *City and County Data Book,* 1954.

16. U. S. Department of Commerce, Bureau of the Census, *Eighteenth Census of the United States: Selected Area Reports,* PC(3) 1-D.

17. U. S. Department of Commerce, Bureau of the Census, *Statistical Abstract of the United States,* 1969.

18. U. S. Department of Health, Education and Welfare, *Comprehensive Studies of Solid Waste Management,* Public Health Service Publication No. 2039, by C. G. Goluuke and P. H. McGauhey (Washington, D.C.: Government Printing Office, 1970).

19. U. S. Department of Health, Education and Welfare, *Proceedings – The Surgeon General's Conference on Solid Waste Management, for Metropolitan Washington,* Public Health Service Publication No. 1729, ed. Leo Weaver (Washington, D.C.: Government Printing Office, 1967).

20. U. S. Department of Health, Education and Welfare, *Preliminary Data Analysis: 1968 National Survey of Community Solid Waste Practices,* Public Health Service Publication No. 1867, by Anton J. Muhich, Albert J. Klee, and Paul W. Britton (Cincinnati: Government Printing Office, 1968).

21. U. S. Department of Health, Education and Welfare, *Solid Waste Management: Abstracts and Excerpts from the Literature,* Public Health Publication No. 2038, Vol. I and II, by C. G. Golueke and the Staff of the

College of Engineering, University of California (Washington, D.C.: Government Printing Office, 1970).

22. U.S. Department of Health, Education and Welfare, *The Role of Packaging in Solid Waste Management 1966-1976* Public Health Service Publication No. 1855, by Arsen Darney and William E. Franklin (Rockville, Maryland: Government Printing Office, 1969).

23. U. S. Office of Science and Technology, *Solid Waste Management: A Comprehensive Assessment of Solid Waste Problems, Practices, and Needs*, Executive Office of the President, by the Ad Hoc Group (Washington, D.C.: Government Printing Office, May 1969).

24. "Wage Rates for Key Construction Trades," *Engineering News-Record*, January 28, 1971.

25. Garret P. Westerhoff and Robert M. Gruninger, "Population Density vs. Per Capita Solid Waste Production," *Public Works*, February 1970.

26. "World Survey Finds Less Organic Matter," *Refuse Removal Journal*, September 1967.

Books

27. M. Clawson, R. B. Held, and C. H. Stoddard, *Land for the Future* (Baltimore: John Hopkins Press, 1960).

28. U. S. Department of Health, Education and Welfare, *Systems Analysis of Regional Waste Handling*, Public Health Service Publication No. 2065, by Norman Morse and Edwin C. Roth (Washington, D.C.: Government Printing Office, 1970).

29. U. S. Department of Commerce, *County Business Patterns* (Washington, D.C.: Government Printing Office, 1959, 1964, and 1969).

Reports

30. Tri-State Transportation Commission, *Water and Waste in the Tri-State Region*, Interim Technical Report 4136-3110, New York, 1969.

Magazine Articles

31. "Forty-nine Municipalities Join in County-wide Incineration Plan," *Public Works*, February 1963.

32. "Refuse Disposal via Railroad," *Public Works*, July 1968.

Newspaper Articles

33. "City Is Urged to Rid Waste in New Ways," *New York Times*, November 30, 1970, p. 45.

34. "Problem of Ridding City of Garbage Eludes a Solution," *New York Times,* March 24, 1970, pp. 49, 95.

35. "The 'Black Mayonaisse' at the Bottom of Jamaica Bay," *New York Times,* February 7, 1971, Sec. VI, pp. 9-11 and 36-40.

36. "The Cost of Litter," *New York Times,* May 25, 1970, pp. 1 and 38.

Other

37. *1968 National Survey of Community Solid Waste Practices.*

24 Solid Waste Disposal in Five Midwestern Cities

HOWARD W. JAFFIE

The five midwestern cities whose solid waste disposal problems are covered in this chapter range in size from the Mobile, Alabama, SMSA with a population of 377,000 to mighty Chicago with nearly 7 million people in the Metropolitan Area [1]. The central cities of the five are all located on water, but land disposal sites are usually available in the surrounding countryside. All but Mobile used substantial incinerator capacity to supplement land disposal (see table 24.1).

Table 24-1
Incinerators in the U.S. Operating in 1966, Selected Midwestern Cities

State	City	Capacity[a] Tons per 24 Hr. Day	Date Installed
Illinois[b]	Chicago	720	1956
		1200	1959
		1200	1963
Louisiana	New Orleans	400	1958
		400	1962
		200	1963
Missouri	St. Louis	500	1950
		500	1958
Ohio	Cleveland	900	1936
		500	1961

Source: U.S. Department of Health, Education and Welfare, Public Health Service, Bureau of Solid Waste Management, TECHNICAL-ECONOMIC STUDY OF SOLID WASTE DISPOSAL NEEDS AND PRACTICES, (Rockville, Maryland: 1969), Section I, pp. 10-15.

[a] Capacity of each incinerator. Add for cumulative capacity.

[b] On Aug. 23, 1971, Chicago Commissioner of Public Works, Milton Pikarsky, informed the author by personal correspondence that a 1,600 ton per day capacity incinerator had been installed for a total capacity of 4,720 tons per day.

Chicago

Chicago disposes of all household wastes collected by the city by incineration. Waste collected by private carters from all buildings containing more than four dwelling units and all commercial and industrial sources is disposed of by landfill, principally at privately controlled sites. That is, all refuse collected from owner-occupied individual private dwellings is incinerated and the remainder is disposed of at landfill sites [8], [9].

Household waste collected by the city is incinerated in four city-owned modern incinerators capable of disposing of 4,720 tons of refuse per day (see table 24.1). Following incineration, metals are reparated from the residue and sold. Steam is generated by the incinerators and used for space heating and processing at the points of disposition. The newest incinerator, with a capacity of 16,000 tons of refuse per day, generates steam at a rate of 440,000 pounds per hour [7]. It is not known whether any expansion of incinerator capacity is planned, but apparently Chicago intends to dispose of virtually all municipally collected wastes by incineration.

If Chicago generates solid waste at the prevailing national average rate of 5 pounds per capita per day, the incinerators would have sufficient capacity to handle about a third of the waste collected. Cook County, in which Chicago is located, responded only partially to the National Survey of Solid Waste Practices. The report indicated that almost three-fourths of the disposal sites were privately owned and 70 percent privately operated (see table 24.2). The anticipated years remaining for these sites averaged near nine. More than three-quarters of the sites were not operated as sanitary landfill and a planned future use was reported for only 30 percent of the sites.

Of the other five counties in the Chicago Standard Metropolitan Statistical Area, only Du Page responded to the National Survey. The anticipated future years remaining for the four disposal sites in Du Page County was six years.

Despite the existence of a modern incineration system that provides for some recycling, there is no evidence of adequate planning for disposal of the future volume of solid waste in the Chicago area.

Cleveland

The Cleveland SMSA encompasses four counties: Cuyahoga, Geauga, Lake, and Medina. The four-county area comprises 1519 square miles and maintains a population of 2.1 million people. The City of Cleveland collects about half of the solid wastes generated in the city, with the other half coming from commercial and industrial establishments not handled by the city.

According to the national average of 5 pounds per person per day, waste generation in Cuyahoga County should have averaged close to 4,300 tons per day in 1970. In that year the Director of Public Service of the City of Cleveland informed the author that the City of Cleveland was collecting about 1,800 tons of waste per day from "250,000 residences and commercial establishments" [10:3]. The same source indicates that Cleveland is incinerating 300 tons per day with the remainder disposed of at private landfill sites [10:5], although it may be noted that table 24.1 suggested an available incinerator disposal capacity of 1,400 tons per day compared to the 300 tons referred to. Perhaps the reason for this discrepancy between incinerator capacity and percent utilization is explained by the fact that incineration costs Cleveland about $9.00/ton compared to $4.00 to $4.60/ton for sanitary landfill disposal [10:5].

Disposal sites in Cuyahoga County reported an average remaining life of only 3.6 years and the average remaining life for surrounding counties in the Cleveland SMSA was only slightly higher at 5.9 years.

St. Louis

The St. Louis SMSA includes a total of 4,119 square miles, of which four Missouri counties—Franklin, Jefferson, St. Charles, and St. Louis—account for 2,652 square miles. The two Illinois counties—Madison and St. Clair—comprise the balance. St. Louis citizens do realize their mounting solid waste problem. In the metropolitan area solid waste disposal has three facets: space, public opinion, and economics. In metropolitan St. Louis refuse is burned or used as landfill. Two ancient incinerators along the banks of the Mississippi burn all the city's trash, polluting its air in the process. New ones are required, but the city must solve the problem of finding some space to dump the trash while the devices are being built.

The counties, meanwhile, are depositing their own trash farther and farther away from its point of collection. Land is getting scarce. In Jefferson County, residents are currently protesting plans to develop a park out of a proposed sanitary landfill [12:9A-17A].

Eighty-seven percent of collection in the SMSA is carried on privately, though under public jurisdiction, in most suburban operations (see table 24.3). Disposal is similarly conducted overwhelmingly by private operations with only limited public jurisdiction exercised.

The lack of forward planning in the St. Louis SMSA is evident from the data in table 24.2. The average anticipated remaining life for disposal sites in St. Louis County is only 5.3 years. Even in the suburban Missouri counties this figure is only 8.5 years. Moreover, there is a planned future use for only five of the twenty-six disposal sites in the Missouri counties of the SMSA. The lack of control over disposal activities is reflected in the fact that none of the disposal sites met sanitary landfill standards. It is evident that the St. Louis SMSA badly needs a planning commission to give guidance to its solid waste disposal activities.

New Orleans

The New Orleans SMSA includes the four parishes of Orleans, Jefferson, St. Bernard, and St. Tammany. The City of New Orleans, located in Orleans Parish, has about 63 percent of the SMSA's population. The city began its program of waste disposal with the installation of its first incinerator in 1916. There were three incinerators operating in 1966 with a combined capacity of 1,000 tons per day, as shown in table 24.1. This incinerator capacity would presumably be capable of handling about 40 percent of the solid waste generated in the SMSA, assuming the national average per capita generation rate of 5 pounds per day.

In both Orleans and the suburban parishes, the public collection and disposal services are limited to household and small commercial installations. Private commercial collectors service multifamily apartments and large commercial and

Table 24-2
Status of Land Disposal Sites in Selected Midwestern SMSAs

	New Orleans SMSA		Cleveland SMSA		Mobile SMSA		St. Louis SMSA		Chicago SMSA	
	Orleans Parish	Suburban Parishes	Cuyahoga County	Suburban Counties	Mobile County	Baldwin County	St. Louis County	Suburban Counties	Cook County	Du Page County[c]
Agency in Charge of Operation										
Public	2	12	12	10	6	6	1	9	9	2
Private	1	2	7	6	1	0	3	13	21	2
Site Owned By										
Public	0	6	13	8	6	4	2	7	8	2
Private	3	8	6	8	1	2	2	15	22	2
Anticipated Years Remaining (Mean)	3	31.2[a]	3.6	5.9[a]	13	12.8	5.3	8.5[b]	9	6
Future Use Planned For Completed Site										
Yes	1	3	2	4	0	0	2	3	9	2
No	2	11	17	12	7	6	2	19	5	0
No Response										16	2

Will Public Agency Control Completed Site										
Yes	0	6	13	8	5	4	3	7	8	2
No	3	8	6	8	2	2	1	15	6	0
No Response									16	2
Is It a Sanitary Landfill										
Yes	0	0	4	4	2	1	0	0	7	1
No	3	14	15	12	5	5	4	22	23	3

Source: Tabulation at Hofstra University from computer printout of U.S. Department of Health, Education and Welfare, Public Health Service, PRELIMINARY DATA ANALYSIS, 1968 NATIONAL SURVEY (Washington, D.C.: 1968).

[a] Average weighted by population.
[b] Weighted average of Missouri counties only.
[c] No response from remaining four suburban counties.

Table 24-3
Community Solid Waste Practices in Selected Midwestern SMSAs[a]

	New Orleans SMSA		Cleveland SMSA		Mobile SMSA		St. Louis Area	
	Orleans Parish	Suburban Parishes	Cuyahoga County	Suburban Counties	Mobile County	Baldwin County	St. Louis County	Suburban Counties[b]
Operate Solid Waste Collection System								
Yes	1	13	40	17	4	4	1	5
No	0	2	12	19	3	2	16	24
Exercise Jurisdiction Over Collection								
Yes	1	13	50	34	6	5	7	26
No	0	2	2	2	1	1	10	3
Operate Disposal Facility								
Yes	1	11	11	12	4	5	5	1
No	0	4	41	24	3	1	12	28
Exercise Jurisdiction Over Disposal								
Yes	1	11	12	15	4	5	7	2
No	0	4	40	21	3	1	10	27

Source: Tabulation at Hofstra University from computer printout of U.S. Department of Health, Education and Welfare, Public Health Service, PRELIMINARY DATA ANALYSIS, 1968 NATIONAL SURVEY (Washington, D.C.: 1968).

[a]Nothing on Chicago SMSA included; no response from 5 of 6 counties in Chicago SMSA, including Cook County in which City of Chicago is located.

[b]Includes only Missouri counties.

industrial establishments. The private collection and disposal services that supplement the public operations involve, in addition to complete contractor and collection services, the operation of on-site incinerators and landfills by individual enterprises for their own use. Finally, there are private dumps operated for profit and financed by a combination of salvage and dumping charges [2:110-11].

As table 24.3 indicates, most communities in the New Orleans SMSA operate collection and disposal facilities as well as exercising control over these activities when privately conducted. However, as table 24.2 makes clear, it does not follow that public control of disposal sites is the rule or that there are planned future uses for most sites. There was no future use planned for two of the three sites in Orleans Parish nor for eleven of the fourteen sites in suburban parishes. None of the landfill sites in the SMSA met standards for a sanitary landfill.

In Orleans Parish, the average anticipated years remaining for the three disposal sites was only three years. However, the urgency that might be suggested by this figure is softened by the average of 31.2 years for the sites in suburban parishes as well as by the existence of substantial incinerator capacity. It appears that despite inadequate planning and poor supervision of disposal sites reflected in the statement that none qualified as sanitary landfill, New Orleans faces no imminent crisis in solid waste disposal.

Mobile

As the data in tables 24.2 and 24.3 suggest, Mobile has few solid waste disposal problems. Most of the communities in both Mobile County and adjacent Baldwin County operate and exercise jurisdiction over both collection and disposal activities and facilities (see table 24.3). Most of the disposal sites in the SMSA are owned and controlled by the public and a public agency will be in control of most sites when disposal operations are completed (see table 24.2).

This is not to say, however, that disposal is carefully planned. On the contrary, there is no planned future use for any of the thirteen sites in the SMSA and ten of the thirteen did not qualify as sanitary landfill (see table 24.2).

Nevertheless, there is no likelihood of any near-term problem of solid waste disposal in the Mobile SMSA. The average remaining years of life was close to thirteen years for all thirteen disposal sites in the SMSA (see table 24.2).

Mobile's one experiment in more sophisticated handling of solid wastes was reluctantly abandoned in 1971 [6]. Mobile established a composting plant in 1965 intending to produce and sell a soil conditioner under the trade name of Mobile Aid. Marketing problems apparently plagued the operation from the outset since freight costs for buyers more than 100 miles from Mobile exceeded the cost of the conditioner. Bags tended to deteriorate if left outside, where many stores kept the stock of Mobile Aid. Almost half the $40,000 annual income from the sale of Mobile Aid went into advertising expenses. In consequence, it was found that at the 1971 rate of sales it was costing the City of Mobile $11 per ton to dispose of waste at the compost plant. This compared to $2 per ton to dispose of solid waste at one of the city's landfills. Commissioner Mims concluded that the city could no longer afford to continue to finance the losing operation [13].

Notes

Government Publications

1. U.S. Department of Commerce, Bureau of the Census, *General Population Characteristics* (Washington, D.C.: Government Printing Office).
2. U.S. Department of Health, Education and Welfare, Environmental Control Administration, *Master Plan for Solid Waste Collection and Disposal, Tri-Parish Metropolitan Area of New Orleans,* 1969.
3. U.S. Department of Health, Education and Welfare, Public Health Service, Bureau of Solid Waste Management, *Technical-economic study of Solid Waste disposal needs and practices* (Rockville, Maryland: 1969), Section I, pp. 10-15.
4. U.S. Department of Health, Education and Welfare, Public Health Service, *Preliminary Data Analysis, 1968 National Survey of Community Solid Waste Practices* (Washington, D.C.: 1968), computer print-out.

Personal Correspondence and Interviews

5. William S. Gaskill, Director of Public Utilities, Cleveland, Ohio. August 30, 1971.
6. Lambert C. Mims, Public Works Commissioner, Mobile, Alabama 36604, August 16, 1971.
7. James J. McDonough, Commissioner, Department of Streets and Sanitation, City of Chicago, Chicago, Illinois 60602, September 7, 1971.
8. Milton Pikarsky, Commissioner, Department of Public Works, City of Chicago, Chicago, Illinois 60602, August 23, 1971.
9. John Rohles, Bureau of Sanitation, City of Chicago, telephone conversation, September 29, 1972.

Lectures

10. Delivered by Ralph C. Tyler, Director, Department of Public Service, Cleveland, Ohio 44114, to Urban Leaders Seminar at Cleveland State University on May 4, 1971.

Books

11. Alfred J. Van Tassel (Ed.), *Environmental Side Effects of Rising Industrial Output* (Lexington, Massachusetts: D.C. Heath and Company, 1970).

Other Publications

12. "Environment," *The Chemical and Engineering News,* March 11, 1968, pp. 9A-17A.

13. Photocopy of newspaper article from *Mobile Press* by Clay Swanzy, Press staff reporter, "City to Close Compost Plant October 1," Reproduction enclosed in letter dated August 16, 1971 from Lambert C. Mims, Public Works Commissioner, City of Mobile, which confirmed that compost plant was "being phased out."

25 Solid Waste Disposal in Five Western Cities

JOHN A. BURNS

In terms of those factors most affecting the generation of solid wastes, the western states have, in recent years, grown faster than any other region in the United States. Population and per capita income gains have exceeded the national average. Total retail sales have risen to new highs as the result of the stimulus provided by consumers with the means and desire to consume.

This chapter deals with five western Standard Metropolitan Statistical Areas (SMSA): Los Angeles, California; San Francisco, California; Seattle, Washington; Denver, Colorado; and Salt Lake City, Utah. It is felt that these cities reflect most of the solid waste disposal problems of other urban areas in the West.

The population growth of the western area during the 1960s is evidenced by the following annual growth rates for the years 1960-68. Los Angeles — 1.6 percent, San Francisco — 1.5 percent, Seattle — 2.3 percent, Denver — 2.4 percent, Salt Lake City — 2.2 percent. During the same period, the average annual growth rate for the United States was 1.3 percent [22:840,856]. Figures outlining specific SMSA increases in population are recorded in tables A25.1 through A25.5 in the Appendix.

Concomitant with the population growth of the western region has been the movement of its population away from the central cities towards the suburbs. Table 25.1 reflects this phenomenon on a national scale as it occurred during the 1960s.

"It is scarcely surprising that most of the growth in metropolitan areas continues to be in the suburbs ... [7:84]. As this population movement towards suburban areas intensifies, availability of suitable land for development in the adjacent counties becomes more acute. Farmland, orchards, and unimproved property all come under the influence of latter-day land speculators. New limited access roads (freeways) stretch out over once rural landscapes that have now been transformed into residential areas complete with homes, shopping centers, parking lots, and other outward signs attributable to "area growth." These developments have become especially critical in the urban West where utilization of the land is virtually the sole means employed for municipal solid waste disposal. Portland, Oregon is the only major western city still operating a municipal incinerator.

Whereas population growth and concurrent movement away from the central city are important factors relating to our study, we must look further to appreciate the total effect these coefficients have on our ability to generate solid wastes. Consideration as to the general makeup of the population, as well as its propensity for consumption, is necessary for lucid comprehension of the problem. Clearly, people alone do not constitute a serious threat to adequate disposal of solid wastes. However, it is "people" with mobility and the means to generate solid wastes, with increased disposable income, that make it imperative provisions be made for ultimate disposal of the solid wastes bound to accrue as a result of the consumption process. The necessity for western urban centers to provide land disposal sites able to accommodate efficiently increased consumption by greater numbers is apparent. As table 25.2 indicates, the expansion and growth of metropolitan areas in the western United States has been coupled with an above-average increase in income, surpassing the national average for the past twenty years.

This increased growth of income contributes more to potential generation of solid waste when vested with those persons in high consumption age brackets. E. B. Weiss, Vice President, Doyle, Dane, Bernbach, has noted that: "By 1975, households between 25 and 34—traditionally the great age of accumulation of possessions—will increase . . . by 50 percent" [1:39]. A review of table 25.1 will confirm that on a national basis, persons in these high consumption groups (ages 20-44) have made up the bulk of those persons moving to the suburbs during the 1960s.

The overall growth of total retail sales, as well as per capita income, is outlined in tables A25.1, A25.2, A25.4, and A25.5 in the Appendix.

In summary, then, we may hypothesize that during the past decade, the following phenomena relating to solid waste coefficients have characterized western urban areas:

1. Movement of population into the counties embracing the SMSA has increased at a rate greater than the national average.
2. This movement is made up of people leaving the central cities as well as those relocating from outside the SMSA.
3. The suburban population growth is centered principally around that segment most likely to exert an upward influence in the consumption process.

Table 25-1
Population by Residence and Age 1960, 1969 (In millions of persons)

1960	Total	Under 20	20-24	25-34	35-44	45-64	65 and over
Residing in Central Cities	57.8	20.2	3.7	7.6	7.8	12.8	5.7
Residing outside Cen. Cities	55.1	22.0	2.9	7.5	8.1	10.4	4.0
1969							
Residing in Central Cities	58.6	21.1	4.9	7.0	6.4	12.6	6.5
Residing outside Cen. Cities	70.6	28.2	5.0	9.0	9.1	14.0	5.4

Source: The Conference Board, Inc., 1970. A GUIDE TO CONSUMER MARKETS, 1970, New York, New York, p. 22.

Table 25-2
Average Annual Growth Rates for Personal Income in Metropolitan Areas

	1950-59	1959-68
All Metropolitan Areas	6.47	6.69
Rocky Mountain Metropolitan Areas	7.96	6.76
Far West (excluding Alaska, Hawaii)	8.30	7.33

Source: The Conference Board, Inc., 1970. A GUIDE TO CONSUMER MARKETS, 1970, New York, New York, p. 91.

4. The growth of personal income in western SMSAs has substantially exceeded the national average for income growth for all metropolitan areas.

These population/income relationships have manifested themselves in a manner so as to make significant solid waste management decisions imperative during the 1970s.

Analysis of Current Solid Waste Conditions in Selected Western SMSAs

Formal recognition of the solid waste problem and an attempt to arrest it first appeared with the passage of the Solid Waste Disposal Act (PL 89-272) in October 1965. Under provisions of this act, the first national survey of community solid waste practices was undertaken in 1968. Before that, no national data had ever been collected. To be sure, the data collected in 1968 are statistically inconsistent and present semantic difficulties. Nonetheless, the survey ranks as a major achievement and has served to point up some of the gross deficiencies in our urban solid waste programs. It is especially useful for our purpose in that the sample population of approximately 92.5 million was 75 percent urban in nature. The survey enabled us in general to observe types and amounts of solid wastes being collected as well as methods and means of disposal. Further information regarding the state of the art, including evaluation of overall resources, was also obtained. The "Community Description" phase of the survey provided information concerning specific community solid waste practices. The "Land Disposal" phase attempted to evaluate the disposal capabilities for all land disposal sites.

At the time of the passage of PL 89-272, there were no solid waste agencies at the state level and only five states had any personnel assigned to solid waste management. In order to undertake the survey, the Secretary of Health, Education and Welfare met with governors of each state, requesting them to designate one person in their state to act as a contact point. Three years later, in 1968, the survey was sent to community officials for completion.

In 1970, a second act, dealing with solid waste, was passed by the Congress. This act (PL 91-512) is entitled The Resource Recovery Act of 1970. It enables the Office of Solid Waste Management to provide regions and municipalities with technical and financial assistance in planning, developing, and conducting a solid waste program. This will enable closer cooperation with the operating sector of these governments and allow utilization of concepts developed at the state level. To augment this, a National Data Network is being established that will function somewhat as a "clearing house" of information and data to local areas.

Los Angeles

In contrast to most urban areas, Los Angeles recognized the potential danger of solid waste pollution shortly after World War II, when its rapid population growth began in earnest. In 1949, codes were enacted regulating disposal sites in then unincorporated areas of Los Angeles County. In 1957, the use of backyard incinerators was prohibited, thereby increasing the volume of waste brought to landfill sites.

Tables 25.3 and 25.4 outline the status of Los Angeles County Community Solid Waste Practices and Land Disposal Sites as recorded by the survey. Table 25.3 shows that twenty cities within the county, including Los Angeles proper, utilized municipal-owned facilities to collect their solid wastes, while fifty-six cities entered into contracts with contract haulers on a competitive bid for franchise basis. Control over collection practices was considerably more evident than over disposal as far as the communities were concerned. However, it should be noted that the Los Angeles County Regional Planning Commission does exercise overall control over waste disposal (and transfer) sites.

Although table 25.4 lists a total of thirty-eight disposal sites, five landfills operated by municipal Sanitation Districts received approximately one-third of all county solid wastes. In addition, five other municipal landfills accommodate another 15 percent.

Sanitary landfill requirements, for the most part, were observed by the land disposal sites. Only one site reported burning of any kind and none reported surface drainage problems. All sites reported that a cover material was used and twenty-seven sites (71 percent) covered their wastes daily, as is required for a sanitary landfill. The degree to which future planning is part of the overall disposal operation is evidenced by the fact that thirty-five out of the thirty-eight sites maintained quantitative records.

Respondents to the survey felt that present areas designated for solid waste disposal (2,597 acres) would be sufficient for their needs through 1980.

Since 1968, the Los Angeles Sanitation Districts have added a new landfill site (Puente Hills) as well as a transfer station. As the result of planning in the late 1940s and throughout the 1950s, Los Angeles appears competent to cope with solid wastes through the 1970s. The establishment of the County Sanitation Districts has been a major factor in enabling this fast-growing area to manage its solid wastes [30:9].

Los Angeles County ranks high in its ability to utilize completed sanitary landfill sites for community improvement. Golf courses and parks have already

Table 25-3
Community Solid Waste Practices—Los Angeles SMSA

	Los Angeles County
Number of Communities Reporting	76
Operate Collection Equipment	
Yes	20
No	56
Exercise Jurisdiction over Collection	
Yes	69
No	7
Operate Disposal Facilities	
Yes	8
No	68
Exercise Jurisdiction over Disposal	
Yes	17
No	59
Planning Agencies Including Solid Waste as part of their planning	
Local	74
County	1
Regional	1
No Agency	0
Collection Principally Enforced by:	
No enforcement	9
Operational Authority	0
Police	1
Health Authority	35
Other	31

Source: Tabulation of Hofstra University from computer printout of U.S. Department of Health, Education and Welfare, Public Health Service, PRELIMINARY DATA ANALYSIS, 1968 NATIONAL SURVEY (Washington, D.C.: 1968).

been thus created. In addition, "a portion of the Palos Verdes landfill has been converted into an Arboretum and Botanic Garden" [13:20]. There is little doubt that with this degree of success, the county will continue with the exclusive use of sanitary landfill as a method of disposal for the foreseeable future [13:17]. In addition to availability of future sites, ease of transportation via existing freeway routes and strict air pollution laws make sanitary landfill both economical and practical.

This sprawling metropolis of the west is sometimes thought of as being besieged with environmental problems. This writer is prone to conclude, however, that whatever their environmental problems may be, solid waste pollution is not one of them.

Table 25-4
Status of Land Disposal Sites—Los Angeles SMSA

	Los Angeles County
Number of Sites Reporting	38
Anticipated Years Remaining (mean)	13
Area to be Used for Land Disposal (acres)	2597
Present Zoning Classification	
Residential	3
Commercial	0
Industrial	18
Agricultural	2
Other	15
Future Use Planned for Completed Site	
Yes	18
No	20
Cover Material	
None	0
Earth	0
Other	38
Frequency of Cover	
None	4
Daily	27
Other	7
Routine Burning Practiced	
None	37
Uncontrolled	1
Planned and Limited	0
Surface Drainage Problems	
Yes	0
No	38
Quantitative Records Maintained	
Yes	35
No	3

Source: Tabulation of Hofstra University from computer printout of U.S. Department of Health, Education and Welfare, Public Health Service, PRELIMINARY DATA ANALYSIS, 1968 NATIONAL SURVEY (Washington, D.C.: 1968).

San Francisco

Virtually all solid waste collection and disposal in the San Francisco Bay area is carried out by private (scavenger) companies under franchise from local governments. Municipal jurisdiction over collection activities is almost universal, with most collection and disposal privately conducted, as shown in table 25.5.

It is the lack of land disposal facilities that prompted a degree of urgency in 1968. San Francisco, San Mateo, and Alameda Counties each estimated that they would not have enough acreage to meet their landfill needs to 1980. This shortage has become even more critical since passage of the McAteer-Petris Act which made permanent the San Francisco Bay Conservation and Development Commission, established in 1965. The Commission has virtually halted all landfill activities in the bay that previously had served as disposal sites for solid waste.

Operating landfill sites reflected inconsistent response with respect to sanitary landfill practices, as shown in table 25.6. No burning was reported at any site (having been banned by the Bay Area Air Pollution Control District). However, over half the sites reporting had surface drainage problems. While all sites reported that a cover was applied, frequency of cover varied and less than half reported they practiced daily covering. Lack of overall planning is evidenced by the fact that quantitative records were maintained by less than half the sites. The majority of sites also stated that no future use was planned for completed sites.

In summary, the solid waste disposal sites in the San Francisco Bay area in 1968 were unsatisfactory. In addition, the Bay area counties with the heaviest concentration of population (San Francisco and San Mateo) were facing the prospect of running out of land for disposal sites by 1980.

The ominous soundings of the 1968 survey became a reality in 1970 when the City of San Francisco found itself without a solid waste disposal site. (The nearby Brisbane site had been completed.) An agreement was reached with the City of Mountain View (approximately 35 miles south of San Francisco) to utilize 544 acres of land in that city for purposes of disposal of solid wastes via planned sanitary landfill program. Cartage of the wastes is currently made to a transfer station near the county line. This station presently has the capacity to accept additional solid wastes generated by other communities if need be. Large container trucks haul the compacted wastes from the transfer station to the sanitary landfill site in Mountain View. Ultimately, this landfill site is planned as a recreational park and boating marina. The project is expected to be complete in the late 1970s.

Although employment of landfill sites away from San Francisco Bay will no doubt be the major method of solid waste disposal for the five-county area in the future, there is reason to believe that the other counties in the SMSA will face a crisis later in the decade similar to that experienced by San Francisco County in 1969-70. Heavily populated San Mateo County outranked only San Francisco County in acreage available for land disposal. In fact, with the inauguration of the Mountain View site, San Mateo County now accepts solid wastes from its own cities as well as San Francisco.

With only two counties (Alameda and Contra Costa) having any appreciable land available for use as solid waste disposal sites, and intercounty movement of wastes becoming more difficult,[a] some means of regional cooperation not now

---

[a]Marin County has enacted a county ordinance "which specifically prohibits importation of garbage and refuse from outside of (the) county." Peter A. Rogers, Bureau of Vector Control, California Department of Health, "San Francisco's Crisis – An Indicator of Bay Area Problems," *The Commonwealth,* Vol. 63, December 22, 1969, San Francisco, California, p. 23.

Table 25-5
Community Solid Waste Practices—San Francisco SMSA

	Counties				
	Alameda	Contra Costa	Marin	San Francisco	San Mateo
Number of Communities Reporting	13	13	11	1	18
Operate Collection Equipment					
Yes	2	0	0	0	3
No	11	13	11	1	15
Exercise Jurisdiction over Collection					
Yes	13	13	10	1	16
No	0	0	1	0	2
Operate Disposal Facilities					
Yes	3	1	0	0	1
No	10	12	11	1	17
Exercise Jurisdiction over Disposal					
Yes	4	2	0	0	2
No	9	11	11	1	16
Planning Agencies Including Solid Waste as part of their planning					
Local	12	12	11	0	16
County	0	0	0	1	0
Regional	1	1	0	0	2
No Agency					0

Collection Principally Enforced by:				
No enforcement	1	2	1	0	2
Operational Authority	0	2	0	0	0
Police	10	6	4	1	13
Health Authority	2	4	6	0	1
Other					

Source: Tabulation at Hofstra University from computer printout of U.S. Department of Health, Education and Welfare, Public Health Service, PRELIMINARY DATA ANALYSIS, 1968 NATIONAL SURVEY (Washington, D.C.: 1968).

Table 25-6
Status of Land Disposal Sites—San Francisco SMSA

	Counties				
	Alameda	Contra Costa	Marin	San Francisco	San Mateo
Number of Sites Reporting	11	6	4	4	13
Anticipated Years Remaining (mean)	12	64	16	5	6
Area to be used for Land Disposal (acres)	1690	1449	575	104	287
Present Zoning Classification					
Residential	1	1	0	0	3
Commercial	0	0	0	0	2
Industrial	2	1	0	1	2
Agricultural	2	1	1	0	1
Other	6	3	5	3	5
Future Use Planned for Completed Site					
Yes	2	1	0	2	4
No	9	5	4	2	9
Cover Material					
None					
Earth	11	6	4	4	13
Other					
Frequency of Cover					
None	1	2	1	0	0
Daily	8	3	1	0	9
Other	2	1	2	4	4

Routine Burning Practiced					
None	11	6	4	4	13
Uncontrolled					
Planned and limited					
Surface Drainage Problems					
Yes	6	4	1	2	7
No	5	2	3	2	6
Quantitative Records Maintained					
Yes	7	3	1	0	6
No	4	3	3	4	7

Source: Tabulation at Hofstra University from computer printout of U.S. Department of Health, Education and Welfare, Public Health Service, PRELIMINARY DATA ANALYSIS, 1968 NATIONAL SURVEY (Washington, D.C.: 1968).

apparent must be devised during the 1970s in order to forestall a crisis of major proportions throughout the Bay area. In comparison with other urban areas covered in this chapter, none approaches the San Francisco SMSA in the degree of urgency currently existing with respect to solid waste.

Seattle

Lack of interest and neglect of future planning appears to be the obvious conclusion to be drawn from the Seattle area response to the survey. Only seven of forty-seven communities in King and Snohomish Counties had planning agencies of any type involved in solid waste planning, as seen in table 25.7. Thirty-six communities (77 percent) reported that they exercised no jurisdiction whatsoever over disposal of solid wastes.

Table 25.8 indicates that the status of land disposal sites did not differ appreciably from the substandard conditions apparent in community collection practices. Twenty-three of thirty sites reporting did not maintain quantitative records. Twenty-six sites reported surface drainage problems. Uncontrolled burning was found to exist at eleven sites. Only ten sites (one-thrid) reported no burning. Further neglect was evidenced by only four sites utilizing daily cover while eleven sites reported no cover whatsoever (in other words, an open dump). In late 1967, "a major Southwest County dump site was closed and . . . the City of Everett and Snohomish County were forced to jointly utilize the Everett site, with an anticipated life of two to five years" [15].

In some ways, it seems perplexing to attempt to reconcile the solid waste situation in Seattle (King County) as it is reflected in the 1968 survey. Four years earlier, a Citizens Advisory Committee warned of approaching problems and recommended "that sanitary landfill be used as the major method of garbage and refuse disposal . . ." [12] in preference to incineration. At the same time, the construction of two transfer stations (now in operation) was recommended. In 1965, a year after the committee's report, the large Ravenna Landfill site was closed forcing the city to make arrangements to use King County facilities. Since this time, the city has acquired an additional canyon site that commenced operation in 1968.

The collection of solid wastes in the county area of Snohomish and King Counties is administered by a state agency, the Washington Utilities and Transportation Commission. This is accomplished by the awarding of franchised service areas by the Commission (some rural areas, however, furnish their own collection service). The bulk of the wastes collected are disposed of at land sites operated by either the county or municipalities.

In King County, collection routes are awarded by the Utilities and Transportation Commission. The regulation of sanitary landfill sites in King County is under the jurisdiction of the county Department of Sanitary Operations. In the city of Seattle, however, a different approach is followed. Operating as a division of the Seattle Engineering Department and on the basis of a functional "utility," the Garbage and Refuse Agency is responsible for the

Table 25-7
Community Solid Waste Practices—Seattle SMSA

	Counties	
	King	Snohomish
Number of Communities Reporting	29	18
Operate Collection Equipment		
Yes	5	8
No	24	10
Exercise Jurisdiction over Disposal		
Yes	4	7
No	25	11
Planning Agencies Including Solid Waste as part of their planning		
Local	3	1
County	2	1
Regional	0	0
No Agency	24	16
Collection Principally Enforced By:		
No enforcement	12	9
Operational Authority	1	6
Police	8	2
Health Authority	4	1
Other	4	0

Source: Tabulation at Hofstra University from computer print-out of U.S. Department of Health, Education and Welfare, Public Health Service, PRELIMINARY DATA ANALYSIS, 1968 NATIONAL SURVEY (Washington, D.C.: 1968).

total management function pertaining to collection and disposal of municipal solid wastes. Collection is accomplished by two private hauling firms on a franchise basis. The utility is responsible for operation of the transfer stations and sanitary landfill site.

As the result of the closing of the waste disposal site in late 1967, a committee was formed in Snohomish County with the purpose of formulating a program for county solid waste disposal. The county has since received a Department of Housing and Urban Development (HUD) grant "to develop plans which can be implemented for water distribution, sewage collection and treatment, solid waste disposal . . ." [15]. This program was scheduled to begin in January 1971.

In conclusion, it would appear that the two counties have begun to make some progress from a point of apparent indifference noted in 1968. However, considerable effort will be necessary to bring existing facilities into sanitary landfill classification. Further, the lack of accurate quantitative records will handicap future development until such time as they can be accumulated to form a basis for future planning.

Table 25-8
Status of Land Disposal Sites—Seattle SMSA

	Counties	
	King	Snohomish
Number of Sites Reporting	13	17
Anticipated Years Remaining (mean)	18	13
Area to be Used for Land Disposal (acres)	605	300
Present Zoning Classification		
Residential	0	0
Commercial	0	1
Industrial	2	2
Agricultural	11	14
Other	0	0
Future Use Planned for Completed Site		
Yes	2	2
No	10	15
Cover Material		
None	2	9
Earth	11	7
Other	0	1
Frequency of Cover		
None	5	10
Daily	3	1
Other	5	6
Routine Burning Practiced		
None	4	6
Uncontrolled	4	7
Planned and Limited	5	4
Surface Drainage Problems		
Yes	2	2
No	11	15
Quantitative Records Maintained		
Yes	4	3
No	9	14

Source: Tabulation at Hofstra University from computer printout of U.S. Department of Health, Education and Welfare, Public Health Service, PRELIMINARY DATA ANALYSIS, 1968 NATIONAL SURVEY (Washington, D.C.: 1968).

Denver, Colorado

In response to the Federal Solid Waste Disposal Act of 1965, the Colorado Department of Health was named as the state agency responsible for the development of solid waste management programs within the state. Inasmuch as there was a complete absence of data of any kind, a state survey questionnaire was submitted to local municipalities. The results of this survey formed the nucleus of the Colorado State Solid Waste Disposal Act of 1967. Under this act, "responsibility for adequate and economical solid waste management . . . rests with the local government officials" [26]. The state survey found that solid waste disposal in the state was considerably below minimum standards. Particularly in the area of land disposal sites (there are no municipal incinerators in Colorado) there was found to be flagrant violation of accepted State Department of Health Standards. "Of the 290 land disposal sites surveyed, only nine were in compliance . . . for a sanitary landfill" [26:II-6].

Conditions found to be prevalent in the state in 1965 were later reflected in the federal survey for the five counties included in the Denver SMSA, as evidenced by table 25.9. All major disposal sites for the Denver SMSA exist outside the county of Denver. Out of a total of twenty-five disposal sites reporting, uncontrolled burning existed at four sites, and on a limited basis at two sites. Surface drainage problems were present at nine sites, including four in Jefferson County, the fastest growing county in the SMSA. This latter point is significant inasmuch as the availability of water is critical to the Denver area. Disposal of solid wastes in areas located in the vicinity of streams can result in water pollution through "leaching." This term is defined as occurring as a consequence of "infiltrating water passing through the landfill . . . causing a discharge of excess water, referred to as leachate . . . which passes out of the bottom of the landfill unseen and often unsuspected" [18:103].

Neglect of future planning is evidenced by the fact that twenty disposal sites (80 percent) did not maintain quantitative records at the time of the survey (see table 25.10).

Since passage of the Colorado State Solid Waste Disposal Act of 1967, there appears to be little if any action at the municipal level to correct disposal deficiencies. Part of the problem must be attributed to lack of knowledge in the form of quantitative data and complete absence of a framework for carrying out an organized solid waste plan. In the Denver area, an association of regional governments has been formed known as the Denver Regional Council of Governments (DRCOG). Under the guidance of this voluntary association, solid waste management as a regional problem has been assumed by the council, which is beset with jurisdictional problems. More specifically, the Denver SMSA has "25 municipalities and hundreds of special districts" [2:3,7]. Partially because of this fragmented authority, compliance with sanitary landfill specifications (made mandatory by the 1967 state Solid Waste Disposal Act) is less than uniform [2:2].

It is interesting to note that automobile bodies, so much in evidence as a prime example of solid waste in other urban areas, were specifically excluded from the 1967 act. This action was the result of an over capacity of auto scrap processing facilities in the Denver area in the late 1960s. Two firms have the capacity to handle 1,000 cars per day. There are also two steel mills nearby in Colorado and Utah that serve as economical outlets for reusable scrap. That portion of the obsolete automobiles not converted to scrap is disposed of at a municipal landfill site in Denver.

Table 25-9
Community Solid Waste Practices—Denver SMSA

	Counties				
	Adams	Arapahoe	Boulder	Denver	Jefferson
Number of Communities Reporting	5	2	2	1	2
Operate Collection Equipment					
Yes	1	0	1	1	0
No	4	2	1	0	2
Exercise Jurisdiction over Collection					
Yes	4	1	1	1	2
No	1	1	1	0	0
Operate Disposal Facilities					
Yes	2	1	1	1	0
No	3	1	1	0	2
Exercise Jurisdiction Over Disposal					
Yes	2	1	2	1	1
No	3	1	0	0	1
Planning Agencies Including Solid Waste as part of their planning					
Local	4	0	2	0	0
County	1	1	2	0	0
Regional	0	0	2	0	0
No Agency	0	1	0	1	2
Collection Principally Enforced By:					
No Enforcement	0	0	0	0	0
Operational Authority	0	0	0	1	0
Police	2	0	0	0	1
Health Authority	0	0	2	0	0
Other	3	2	0	0	1

Source: Tabulation at Hofstra University from computer printout of U.S. Department of Health, Education and Welfare, Public Health Service, PRELIMINARY DATA ANALYSIS, 1968 National Survey (Washington, D.C.: 1968).

The major decisions facing the solid waste planners for the Denver SMSA are: (1) bringing together and coordinating the activities of the many political entities in the area, and (2) making a decision on the most efficient manner to transport and dispose of solid wastes to the area's largest land disposal site, the former Lowry bombing range. The site is 22 miles from the center of Denver and if it is to be used to accommodate increased amounts of waste as other landfills are exhausted, additional transfer stations (presently there is one such facility) will be needed.

Salt Lake City, Utah

The least populous of the five SMSAs under review in this chapter, Salt Lake City, reflects a perceptible unconcern over solid waste. Only one community in

Table 25-10
Status of Land Disposal Sites—Denver SMSA

	Counties				
	Adams	Arapahoe	Boulder	Denver*	Jefferson
Number of Sites Reporting	5	8	6		6
Anticipated Years Remaining (mean)	20	10	5		5
Area to be Used for Land Disposal (acres)	92	1753	14		98
Present Zoning Classification					
Residential	0	0	1		0
Commercial	0	0	0		0
Industrial	2	3	1		2
Agricultural	3	5	3		3
Other	0	0	1		1
Future Use Planned for Completed Site					
Yes	3	5	1		3
No	2	3	5		3
Cover Material					
None	0	0	2		0
Earth	4	7	4		6
Other	1	1	0		0
Frequency of Cover					
None	0	0	2		0
Daily	2	5	3		2
Other	3	3	1		4
Routine Burning Practiced					
None	3	6	6		4
Uncontrolled	2	2	0		0
Planned and Limited	0	0	0		2
Surface Drainage Problems					
Yes	0	0	5		4
No	5	8	1		2
Quantitative Records Maintained					
Yes	2	1	1		1
No	3	7	5		5

Source: Tabulation at Hofstra University from computer printout of U.S. Department of Health, Education and Welfare, Public Health Service, PRELIMINARY DATA ANALYSIS, 1968 National Survey (Washington, D.C.: 1968).
*Not reported.

the two-county SMSA included solid waste as part of their overall planning. All communities did, however, state that they exercised jurisdiction over collection practices (table 25.11).

Table 25.12 shows that with the exception of surface drainage and open burning problems at two sites, in rural Davis County, all other sites, including all four landfill sites in Salt Lake County complied in total with sanitary landfill

Table 25-11
Community Solid Waste Practices—Salt Lake City SMSA

	Counties	
	Davis	Salt Lake
Number of Communities Reporting	10	6
Operate Collection Equipment		
Yes	3	5
No	7	1
Exercise Jurisdiction Over Collection		
Yes	10	6
No	0	0
Operate Disposal Facilities		
Yes	2	4
No	8	2
Exercise Jurisdiction over Disposal		
Yes	4	5
No	6	1
Planning Agencies Including Solid Waste as part of their planning		
Local	0	1
County	0	0
Regional	0	0
No Agency	10	5
Collection Principally Enforced By:		
No enforcement	0	0
Operational Authority	0	0
Police	0	0
Health Authority	10	5
Other	0	1

Source: Tabulation at Hofstra University from computer printout of U.S. Department of Health, Education and Welfare, Public Health Service, **PRELIMINARY DATA ANALYSIS**, 1968 National Survey (Washington, D.C.: 1968).

standards. In addition, quantitative records were maintained at three of the Salt Lake County sites. Future use (probably agricultural) is planned for all Salt Lake County sites. There is no stated end use for the Davis County sites, probably due to the basically agrarian status of the county area.

There have been no significant developments since 1968 with respect to solid waste disposal in the Salt Lake City SMSA. According to the survey, metropolitan Salt Lake County employed four landfill sites, all in full compliance with sanitary landfill requirements. Collection is enforced by the County Health Department throughout the county. Ultimately, a planning commission to deal with future solid waste needs will be necessary. At present, there is no provision at county level to perform this function. The present status of solid waste should continue through 1980.

Table 25-12
Status of Land Disposal Sites—Salt Lake City SMSA

	Counties	
	Davis	Salt Lake
Number of Sites Reporting	3	4
Anticipated Years Remaining (mean)	35	6
Area to be Used for Land Disposal (acres)	168	296
Present Zoning Classification		
Residential	0	0
Commercial	0	1
Industrial	0	0
Agricultural	3	3
Other	0	0
Future Use Planned for Completed Site		
Yes	0	4
No	3	0
Cover Material		
None	0	0
Earth	3	4
Other	0	0
Frequency of Cover		
None	0	0
Daily	1	4
Other	2	0
Routine Burning Practiced		
None	1	4
Uncontrolled	1	0
Planned and Limited	1	0
Surface Drainage Problems		
Yes	2	0
No	1	4
Quantitative Records Maintained		
Yes	1	3
No	2	1

Source: Tabulation at Hofstra University from computer printout of U.S. Department of Health, Education and Welfare, Public Health Service, PRELIMINARY DATA ANALYSIS, 1968 NATIONAL SURVEY (Washington, D.C.: 1968).

Future Projections for Solid Waste Needs Through 1980

As mentioned earlier, the 1968 federal survey did have statistical as well as semantic limitations. However, the sample population was sufficient and more important, oriented towards urban areas.

For western SMSAs, the degree of urgency toward solid waste bears a direct relationship to the amount of land available for future landfill sites. The extent to which existing landfill sites can be brought within sanitary standards and efficiently managed will forestall the use of distant landfill sites. These sites are naturally more costly and involve large capital expenditures for massive collection vehicles and transfer stations.

A review of tables A25.1 through A25.4 in the Appendix will reveal that solid waste coefficients for these SMSAs will continue to grow through the 1970s. If we accept these growth patterns and future projections as realistic, we may express them in a model and develop an aggregate growth factor (AGF) for each SMSA. The AGF, as shown in table 25.13, is a composite of three coefficients of solid waste: population, income, and volume. For anticipated annual population gain through 1980, the average annual change for the years 1960-68 was used. The average annual rate of growth in personal income for the years 1960-68 was reduced 80 percent of the calculated figure to allow for income not channeled into the consumption process[b] but held at the discretion of the consumer for savings or investment. The volume factor (perhaps the most discretionary) reflects the continued growing use of packaging materials in the modern-day marketing process. Consumption of paper, paperboard, and metal cans each is presently growing at an annual rate in excess of 3 percent. Current publicity concerning recycling, etc., notwithstanding, the use of nonreturnable bottles was expected to grow at a 12 percent annual growth rate through 1970 [28:41]. We can do no less than assume continued growth (although not necessarily at this rate) through the decade.

Given these growth rates, the Annual Aggregate Growth Factor for each SMSA is shown in table 25.13.

Tables 25.14 through 25.18 detail expected land requirements, in acres, necessary for adequate solid waste disposal in 1980 given these AGF rates for each SMSA. A standard requirement of one acre for every 10,000 population (7 foot lift) was used to develop acreage required in 1980.

It should be realized that land projections for 1980 are estimates of land that will be needed and are based on original data obtained from the 1968 survey, which may not have been entirely accurate with respect to detailing acres budgeted for solid waste in 1968. In any event, the reader can readily determine "crisis areas" as opposed to areas that can be expected to accommodate their solid wastes through 1980.

Los Angeles

An estimated population in 1980 of over 8.3 million will require in excess of 14,000 acres for solid wastes (see table 25.14). Acres now budgeted (2,597) will be exhausted in 1980 and over 11,000 new acres must be found. Due to continuing and advanced planning within the county along with an abundance of optimal landfill sites such an canyons and gravel pit excavations, Los Angeles

---

[b]Long-range income growth for each SMSA as shown in the Appendix tables, however, assures that we can expect consumption to rise at least proportionally over the decade.

Table 25-13
Annual Aggregate Growth Factor for Western SMSAs Period 1968-80 (expressed in percentage)

SMSA	Population	Income	Annual Increase Volume	Total
Los Angeles	1.6	5.2	2.0	8.8
San Francisco	1.5	5.8	2.0	9.3
Seattle	2.3	6.2	2.0	10.5
Denver	2.4	5.7	2.0	10.1
Salt Lake City	2.2	5.1	2.0	9.3
Average ($\overline{X}$)	2.0	5.6	2.0	9.6

Source: Population and income percentages from United States Bureau of the Census, Statistical Abstract of the United States: 1970, Washington, D.C. 1970, pp. 840, 856 (population), pp. 842, 858, 859.

will be well able to set aside sufficient lands enabling it to cope with its solid wastes through 1980. Los Angeles County is fortunate in that there are still available ample potential landfill sites to easily meet the 50 year requirement . . ." [13:17].

San Francisco

Analysis of acreage data for San Francisco SMSA reveals an expected deficit of over 2,000 acres for the five-county Bay area. As is currently the case, the counties of San Francisco and San Mateo will present the most critical areas. It was pointed out earlier that San Francisco utilizes a site over thirty miles away from its county line, and this sanitary landfill is expected to be complete well before the end of the decade—probably in 1977.

It is conceivable that improved techniques such as compaction and the use of additional transfer stations could extend the life of some inland sites, as well as Mountain View. However, the sheer weight of increasing population will force some decision before the mid-1970s. Providing intergovernmental jurisdiction problems can be resolved, the prospect of rail haulage of solid wastes to rural areas is well within the realm of possibility.

There appears to be some uncertainty, however, as to just what the next step will be. Feasibility studies have been undertaken concerning disposal at sea as well as transporting solid wastes by rail to distant counties. However, as far as the writer can determine, no serious effort is currently being applied to develop these proposals. It is not inconceivable, pending a successful landfill operation at Mountain View, that another peninsula city will reach agreement to accept San Francisco's solid wastes under similar conditions. This would only be a temporary measure, however, and would not change the overall outlook suggested by table 25.15.

Table 25-14
Summary of Current and Future Land Requirements for Solid Waste Disposal, Los Angeles SMSA

County	Population January, 1958	Acres Budgeted for Solid Waste 1968[a]	Estimated Population 1980	Minimum Acres Required for Solid Waste 1980	Projected Surplus (deficit) in Acres 1980
Los Angeles	7,111,400	2,597	8,334,703	14,148	(11,551)

Source: Tabulation at Hofstra University from computer printout of U.S. Department of Health, Education and Welfare, Public Health Service, PRELIMINARY DATA ANALYSIS, 1968 NATIONAL SURVEY (Washington, D.C.: 1968).

[a]Estimates for future land requirements based on a standard of one (1) acre per 10,000 population per year (7 foot lift). Population figures for 1968 from SALES MANAGEMENT ANNUAL SURVEY OF BUYING POWER (New York, New York: Sales Management, Inc.) p. D-10.

Table 25-15
Summary of Current and Future Land Requirements for Solid Waste Disposal, San Francisco SMSA

County	Population January, 1968	Acres Budgeted for Solid Waste 1968[a]	Estimated Population 1980	Minimum Acre Required for Solid Waste 1980	Projected Surplus (deficit) in Acres 1980
Alameda	1,069,600	1,690	1,241,303	2,184	(494)
Contra Costa	550,500	1,449	628,872	1,125	324
Marin	204,500	575	237,328	420	155
San Francisco	724,300	104	840,572	1,482	(1,378)
San Mateo	555,600	287	644,790	1,137	(850)
SMSA Total	3,104,500	4,105	3,602,865	6,348	(2,243)

Source: Tabulation at Hofstra University from computer printout of U.S. Department of Health, Education and Welfare, Public Health Service, PRELIMINARY DATA ANALYSIS, 1968 NATIONAL SURVEY (Washington, D.C.: 1968). Population figures for 1968 from SALES MANAGEMENT ANNUAL SURVEY OF BUYING POWER (New York, New York: Sales Management, Inc.) pp. D-9, D-10, D-14, D-18, D-20.

[a]Estimates for future land requirements based on standard of one (1) acre per 10,000 population per year (7 foot lift).

Seattle, Washington

The two-county Seattle SMSA will require an additional 1,900 acres for solid waste disposal by the year 1980. The AGF factor for the Seattle SMSA may be subject to abnormal variation due to economic conditions in the area. Unemployment in the Seattle SMSA, for example, averaged 4.94 percent during the 1960s. Preliminary figures for the first five months of 1970 reflect an average unemployment rate of 7.3 percent [37:4]. In any event, table 25.16 indicates that additional acreage will have to be obtained, particularly in more densely populated King County.

Denver, Colorado

Table 25.17 suggests that the Denver SMSA will need to obtain an estimated 456 acres by 1980 to meet its aggregate growth needs of the 1970s. Fortunately, almost 90 percent of the acreage currently budgeted for solid waste is in Arapahoe County, which, with the exception of outlying Boulder County, is the most sparsely populated county in the SMSA. Further, metropolitan Denver County, which does not contain municipal landfill sites, does have access to sufficient landfill acreage to assimilate its solid waste needs through the 1970s. "The Lowry dump is big enough to handle everything Denver could haul there for at least 30 years," said Charles Roos of *The Denver Post* in 1969 [2:5]. About the most pressing problem that can be foreseen for this area centers around determination of the most efficient way to transport solid wastes from population centers to outlying disposal areas such as the Lowry site. In 1968, the City of Denver commissioned a private consulting firm to undertake a study concerning solid waste disposal. One recommendation was for the construction of an additional transfer station, which, according to the study, would be less costly than rail haulage also under consideration at the time.

Salt Lake City

Although table 25.18 indicates that Salt Lake SMSA must secure an additional 670 acres to satisfy its needs for 1980, the outlook is not cause for apprehension. This forecast is the result of a relatively moderate Aggregate Growth Factor combined with abundant land that can be utilized for future landfill sites. Compared to other areas surveyed in this chapter, Salt Lake City SMSA does not appear to have cause for concern regarding disposal of solid wastes through 1980. Larry N. Jensen, Environmental Coordinator, Salt Lake County Health Department, told the author in a personal letter in May 1971 that: "In Salt Lake County, being located in the valley floor, there are hundreds of acres of land that are either alkaline or salty and not fit for agriculture, making an excellent area for sanitary landfill and land reclamation use. At the present time there is not a crisis for solid waste disposal areas within the Salt Lake Valley, nor will there be in the near future" [40].

Table 25-16
Summary of Current and Future Land Requirements for Solid Waste Disposal, Seattle SMSA

County	Population January, 1968	Acres Budgeted for Solid Waste 1968[a]	Estimated Population 1980	Minimum Required for Solid Waste 1980	Projected Surplus (deficit) in Acres 1980
King	1,065,300	605	1,337,292	2,340	(1735)
Snohomish	225,300	300	284,824	504	(204)
SMSA Total	1,290,600	905	1,620,116	2,844	(1939)

Source: Tabulation at Hofstra University from Computer printout of U.S. Department of Health, Education and Welfare, Public Health Service, PRELIMINARY DATA ANALYSIS, 1968 NATIONAL SURVEY (Washington, D.C.: 1968). Population figures for 1968 from SALES MANAGEMENT ANNUAL SURVEY OF BUYING POWER (New York, New York: Sales Management, Inc.).

[a]Estimates for future land requirements based on standard of one (1) acre per 10,000 population per year (7 foot lift).

Table 25-17
Summary of Current and Future Land Requirements for Solid Waste Disposal, Denver SMSA

County	Population January, 1968	Acres Budgeted for Solid Waste 1968[a]	Estimated Population 1980	Minimum Required for Solid Waste 1980	Projected Surplus (deficit) in Acres 1980
Adams	165,400	92	209,668	354	(262)
Arapahoe	139,000	1,753	176,202	297	1,456
Boulder	111,100	14	140,834	237	(223)
Denver	494,100	N/R[b]	626,341	1,056	(1,056)
Jefferson	219,400	98	278,122	469	(371)
SMSA Total	1,129,000	1,957	1,431,167	2,413	(456)

Source: Tabulation at Hofstra University from computer printout of U.S. Department of Health, Education and Welfare, Public Health Service, PRELIMINARY DATA ANALYSIS, 1968 NATIONAL SURVEY (Washington, D.C.: 1968). Population figures for 1968 from SALES MANAGEMENT ANNUAL SURVEY OF BUYING POWER (New York, New York: Sales Management, Inc.) pp. D-24, D-25.

[a]Estimates for future land requirements based on standard of one (1) acre per 10,000 population per year (7 foot lift).
[b]Not reported.

Table 25-18
Summary of Current and Future Land Requirements for Solid Waste Disposal, Salt Lake City SMSA

County	Population January, 1968	Acres Budgeted for Solid Waste 1968[a]	Estimated Population 1980	Minimum Required for Solid Waste 1980	Projected Surplus (deficit) in Acres 1980
Salt Lake	460,000	296	571,826	941	(645)
Davis	94,500	168	117,472	193	(25)
SMSA Total	554,500	464	689,298	1134	(670)

Source: Tabulation at Hofstra University from computer printout of U.S. Department of Health, Education and Welfare, Public Health Service, PRELIMINARY DATA ANALYSIS, 1968 NATIONAL SURVEY (Washington, D.C.: 1968). Population figures for 1968 from SALES MANAGEMENT ANNUAL SURVEY OF BUYING POWER (New York, N.Y.: Sales Management, Inc.) p. D-179.
[a]Estimates for future land requirements based on standard of one (1) acre per 10,000 population per year (7 foot lift).

Conclusion

It is apparent from the foregoing discussion that additional acreage for disposal of solid wastes will be required for each of the SMSAs covered in this chapter. However, only the San Francisco SMSA appears to have the potential to develop into a critical situation by 1980. In addition to land shortages, the region is beset with jurisdictional problems that make any area solution within the decade almost too much to expect.

There are those who feel that a regional authority is the answer to urban solid waste problems. Unfortunately, as we have seen in the Denver and San Francisco SMSAs, the existence of a regional authority results in a minimum of authority and a maximum of political indifference. Only in Los Angeles has there been any effective regional solid waste management. This is not to say that regional planning cannot be useful, but rather that it cannot transcend the inevitability of local politics.

Unlike its sister city to the north, Los Angeles, while continuing to grow in terms of solid waste coefficients, appears to have established the framework to handle its solid wastes through the 1970s. This can be attributed to the establishment of Sanitation Districts during the 1940s as well as the availability of adequate surrounding land suitable for disposal sites.

The Seattle and Denver SMSAs have taken important steps towards making provisions to solve their currently substandard solid waste disposal practices. The outlook for each of these SMSAs is optimistic for continued progress through the 1970s. The condition of solid waste in the Salt Lake City SMSA should remain well within the capability of existing facilities through the decade.

Concurrent with the need for each of these SMSAs to secure more land for disposal will be the necessity to achieve more efficiency in the collection sector. Only Salt Lake City approaches the national urban average of four dollars spent on collection for every dollar spent on disposal. Los Angeles and Denver spend in excess of this ratio. While results of the San Francisco SMSA are inconclusive, it is safe to assume that they too have at least a four-to-one collection/disposal ratio.

There are those who suggest that this national disproportionate collection/disposal ratio[c] exists because of a historical emphasis on collection practices rather than disposal facilities and sites. Indeed, the current state of neglect of the western urban landfill sites investigated seem to bear this out.

It is in the area of municipal collection practices, however, where major contributions to efficiency are ripe for innovation. Here is the fertile area where solid waste management can forcibly assert itself. The duties of collection (garbage) men are presently looked upon as menial and degrading. The accident rate is high and the social-political relationships are complex. The increasing number of sanitation workers who are becoming unionized serves to heighten the complexity of the problem.

---

[c]Anton J. Muhick, Chief, Systems and Operations Planning, Solid Waste Program, *An Interim Report 1968 National Survey of Community Solid Waste Practices,* address presented at 1968 meeting Institute for Solid Wastes, Miami Beach, Florida, October 1968.

This writer does not believe there exists a single solid waste solution that can be applied to all urban areas. Local government must find the combination that will best serve their locale. One concept that has merit is the "public utility" concept of managing our solid wastes. This approach has been used successfully in the State of Washington. The City of Tacoma has utilized this approach for many years on a "pay as you go" basis. It offers a means whereby rigid business practices may be employed and overall control exercised for the benefit of the community.

In order for us to obtain some degree of priority with respect to our urban solid wastes, we must view the problem as an opportunity for management innovation and leadership. In short, it must be considered and acted upon as a business problem. The pursuit of sound business practices will do much to uplift the image of the line sanitation worker. This type of encouragement can ultimately lead to more efficient and productive output. It does not seem far-fetched to envision private capital actively engaged in collection and disposal of solid wastes with incentive plans, etc., that are now a standard part of industrial compensation.

In conclusion, it must be said that our solid waste problem—this "anonymous pollution"—will not be solved per se in the 1970s. It must, however, be arrested and ultimately managed. We are beginning to proceed to arrest the problem in the 1970s. In the final analysis, our ability to manage will determine the outcome.

Appendix

Population, Per Capita Income, Retail Sales

Los Angeles	Table A25.1
San Francisco	Table A25.2
Seattle	Table A25.3
Denver	Table A25.4
Salt Lake City	Table A25.5

Table A25-1
Population, Income and Retail Sales Statistics for Los Angeles SMSA, Years 1959, 1964, 1968

	Los Angeles County
Population:	
1980 (estimated)	8,603,514
1968	7,111,400
1964	6,658,900
1959	5,857,600
Per Capita Income:	
1980 (estimated)	$ 7,825
1968	3,634
1964	3,018
1959	2,204
Percentage Households Earnings $10,000 or More Annually:	
1980 (estimated)	63
1968	33
1964	26
1959	9
Total Retail Sales (In Thousands of Dollars):	
1980 (estimated)	22,963,223
1968	14,342,871
1964	10,713,183
1959	8,346,378

Source: SALES MANAGEMENT SURVEY OF BUYING POWER, published by Sales Management, Inc., New York, New York.

1959 Data; pages 230, 248

1964 Data; page 244

1968 Data; page D-10

Note: Estimates for 1980 in each category except "Households earning $10,000 or more annually" based on average annual change for years 1960-68.

Source: United States Bureau of the Census, Statistical Abstract of the United States: 1970, Washington, D.C., 1970, pp. 840, 856. Estimates for "households" based on straight line projections.

Table A25-2
Population, Income and Retail Sales Statistics for San Francisco-Oakland SMSA, Years 1959, 1964, 1968

	Alameda	Contra Costa	Marin	San Francisco	San Mateo
Population					
1980 (estimated)	1,278,824	658,183	244,502	865,803	664,281
1968	1,069,600	550,500	204,500	724,300	555,600
1964	982,500	470,700	176,500	745,300	515,100
1959	901,400	382,600	133,400	801,500	400,000
Per Capita Income:					
1980 (estimated)	$ 7,628	7,733	9,184	9,011	9,067
1968	3,275	3,320	3,943	3,869	3,893
1964	2,555	2,428	3,153	3,198	3,041
1959	2,240	2,005	2,288	2,628	2,464
Percentage Households Earning $10,000 or More Annually:					
1980 (estimated)	62	78	86	54	90
1968	31	39	43	27	45
1964	23	27	35	24	34
1959	9	9	13	11	14

Total Retail Sales
(In Thousands of Dollars):

1980 (estimated)	$ 3,687,250	1,628,912	657,140	3,384,055	1,857,097
1968	2,053,206	907,042	365,365	1,884,375	1,034,556
1964	1,490,615	583,707	233,775	1,408,645	723,463
1959	1,159,677	417,306	152,332	1,157,612	523,083

Source: SALES MANAGEMENT SURVEY OF BUYING POWER, published by Sales Management, Inc., New York, New York.

1959 Data, pages 228, 229, 234, 240, 246, 256, 258
1964 Data, pages 244, 246, 250, 254, 257
1968 Data, pages D-9, D-10, D-14, D-18, D-20

Note: Estimates for 1980 in each category except "Households earning $10,000 or more annually" based on average annual change for years 1960-68.

Source: United States Bureau of the Census, Statistical Abstract of the United States: 1970, Washington, D.C., 1970, pp. 840, 856. Estimates for "households" based on straight line projections.

Table A25-3
Population, Income and Retail Sales Statistics for Seattle SMSA Years 1959, 1964, 1968

	Counties	
	King	Snohomish
Population:		
1980 (estimated)	1,399,516	295,984
1968	1,065,300	225,300
1964	1,015,900	194,600
1959	882,100	142,100
Per Capita Income:		
1980 (estimated)	$ 9,432	7,430
1968	3,830	3,017
1964	2,644	1,889
1959	2,105	1,635
Percentage Households Earning $10,000 or More Annually:		
1980 (estimated)	87	66
1968	41	31
1964	24	13
1959	8	4
Total Retail Sales (In Thousands of Dollars):		
1980 (estimated)	$ 5,369,807	928,388
1968	2,132,433	368,677
1964	1,662,634	213,000
1959	1,201,746	144,797

Source: SALES MANAGEMENT SURVEY OF BUYING POWER, published by Sales Management, Inc., New York, New York.

1959 Data; pages 704, 706, 710, 711
1964 Data; pages 562, 564, 565
1968 Data; page D-188

Note: Estimates for 1980 in each category except "Households earning $10,000 or more annually" based on average annual change for years 1960-68.

Source: United States Bureau of the Census, Statistical Abstract of the United States: 1970, Washington, D.C., 1970, pps. 840, 856. Estimates for "households" based on straight line projections.

Table A25-4
Population, Income and Retail Sales Statistics for Denver SMSA Years 1959, 1964, 1968

	Adams	Arapahoe	Counties Boulder	Denver	Jefferson
Population:					
1980 (estimated)	219,853	184,762	147,676	656,768	291,631
1968	165,400	139,000	111,100	494,100	219,400
1964	149,100	135,000	88,300	523,800	167,300
1959	71,700	92,400	58,900	527,500	97,400
Per Capita Income:					
1980 (estimated)	$ 5,899	7,555	6,910	7,674	7,974
1968	2,561	3,280	3,000	3,332	3,462
1964	1,648	2,319	2,197	2,620	2,380
1959	1,666	1,691	1,763	2,109	1,712
Percentage Households Earning $10,000 or More Annually:					
1980 (estimated)	55	77	52	54	77
1968	28	39	26	27	39
1964	14	25	17	20	25
1959	6.7	6.7	6.2	9.9	7.3

Table A25-4 (cont.)

	Adams	Arapahoe	Counties Boulder	Denver	Jefferson
Total Retail Sales (In Thousands of Dollars):					
1980 (estimated)	$ 283,832	520,425	404,131	2,073,246	730,422
1968	158,049	289,793	225,036	1,154,465	406,728
1964	83,271	167,040	140,163	961,333	214,661
1959	72,421	106,562	79,221	879,558	117,315

Source: SALES MANAGEMENT SURVEY OF BUYING POWER, published by Sales Management, Inc., New York, New York.

1959 Data; pages 260, 264
1964 Data; pages 262, 264, 265
1968 Data; pages D-24, D-25

Note: Estimates for 1980 in each category except "Households earning $10,000 or more annually" based on average annual change for years 1960-68.

Source: United States Bureau of the Census, Statistical Abstract of the UniteStates: 1970, Washington, D.C., 1970, pps. 840, 856. Estimates for "households" based on straight line projections.

Table A25-5
Population, Income and Retail Sales Statistics for Salt Lake City SMSA, Years 1959, 1964, 1968

	Counties	
	Salt Lake	Davis
Population:		
1980 (estimated)	597,264	122,699
1968	460,000	94,500
1964	431,600	81,300
1959	371,100	56,500
Per Capita Income:		
1980 (estimated)	5,539	4,992
1968	2,661	2,398
1964	2,090	1,734
1959	1,657	1,395
Percentage Households Earning $10,000 or More Annually:		
1980 (estimated)	46	53
1968	25	29
1964	18	17
1959	6.1	3.5
Total Retail Sales (In Thousands of Dollars):		
1980 (estimated)	$ 1,148,852	127,298
1968	805,788	89,285
1964	645,927	43,040
1959	476,733	30,894

Source: SALES MANAGEMENT SURVEY OF BUYING POWER, published by Sales Management, Inc., New York, New York.

    1959 Data; pages 684, 685, 686
    1964 Data; pages 546, 547
    1968 Data; page D-179

Note: Estimates for 1980 in each category except "Households earning $10,000 or more annually" based on average annual change for years 1960-68.

Source: United States Bureau of the Census, Statistical Abstract of the United States: 1970, Washington, D.C., 1970, pps. 840, 856. Estimates for "households" based on straight line projections.

Notes

Newspapers

1. *Advertising Age,* February 8, 1971, p. 39.
2. *The Denver Post,* December 9, 1969, pp. 2-7.
3. *The Sacramento Bee,* November 12, 1970, p. A-9.
4. *San Francisco Examiner,* January 10, 1971, pp. 1, A-16.
5. *San Francisco Examiner,* July 18, 1970, pp. 1, 4.
6. *Wall Street Journal,* October 22, 1970.

Periodicals

7. "New Questions About the U. S. Population," *Fortune,* February 1971, pp. 80-85.
8. "Refuse Collection and Disposal Practices in 99 Western Cities," Reprinted from *Western City* magazine, February-March, 1969, 11 pages.
9. "Revolution in Suburbia," *Forbes,* April 1, 1970, pp. 24-26, 31-32.
10. "Welcome to the Consumption Community," *Fortune,* September 1967, p. 118.

Unpublished Materials

11. City and County of Denver, Sanitary Services Division of Public Works Operations, *Annual Report,* 1969, Denver, Colorado, 1969.
12. City of Seattle, *History of the City of Seattle's Solid Waste,* undated.
13. John D. Parkhurst and John A. Lambie, *Solid Waste Disposal in Los Angeles Metropolitan Area,* a paper presented to a Joint Committee of Governmental Agencies, August 1968, Los Angeles, California.
14. Salt Lake County, *Cost of Operations for the Salt Lake County Sanitary Landfill 1968-1969,* Midvale, Utah, 1970.
15. Snohomish County Planning Department, *The Status of the Snohomish County Solid Waste Management Program,* September 1, 1970.

Government Publications

16. Association of Bay Area Governments, *Bay Area Regional Planning Program;* Refuse Disposal Needs Study, July, 1965. Association of Bay Area Governments, Berkeley, California.
17. California State Department of Public Health, *California Vector Views,* December, 1968, Berkeley, California.

18. California State Department of Public Health, *California Vector Views,* November, 1969, Berkeley, California.

19. U. S. Department of Health, Education and Welfare, *Preliminary Data Analysis: 1968 National Survey of Community Solid Waste Practices – An Interim Report,* 1970, Cincinatti, Ohio (computer printout).

20. U. S. Department of Health, Education and Welfare, *Policies for Solid Waste Management:* 1970 (Washington, D.C.: Government Printing Office, 1970).

21. U. S. Department of Commerce, Bureau of the Census, *County and City Data Book,* 1962 and 1967 editions (Washington, D.C.: Government Printing Office, 1962, 1967).

22. U. S. Department of Commerce, Bureau of the Census, *Statistical Abstract of the United States,* 1970 (Washington, D.C.: Government Printing Office, 1970).

23. U. S. Department of Health, Education and Welfare, *Summaries Solid Wastes Demonstration Grant Projects 1969* (Washington, D.C.: Government Printing Office, 1969).

24. U. S. Department of Commerce, *Pulp, Paper and Board Quarterly Industry Report,* October, 1970 (Washington, D.C.: Government Printing Office, 1970).

25. California State Department of Health, *A Program for Solid Waste Management in California,* Volume II, January 1970.

Reports Published

26. Colorado Department of Health, *Solid Waste Management Plan: June 1970,* Denver, Colorado, 1970.

27. Commonwealth Club of California, *What Methods of Solid Waste Disposal Should Be Used in San Francisco Bay Area – Landfill, Incineration, Export Out of Area, Other?* Commonwealth Club of California, San Francisco, California, December 22, 1969.

28. Arsen J. Darnay, Jr., *The Role of Packaging in Solid Waste* (Kansas City, Missouri: Midwest Research Institute, 1968).

29. Los Angeles Area Chamber of Commerce, *Southern California Business,* Los Angeles, California, February 24, 1970.

30. Los Angeles County Sanitation Districts, *Resume of Sewerage and Refuse Disposal Practices of Sanitation Districts of Los Angeles County,* Los Angeles, California, June 1970.

31. Oakland Chamber of Commerce, *1969 Handbook, Comparative Data Bay Area Counties,* Oakland, California, September 1969.

32. The Conference Board, Inc., *A Guide to Consumer Markets 1970,* New York, New York, 1970.

33. Sales Management, Inc., *Sales Management Survey of Buying Power. 1960, 1965, 1969 editions,* New York, New York.

34. San Francisco Bay Conservation and Development Commission, *Refuse Disposal and San Francisco Bay,* San Francisco, California, October 1966.

35. Snohomish County Planning Department, *Survey of Disposal Sites – Snohomish County,* Everett, Washington, 1970.

36. The Regional Planning Commission, County of Los Angeles, California, *Quarterly Bulletin,* Los Angeles, July 1, 1970.

37. The Seattle Area Industrial Council, *Seattle, Tacoma, Everett, Metropolitan Area Economy,* Seattle, Washington, August 1970.

Personal Interviews

38. Robert Clark, Chief, Urban Data Section, Bureau Solid Waste Management, Cincinnati, Ohio. Interview January 12, 1970.

39. Richard Toftner, Chief Planning Section, Bureau Solid Waste Management, Cincinnati, Ohio. Interview January 12, 1970.

Personal Letters

40. Larry N. Jensen, personal letter, October 20, 1970; May 17, 1971.
41. Lamont B. Gundersen, personal letter, November 15, 1970.
42. Larry Martin, personal letter, December 9, 1970.
43. Leonard Stefanelli, personal letter, November 12, 1970.
44. Jack Coyne, personal letter, October 19, 1970.
45. M. Wetzel, personal letter, November 20, 1970.

26 Summary of Findings Concerning Solid Waste

ALFRED J. VAN TASSEL

There is a natural tendency to associate solid waste with household garbage, but this latter is actually a small part of the problem. Retail stores and other commercial establishments generate large amounts of solid waste, especially packaging materials. Industry generates still more, as the slag heaps of a steel town will testify, but its treatment is so varied as to defy generalization. Household garbage is that portion usually collected by the municipality; private firms sometimes collect household garbage, but they play a larger role in collecting commercial waste. Whoever collects it, and except for industrial waste, disposal is usually on property controlled by the municipality.

We can dispose of the question of industrial wastes promptly. Each industry generates a waste peculiar to that industry and much of this is sorted so as to recover valuable portions by recycling. Since such industrial waste is concentrated at the point of origin the opportunities for by-product recovery are at a maximum. However, each situation is a case by itself and there is little basis for a general study. This chapter will deal principally with household garbage and will not attempt to discuss industrial wastes further. Neither will it deal extensively with such other forms of waste originating from agricultural or sewage treatment residues.

Sources of Household Waste

A major and growing source of household waste in the past decade has been packaging materials. The use of paper and paper board, for example, has been growing at an annual rate of 4.5 percent. The use of metal cans for food products was expected to grow through 1976 at an annual rate of 3.4 percent. In general, nonreturnable containers increased in popularity during the 1960s and the trend was expected to continue. Output of nonreturnable bottles was expected to more than triple from 1966 to 1976, while production of cans was approximately doubled in the same period. The impact on disposal is evident when it is realized that each returnable bottle may make nineteen trips to the market before entering the garbage heap, which is the immediate end of the line for a nonreturnable. Plastics, usually non-biodegradable, have entered to complicate disposition.

As a consequence of such consumption trends, the volume of household waste per person has tended to increase from an estimated 2.8 pounds per person per day in 1920 to 4.8 pounds in 1965 with a further increase to 5.5

pounds projected for 1980. Interestingly, this consequence of the packaging revolution reverses a downward trend in household waste per capita occasioned by a shift in preferred fuels. As fuel oil and, more recently, natural gas replaced coal for household space heating, the generation of ash in the household declined precipitously.

However, it must be noted that the rapid increase in the use of paper and other relatively light packaging materials means that the bulk of solid waste has increased faster than its mass. Hence garbage trucks with compactors have come to play a strategic role. Moreover, new sources of denser waste are increasing in importance as, for example, when efforts to reduce water pollution lead to accelerated creation of sewage residues.

Population, Affluence, and Solid Waste

In general, we may expect the volume of solid wastes generated from household and commercial sources to rise in some rough proportion to consumption of goods and services. Also, as people are more crowded in apartments as against single homes with attics, discards find their way more promptly to the trash heap. Summarizing, our investigators found a likelihood, in the years ahead, of a faster increase in per capita waste generation than had been anticipated and an even faster rise in the volume of trash.

Open Dump or Sanitary Landfill

The preferred method of disposal was by deposit on open land. If done in accordance with best practices, the procedure is designated as sanitary landfill and can lead in time to the creation of areas suitable for recreational purposes. However, the 1968 national survey found true sanitary landfill was practiced on only about 360 of 6,000 disposal sites surveyed and many of the others were no better than open dumps and likely sources of vermin, odors, and water pollution. Because of this widespread failure to observe good standards, "sanitary landfill" has a bad reputation that arouses local opposition near any site for which it is proposed. Most importantly, it has become progressively more difficult to find sites for land disposal and several cities are within a few years of exhausting the capacity of currently available sites.

Reducing Bulk: Incineration and Other Means

One way of overcoming the rise in volume of trash is to find means of reducing its bulk. Since much of this comes from paper, a favorite method has been through incineration, which declined in popularity to about 1950. Thereafter, installed incineration capacity increased by 3¼ times to 1966. Through compaction and incineration it is possible to reduce the bulk of trash to about one-seventieth of its original volume with a corresponding saving in disposal area. But just as careless performance has given "sanitary landfill" a bad reputation, so smoky incinerators have led to widespread opposition to this solution, even though authorities point out that emissions from burning trash are low in the dangerous pollutants. Air pollution devices involve large additions to capital costs as removal efficiency reaches optimum limits.

Since incineration is more expensive, to begin with, than landfill as a disposal method, such pollution control expenses increase its competitive disadvantage. Such added costs may be partially offset if it proves possible to sell steam or energy produced by incinerators, as is done in Chicago and several European cities. The recovery of recycled materials from the waste is another source of offsetting income. A machinery manufacturer has offered to build a $30 million plant for Nassau County, New York, at its own expense to generate steam for electric power, and to recover recycled by-products. Profits from the sale of steam and by-products would be shared with the county, provided the county undertakes to supply 1,200 tons of refuse daily [1].

Incinerators that generate steam are water-cooled with metal walls. Polyvinyl chloride (PVC) plastics—about 2 percent of all unsorted refuse—form hydrogen chloride gas upon combustion, which is highly corrosive to the metal walls of the incinerator.

The Political Context

Collection accounts for four-fifths of the cost of dealing with solid wastes, but disposal is the ultimate problem. Much of the collection cost is associated simply with the problem of litter, but such difficulties can be dealt with; for some cities the question of disposal appears more intractable. Incineration is one way of reducing bulk, but it is expensive relative to landfill and may complicate air pollution difficulties. Compaction and composting are other ways of treating solid waste, but they have hardly received more than experimental attention.

Since land for ultimate disposal—whether of untreated refuse, incinerator residue, or compacted garbage—is the ultimate problem, the desirability of a regional solution is obvious. But political jealousies stand in the way of such a solution, however logical. No one wants to share the burden of his neighbor. How much more pressing the problem must become to overcome these barriers is hard to say. Meanwhile, the problem manifests itself with varying severity in specific localities.

The Cities

New York City

As in every environmental area, solid waste disposal problems are especially severe in the nation's largest city. Based on National Survey data, the conclusion is reached that current landfill areas will be exhausted by 1977 at the latest.

Collection costs accounted for 82 percent of the total New York expenditures for solid waste disposal. To save precious landfill space, New York has considered adding to its 9,500 tons per day incinerator capacity, but this solution has been delayed and made more costly by the need to consider the impact on air pollution.

New York City's sanitary landfill sites are well planned and suitable for conversion, but they are located at long distances from points of waste origin. Elimination of apartment house incineration to reduce air pollution would further burden inadequate disposal sites.

Private firms have demonstrated flexibility in using regional rather than local disposal sites, and there is some evidence that private operations are more highly mechanized and hence presumably more efficient. It is not clear, however, to what extent this is because the private carters enjoy the economies of scale associated with removal of commercial refuse as against serving individual households.

Despite generally good planning, New York faces a more imminent solid waste crisis than any other city studied.

Philadelphia

Unlike New York, Philadelphia is unable to extend its land area and gain disposal space by landfill at shore areas. Although its population is growing at a moderate rate, Philadelphia will have exhausted its own disposal sites within a few years. However, there is a good deal of cooperative regional planning and sufficient disposal sites should be available in surrounding rural areas. Disposal costs will rise as more distant sites require longer trucking. Consideration is being given to rail haul of refuse to upstate strip mines.

Atlanta

Atlanta has been growing rapidly in all respects, including the generation of solid waste. The outlook is for continued rapid growth to 1980. Forward planning re disposal has not been especially notable and the city is likely to have exhausted its present sites by 1975. None of its present sites can be considered sanitary and only one site has a planned after-use. Meanwhile, "urban sprawl" in the suburban areas surrounding Atlanta may be reducing the number of potential disposal sites outside the city.

Pittsburgh

Pittsburgh is cited as a prime example of a city where a large gap exists between the quantity of wastes generated—particularly by the industrial sector—and the wastes collected. Its disposal sites do not meet sanitary landfill standards and there is little evidence of effective regulation. Nevertheless, Pittsburgh is well endowed with land area for disposal and, unlike New York, faces no imminent disposal crisis.

Chicago

Chicago has four advanced incinerators that handle all household waste generated in individual dwellings in the city. Metals are recovered for recycling from these incinerators and steam is generated for sale as well. Other wastes are disposed of by landfill, with most sites not classed as sanitary landfill and with an average remaining life of about nine years in 1968. Lack of response to the 1968 survey clouds the outlook, but there is little evidence of careful advance planning.

Cleveland

Cleveland uses a combination of incineration and land disposal to handle a mounting volume of solid waste. The short average remaining life for disposal sites, both in the central and in suburban counties, suggests that Cleveland may face a critical situation shortly.

St. Louis

St. Louis uses old incinerators and landfill operations not qualifying as sanitary to dispose of the refuse of about two million people. There is little evidence of advance planning, and present sites in 1968 had average remaining life of 5.3 years in St. Louis County and 8.5 years in the suburbs.

New Orleans

The New Orleans SMSA employs incinerators and landfill to dispose of the solid waste of over one million people. None of the landfill sites qualified as sanitary. Although the average remaining life in 1968 for sites in the central parish was only three years, the average remaining life in the suburbs was thirty-one years. It does not appear that the New Orleans SMSA will have a disposal crisis as a penalty for the lack of planning.

Mobile

Mobile has abundant land available for disposal at sites that do not qualify as sanitary for the most part. The low cost of such disposal doomed an experiment in the production for sale of a soil conditioner by the composting of solid waste.

Los Angeles

Los Angeles has been planning, since the end of World War II and the beginning of its rapid population growth, to meet well in advance its needs for land disposal sites. Need for additional space was recognized when the prohibition of backyard incineration in 1957 as an air pollution control measure added to the solid waste burden.

High standards of landfill practices obtain at most Los Angeles disposal sites and quantitative records are maintained as an aid to planning. Los Angeles has had notable success in converting completed landfill sites to recreational and other park purposes. There are plans for use for nearly half the present disposal sites in Los Angeles County. Thus although rapid growth of a generally affluent population is projected for the future, it is expected that the sizable requirements for new disposal sites will be met as needed. Good forward planning is expected to yield without stress or strain the projected 11,000 plus acres required for solid waste disposal in Los Angeles by 1980.

San Francisco

A crucial turning point in solid waste disposal in the San Francisco Bay area came with the series of decisions beginning in 1965 to curtail landfill in San Francisco Bay. This turn in public policy designed to correct a course which it was felt would have led to the inevitable pollution of bay waters to an unacceptable degree induced a solid waste disposal problem that shows no sign of going away. The 1968 National Survey of Bay Area disposal revealed that the anticipated years remaining for disposal sites ranged from four for San Francisco County to six for adjoining peninsular San Mateo County to twelve for Alameda County, which is the seat of Oakland and Berkeley. Other Bay area counties, Marin and Contra Costa, found no immediate crisis. Other responses to the 1968 National Survey did not argue an especially planful or technically progressive handling of land disposal by Bay area counties.

In general, John Burns concluded that of all the western urban areas studied, "none approaches the San Francisco SMSA in the degree of urgency currently existing with respect to solid waste."

The outlook for the future was no more promising. Forced to truck its refuse 35 miles south by the end of the 1960s, San Francisco faced a projected deficit of 1,378 disposal acres by 1980. Adjoining San Mateo faced a deficit of 850 acres needed for 1980 disposal, and Alameda (Oakland, Berkeley) 494, for a combined deficit for the three counties of 2,722 acres. Only the projected surplus of 324 acres in Contra Costa County was likely to be of any value in easing this pinch.

Seattle

Relying almost entirely on landfill for disposal, the communities of the Seattle area gave every indication of a very casual, even careless and certainly unplanned, approach to solid waste disposal. Most of sites in the Seattle area covered by the 1968 survey did not maintain quantitative records, most reported surface drainage problems, and over a third reported uncontrolled burning. With the closing of a dump site in 1967 the City of Everett and Snohomish County were forced to turn to a dump site for both with a life expectancy of only two to five years. Despite this apparent lack of planning the two major counties of the area—King (Seattle) and Snohomish (Everett)—possessed sites with life expectancies of eighteen and thirteen years respectively.

The projection of economic growth in the Seattle area has become hazardous because of the economic difficulties experienced by its aerospace industry. However, if economic growth is projected at the same rate as that postulated for other areas, Seattle may face a considerable solid waste crisis by 1980. A land-for-disposal deficit of 1,735 acres is projected for 1980 in King County and 204 acres for Snohomish. If this comes to pass it will be a clear example of a lack of planning coming home to roost.

Denver

Denver does not display an especially outstanding record in the field of solid waste management, but neither is it likely to face an imminent crisis. While Denver County itself faces a probable deficit in disposal area of more than a thousand acres, this is likely to be largely offset by surplus acreage available elsewhere in the SMSA. Specifically, Fort Lowry in nearby Arapahoe County is viewed as likely to be able to handle all wastes generated in Denver for the next twenty years. Thus, although growth projections suggest a deficit in disposal acres of 456 for the SMSA as a whole, local observers might regard as the most pressing problem the determination of the most efficient way to transport wastes from population centers to outlying disposal sites such as Fort Lowry.

Salt Lake City

The least populous of the five SMSAs surveyed in the western states, Salt Lake appears to have no future disposal problem of consequence. Its existing sites are all managed in accordance with sanitary landfill standards and there is abundant land available when needed. Thus there is no pressure for extensive forward planning.

Notes

1. Carole Ashkinaze, "County Gets a Cash-for-Trash Offer," *Newsday* (Garden City, N.Y.), July 6, 1972, p. 19.

**Part 4
Our Environment: Conclusions**

27 Conclusions

ALFRED J. VAN TASSEL

For the study of air pollution and solid waste disposal, the Hofstra study covered fifteen cities or metropolitan areas in the United States. In addition, nine of these cities were involved in the study of water pollution problems or recycling possibilities. Finally, twelve of the cities were covered in the study of noise pollution due to jet transport operations.

The fifteen cities had a combined population of 47 million or 38 percent of total U. S. population in cities within the same size range.[a] The percentage of total U. S. metropolitan area population included in the sample is 31 and of all U. S. population, urban and rural, is 23. The nine cities studied for water pollution or recycling possibilities constituted 77 percent of the total population in the fifteen [5]. Statisticians know that the notion of a "representative" sample is a slippery concept and no such claim is made for this one. Nevertheless, the portion of urban America covered with respect to selected environmental problems is substantial, and a summary of findings with respect to these urban areas should throw some light on the overall outlook for the environment in the America of 1980.

[a]Throughout this chapter when reference is made to the populations of cities, the data employed are the populations of the metropolitan areas, as defined by Standard Rate and Data Service, that include the central cities.

New York

What will be the net effect of the forces at work on the New York environment of 1980 in terms of general quality? Is the question a meaningful one? Is it possible to put in the balance the advances that seem certain in some areas and weigh these against the effects of increased output and population? Certainly, no quantitative answer is possible. We may be able to approach some quantitative judgment concerning the state of air pollution *or* of water pollution, but value judgments must enter into any weighting system designed to reach an overall judgment. However, let's make the effort.

For the New York City area as a whole, all projections agree on a moderate percentage increase in population, which amounts to a very sizable absolute increase.

In terms of the major waterways in the New York City area, there is little reason to anticipate any substantial improvement by 1980. The lower Hudson River is so heavily polluted at present that any improvements that seem

557

achievable by 1980 are likely to be overbalanced by growth of population and perhaps industrial output. The more pressing question is likely to be whether through tidal action, especially in the summer months when stream flow is sluggish, the pollution of the lower Hudson could reverse the progress that seems likely in the mid-Hudson area.

So far as Long Island Sound is concerned, there is nothing currently planned to reduce the buildup of nutrient concentrations that has proceeded despite some advances in sewage treatment. For the western sound in particular, this growing concentration of nutrients poses the danger of a sudden deterioration of water quality occasioned by algae growth and decay depleting dissolved oxygen.

In terms of water recreation for the New York area, a great deal will depend on the decisions reached by Long Island communities concerning the interlocking questions of sewage disposal and water supply. Most of the homes in Nassau and Suffolk Counties, the two Long Island counties lying outside New York City, were not served by sewers in the 1960s. The proportion sewered is expected to be 85 percent in Nassau by 1980 and a majority in Suffolk by 1980. In terms of recreational opportunities for residents of the New York area much will depend on what is done with the effluent of the sewage treatment plants. So far, on the North Shore of the Island effluent has been discharged to Long Island Sound without any deleterious effect on open waters of the sound except for an increase in nutrient concentrations. However, a barrier reef along the South Shore encloses a series of bays, and discharge of sewage effluent might have a serious deleterious effect on water quality and as a minimum would involve a comparatively rapid buildup of nutrient concentrations. The only alternative extensively considered so far would involve expensive ocean outfalls. Either bay or ocean discharge results in a decline of the unique and extensive underground water reservoir that exists under Long Island. Though the economic and other advantages seem evident, little serious consideration has been given to water recycling as a solution, in part because of major technical difficulties such as that of removing dissolved nitrates. In view of the diffident and tentative approach to water recycling that has characterized the Long Island scene to date, it seems unlikely that there will be any substantial adoption of this approach by 1980.

The quality of the waters of the Hudson, and to a much smaller extent those of Long Island Sound, strongly affect the quality of the waters of New York Harbor. The quality of the waters of New York Harbor is powerfully affected by forces that were not covered in the Hofstra study, such as the direct discharge of the effluent of sewage treatment plants (such discharges are usually to waterways such as the East River, a tidal estuary, or to Jamaica Bay, and the extent of pollution depends, of course, on degree of treatment). Important also is the continued dumping of sewage sludge and industrial wastes in the near offshore waters; there are no current plans to curtail such dumping, and Jerome Kretchmer—the Environmental Protection Agency administrator for the city— has even proposed dumping garbage at sea as a solution to the shortage of solid waste disposal sites. The volume of solid waste to be disposed of is likely to increase if apartment incinerators are banned in the interest of improved air quality, as has been discussed for some time. The capacity to absorb additional solid waste was within nine years of exhaustion at city-operated sites in 1968.

The one aspect of the environment with respect to which some improvement seemed certain was the quality of the air. Total air pollutants emitted in the New York Standard Metropolitan Statistical Area were projected to decline by 58 percent between 1969 and 1980. Most of this was attributable to a reduction in pollutants coming from automobiles; emission of hydrocarbons, mostly from this source, was projected to decline by at least 63 percent from 1969 to 1980, and of carbon monoxide by 58 percent. Not so nitrogen oxides, which were projected to rise by 110 percent even if the highest automobile pollution control levels were achieved and by 250 percent if only the 1971 standards proved feasible.

In terms of total tonnage of air pollutants emitted to the skies of the New York area, the outlook is for a reduction of 45 to 58 percent depending on whether or not it proves feasible to meet the Muskie bill standards. Since nitrogen oxide emissions, the principal factor in smog formation, are projected to increase sharply in either event, we are forced to the paradoxical conclusion: the New York air of 1980 will be cleaner with more frequent smog.

An advance was also expected in the New York area with respect to the dangerous sulfur oxide emissions that were projected to decline by 74 percent from 1969 to 1980. However, the emission of soot (particulates) was projected to rise by 6 percent from 1969 to 1980.

Finally, whatever happens to the quantity of physical air pollutants in the New York area, the volume of noise pollution attributable to jet planes seems certain to increase. The number of takeoffs and landings by jet planes in the New York area was projected to increase by 70 percent between 1969 and 1980. Moreover, it was to be expected that the number of takeoffs and landings would be performed by larger and noisier planes. Accordingly, the outlook in New York might be for cleaner skies, but emphatically noisier ones.

There are, of course, many more aspects to the quality of life and the environment in New York City than those covered in the Hofstra study. In particular, the outlook for mass transit is of decisive importance for millions of citizens of the area. And no effort was made by the study to consider such things as the state of law and order. Within the limits of the scope of the study, the outlook for the New York area is for some considerable improvement in air quality, but otherwise a negligible change in the overall quality of the environment. The possibility of a decline in the quality of the waterways in the New York vicinity cannot be excluded despite the strenuous efforts to improve it.

Savannah

The sharp difference in the 1980 environmental outlook between Savannah, Georgia, and New York City emphasizes the distinction between the prospects of small and large cities. If Savannah achieves half her goals for 1980 for the improvement of the state of the Savannah River, there will be a decisive advance in environmental quality. With the quantity of Savannah air pollutants always insignificant relative to the mass of receiving air, dilution could always be counted on to reduce any air pollution problem to negligible proportions. With net overall advances expected, especially in pollutants from automobiles, the prospect is for making a generally satisfactory situation better. With respect to solid waste, the problem is so trivial that the analyst to whom Savannah was assigned decided not to include it in his analysis.

The big question, then, is whether or not the major concerns responsible for most of the water pollution can be prevailed upon to carry through the measures needed to reduce decisively the pollution of the Savannah River. If they can, the 1980 environment of Savannah will be greatly improved.

Atlanta

Atlanta and Savannah are both southern cities, but there the similarity ends. Water pollution is of little concern to Atlanta since it has no major waterways, but it confronts the problem of air and noise pollution associated with active growth. Emissions of nitrogen oxides and sulfur oxides are projected to increase in Atlanta between 1969 and 1980 by 131 and 49 percent respectively. Although no imminent crisis is foreseen, Atlanta will no doubt be forced to make arrangements with surrounding areas to dispose of its burgeoning volume of solid waste. The most difficult adjustment may be with respect to noise pollution. Atlanta, without major waterways, was established to serve as a transportation hub in the era when the railroad was coming into its own; it has made the transition to the air age gracefully, and is now the most rapidly expanding air hub in the nation. Between 1969 and 1980, the number of takeoffs and landings in Atlanta is projected to increase by 143 percent, a more rapid pace than for any other air hub covered in this study. In total takeoffs and landings, Atlanta will rank just behind Los Angeles. This means a vast increase in the amount and intensity of noise in the vicinity of its airport.

The Atlanta air will be noisier in 1980 than it was in 1969 and it seems unlikely that it will be much cleaner. If the Muskie bill standards for automobile emissions are met, the total weight of pollutants emitted to Atlanta air is projected to decline by 11 percent. If only the 1971 standards for automobiles are achieved, the total of pollutants in Atlanta air will increase slightly by 3 percent. In either case, sulfur oxides in Atlanta skies are projected to increase substantially by 49 percent between 1969 and 1980 due almost entirely to expanded output of electricity from coal-burning generating stations. Emission of nitrogen oxides in Atlanta will also increase rapidly between 1969 and 1980, by 131 percent if Muskie bill standards are met and by 168 percent otherwise.

Whether Atlanta will also pay the price of expansion with a solid waste crisis is not clear at this time. The city is likely to have exhausted its present solid waste disposal sites by 1975 and urban sprawl may eliminate some suburban replacements.

Philadelphia

Philadelphia draws its drinking water from the heavily polluted Delaware River, and in one severe drought its supply suffered from the competition of New York

City, which draws its water from the unpolluted upper reaches. Extensive chlorination has been required to make Delaware River water safe for drinking. Philadelphia water has been described as a chlorine cocktail and also as the only water in the world that one can chew. The Delaware was not among the waterways included in the Hofstra study, but it is not unreasonable to expect some considerable improvement in the state of Delaware waters by 1980 in the light of national trends in water pollution control.

Philadelphia's air should be very much cleaner in 1980 than it was in 1969 if the Muskie bill standards for automobiles can be met, with the overall tonnage of pollutants reduced by 30 percent, accomplished mainly through a near 90 percent reduction in carbon monoxide emissions. Even if only the 1971 standards prove feasible, the reduction in overall tonnage of pollutants emitted to the air in Philadelphia is projected to decline by 14 percent between 1969 and 1980.

However, emissions of sulfur oxides are projected to increase by 62 percent from 1969 to 1980 and nitrogen oxides by 39 percent even if 1976 standards are achieved by automobiles; if only the 1971 standards prove attainable the projected increase in nitrogen oxides will be nearly 90 percent. Thus, the outlook may be for more frequent smog despite the generally cleaner air.

Whether cleaner or not, the air in the vicinity of Philadelphia airports promises to be noisier in 1980 than it was in 1970. Jet operations in Philadelphia are projected to increase to 210 percent of the 1970 level by 1980.

Philadelphia is not expected to face an especially critical situation with respect to solid waste disposal in 1980. Although disposal sites within the city may well have been exhausted before that date, there is every indication that sites in surrounding rural areas will be available, thanks to a substantial degree of cooperative regional planning.

Pittsburgh

When Alfred Marshall, the English economist, wanted an example of an external cost he chose the skies of Pittsburgh, made smoky by the manufacture of steel. Progress in Pittsburgh in cleaning up the sources of industrial pollution has been such that it long ago lost its reputation for having the dirtiest air, a doubtful distinction that passed to Los Angeles, whose smog-ridden skies come not from production but from consumption—the use of private automobiles. While industrial air pollution, except that from power stations, was not included in the Hofstra study, it still seems safe to predict that the air in Pittsburgh will be cleaner in 1980 than it was in 1969. Total tonnage of air pollutants from the sources covered is projected to decline by 45 percent from 1969 to 1980 if the Muskie bill standards can be met and by 29 percent even if they can't. Nitrogen oxide emissions are expected to increase by 11 to 51 percent depending on which automobile control standards are achieved, and sulfur oxides by 25 percent. However, these increases are more than outweighed by large projected reductions in carbon monoxide emissions.

The level of jet operations at Pittsburgh airports is expected to more than double from 1969 to 1980, but to still be at the lowest volume of any major eastern city.

Although characterized by lax disposal practices and little evidence of much forward planning, there seems to be little prospect of any serious difficulties in the disposal of solid wastes in Pittsburgh.

The rivers flowing through Pittsburgh were not included in the Hofstra study, but as in Philadelphia it is not unreasonable to project some improvement in water quality by 1980.

Chicago

It appears likely that New York City in 1980 will pay many environmental penalties for its bigness, whether judged in terms of the state of its air, its waterways, or its problems of garbage disposal. For the nation's second city, the 1980 outlook for the environment is considerably brighter. In part this is because Chicago has long exported its major water pollution problems southward to the Mississippi via the Chicago River. The export will be made even more complete by 1980 through the addition of the effluent of some sewage systems north of Chicago to the Chicago River flow.

With respect to air pollution, the 1980 projections indicate that the total tonnage of pollutants from the sources covered in the Hofstra study will be reduced by 22 to 40 percent, depending on whether or not Muskie bill standards for automobile emissions are achieved. Reductions in emissions are projected to be dramatic with respect to carbon monoxide, hydrocarbons, and particulates. However, sulfur oxides are projected to increase by a modest 6 percent and nitrogen oxides by 15 to 66 percent, depending on whether or not the Muskie bill standards prove to be attainable. Chicago's air quality benefits from favorable atmospheric conditions much of the time.

The vicinity of O'Hare Airport, already noisy at present, promises to be still noisier in 1980. Jet operations from Chicago airports are projected to increase by 81 percent between 1970 and 1980. Operations tend to be more concentrated at O'Hare than is the case for any single airport in the New York area.

No particular advance is foreseen in the field of solid waste disposal. For some time Chicago has relied on moderately advanced incinerators for the disposal of those solid wastes collected by the city, relying on land disposal for the considerable portion collected from commercial and industrial sources by private carters. Sites for the latter were approaching exhaustion, with little advance planning in evidence.

Cleveland

The environmental outlook for Cleveland in 1980 was dominated by the relative intractability of the Lake Erie water pollution problem. Since the basic difficulty with Lake Erie rests with the relatively high concentration of nitrogenous and phosphatic nutrients, there was no likelihood of a rapid reversal of the lake's condition in the unlikely event of a sharp reduction in the input of pollutants attributable to industry and population. Actually, inputs of pollutants from both sources were likely to increase between 1969 and 1980 at least in both the western and central basins of Lake Erie, those portions already the most overloaded.

If there was no sound prospect of improvement in the quality of the waters off Cleveland, there was good reason to expect cleaner air above it. Projections for 1980 pointed to a 20 to 37 percent reduction in total tonnage of air pollutants, depending on whether or not the Muskie bill automobile emissions were achieved. In either event, some increase in nitrogen oxides was to be expected, a sizable 59 percent rise if the Muskie bill standards proved to be out of reach. A 54 percent increase in sulfur oxide emissions was projected for 1980 over 1969. However, these increases were more than outweighed by sharp reductions in hydrocarbons, carbon monoxide, and particulates.

Also projected to increase sharply was the scale of jet operations at Cleveland airports, which was expected to rise by 97 percent. However, this expanded scale of operations would only be about half the present load at Chicago airports and less than 30 percent of that projected for Chicago in 1980.

Although data were scanty, the indications were that Cleveland faced early difficulties in finding sufficient land for solid waste disposal.

St. Louis

St. Louis has had the doubtful distinction of being the dirtiest city in the midwest in terms of its air. Judged on a basis of total tonnage of air pollutants emitted per year per square mile of the central city, St. Louis in 1969 had 16,134 tons per square mile per year, or 22 percent more than its closest rival, Chicago. The outlook is for considerable improvement by 1980, with total pollutants emitted in St. Louis reduced by 27 to 43 percent from the 1969 levels, depending on whether or not the Muskie bill standards are attained. Nitrogen oxides will go up by only 6 percent if Muskie bill standards are attained, but will jump by 52 percent in St. Louis if only the 1971 standards prove attainable by 1980. In either case, sulfur oxides are expected to increase by only 7 percent and particulates to be reduced by 83 percent, reflecting a continued shift to natural gas for space heating. Meanwhile, since Chicago is likewise projected to reduce its emissions of air pollutants, St. Louis, though much cleaner, will remain the midwestern city with the highest level of air pollutants entering its air. In terms of noise pollution from jet planes, St. Louis compares much more favorably. Thus, the projected 1980 level of jet operations at St. Louis airports is only 24 percent of that in Chicago, even though the 1980 figure for St. Louis represents a 22 percent increase over 1970.

Solid waste disposal activities in St. Louis appear to proceed with a minimum of forward planning. Average remaining life of disposal sites in St. Louis county in 1968 was only 5.3 years and even in the surrounding suburban Missouri counties was only 8.5 years.

Similar casual treatment characterizes St. Louis in relation to the mighty river that first gave the city its importance. The potential BOD load of industrial and municipal wastes discharged to the Mississippi River in the St. Louis area was 350 million pounds in 1970, and this figure was projected to increase to over 400 million pounds in 1980. These discharges included effluents from such heavily oxygen-demanding industrial processes as those in paperboard mills, chemical plants, and petroleum refineries. Dissolved oxygen levels have dipped dangerously low at various times and there have been periodic color and odor problems. Generally, the tremendous rate of flow of the river—174,000 cubic feet per second at St. Louis—overwhelms such pollution problems.

New Orleans

The rate of flow of the Mississippi is even more immense at New Orleans, at 406,000 cubic feet per second, than it is upstream in St. Louis. Accordingly, although both the industrial and nonindustrial load are increasing more rapidly than in St. Louis, it is expected that dilution will serve to obscure any major new water pollution problem in the New Orleans area.

New Orleans has the least problem of air pollution of any city of its size covered in the Hofstra study. Natural gas is the major fuel employed, so there is little pollution from this source. The automobile is the major source of air pollutants, but since the car population is relatively small the load is not excessive. The total tonnage of pollutants emitted is projected to decline from 1969 to 1980 by 40 to 61 percent, depending on whether or not the Muskie bill standards are achieved. Nor does noise pollution from jet planes loom as a major problem although operations are projected to increase by 123 percent between 1970 and 1980. Still, the scale of operations will be only 22 percent of that in Chicago.

Similarly with respect to solid waste disposal, although there is no evidence of careful planning, it appears that New Orleans will find adequate land for disposal in nearby suburbs.

Mobile

Few of the environmental problems that plague the larger centers have much effect on life in Mobile. Air pollution in Mobile, measured by total tonnage of pollutants emitted from sources covered in this study, was the lowest in 1969 for any of the five midwestern cities studied. Tonnage of pollutants emitted in Mobile was projected to decline further in 1980 by 17 to 36 percent. The quality of Mobile's waterways was not covered in the Hofstra study.

Solid waste disposal at one of the city's landfill sites costs Mobile only $2 per ton. It was this low cost that doomed an experiment in more sophisticated methods of solid waste disposal. For several years, Mobile operated a composting plant where solid waste was converted to soil conditioner for commercial sale with the hope of defraying some part of disposal costs. However, it was found that the net cost to Mobile was $11 per ton for solid waste disposal by composting.

Los Angeles

With respect to the environment and especially with regard to air pollution, the nation stands in debt to Los Angeles. Los Angeles officials at an early stage foresaw the rise of the automobile as the main factor in U. S. transportation and set out to welcome the automobile's ascendancy. With an elaborate system of limited-access highways, Los Angeles was prepared to make the automobile work effectively if it ever could anywhere. In consequence, the automobile population of Los Angeles soared. This fact, together with local atmospheric conditions that made inversions frequent, meant that Los Angeles was the place where trouble in the automobile paradise first became evident. A new word, "smog," entered the vocabulary to describe the haze produced by the reactions of automobile pollutants.

Los Angeles became a laboratory for experiments in air pollution control as air pollutants from industry, home incinerators, and other sources were brought under control. But the finger of blame still pointed to the automobile, and Los Angeles was the scene of the first experiments in control of automobile emissions. As it became clear that Los Angeles was simply an extreme case of a general national problem, federal legislation and regulation entered, as noted in Chapter 15.

For Los Angeles, the 1970s should bring a turn in the deterioration of air quality that has proceeded with only slight checks since World War II. By 1980 the total tonnage of air pollutants released to Los Angeles air should be 70 percent below the 1969 figure, if it proves feasible to meet Muskie bill standards. Even if only the 1971 standards prove attainable, the 1980 reduction from 1969 should be 55 percent in total tonnage of air pollutants emitted. Most of this decline in total pollutants comes from reduction of emissions of carbon monoxide and hydrocarbons. It is expected that improved automobile pollution control will be aided by a continued shift to natural gas for space heating and even power generation.

However, Los Angeles may still confront a worsening of the situation with respect to nitrogen oxides—a major contributor to the smog that has become Los Angeles' unwelcome trademark. If the Muskie bill standards prove attainable, the emission of nitrogen oxides in the Los Angeles of 1980 will be 27 percent below 1969 levels. But if only the 1971 standards can be achieved, Los Angeles air in 1980 will have 43 percent more nitrogen oxides than in 1969.

Los Angeles depends for its existence on the automobile and there is a limit to how far it can go in efforts at unilateral control of automobile emissions. Its regulations in other areas are accordingly stricter and this may amount to, in some instances, an export of pollution almost as definite and clear-cut as that practiced by Chicago in the diversion of its sewage effluent southward. In this day of interlocking electric power networks where utilities in adjacent areas supply one another with power to meet peak demands, it is difficult to say precisely where the power a city employs is generated, though usually it will be as close to home as possible. Southern California Edison, the electric utility for the Los Angeles area, together with several Southern California cities, is a member of Western Energy Supply and Transmission (West) Associates, a combine established to generate and transmit the power needed to meet the burgeoning demands of the southwest states [1:35, note 2]. Among the projects undertaken by West Associates is the erection and operation of a 2 million kilowatt coal-burning plant known as the Four Corners Plant because of its desert location near the meeting place of the states of Utah, Colorado, Arizona, and New Mexico. One attraction of this desolate site was that it permitted freedom from air pollution restrictions existing in Southern California [1:22]. That West Associates exploits this freedom to the fullest was evidenced in 1970, when NBC News pointed out that the only man-made feature on the surface of the earth which was distinguishable to astronauts returning from the moon was the Four Corners Power Plant's plume of smoke [1:29]. Four more plants in addition to the two now in operation are planned by West Associates for this scenic desert area.

However, the bulk of Los Angeles' air pollution arises from its millions of automobiles. The fate of their emission controls, rather than that coming from power generation, is the key to the future.

Whatever the state of Los Angeles air, its International Airport has experienced the nation's most expensive reaction to noise pollution. Los Angeles has only about half the total jet takeoffs and landings of the New York airports, but it is second only to Chicago's O'Hare for a single airport. In 1971, Los Angeles airport commissioners felt forced to spend more than $200 million to buy and remove nearly 2,000 homes located near the airport [2]. With jet operations projected to increase by 70 percent between 1970 and 1980 and with court rulings supporting homeowner claims, there were indications that the troubles of Los Angeles airport officials were just beginning [3].

Los Angeles recognized its potential solid waste disposal problem at an early date and took the necessary steps to assure adequate sites for disposal. There is every indication that this forward planning has provided for the future as well. This recognition of the need for planning may well have been advanced by the decision in 1957 to ban home incineration as an air pollution control measure and the realization that this would impose an added burden on disposal facilities. Los Angeles has succeeded in converting completed landfill sites to recreational and park purposes.

The perceptive forward planning characterizing the solid waste disposal program in Los Angeles has not carried over to the field of water supply and sewage disposal. Water supply has, from the first, been dealt with by a mobilization of Los Angeles County's political muscle. A series of boldly conceived and enormously expensive aqueducts have assured this semi-arid region an ample supply of water for one-time usage, sometimes at the expense of adequate water for other areas. Latest of these constructions is the California Aqueduct, which draws on the abundant water resources of Northern California. The 1971 cost per acre foot was about 350 percent of that in 1941.

Meanwhile, the Los Angeles Sanitation Districts have been faced with the problem of disposing of the mounting volume of sewage generated by a rising population. In experimental efforts to offset such costs, the sanitation districts have been able to reclaim and sell small quantities of water at a cost of less than 21 percent of that of imported water.

San Francisco

An early reaction by Los Angeles to the threat of air pollution was to ban backyard incinerators and make plans to dispose of the increased volume of solid waste. When San Francisco became aware that its bay was threatened with irreparable damage if landfill continued, it banned further operations of this sort and faced the solid waste disposal problem that was sure to follow. Since then, the San Francisco Bay region has developed a far-reaching plan for cleaning up the waters entering the bay involving expenditures of $1 billion by 1980. However, the prospect is for only slight improvement or maintenance of present quality levels of bay waters. A great deal depends on whether or not substantial progress can be made in treating the increased volume of pollutants from industrial sources that are expected to approximately double by 1980. In the final phases of the plan after 1990 consideration will be given to water reclamation.

The Bay Area Rapid Transit System (BART) is San Francisco's most solid claim to advanced planning for environmental quality. Faced with the prospect of a wholesale remolding of the city to accommodate additional limited-access highways, San Francisco chose another course and voted a massive bond issue to finance construction of a modern transit system. First units of the new system began operating in 1972. Part of the reason for BART was to reduce the number of automobiles entering San Francisco and thereby reduce air pollution, but the indications are that the statistical impact on air pollutants will be trivial. Nevertheless, BART should contribute to more gracious living.

Air pollution has not been a major problem in San Francisco. Total tonnage of air pollutants in 1969 in San Francisco County was only about 8 percent of that in Los Angeles County and San Francisco is favored with much better atmospheric conditions. Total tonnage of pollutants emitted in San Francisco is expected to decline from 1969 to 1980 by 45 to 62 percent, depending on whether or not Muskie bill standards are achieved.

The air in the vicinity of San Francisco airports will be roughly twice as noisy in 1980 as it was in 1970. San Francisco/Oakland ranked fifth among the nation's air hubs in projected level of 1980 jet operations.

San Francisco faces its most immediately pressing environmental problems in the field of solid waste disposal. The decision to cease landfill operations in San Francisco Bay precipitated an immediate crisis for San Francisco and adjoining San Mateo County. Other Bay area counties are not faced with such imminent difficulties.

Seattle

Seattle was the scene of one of the most successful environmental campaigns in the 1960s when the installation of a sewer system and the diversion of effluent reversed the growing eutrophication of Lake Washington, a major recreational asset, which was threatened with excessive growth of algae. The Lake Washington program may have supplied the stimulus, but, in any event, the State of Washington has developed an advanced pollution control program. Vigorous enforcement of pollution control measures is essential because the pulp and paper industry is a potential source of enormous pollution for the waters of Puget Sound. If pollution control enforcement threatens the continued existence of any pulp and paper mills in the Seattle area, it seems likely that the responsible officials might be inhibited by a consciousness of the depressed condition of the region's economy.

Air pollution has not been a major problem in Seattle. Total tonnage of pollutants is projected to decline from 1969 to 1980 by 30 to 50 percent, depending on whether or not the Muskie bill standards are achieved.

Although solid waste disposal in Seattle appears to be characterized by a casual and unplanned approach, there does not seem to be evidence of the sort of immediately critical situation such as that facing San Francisco.

Denver

Denver's major environmental problem arises from its atmospheric conditions. It has the highest rate of inversions of any of the cities covered in this study except for Los Angeles, and the precipitation rate is low. Coal is used extensively as fuel for power generation, and this contributes to a moderately high rate of generation of sulfur oxides. Emission of nitrogen oxides is projected to increase from 1969 to 1980. However, total tonnage of pollutants should decline between 1969 and 1980 by 21 to 35 percent, depending on whether or not the Muskie bill standards are achieved.

Denver jet operations are projected to increase by 84 percent between 1969 and 1980, but will remain at a fraction of the level of intensity at the major hubs.

Solid waste disposal is not expected to pose any major problems in Denver. Water pollution questions were not covered in the study.

Salt Lake City

The total tonnage of air pollutants emitted in Salt Lake by sources covered in the Hofstra study were projected to decline between 1969 and 1980 by 39 to 61 percent depending on whether or not the Muskie bill standards are achieved. A nearby copper smelter emits 177,000 tons of sulfur oxides annually, or more than any of the five western cities studied.

No solid waste disposal difficulty is anticipated for Salt Lake City. Water pollution problems were not studied.

General Conclusions

Out of all this mass of facts, what general conclusions can we draw? To be sure, our study did not include all forms and sources of pollution, nor all cities in the country, and certainly none of the rural areas. But it did cover a reasonably representative sample of the nation's cities. Nobody else has attempted to survey as large a nationwide sample of locations using a common methodology which yields at least semi-quantitative answers.

The first general conclusion relates to the small cities. Our sample included only three "smaller" cities—Savannah, Mobile, and Salt Lake, with populations ranging from 141,000 to 255,000. No smaller cities or less densely populated areas whatsoever were included in our study. In 1970, about 40 percent of the people in the United States lived in rural areas or in populated places smaller than the smallest city covered in the Hofstra study (which happened to be Savannah, Georgia [5]). Does this mean the results of the Hofstra study tell us nothing about what happened to the environment for 40 percent of the people of the United States? This is plainly not so. In the field of solid waste disposal, for example, not one of the smaller cities covered in the study anticipated any difficulty in finding a place in which to dispose of solid waste. This was not true for the larger cities, most of which could foresee a time when land disposal sites reasonably near the city would be exhausted. The immediacy of this latter increased roughly in proportion to the size of the city, though this relationship was altered for some, such as Chicago and Cleveland, by use of incinerators. The conclusion is clear, nevertheless, that there is no widespread imminent crisis in finding a place to dispose of solid waste in U.S. SMSAs with central cities of less than 150,000 population and most smaller suburban and rural areas. Litter may be ubiquitous as nonreturnable containers and lavish packaging deface the landscape. Collection costs, which are 80 percent of all collection and disposal costs, may escalate as an increasingly affluent America disposes of more wastes, usually with more carelessness. But it may be safely concluded that the percentage of U.S. peoples living in cities with less than 150 thousand inhabitants or in smaller places will find little difficulty locating a disposal site once their solid waste is collected. This relationship might change radically for a given city if it had a few largely uncontrolled sources of industrial waste to contend with.

A somewhat similar relationship holds for those sources of air pollution covered in the Hofstra study. Considering only air pollutants related to automobiles, space heating, or power generation, the smaller cities covered all had reasonably satisfactory air in 1969 and good prospects for generally cleaner air by 1980. Again this relationship was roughly proportional to the size of the city, with air quality tending to deteriorate as city size increases.

This tidy and convenient generalization may be shattered by the presence in a small city or town of an industry that produces an extraordinarily large amount of a particular air pollutant from an industrial source. In Salt Lake City, for example, there is a single copper smelter that would produce over 581,000 tons annually of sulfur oxides if emission were *not* controlled, and actually produced 177,000 tons in 1969 with relatively satisfactory controls employed. The conclusion is clear that in the small cities and rural areas, the air quality in 1980 is likely to be decisively better than it was in 1969. It also seems likely that air quality will tend to improve as city size declines. This relationship may be effectively and dramatically altered by the presence of major industrial air polluters as in Salt Lake City.

No such precautionary qualification is necessary with respect to the relationship between size of city and noise pollution due to jet aircraft. To be sure, both Atlanta and San Francisco/Oakland had a level of jet operations out of proportion to their populations in 1969 and the relationship was not precise in any case. But, by and large, there appeared to be a rather clear and definite relation between size of city and extent of noise pollution attributable to jet air operations.

The reaction to this might well be: so what? It comes as no surprise that the larger a city, the more it tends to be cursed with air and noise pollution and difficulties with solid waste disposal. Before accepting this dismissal, it might be well to note that no one ever suggested that the relationship is a simple linear one, as witness Los Angeles, a giant city that has successfully handled solid waste disposal; New Orleans, a metropolitan area of a million people with negligible air pollution; and Atlanta, with a noise pollution from jet operations 14 percent greater and much more concentrated than that of San Francisco/Oakland with twice the population.

The disparity between size of population and extent of pollution becomes much more evident when we consider water pollution. In 1969, rivers so polluted that they were unfit for any but the lowest uses flowed past the largest (New York) and the smallest (Savannah) cities covered in the Hofstra study. Cleveland, a major city of 790,000 on Lake Erie, found many of its beaches unsafe for swimming, but then so did Green Bay, Wisconsin, a city of 97,000 on Lake Michigan. It is evidently not safe to conclude without question that the large cities bear the brunt of America's pollution problems. It is worth noting, however, that Savannah's poor water quality derives from the effluents of a handful of industrial polluters who are publicly pledged to correct the situation, and Green Bay's problems could also be dealt with by attack on a limited front. In contrast, New York's involves many sources of pollution. The problem for Cleveland may be even more intractable because it relates to the excessive growth of vegetation in Lake Erie, which is in turn related to a long-term buildup in the concentration of plant nutrients.

Nevertheless, it does appear that improvement in water quality will generally be greater in the vicinity of small cities than of large. What can be said more positively is that in regard to water pollution, industrial sources exert a decisive influence for any specific locality. There will be some sharp contests in dealing with industrial water polluters between now and 1980.

One method used by giant cities which pack heavy political and economic muscle is to turn its problems over to others, to export them. There are a number of examples: (1) Chicago exports its sewage south to the Mississippi River and enjoys pollution-free swimming in Lake Michigan; (2) New York disdains the waters of the Hudson River as a source of drinking water since it pollutes them and draws instead on the Delaware River at the expense of Philadelphia in drought periods; (3) Los Angeles sets such strict limits on pollution from power plants within its city limits that its electric utilities join combines free to befoul the desert air hundreds of miles away; (4) Los Angeles also draws on waters from the Colorado River in competition with Arizona as well as distant California sources and then discards such waters although recycling for some purposes is plainly feasible.

Such examples seem sporadic no matter how individually persuasive. Overall, America's smaller communities stand to gain more with respect to quality of life in the next decade than the larger cities do. The Hofstra study covered with some degree of thoroughness only those situations summarized above for individual cities. Our study did not cover urban transportation, which appears to be especially crucial for the environment in larger cities. There is good reason to believe that the Interstate Highway System, which will be substantially completed during the 1970s, has effected a considerable net improvement in the ease and convenience of transportation in and between America's smaller cities and rural areas. Simultaneously, the development of the Interstate Highway System has tended to unload an increasing volume of rubber-tired traffic on the already overloaded arteries of the larger cities, sometimes to a point approaching strangulation. Still, any suggestion that some part of the gasoline taxes collected from this traffic be devoted to investment in mass transit facilities to ease this large-city congestion is met with instant rebuff by the guardians of the Highway Trust Fund, the carefully husbanded depository for gasoline taxes, whose resources may be employed solely for highway construction. Thus, the outlook for the big cities appears to favor increased air pollution and traffic congestion.

Nor is there any solid expectation of major assistance in solid waste disposal for the large centers. This is not an especially troublesome problem for most smaller communities, though it tends to command increasing attention.

Water pollution control is a sector in which the political economy of environmental measures may not be entirely stacked against the large city. Small town and suburban members of Congress, and more importantly members of key congressional committees, may be prepared to support assistance to large cities in water pollution control measures as a way of cleaning up waterways in their own vicinity. No such compelling arguments are likely to be available in behalf of air pollution control or solid waste disposal measures.

Meanwhile, a portion of the pollutants generated in cities large and small finds its way into the ultimate sinks, the upper atmosphere and the oceans. Since mankind does not fully comprehend the effect of recent developments on the environment of the continents it inhabits, it is not surprising that many of the effects of pollutants on the seas and on the upper atmosphere are only beginning to be explored. Some of the possibilities are sufficiently menacing to encourage more extensive worldwide monitoring to detect any harmful effects before they are out of hand.

Returning to earth, and on balance, the outlook is for improvement by 1980 in many, perhaps most, aspects of the environment in America in the small cities and rural areas. In the larger cities, improvement by 1980 in some aspects of the environment seems certain, but the total impact is less clear, due especially to reliance on modes of transportation that may be partially obsolete.

Notes

1. Roy Craig, "Cloud on the Desert," *Environment* 13, 6 (July/August, 1971), pp. 21-24, 29-35.

2. Robert Lindsey, "Jet Noise in Los Angeles Is Dooming 1,994 Homes," *New York Times,* July 21, 1971, p. 1.

3. Robert Lindsey, "California Jetports Facing Dual Problems with Noise," *New York Times,* July 23, 1972, Section S, p. 24.

4. Alfred J. Van Tassel (Ed.), *Environmental Side Effects of Rising Industrial Output* (Lexington, Mass.: D.C. Heath and Company, 1970), Tables 8.25, 8.26.

5. Standard Rate and Data Service *Newspapers* (Skokie, Ill.: October 22, 1972) Metro Area Population.

Glossary

AGF	– average growth factor
AOCI	– Airport Operators Council International
BOD	– biological oxygen demand
BTU	– British thermal unit
DO	– dissolved oxygen
EBI	– effective buying income
EPA	– Environmental Protection Agency
EPNL	– effective perceived noise level (see Chap. 20)
gpm or GPM	– gallons per minute
kwh or KWH	– kilowatt hour(s)
LOT	– load on top (see Chap. 14)
mgd or MGD	– million gallons per day
ml	– milliliter
MPN	– most probable number
NEF	– noise exposure forecast (see Chap. 20)
OBW	– oversize burnable waste
pH	– a measure of acidity
ppb	– parts per billion
ppm	– parts per million
ppt	– parts per trillion
SIC	– Standard Industrial Classification
SMSA	– Standard Metropolitan Statistical Area
SWL	– sulfite waste liquor
VFR	– visual flight rules (see Chap. 20)

Index

Index

Abandoned vehicles, 457
Abrahamson, Dean E., 70n
Absorbed radiation, 397
ABS (Akyl Benzene Sulfurate), 217, 218
Acid formation, 412
Acid wastes, 198
Ackerman, Joseph, 446n, 447n
Ackerman, Joseph; Clawson, Morrison; Harris, Marshall, 446n, 447n
Ackley, C. *see* Magill, P.L.
Adams County, Colorado, 524, 533
Adirondack region, 149, 160
Aerosols, 397, 398, 450
Agricultural wastes, 103, 104
Airport Operators Council International, 430, 430n, 433, 436, 438, 449n
Air transportation, 417, 451
Akron, Ohio, 56, 64
Alameda, Calif., 101, 105, 108, 116, 285
Alameda County, Calif., 515, 516, 518, 530, 552
Albany County, N.Y., 153, 154, 158, 163
Albany, N.Y., 152, 157, 286
Albedo or reflective power, 400
Alcohol sulfates, 218
Aldehydes, 304, 341, 355
Alewives, 47, 49
Algae, 44, 56, 60, 96, 110, 139, 273, 276, 558, 567
Allegheny County, Pa., 319, 475, 492, 493
Altman, Lawrence, K., 274n
Altschul, Selig, 418n
American Can Co., 46
American Cyanamid Co., 195, 197, 198, 199, 200, 206
American Gas Association, 306, 306n, 307, 343, 344, 345
American Petroleum Institute, 264, 306, 346
American Public Works Assoc., 457, 460, 462, 499n
Anacortes, Washington, 76, 86, 87
Anaheim, Calif., 238
Anderson, Homer, 433
Antarctic continent, 271
Antioch, Calif., 95, 98, 100
Appalachian Mountains, 175, 316
Apparel, 158, 160
Arapahoe County, Colorado, 524, 531, 533, 553
Arizona, 244, 253, 292, 566, 570
Arnold, C.E., 246
Artesian well, 237
Ash, 461

Ashkinaze, Carole, 549n, 553
Ashtabula River, 64
Asphyxiation, 272
Assimilative capacity, 284
Astoria, Queens, N.Y., 314
Atlanta, Ga., 305, 321, 322, 323, 324, 325, 331, 332, 344, 424, 449, 475, 477, 480, 489, 490, 491, 492, 550, 560, 570
Atlantic Coast, 28
Atlantic Ocean, 127, 149, 150, 187, 199, 211, 229
Atmospheric particulate loading, 398, 568
Atmospheric Pollutants, 401
Atomic fuels, 12
Atomic oxygen, 412
Automobile emission coefficients, 299
Automobile emissions, 301, 302, 303, 312, 313, 314, 316, 317, 320, 323, 324, 325, 336, 337, 346, 349, 352, 357, 385, 401, 405, 412, 445, 560
Automobile Manufacturers Assoc., 303n, 313, 317, 320, 322, 325, 337n, 346, 349, 352, 354, 357

Babylon, N.Y., 218, 228
Back River, 202
Bacterial concentrations, 60
Bacterial growth, in water pumped from injection well, 227
Baghouse, 409
Baier, Joseph H., 217n
Bailly, Indiana, 35
Baking, 305
Baldwin County, Ala., 502, 504, 505
Bank of America, 96
Barnett, H.J., 2
Barrow, T.D., 263n
Baxley, Ga., 324
Bay Area Air Pollution Control District, 515
Bay Area Rapid Transit System (BART), 567
Bay City, Mich., 47
Bay Park, N.Y., 226, 227, 230
Bay Park Treatment Plant, 229
Beaver County, Pa., 319
Beckman, Wallace J., 230n
Beer and ale, 459
Bellingham Bay, 77, 86, 87
Bellingham, Wash., 76, 78, 86, 87, 88
Bennett, G.D., 227n
Bergen County, N.J., 152, 153, 154, 168
Bering Sea, 274
Berlin, Germany, 460
Besselievre, E.B., 161n

577

Bethlehem Steel Co., 65
Beverly Hills, Calif., 238
Big Rock nuclear plant, 48
Biodegrade, 27, 219
Biostimulation, 96, 97, 110, 115
Birnbaum, B., 257n, 267, 268, 273n, 274n
Bishop, Dwight E., 427, see also Galloway, William J.
Bjorkman, Aki, V., 466n
Black, R.J.; Muhich, A.J.; Klee, A.J.; Hickman, H.L., Jr.; and Vaughan, R.D., 460n
Black River, 64
Blanchard River, 64
Block Island Sound, 128, 129
Boeing Co., 76, 88
Boating discharges, 131
BOD (biological oxygen demand), 11, 18, 27, 28, 36, 38, 41, 42, 44, 45, 84, 91, 112, 178
  of industrial discharges, 30, 31
  of municipal sewage, 32, 39, 40
Bottles, 459
Boulder County, Colorado, 524, 531, 533
Bolso Island, 243, 244
Boston, Mass., 335, 433
Boulding, Kenneth, 391
Bower, Blair T., 18
Boyle, Robert H., 163n
Brine, 226
Bronchitis, 447
Bronx, N.Y., 152, 154, 168
Brookhaven, N.Y., 218
Brooklyn Navy Yard, 484
Brooklyn, N.Y., 152, 212, 214
Brown, H., 391n, 402n, 408n
Brown, Secor D., 439
Brown and Caldwell Civil and Chemical Engineers, 76n
Bruce, Ronald, 417, 451
Bryson, Dr. Reid, 400
BTUs (British Thermal Units), 33
Bucks County, Pa., 315
Budyko, M.I., 398
Buffalo, N.Y., 58, 60, 293
Buffalo River, 58, 65
Bulky wastes, 457
Burbank, Calif., 238
Burlington County, N.J., 315
Burns, John A., 455, 509, 552
Burns, William, 425n, 426n
By-product, 11, 107, 449

Cairo, Ill., 4, 5n
California, 95, 231, 244, 274
  northern, 292, 566
  southern, 227, 288, 566
California Aqueduct, 244, 566
California Dept. of Public Health, 523n, 545

California Institute of Technology, 248
California Water Project, 249
California Water Resources Control Board, 95n, 99, 100, 102, 110n, 111, 112, 116, 117, 118
Calumet Conference, 35
Calumet, Indiana, 22, 32, 38, 39, 42, 43, 48
Canned fruits and vegetables, 30, 31, 39, 40, 44, 74, 78, 79, 104, 105, 120, 162
Cans, nonreturnable, production of, 459
Cape Girardeau, Missouri, 176
Caponi, Robert, 127, 285
Carbon absorption, 231
Carbon-based fuels, 393
Carbon Dioxide, 395, 396, 410, 450
  runaway, 411
Carbon monoxide, 298, 300, 301, 304, 306, 313, 314, 316, 317, 318, 320, 321, 322, 324, 325, 327, 328, 329, 330, 331, 332, 333, 336, 341, 346, 347, 348, 349, 350, 352, 355, 357, 358, 368, 369, 370, 373, 374, 375, 377, 378, 379, 381, 382, 383, 384, 385, 386, 387, 412, 445, 446, 559, 561, 562
Carmichael, James, 237, 288
Carr and Hiltunen, 67
Carr, James, 433
Carry, C.W., 249n
Carter, Sylvia, 285n
Cascade mountain range, 75, 376
Catalytic converters, 312
Catskill Mountains, 149
Cattaraugus Creek, 65
Carlson, Frank, 305, 335, 450
Center for Study of Responsive Law, 287
Central Valley, Calif., 95
Certain-Teed Prod. Corp., 194, 195, 196, 200, 206
Cesspools, 211
Chagrin River, 54
Changnon, S.A., 413
Charlevoix, Michigan, 48
Charmin Paper Products Co., 46
Chatham City, Ga., 188
Chatham County, Ga., 187, 197, 203, 324
Chatham County Health Dept., 202
Chemical and allied products, 27, 29, 30, 44, 60, 96, 160, 184, 225, 564
Chemical Economics Handbook, 267n
Chester County, Pa., 315
Cheswick Power Station, 320
Chicago, Ill., 27, 42, 283, 292, 344, 345, 346, 359, 360, 420, 424, 446, 449, 450, 463, 499, 500, 502, 551, 562, 563, 564, 565, 569, 570
Chicago River, 283, 292, 293, 562
Chipps Island, 95
Chlorination, 34

Chub, 28
Clarifier, 197
Clark Hill Dam, 202
Clawson, M.; Held, R.B.; and Stoddard, C.H., 477n, 490n
Clawson, Marion, *see* Ackerman, Joseph
Clayton County, Ga., 321
Clean Air Act of 1970, 300, 445, 446
Cleveland, 56, 60, 64, 292, 293, 344, 345, 351, 359, 360, 424, 449, 450, 499, 502, 504, 551, 562, 563, 569, 570
Cleveland Harbor, 56
Cleveland Waterworks, 56
Climatological effects, 400
Clinton River, 54, 64
Cloverdale, Ga., 188
Cloud formation, 400
"Club of Rome," 4
Coagulation, 231
Coal, 395, 404
Coal, gas and oil, 303, 339, 384
Cobb County, Ga., 321
Cohen, Phillip, 211n, 212n, 213n, 215n, 217n, 220n
Coho salmon, 29, 49, 163
Cold trap, 400
Colfax Power Station, 320
Coliform bacteria, 14, 36, 42, 44, 77, 82, 96, 109, 110, 115, 190, 191
Colorado, 566
Colorado Dept. of Health, 522, 523n, 545
Colorado River, 237, 238, 244, 245, 253, 292, 570
Colorado River Aqueduct, 237, 243, 288
Colorado Solid Waste Disposal Act of 1967, 523
Columbia County, N.Y., 152, 153, 166
Combustion Engineering Co., 474, 475, 478, 483n, 486n, 490n, 494n
Commencement Bay, 77, 89
Commercial fishing, 204, 205
Commoner, Barry, 1n, 51n, 56, 59, 284, 293, 293n
Commonwealth Edison Co., 347
Composting plant, 505
Compton, Calif., 238
Conference Board, Inc., 510, 511
Connecticut, 127, 129, 131, 135, 143, 144, 145, 149, 224
Connolly, Hugh H., 467n
Construction, 241, 457
Consumers Power Company, 35
Consumption wastes, 473, 474
Continental Can Co., 194, 195, 196, 200, 206
Continental shelves, 258
Contra Costa County, Calif., 105, 107, 108, 116, 285, 515, 516, 518, 530, 552

Contrails, 400
Conway, N.M., 329n, 331n
Cook County, Ill., 500, 502
Coosa River, 282
Cooling effect, 411
Cooling ponds, 35
Cooling towers, 35
Cooling water, 31
Copper, 380, 448, 568, 569
Corino, Edward R., 259, 261, 262n, 263n
Counts, Harlan B., *see* McCollum, M.J.
Coyote Creek, 95
Craig, Roy, 566n, 571
Cuyahoga County, Ohio, 501, 502, 504
Cuyahoga River, 56, 64, 66, 500, 501

Dales, J.H., 60n
Darnay, Arsen J., Jr., 528n
Darney and Franklin, 468, 468n
Davis County, Utah, 525, 526, 534
DDT, 267, 271, 275, 276, 277, 278
Dead animals, 457
Deem, Warren H. and Reed, John L., 434n, 435n
DeKalb County, Ga., 321
Delaware County, Pa., 315, 488
Delaware River, 211, 224, 292, 316, 560, 561
Denver, Colorado, 335, 344, 381, 384, 385, 386, 387, 424, 463, 509, 522, 523, 524, 525, 529, 531, 533, 535, 541, 553, 568
Denver County, Colorado, 524, 531, 533
Denver Regional Council of Government (DRCOG), 523
Desalination of Water, 224, 225, 231, 253
  Projected costs, 225
  Standards for drinking, 216, 227
Detroit, Mich., 47, 60, 66, 283, 359, 445
Detroit River, 47, 52, 54, 64
Dilution, 11, 179, 287, 291
Dissolved metals, 34
Dissolved oxygen, 39, 42, 109, 135, 136, 203, 281
Distillation, 225
Distilled liquors, 104, 107, 121, 162
Donner-Hanna Coke, 65
Drainage basins, 20, 21, 75
Drinking water, 9, 51, 175, 176, 179, 293, 560
Duke, T.W., 275n
DuPage County, Ill., 500, 502
Dutchess County, N.Y., 149, 152, 154, 159, 166
Duquesne Light Co., 320, 329
Duwanirsh River, 77

Earth surface temperature, 398
East Chicago, Ind., 41

East Hampton, N.Y., 223
East River, 145, 160, 558
Eckenfelder, W.W., 18
Ecological imbalances, 220
Egbert, Col. John S., 202n
Eisenbud, M., *see* Howells, G.P.
Electric power generating plants, 12, 25, 31, 35, 70, 131, 137, 303, 316, 320, 323, 326, 338, 351, 381, 401, 405, 569
Electrical and nonelectrical machinery, 158
Electrodialysis, 225
Electrostatic precipitators, 314, 329, 409, 465
Elliott Bay, 77
English, J.N., *see* Parkhurst, J.D.
Erie, Pa., 56, 58, 65
Essex County, N.J., 153, 154, 168
Essex County, N.Y., 152, 153, 163
Esso Research and Engineering Co., 261
Eutrophication, 12, 36, 47, 59, 66, 139, 150, 171, 179, 278, 289, 291, 292, 294
Evapotranspiration, 215
Everett Harbor, 77, 88
Everett, Wash., 76, 78, 552

Fairport, Ohio, 64
Feather River, 249
Feather River Project, 244
Federal Water Pollution Act of 1965, 278
Federal Water Quality Improvement Act of 1970, 71
Feed-lot runoff, 28
Fennimore, George, 450
Feth, J.H., 150n
Finch, Robert H., 467
Fisher, J.L., and Potter, N., 391n
Flint, Mich., 47
Flood, Francis J.J., 217
Flood-control methods, 176
Florida, 270
Fluid milk, 30, 31, 39, 40, 74, 78, 79, 104, 120, 162
Flushing action, 76, 150
Food and kindred products, 30, 96, 160, 164, 166, 182
Food Chain, 275, 276
Ford, Henry, II, 302, 328
Ford Motor Co., 302
Fort Jackson, Ga., 202
Fort Lowry, Colorado, 553
Fossil fuel demand, 394
Fossil fuel for electricity generation, 33, 35, 402
Four Corners Plant, 566
Four-State Enforcement Conference, 34, 35, 37, 48
Fox River, 27, 36, 45

Foxworthy, B.L., 211n, 212n, 213n, 215n, 217n, 220n
Franke, O.L., 211n, 212n, 213n, 215n, 217n, 220n
Franklin County, Mo., 501
French, Rogers, 187, 287
Frozen fruits and vegetables, 74, 78, 79, 89, 104, 107, 120, 162
Fuel consumption, 299, 304, 368, 372, 377, 381, 384
Fuel oil, 404
Fullerton, Calif., 238
Fulton County, Ga., 321, 489, 490, 491, 492
Fulton County, N.Y., 152, 153, 159, 165, 475

GAF (formerly Ruberoid), 195, 196, 200, 206
Galloway, William J., and Bishop, Dwight E., 426n, 428n
Garden City, N.Y., 187, 188, 207
Gary-Hammond-East Chicago, Ind., 27
Gary, Ind., 41, 347
Gas consumed for space heating, 386
Gas turbine engine, 312
Gasoline consumption, 405, 406,
Geauga County, Ohio, 500
Georgia, 187, 190
Georgia-Pacific Co., 86, 87, 88, 92
Georgia Power Co., 323, 331
Georgia Health Dept., 190
Georgia Water Quality Control Board, 195, 198, 199, 202, 203, 209
Geraghty, J.J., 213n
Geyer, John C., 34n
Glacial aquifer, 215, 223
Glendale, Calif., 238, 251
Gloucester County, N.J., 315
Goffin, Peter L., 311, 450
Golegar, W.C., *see* Poston, N.W.
Goodrich, W.F., 461
Gordon, Fair M., 34n
Gould, Dr. Jay, 180
Governor's Island, 480
Graham, Frank, Jr., 72
Grand River, 27, 64
Grand River Basin, 32, 39, 40, 46
Grand River, Mich., 36
Gray iron founderies, 181
Great Britain, 263, 270
Great Lakes-Illinois River Basin Project, 28
Green Bay, Wis., 27, 36, 38, 45, 47, 282, 289, 570
Green Bay Metropolitan Sewage District, 32, 39, 46
Greene County, N.Y., 152, 154, 166

Greenhouse effect, 400, 410
Gross, M. Grant, 285
Groundwater, 99, 211, 242, 249
Gruninger, Robert M., *see* Westerhoff, Garrett P.
Guemes Channel, 77, 86, 87
Gurnham, C.F., 18
Guyol, N.B., 401n, 404
Gwinnett County, Ga., 321

Handbook of Analytical Toxicology, 277n
Hansen, David J., 221
Harbor dredgings, 28, 47
Hardenbergh, W.A., 14n
Harris, Marshall, *see* Ackerman, Joseph
Hartwell Dam, 187
Harwood, Michael, 481
Haze, 412
Heat exchange medium, 9
Heath, R.C.; Foxworth, B.L.; and Cohen, Phillip, 212n
Heating degree days, 344
Heavy metals, 38, 44
Hedger, H.E., 246
Held, R.B., *see* Clawson, M.
Heller, Austin, 312
Hempstead Harbor, 138, 229
Hempstead, N.Y., 223
Henzey, William V., 418n
Hercules, Inc., 194, 195, 196, 200, 206
Herkimer County, N.Y., 152, 153, 165
Hickman, H. Lanier, Jr., 462 *See also* Black, R.J.
Highway Trust Fund, 571
Hofstra, 1, 22, 298, 482, 487, 491, 494, 503, 504, 513, 514, 517, 519, 521, 522, 524, 525, 526, 527, 530, 532, 533, 534, 557, 558, 559, 568, 569, 570, 571
Hofstra, School of Business, 1
Hofstra study, 561, 562, 564, 569
Hofstra Yearbooks of Business, 1
Holden, F.R., *see* Magill, P.L.
Holding ponds, 197
Holzmacher, McLendon, and Murrell, Consulting Engineers, 212n, 217n, 218n, 223, 224n, 225n, 228, 231n, 232n
Hoover Dam, 238
Hoover, Isaac H., and Cochran, Donald G., 435n
Howard, R.S., 193, 200
Howells, G.P.; Eisenbud, M.; and Kneip, T.J., 151n
Hubbert, King M., 393
Hudson Corridor, 152, 153, 166
Hudson County, N.J., 152, 154, 168
Hudson River, 3, 149, 150, 153, 162, 211, 281, 286, 289, 290, 291, 292, 293, 294, 557, 558

Upper, 153, 154, 159, 161, 163, 164 171
Mid, 152, 153, 154, 159, 161, 162, 166, 171, 172
Lower, 152, 153, 154, 156, 159, 160, 161, 163, 167, 168, 169, 171
Hudson River Valley Commission, 150n
Hudson Canyon, 149
Hudson Valley, 149, 152, 156, 286
Humidification, 225
Hunt-Wesson Foods, Inc., 195, 196, 200, 206
Husson, Frank, Jr., 257, 290, 294
Hydrocarbons, 266, 298, 300, 301, 304, 306, 313, 314, 316, 317, 318, 320, 321, 322, 324, 325, 327, 328, 330, 331, 332, 333, 336, 341, 346, 347, 348, 349, 350, 352, 354, 355, 357, 358, 368, 369, 370, 373, 374, 375, 377, 379, 381, 382, 384, 385, 386, 387, 412, 445, 446, 559, 562, 563
Hydrogen ion concentration, 109, 110
Hydrogeological setting, 213
Hydrologic conditions, 43
Hydrological imbalance, 212, 223, 228

Illinois, 21, 33, 36, 42
Illinois River, 27, 38, 43, 47
Illinois Sanitary Water Board, 43n
Imperial County, Calif., 241
Importing fresh water, 224, 231, 243, 253
Incineration, 464, 477, 478, 484, 490, 494, 500, 501, 505, 548, 562, 569
Indiana, 21, 27, 31, 33, 36, 42, 47, 54
Indian Point, N.Y., 151
Indra, R. Raja, 40
Industrial applications of reclaimed water, 249
Industrial processes, as contributors of pollutants to upper atmosphere, 401, 407
Industrial wastes, 11, 25, 34, 62, 64, 65, 66, 83, 99, 103, 104, 117, 118, 129, 135, 144, 160, 164, 166, 176, 194, 196, 200, 206, 239, 259, 265, 282, 283, 286, 289, 547
Ingram, Dr. Hollis S., 170
Inland Steel Co., 41
Inorganic chemicals, 181, 183
Internal-combustion engines, 339
International Joint Commission, 70n, 71
Intrusion of salt water, 226
Interstate Highway System, 571
Islip, N.Y., 228
Inversion conditions, 319, 565, 568
Iron and manganese, 217
Irrigation systems, 175, 249

Jaffie, Howard W., 499

Jamaica Bay, 558
Japan, 263, 409
Jasper County, S.C., 187
Jefferson County, Colorado, 523, 524, 533
Jefferson Parish, La., 501
Jensen, Larry N., 531, 531n
Jet-powered aircraft, 423
Johnels, Dr. Alf, 273
Junge, Christian E., 398n, 399n, 409n, 412n
Junge dust cloud, 399

Kalamazoo County, Mich., 27
Kalamazoo River, 27, 46
Kalamazoo River Basin, 32, 39, 40, 46
Katz, Carl A., 391, 450
Kennedy, F.S., *see* J.M. Wood
Kenosha County, Wis., 44
Keowee River, 187
Kern County, Calif., 241
King County, Wash., 76, 77, 79, 80, 81, 88, 89, 154, 284, 288, 376, 520, 521, 531, 532, 553
Kings County (Brooklyn), N.Y., 168, 211, 212, 213, 215, 217, 220, 222, 230
Kingston, New York, 150
Klee, A.J., *see* Black, R.J.
Kneese, Allen V., 18
Kneip, T.J., *see* Howells, G.P.
Kohn, R.E., 2
Kraft mill, 196, 197, 201
Kretchmer, Jerome, 558

Lake Erie, 3, 28, 35, 47, 51, 54, 59, 60, 62, 64, 65, 66, 68, 70, 72, 283, 289, 290, 291, 292, 293, 562, 563, 570
Lake Erie Basin, 60, 61, 62
Lake Erie, polluted beach areas, 69
Lake herring, 28
Lake Michigan Inter-League Group, 33
Lake Huron, 28, 54
Lake Itasca, 175
Lake Michigan watershed, 25, 30, 31, 39, 40
Lake Michigan basin, 26, 32, 39, 40, 43, 48
Lake St. Clair, 47, 54, 64
Lake Superior, 28
Lake Tahoe, 113
Lake Tear-in-the-Clouds, 149
Lake Washington, 80, 81, 284, 292, 567
Lake Washington Ship Canal, 77, 78
Lake Winnebago, 36
Lambie, John A., 544
Land and Right of Way, 99, 117, 118
Land disposal sites, 502, 514, 522, 525, 527
Landfill, 285, 478, 480, 482, 494
Landsberg, Hans H., 2n, 306, 306n, 342
Landsberg, Hans H.; Fishman, Leonard L.; Fisher, Joseph L., 422n

Lang, Henry, 211, 288
Lamp, The, 264n, 272n
LaSalle, 175
Leaching, 217
Lead dioxide, 413
League of Woman Voters, 33
Leather tanning, 27, 30, 31, 39, 40, 44
Legislative Reference Service of the Library of Congress, 412n
Life Chain, 294
Lignite, 395, 404
Lindsey, Robert, 566n, 572
Liquid hydrocarbon, 395
Little, Arthur D., Inc., 265
Little Neck Bay, 229
Little Ogeechee Rivers, 188
Livermore County, Calif., 101
Lloyd's Register of Shipping Casualty Reports, 259n, 260, 263n
Lloyd sand, 216
London, England, 433, 460
Long Beach, N.Y., 216, 237, 238
Long Island groundwater pollution study, 217n, 219n, 221n, 222, 223
Long Island, N.Y., 127, 137, 211, 214, 215, 274, 288, 558
Long Island Lighting Co., 314
Long Island Sound, 3, 160, 285, 289, 290, 291, 292, 558
Long Island State Park Commission, 212
Long Island's underground fresh water supply, 215
Los Angeles Aqueduct, 237, 243, 288
Los Angeles Area Chamber of Commerce, 242, 243, 252
Los Angeles Bureau of Engineering, 251
Los Angeles, Calif., 237, 240, 241, 242, 243, 254, 292, 298, 305, 335, 344, 367, 368, 369, 370, 371, 374, 378, 420, 424, 446, 463, 509, 512, 528, 529, 530, 535, 537, 551, 552, 560, 561, 565, 566, 567, 570
Los Angeles County, Calif., 237, 238, 239, 246, 512, 529, 552, 566
Los Angeles County flood control, 246
Los Angeles County Regional Planning Commission, 512
Los Angeles County Sanitation Districts, 512n, 545, 566
Los Angeles Int. Airport, 432
LOT (Load on Top), 261, 262, 277
Louisiana, 263
Ludwig, J.H., 398

McCollum, M.J., and Counts, Harland B., 201n
McCormick, R.A., 398

McDonough, James J., 506
McGuinnes, C.L., 201n
McKee, Jack E., 245
Madison County, Ill., 501
Magill, P.L.; Holden, F.R.; and Ackley, C., 408n
Magothy and Jameco aquifers, 211, 219, 223
Magothy stratum, 219
Mallard, 274
Malthus, 2
Mamaroneck, N.Y., 141
Manobe, S., et al., 400, 400n, 401
Manhasset Bay, 138, 229, 285
Manhattan, N.Y., 170, 480
Manistee River Basin, 32, 39, 40
Marin County, Calif., 102, 515, 516, 518, 530, 552
Marine Sciences Research Center, 139, 285
Marshall, Alfred, 561
Martin, Richard, 149, 286, 291
Massachusetts, 149, 157, 265
Masse, A.N., *see* Parkhurst, J.D.
Matern, Raymond, 9, 95, 285, 291
Mathews, Lake, 238
Maumee River Basin, 51, 52, 54, 64, 65
Memphis, Tenn., 176, 287
Marshes, 201
Mayer, Lawrence A., 422n
Meat processing plants, 30, 31, 39, 40, 74, 78, 79, 89, 104, 119, 162, 181, 183
Meddin Packing Co., 194, 195, 200, 206
Medina County, Ohio, 500
Melting of the Antarctic Ice Cap, 411
Melt Zit, 466
Menominee River, 45
Mercury pollution, 266, 267
Metallic mercury, 273, 274
Methane, 447
Meteorological conditions, 409
Methano–basterium omelanskii, 273, 276
Methylene blue active substances, 218
Methyl mercury, 274, 276, 277, 294
Metropolitan Water District of Southern California, 237, 238n, 240, 242n, 245n, 247, 248, 288
Gulf of Mexico, 4, 175
Miami, Fla., 424, 433
Michaels, Abraham, 464n, 500n
Michigan, 21, 27, 31, 33, 54
Michigan, Lake, 3, 21, 22, 25, 27, 28, 29, 31, 33, 34, 35, 36, 37, 41, 42, 45, 46, 49, 282, 283, 289, 290, 291, 292, 570
  See also Lake Michigan Basin
Middlesex County, N.J., 152, 153, 154, 168
Milk River, 64
Milpitas, Calif., 98, 101
Milwaukee County, Wis., 44

Milwaukee Harbor, 283
Milwaukee River, 36, 44
Milwaukee, Wis., 27, 31, 32, 36, 38, 39, 44, 283, 289
Millwork plants, 104, 122
Mims, Lambert C., 505n, 506
Minamata Bay, 274
Minamata disease, 257, 266
Mining of water, 222, 230
Minnesota, 175
Mississippi, 292, 293
Mississippi River, 3, 175, 287, 564, 570
  Average water discharge, 184
Mississippi River Valley, 176
Mitchell, Murray J., 410
Mitsubishi Heavy Industries, 263
Mobile, Ala., 344, 345, 359, 360, 499, 502, 504, 505, 551, 564, 569
Mobile County, Ala., 502, 504, 505
Mobil Oil Corp., 65
Modena Creek, 159
Mohawk River, 149, 150, 156, 159, 163
Mohawk Valley, 150, 152, 153, 158, 159, 161, 163, 164, 171
Monarch Tanning Co., 65
Monitoring program, 89, 171, 202, 279
Monongahela River, 319
Montgomery County, N.Y., 152, 153, 165
Montgomery County, Pa., 315
Morgan, G.B., 398
Morris County, N.J., 152, 153, 154, 168
Mossbunker, 139
Mt. Marcy, 149
Mountain View, Calif., 515, 529
Muhich, Anton J., 535 See also Black, R.J.
Municipal sewage, 10, 25, 28, 29, 46, 57, 61, 99, 115, 160, 289, 290
Municipal wastes, 63, 103, 104, 130, 132, 137, 176, 193
Muskegon River basin, 32, 39, 40, 46, 47
Muskie Bill, 560, 563, 565, 567, 568
Muskie, Senator Edmund S., 312, 445, 446

Nader, Ralph, et al., 195, 196n, 201n, 204n, 206
Napeaque, N.Y., 223
Nassau County, N.Y., 131, 138, 142, 211, 212, 213, 215, 217, 218, 219, 220, 221, 222, 223, 224, 226, 227, 228, 229, 230, 231, 233, 288, 312, 549, 558
Nassau County Dept. of Health, 218, 219
Nassau County Dept. of Public Works, 226, 227, 229n
Nassau-Suffolk Regional Planning Board, 220
National Coal Association, 303, 314, 318, 321, 324, 327, 340, 347, 352, 355, 357
National data network, 512

National Envrionmental Policy Act, 432, 433
National Petroleum News, Annual Factbook Issue, 307
National Survey of Community Solid Waste Practices, 460, 462, 478, 483n, 486n, 500
Natural gas, 314, 315, 343, 395, 404, 446, 449, 563, 564
Nethemoglobinemia, 218
Netherlands, 466
Nevada, 244
Newark Bay, 168
Newark, N.J., 420, 424
Newell, R.E., 398n, 399n, 400, 411n, 412n
New Haven, Conn., 129
New Jersey, 143, 149, 152, 168, 172, 273, 286, 488
New Jersey Dept. of Conservation and Economic Development, 152, 154
New Jersey Meadowlands, 481, 484
New London Harbor, 131, 135
New Mexico, 566
New Orleans, 4, 5n, 176, 179, 185, 287, 344, 345, 355, 359, 360, 424, 499, 501, 502, 504, 505, 551, 564, 570
New York, 54, 129, 143, 145, 314, 316, 321, 324, 328, 331, 335, 344, 420, 424, 433, 446, 448, 449, 460, 570
New York basin, 65
New York City, 127, 145, 152, 154, 159, 160, 168, 172, 211, 213, 214, 224, 228, 229, 231, 286, 288, 292, 293, 312, 323, 476, 479, 480, 482, 483, 484, 486, 487, 488, 549, 550, 557, 558, 559, 560, 562
New York Harbor, 150, 151, 160, 167, 168, 169, 172, 290, 558
Narrows, 160
New York metropolitan area, 129, 151, 166
New York State, 10, 149, 151, 154, 157, 158, 159, 168, 170, 211, 290
  Conservation Dept., 213n, 215n, 220, 221, 22n, 222, 223, 224n, 225n, 227n
  Dept. of Commerce, 157n, 158n
  Dept. of Environmental Conservation, 142n, 151n
  Dept. of Health, 140n, 143n, 144n, 170, 229
  Division of the Budget, 152, 156, 159n
  Legislature, 170
  Office of Planning Coordination, 152, 152n, 154, 220
  Pure Waters program, 143, 170
  Water Resources Commission, 213, 216, 223n
Niddrie, Donald, 9, 75, 284
Northern Public Service Co., 35

Nitrates, 151, 218, 288
Nooksack River, 77
North America, 149
North River project, 170, 172
Nitrogen, 31, 97, 111, 217
Nitrogen oxides, 298, 300, 301, 302, 304, 306, 313, 314, 316, 317, 318, 320, 321, 322, 324, 325, 327, 328, 329, 330, 331, 332, 333, 336, 341, 346, 347, 348, 349, 350, 352, 354, 355, 357, 358, 367, 368, 369, 370, 373, 374, 375, 377, 378, 379, 381, 382, 383, 384, 385, 386, 387, 412, 446, 448, 559, 560, 561, 562, 563, 565
Noise
  abatement, 451
  intensities, 426
  pollution, 559, 560
Nuclear power plants, 33, 35, 315, 323, 339, 369
Nutrients, 31, 48, 151, 558

Oakland, Calif., 98, 424, 538
OBW (oversized burnable wastes), 459
Ocean outfalls, 230, 251
O'Connor, William F., 140
Oconto River basin, 45
Odor, 42, 44
Off-shore drilling, 259
Off-shore wells, 263
Oglethorpe, James, 188, 204
O'Hare Airport, Chicago, 562, 566
Ohio, 54
Oil and grease, 38, 39, 42, 96
Oil refineries, *see* Petroleum refineries
Oil spills, 28
Olefins, 412
Olsson, M., 273
Olympic Mountains, 75, 376
Oneida County, N.Y., 152, 153, 158, 159, 165
Ontario, Canada, 54
Ontario, Lake, 28
Orange County, Calif., 154, 238, 239, 240, 241, 242, 243
Orange County, N.Y., 152, 159, 166
Orleans, Parish, La., 501, 502, 504, 505
Oroville Dam, 244
Ottawa River, 64
Organic Chemicals, 181, 183, 304, 341
Output per employee, 21
Owens Valley, 237
Oxygen, consuming potential of wastes, 27
Ozone depletion, 400
Ozone oxygen, 412

Pacific Northwest, 76
Painsville, Ohio, 64

Palos Verde, Calif., 513
Panofsky, H., 409n
Pulp and Paper Industry, 27, 29, 30, 46, 47, 48, 86, 88, 89, 90, 164, 282, 284, 461
Paperboard mills, 27, 30, 31, 39, 40, 74, 78, 79, 89, 105, 162, 181, 183, 196, 564
Paper coating and glazing, 30, 31, 39, 40, 104, 122, 184
Parker Dam, 238
Parker, Jay, 9, 25, 282, 291
Parkhurst, J.D., 249n, 251n, 513n, 529n, 544
Particle dispersion, 409
Particulates, 298, 304, 306, 313, 314, 316, 317, 318, 320, 321, 322, 324, 325, 327, 328, 330, 331, 332, 333, 337, 341, 346, 347, 348, 349, 350, 352, 354, 355, 357, 358, 373, 377, 378, 379, 381, 382, 383, 384, 385, 387, 412, 446, 449, 450, 562, 563
Pasadena, Calif., 237, 238
Passaic County, N.J., 152, 154, 168
Passaic River, 168
Pearson, E.A., 95n, 97n, 109, 110n
Pearson, F.J. 227n
Pennsylvania, 54
Pennsylvania basin, 65
Percolation, 288
Perlmutter, N.M., 212n, 213n, 227n
Peshtigo River basin, 45
Pesticides, 28
Peter Cooper Corp., 65
Peters, John H., 228n, 230n
Petroleum products, 30, 162, 184, 265
Petroleum refining, 27, 30, 31, 38, 39, 40, 48, 74, 78, 79, 104, 107, 123, 160, 181, 183, 407, 564
Philadelphia Electric Co., 316, 317, 323
Philadelphia, Pa., 224, 315, 316, 317, 318, 321, 324, 329, 330, 331, 344, 424, 449, 475, 476, 481, 485, 487, 488, 550, 560, 561, 570
Phosphate, 31, 34, 44, 151
Phosphatic detergents, 13
Phosphorus content, 41, 97
Photodissociates, 412
Phytoplankton, 275
Pierce County, Wash., 76, 77, 79, 80, 81, 89, 284
Pikarsky, Milton, 499, 506
Pittsburgh, Pa., 319, 321, 322, 323, 324, 329, 330, 331, 344, 424, 449, 475, 481, 493, 494, 550, 561, 562
Pollutional coefficients, 78, 404
Poole, Robert, Jr., 419n
Port Chester, N.Y., 140, 141, 143
Port Rowan, Ontario, 59

Port Susan, Wash., 77
Port Wentworth, Ga., 187, 188, 207
Portage River, 64
Portland, Oregon, 509
Possession Sound, 77
Poston, N.W., and Golegar, W.C., 184
Potter, N., see Fisher, J.L.
Poultry processing plants, 78, 79, 162
Precipitation, 214, 215
Primary metals industries, 27, 184
Printing and Publishing, 96, 159, 160
Puget Sound, 3, 75, 76, 91, 284, 289, 376, 568
Pulp and paper mills, 30, 31, 37, 39, 40, 44, 45, 74, 78, 79, 80, 87, 104, 122, 162, 163, 183
Putnam County, N.Y., 152, 154, 166
Puyallup River, 77

Quantitative methodology, 1, 281
Queens, N.Y., 152, 168, 211
Queens County, N.Y., 154, 212, 213, 214, 215, 217, 220, 222, 230, 288, 487
Quesada, E.R., 417

Racine, Wis., 44
Raisin River, 54, 64
Rawn, A.M., 246
Reagan, Gov. Ronald, 369
Recharge basins (sumps), 215
Recharge operations, 248
Reclaimed water, 246, 285, 289
Reclamation plant, 251
Recreation, 175, 249, 480, 515
Recycling of sewage effluent, 231, 475, 495
Redwood City, Calif., 98, 101
Reed, John L., see Deem, Warren
Regional Plan Association, N.Y., 18, 162, 178n, 469n, 500
Rensselaer County, N.Y., 152, 153, 158, 163
Resources for the Future, Inc., 2
Resource Recovery Act of 1970, 512
Respiratory disease, 447
Revelle, Roger, 270n
Reverse osmosis process, 225
Richmond (borough), N.Y., 152, 154, 168, 480, 482, 484, 487
River Rouge, 47
Riverhead Sewer District, 227
Riverside County, Calif., 238, 239, 240, 241
Robbins, R.C., 405, 407n, 411n, 412n
Robinson, E., 405, 407n, 411n, 412n
Rockefeller, Gov. Nelson, 170
Rockland County, N.Y., 152, 153, 154, 159, 160, 168, 312
Rocky Mts., 175, 384

Rocky River, 56
Rodie, Edward B., 14n
Rogers, Peter A., 515
Rogus, Casimir A., 445n, 465n, 466
Rohles, John, 506
Rohr Corp., 438, 439
Roos, Charles, 531
Rose, John L., 228n, 230n
Rosen, Dr. Carl, *see* J.M. Wood
RPA, *see* Regional Plan Association
Rubbish, 456
Ruberoid Co., 194
Ruckelshaus, William D., 315

Sacramento, Calif., 95, 104
Safe yield, 220, 230
St. Bernard Parish, La., 501
St. Charles County, Mo., 501
St. Clair, Ill., 501
St. Francisville, La., 176, 287
St. Joseph River basin, 27, 32, 39, 40, 46
St. Joseph River, Indiana, 36
St. Joseph River, Michigan, 36
St. Lawrence River, 28
St. Louis County, Mo., 501, 502, 504
St. Louis, Mo., 175, 176, 179, 185, 287, 344, 345, 349, 359, 360, 424, 449, 499, 501, 502, 504, 551, 563, 564
St. Tammany Parish, La., 501
Sales Management, 476, 479, 481, 485, 486n, 489, 490n, 492n, 537, 539, 540, 542, 543
Salt Lake City, Utah, 380, 381, 382, 383, 448, 509, 524, 525, 527, 529, 531, 534, 535, 543, 553, 568, 569
Salt Lake County, Utah, 526, 534
Salt water intrusion, 212, 222, 230
Samish Bays, 77
San Bernardino County, Calif., 239, 240, 241, 242, 243
San Diego, Calif., 239, 240, 241, 242, 243, 249, 252, 289
Sandusky Bay, 56
Sandusky, Ohio, 54, 283
Sandusky River, 64
San Francisco, Calif., 293, 344, 369, 372, 373, 374, 375, 376, 424, 509, 514, 516, 518, 529, 530, 535, 538, 552, 567
San Francisco Bay, 3, 95, 96, 97, 98, 102, 109, 113, 284, 289, 290, 291, 292, 514, 515, 552, 567
San Francisco Bay Conservation and Development Comm., 515
San Francisco County, Calif., 515, 516, 518, 529, 530
San Francisco International Airport, 96, 432

San Francisco/Oakland, 567, 570
San Gabriel River, 237, 251
San Gabriel Valley, 248
San Joaquin, Calif., 95, 104, 105, 107, 108, 110, 116, 285
San Jose, Calif., 95
San Marino, Calif., 238
San Mateo, Calif., 113
San Mateo County, Calif., 515, 516, 518, 529, 530, 552, 567
San Pablo, Calif., 99, 110, 111, 285
San Luis Obispo County, Calif., 241
Santa Anna, Calif., 238
Santa Barbara, Calif., 272
Santa Barbara channel, 264
Santa Barbara County, Calif., 241
Santa Clara, Calif., 105, 108, 116, 285
Santa Monica, Calif., 238
Santee, Calif., 252, 289
Santee County Water District, 249
Sanitary food containers, 104, 162
Sanitary landfill, 462, 477, 501, 503, 505, 512, 515, 548
Saratoga County, N.Y., 152, 153, 158, 163
Savannah Beach, Ga., 187, 188, 207
Savannah Electric & Power Co., 326
Savannah, Ga., 187, 188, 190, 192, 204, 207, 324, 325, 327, 332, 344, 559, 560, 569, 570
Savannah Harbor, 202
Savannah River, 3, 281, 289, 290, 292, 559, 560
Savannah River Enforcement Conference, 192
Savannah Sugar Refinery, 194, 195, 196, 200, 206
Schaeffer, Vincent, 398, 409n, 412n
Schenectady County, N.Y., 152, 153, 163
Schenectady, N.Y., 157
Schmid, Allan A., 431n, 436n
Schoharil County, N.Y., 152, 153, 163
Schreiver, Bernard A., and Seifert, William W., 422n
Schurr, Sam H., 306, 306n, 342
Schwarts, Robert, 175, 287
Scott Paper Co., 86, 88, 92
Sea lamprey, 29
Sea water encroachment, *see* Salt water intrusion
Seaburn, G.E., 215n
Seaman, Michael J., 455, 473
Seattle, Wash., 76, 80, 89, 284, 292, 344, 376, 377, 378, 379, 380, 424, 509, 520, 520n, 521, 522, 529, 531, 532, 535, 540, 544, 552, 553, 567, 568
  METRO, 81, 81n, 88n, 89n
Sedimentation, 175

Seifert, William W., *see* Schreiver, Bernard A.
Seismograph surveying, 263
Septic tanks, 29, 137, 138
Sewage Treatment, 56, 59, 113, 114, 229, 250, 285, 457
  primary treatment, 10, 30, 34, 77, 82, 135, 136, 140, 141, 188, 190, 250, 251
  Secondary treatment, 10, 34, 44, 61, 82, 98, 135, 138, 141, 190, 192, 197, 251
  Tertiary treatment, 10
Sheboygan County, Wis., 44
Shellfish, 110, 141, 142, 190
Shipping, 259, 260
Shortly, George, 427
Sierra Nevada Mountains, 237
Simpson-Lee Co., 88, 92
Simpson, Myles A., 427
Singer, S.F., 267n, 268, 275n, 397, 398n, 400n, 405n, 410n
Skagit, Wash., 76, 77, 78, 79, 80, 86, 92, 284
Skagit Bay, 77, 87
Skagit River, 77
Sludgeworms, 67
Smelt, 28
Smelting, 408
Smith, Sheldon O., 212n, 216n, 217n
Smog, 313, 369, 412, 446, 447, 448, 559, 561
Smokestack height, 409
Snohomish County, Wash., 80, 87, 88, 92, 520, 521, 532, 552, 553
Snohomish County Planning Dept., 521n, 544
Snohomish River, 77, 88
Snohomish, Wash., 76, 77, 79, 284
Soap and Detergent Association, 219
Society of Automotive Engineers, Inc., 425n
Solid Waste Disposal Act, 511
Somerset County, N.J., 152, 153, 154, 168
Sonoma County, Calif., 98, 102
Soren, Julian, 212n, 213n
South Atlantic Fisheries, 205
South Bay, 99, 110, 111
South Carolina, 187, 190
South Carolina Health Board, 190
South Haven, Mich., 35
Southern California Edison, 565
Southern Company System, 323
Space heating, 305, 315, 316, 317, 318, 321, 322, 324, 325, 326, 341, 348, 350, 355, 358, 370, 374, 375, 378, 379, 383, 386, 401, 569
Spectrophotometer, 89
SSTs, 399
Standard Industrial Classifications, 78, 162

Standard Rate and Data Service, 298, 298n, 317, 320, 322, 325, 346, 348, 350, 353, 355, 357, 370, 557, 557n, 569n
Standard Oil of New Jersey, 263, 264
Steel Production, 30, 38, 41, 181, 409
Stockholm, Sweden, 433
Stockton, Calif., 110
Stoddard, C.H., *see* Clawson, M.
Stoney Point, N.Y., 149
Storm water recharge basins, 228
Storm water runoff, 218
Stormont, D.H., 18
Straits of Juan de Fuca and Georgia, 76
Suez Canal, 263
Suffolk County, N.Y., 131, 211, 212, 213, 214, 215, 217, 218, 219, 220, 221, 222, 223, 224, 228, 229, 230, 231, 234, 288, 312, 558
Suffolk County Detergent Ban, 218
Suffolk County Health Dept., 228
Suffolk County Water Authority, 212, 229
Sugar Refining, 30, 103, 105, 162, 183
Suisun Bay, Calif., 99, 110, 111, 285, 291
Sulfates, 398
Sulfite paper mill, 46, 48
Sulfur dioxide, 398, 407, 408, 411
Sulfur oxides, 298, 304, 306, 313, 314, 316, 317, 318, 320, 321, 322, 324, 325, 327, 328, 329, 330, 332, 333, 337, 341, 346, 347, 348, 349, 350, 352, 354, 355, 357, 358, 367, 368, 373, 375, 377, 378, 379, 380, 381, 382, 383, 384, 385, 387, 446, 448, 561, 562, 563, 568, 569
Sulfur Trioxide, 304, 341
Sulfuric acid, 198, 199, 409, 449
Swanzy, Clay, 507

Tacoma, Wash., 76, 81, 89, 284, 424, 536
Taconic Mountains, 149
Tanker accidents, 262, 263 *See also* Oil spills
Tanker operation, 260, 262
Tappan Zee Bridge, 149
Tarrant, K.R., 270n
Thames River, 135, 145
Temperature, 42, 380
Terpenes, 447
Thermal pollution, 12, 16, 31, 33, 68, 137, 283
Thermal stratification of lake waters, 56, 59
Thoman, John R., 193, 200
Tokyo, Japan, 433
Toledo, Ohio, 54, 60, 66, 283, 293
Torrey Canyon, 272
Torrance, Calif., 238
Total potential pollution, 405
Total yearly atmospheric pollution, 401
Toxicity, 96, 97, 115

Travis Field, Ga., 188
Tri-State Transportation Comm., 464n, 480n
Troy, N.Y., 149, 150, 157
Tugaloo River, 187
Turbidity, 36, 42, 78, 113
Twain, Mark, 4n
Two-State Enforcement Conference, 35
Tyler, Ralph C., 506

Ulster County, N.Y., 149, 152, 154, 159, 166
Underground basins, 242
Underground water storage, 558
Union Bag-Camp Paper Corp., 192, 194, 195, 196, 197, 198, 200, 206
Union County, N.J., 152, 154, 168
United Kingdom, Dept. of Scientific and Industrial Research, 150n
United Nations Conference on the Human Environment, 264n, 265n, 269, 451
United Nations Statistical Office, 259n, 262n, 265n, 268, 392, 402, 404, 406, 408, 410n
United States, 75, 90, 158, 228, 265, 267, 274, 275, 278, 281, 297, 569
  Atomic Energy Commission, 33, 34
  Civil Aeronautics Board, 418n, 419, 420, 421, 439
  Department of Agriculture, 175
  Department of Commerce
    Bureau of the Census, 10n, 20, 71, 154, 156, 178, 207, 299, 301, 492, 505n, 506, 529, 537, 539, 540, 542, 543, 545
    Bureau of the Census, *Census of Housing*, 308
    Bureau of the Census, *Census of Manufactures*, 20, 21, 30, 31, 32, 40, 78, 79, 80n, 85, 123, 124, 164, 165, 167, 169, 180, 184
    Bureau of the Census, *County Business Patterns*, 20, 30, 31, 32, 38, 39, 40, 79, 85, 164, 165, 167, 169, 180, 350, 355, 476, 479, 485
    Bureau of the Census, *Statistical Abstract of the United States*, 299n, 301, 405n, 406, 431n, 486n, 509n
  Department of Defense
    Army Corps of Engineers, 187, 198, 201, 202,
    Naval Institute, 260n, 264n
  Department of Health, Education and Welfare, 503, 504, 506, 513, 514, 517, 519, 521, 522, 524, 526, 527, 530, 532, 533, 534

Compilation of Air Pollutant Emission Factors, 306, 405, 409n, 410n, 458n
Conference on Pollution of the Interstate Waters of the Lower Savannah River, 188n, 189, 190n, 191, 192n, 194, 201n, 203n, 206, 207
Report on Pollution of the Hudson River and its Tributaries, 150n
  Public Health Service, 195, 301, 304, 335n, 336n, 337, 338n, 341, 342n, 376, 455, 462n
  Office of Solid Waste Management, 5, 12
  Department of Housing and Urban Development, 521
  Department of the Interior, 26n, 27n, 28n, 36n, 87, 170n
  Department of Interior, Bureau of Mines, 267, 268, 306n, 307, 343, 345, 346
  Federal Water Pollution Control Administration, 2, 26, 27n, 30, 31n, 33n, 44n, 50n, 60n, 61n, 62, 62n 64, 65, 65n, 67, 68, 69, 80n, 87n, 88n, 91, 92, 128, 170, 185n, 196, 196n
  Geological Survey, 175, 225, 226
  National Marine Fisheries Services, 286
  Department of Transportation
    Federal Aviation Agency, 399, 417n, 418, 419, 419n, 420, 421, 423, 423n, 424, 426, 428, 430n, 432, 433, 434n, 437, 437n, 438, 438n, 439, 439n
    Federal Highway Administration, Bureau of Public Roads, 299, 299n, 301, 337
  Environmental Protection Agency, 130, 302, 317, 321
  Federal Power Commission, 304, 305, 305n, 339, 340, 341, 376, 368, 449
  General Accounting Office, 2, 12, 23n
  Office of Science and Technology, 300, 300n, 302, 302n, 431n, 432n, 434n, 436n, 445, 447, 459n, 474
U.S. Steel Corp., 41
University of California, 96, 97
University of Illinois, 273
Upper glacial aquifer, 211
Urban waste generation, 474
Urethane, 272
USSR, 409
Utah, 566,
Utica-Rome, N.Y., 158

Vacuum Freezing vapor compression, 225
Van Tassel, Alfred J., 1, 2n, 12n, 13n, 14n, 19n, 163n, 281, 297, 297n, 298n, 300n, 445, 547, 557

Vaughan, R.D., *see* Black, R.J.
Vecchio, John, 227n
Ventura County, Calif., 239, 240, 241, 242, 243
Vermont, 149, 157
Vernon River, 188
Vicksburg, Miss., 176, 287
Volcanic activity, 450
Volkswagen Works, 466
Vollkommer, Robert, 51

Wankel engine, 338
Wantagh, N.Y., 230
　Water Pollution Control Plant at, 229
Warren County, N.Y., 152, 153, 158, 163
Washington County, N.Y., 81, 152, 153, 163
Washington County, Wis., 44
Washington Utilities and Transportation Commission, 520
Washington State, 75, 76, 80, 81, 82, 83, 90, 91, 567
　Department of Ecology, 77, 81, 82, 82n, 83, 83n, 84, 86, 86n, 87, 88n, 89, 89n
　Department of Commerce and Economic Development, 76n
Waste Management, 20, 21, 28n, 90, 105, 106, 107, 179
Waste disposal permits, 84, 482, 494, 561, 562, 564
Waste heat, 35
Watercraft pollution, 49, 115
Water Quality Control Act of 1965, 41, 170
Water Quality Improvement Act of 1970, 60
Water recycling, 226, 246
Water renovation, 231, 249
Waukegan, Ill., 43
Waukiesha County, Wis., 44
Weiss, E.B., 510

Welder, B.Q., *see* Paulson, E.G.
Wesson Division, Hunt Food and Industry, Inc., 194
Westchester County, N.Y., 131, 140, 142, 149, 152, 153, 154, 159, 160, 168, 229, 312
Westerhoff, Garret P. and Gruninger, Robert M., 461n
Western Energy Supply and Transmission (West) Associates, 565, 566
West Falmouth, Mass., 272
Westmoreland County, Pa., 319
West Memphis, Arkansas, 176
Wet scrubber system, 330, 465
Wet tower, 35
Weyerhauser Co., 88, 92
Whatcom County, Wash., 76, 77, 79, 80, 86, 87, 92, 284
Whittier Narrows, Calif., 246, 248, 252
Whittier Reclamation Facility, 252
Whyte, William, 490
Williams, Dudley, 427
Williams Research Corp., 312
Willoughby, Ohio, 56
Wilmington River, 188
Wilson, Alfred J., Jr., *see* Hansen, David J.
Wilson, Carroll L., 392n, 397n, 398n
Wingo, Lowdon, Jr., 420n
Wisconsin, 21, 27, 31, 33
Wood, Dr. J.M., 273
Wood, John Walter, 425n
World electric production, 404
World fossil fuel consumption, 394, 395
World War II, 80, 152

Yellow perch, 28
Yolo County, Calif., 102
Yonkers, N.Y., 151
Youngstown Sheet and Tube Co., 41

About the Editor

Alfred J. Van Tassel is a member of the faculty of the School of Business, Hofstra University, Hempstead, New York. Professor Van Tassel became interested in resource questions in 1949 when he served as Executive Secretary of the first United Nations Scientific Conference on the Conservation and Utilization of Resources. He holds a bachelor's degree in chemistry from the University of California at Berkeley and a PhD degree in economics from Columbia University. In 1970 Lexington Books published the results of research he had directed in a volume entitled *Environmental Side Effects of Rising Industrial Output*.